Fire Performance Analysis for Buildings

Fire Performance Analysis for Buildings

Second Edition

Robert W. Fitzgerald and Brian J. Meacham

Worcester Polytechnic Institute, MA, USA

Library of Congress Cataloging-in-Publication Data

Names: Fitzgerald, Robert W., author. | Meacham, Brian J., author.
Title: Fire performance analysis for buildings / Robert W. Fitzgerald, Brian J. Meacham.
Other titles: Building fire performance analysis
Description: Second edition. | Chichester, UK ; Hoboken, NJ : John Wiley & Sons, 2017. |
   Revised edition of: Building fire performance analysis. 2004. | Includes index.
Identifiers: LCCN 2016039775 (print) | LCCN 2016054258 (ebook) | ISBN 9781118657096 (cloth) |
   ISBN 9781118926499 (pdf) | ISBN 9781118926338 (epub)
Subjects: LCSH: Building, Fireproof. | Fire prevention–Inspection.
Classification: LCC TH1065 .F574 2017 (print) | LCC TH1065 (ebook) | DDC 693.8/2–dc23
LC record available at https://lccn.loc.gov/2016039775

A catalogue record for this book is available from the British Library.

Set in 10/12pt Warnock by SPi Global, Pondicherry, India

Cover image: Petr Student/Shutterstock
Cover design: Wiley

Printed and bound in Malaysia by Vivar Printing Sdn Bhd

10 9 8 7 6 5 4 3 2 1

Contents

Preface

This book describes a framework to analyze the fire performance for any building – in any location, under any regulatory system, and constructed in any regulatory era.

The book is intended for any individual who wants to understand the fire performance of buildings. The approach allows one to examine the performance of specific components in isolation or to integrate them to describe holistic behavior.

It is anticipated that readers will have varied backgrounds and levels of knowledge in the subject. The book enables a reader to obtain specific information in isolation. For example, an individual may wish to increase their knowledge of fire behavior or the operational details of a specific fire defense. When such background knowledge is already known, the reader may go directly to the analytical techniques. Although a reader may move through specific topics in isolation, the content is structured in a logical progression.

Unit One describes the foundation on which the analytical framework is organized. Theory and practice are based on well-established techniques. Because fire performance in buildings is dynamic, the Interactive Performance Integration (IPI) chart is given special attention. This chart is an essential tool to relate the fire and the phasing in and phasing out of fire defenses and risk characterizations.

Unit Two explains each part of the system of fire and buildings. Fire department operations are of particular interest in building analysis. The procedures are described for fire safety professionals with no experience in fire ground operations. Techniques relate the fire size to the time of water application and to damage estimates at eventual extinguishment.

Modern structural analysis and design for fire conditions is another important part of fire performance in which many fire safety professionals have little knowledge. Chapter 18 describes the evolution of structural requirements while Chapter 19 makes the transition from traditional regulations to modern calculation methods. The information enables a fire safety professional to work with a structural engineer to establish performance understanding of the building's structural system.

Unit Three identifies the analytical framework for each component and for holistic performance. The organization is based on the framework of Unit One and the component behavior of Unit Two. The IPI chart is an essential tool for ordering the time-related phases of fire and fire defenses.

Fire safety engineering is an evolving discipline. Although some components are now reaching early maturity, others are making a transition from infancy into adolescence. Uncertainty is inherent to all analysis and design. *Unit Four* describes ways to manage uncertainty and communicate credible knowledge to other individuals who are involved in the built environment.

Acknowledgements

Harold E. (Bud) Nelson created the foundation of performance analysis and design nearly half a century ago. This book represents the current status of the "Nelson Method."

The acknowledgements in the first edition identified many of the pioneers who contributed significantly to the maturation of this structure for fire safety performance. Although their names are not repeated in this edition, their contributions should not be forgotten. Nevertheless, the names of Rexford Wilson and Rolf Jensen are again recognized because of their significance to the development of these procedures and to the history of performance based fire safety engineering.

The first edition attempted to describe performance analysis for unique, site-specific buildings. Unfortunately, recognition of the analytical framework was obscured by emphasis of probabilistic performance descriptors that were used to sort out complicated interactions. This second edition emphasizes state-of-the-art deterministic fire science and engineering in performance quantification. The role of the Interactive Performance Information (IPI) chart has been expanded to describe dynamic interactions.

The role and depiction of the framework and quantitative measures have been reorganized in this edition. Techniques for evaluating a building design for fire department extinguishment and analyzing structural performance have been upgraded. Essentially, this second edition is an entirely new book that is based on concepts of the first edition.

One of the important new techniques involves building analysis for fire department suppression. James F. Callery, District Chief (ret) Worcester (MA) Fire Department, Clifford S. Harvey, Assistant Chief (ret), Boulder (CO) Fire Department, Peter V. Mulvihill, Nevada State Fire Marshal, and Matthew T. Braley, District Chief, Worcester (MA) Fire Department have made valuable contributions. Professor Guillermo F. Salazar (WPI) provided support for BIM drawings and construction management procedures.

The state of the art of structural design for fire conditions has progressed significantly in recent years. The fire safety engineer and the structural engineer have interactive roles in understanding structural performance for fire conditions. Professor Leonard D. Albano (WPI) and Roger Wildt, P.E., gave valuable support for the structural engineering documentation.

The Society of Fire Protection Engineers (SFPE) provided important support for this edition. Professor Tahar El Korchi of the WPI FPE Department funded students to test practices, develop numerical examples, draw figures, and format the product. Professor Roberto Pietroforte guided the architectural interface. Professor Robert C. Till (John Jay) used early drafts of the text to provide useful feedback. We are very grateful for the support of SFPE, WPI, and the following students: Ian Jutras, Drew Martin, Yu Liu, Yecheng Lyu, Young-Geun You, Milad Zabeti Targhi, and Camille Levy.

A book of this type requires an enormous amount of time to organize, discuss, and prepare. We appreciate the tolerance and sacrifice given by our wives, Margaret and Sharon. Their support has been important to the completion of this project.

Robert W. Fitzgerald
Brian J. Meacham

1

Fire Performance and Buildings

1.1   The Dynamics of Building Fire Performance

A building fire is dynamic because hostile fire characteristics change minute by minute. The dynamic fire produces products of combustion that affect the building and its fire defenses. The continually changing building environment influences time relationships for risk characterizations involving occupants and building functions. These actions occur in a variety of sequences and ways for different buildings.

During a fire, some components complete their roles and become inactive before other components become operational. Additionally, actions of some parts of the system depend on the status and sequential phasing of other components. Performance evaluations analyze interactions that combine time-dependent changes in the fire, building fire defenses, and people.

The goal of this book is to organize the complicated process into an analytical framework with which an engineer can evaluate fire performance. A performance evaluation enables one to understand specific component behavior as a part of holistic building performance.

Time is the common factor that links all of the important events.

1.2   The Anatomy of Building Fire Safety

Figure 1.1 shows the major parts of the complete system of fire performance for buildings. Initially, the system is organized into three major groups:

1) The *composite fire* combines a diagnostic fire and the active extinguishment actions provided by local fire department manual extinguishment and automatic sprinkler suppression, if present.
2) The *building response* is based on the flame-heat and smoke-gas products of combustion produced by the composite fire and their movement through the building. The process continues from ignition to extinguishment.
3) The *risk characterizations* for exposed people, property, and functions are based on the building's response.

Figure 1.1 is a static representation of the major parts. At each minute into the fire, the status of each part changes.

Fire Performance Analysis for Buildings, Second Edition. Robert W. Fitzgerald and Brian J. Meacham.

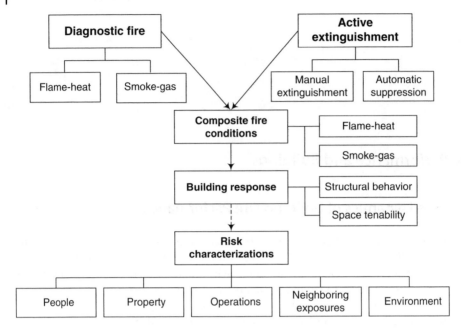

Figure 1.1 System components.

The analytical framework decomposes each part into components that can be evaluated separately. The components are recombined to incorporate the influences of time, fire conditions, and other components within the system. This allows each component to be evaluated as an independent unit and the effects combined to describe holistic performance.

1.3 Analysis and Design

Analysis and design are two sides of the same coin. In its most basic form, all design involves trial and error. For example, a design process starts by gathering information about a building's function, the design objectives, hazards to which the building will be subjected, the dimensional, material, economic and site constraints, and regulatory expectations. An initial trial design is formulated and then analyzed to evaluate the extent to which function, economics, and safety are acceptable. The design is then updated by changing parts of the trial design that did not perform in an acceptable manner. The iterative process of design–analyze–redesign continues until an updated design produces acceptable conditions for function, safety, and economy.

This book does not address building design, nor does it use any specific code or design standard. Rather, it describes how to analyze a building for a hostile fire. The results of the fire analysis provide a basis to characterize risk for people, property, and function. The goal is to describe a way to understand fire performance and risk characterizations for any existing building or proposed new building design. Although the book does not describe conventional procedures to accomplish design objectives, a performance analysis will give an insight into effective ways to achieve stated objectives.

1.4   Performance Analysis

A performance analysis creates an understanding of what to expect during a building fire. After evaluating the building's performance, one can identify associated risk characterizations to people, property, operational continuity, neighbors, and the environment.

Evaluation procedures integrate two distinct parts:

1) An analytical framework to provide systematic, methodical procedures to structure individual component behavior and integrate all parts into a holistic entity.
2) Quantification to provide numerical measures of performance.

The primary goal of this book is to identify a framework for analyzing fire performance in buildings. However, a framework is sterile without ways to quantify the critical events. One cannot exist without the other.

Fire safety is an emerging engineering discipline. Consequently, all numerical measures for component quantification do not have the same level of development. Some components, such as structural fire analysis and detector actuation, are relatively well developed and one can have confidence in calculations. Room fire models can provide accurate representations of behavior within their limits of theory and input knowledge. On the other hand, certain aspects of manual fire extinguishment, automatic fire suppression, and barrier effectiveness are inadequate for comprehensive numerical analyses. Nevertheless, the framework uses existing knowledge for quantification and developing a performance understanding.

Quantification uses any information or calculation tool that is relevant and seems appropriate to obtain the necessary numerical measures. Sources such as computer programs, experimental data, calculated values, observed information, and failure analyses become resources for quantification. Quantification procedures may be viewed as a set of tools. An engineer selects appropriate and available tools for each need. The framework organizes the analysis to incorporate quantitative measures of performance.

1.5   Quantification

The goal of a performance analysis is to understand expected building behavior and the associated risk characterizations during a fire. Building evaluations use specific fire scenarios to acquire this understanding.

A scenario evaluation uses three types of analysis. A *quantitative analysis* calculates outcomes using available information. Fire safety has not yet evolved to provide reliable, unique source quantification for the range of conditions routinely encountered in buildings. Therefore, the quantitative analysis is augmented by a *qualitative analysis* to provide a sense of proportion for expected behavior. A qualitative analysis incorporates many features that affect outcomes for interpreting numerical calculations.

Quantitative analyses and qualitative analyses are used together in an evaluation. Often, a quantitative analysis is a primary source for performance measures. The qualitative analysis helps to ensure that an outcome incorporates all of the important features and provides reasonable values. At other times, a qualitative analysis is the dominant evaluative tool and quantitative information is used to augment or give confidence to the estimates. Qualitative analyses are often used to select initial scenarios that become the basis for a performance analysis.

Both quantitative and qualitative analyses are sensitive to changes in condition. For example, the status of a door being open or closed may significantly affect performance. Fuel packages may use differing construction materials that can have significantly different burning characteristics. This produces different time-related outcomes that, in turn, may affect the performance of other components.

Often, "what if" questions become evident during the decision-making function of scenario identification and one may wish to examine performance differences that could occur. A *variability analysis* provides a basis for ascertaining if possible changes will significantly affect performance outcomes or will have only a relatively benign influence. A variability analysis examines important questions that could affect quantitative or qualitative outcomes. Variability analyses establish "windows of behavior" to better understand building features that affect fire safety.

1.6   The Organization

This book organizes the complex system of fire in buildings in a way that enables one to understand both an individual component's behavior and its effect on holistic building performance. This involves:

1) Identification of a comprehensive analytical framework. This framework is logically structured and consistent to be adaptable for any building and geographical location.
2) Use of deterministic component evaluations that combine state-of-the-art fire science with engineering knowledge and information.
3) Use of organizational charts to record key information and to visualize time-related complexity in a way that performance expectations may be explained to other professions.

The analytical framework is the primary focus of attention, and different aspects of the framework and its quantification are presented in each of the four units of this book:

Unit One: The Foundation. This unit describes the structure of the organizational framework. The framework adapts established techniques of other disciplines for fire safety evaluations. The Interactive Performance Information (IPI) chart becomes the central tool to relate time sequencing with critical events for performance evaluations.

Unit Two: The Parts. The functional behavior and operation of the major components are described in the context of the analytical framework. Functional and quantitative relationships provide guidelines for evaluation.

Unit Three: The Analysis. The descriptive base for the components of Unit Two is organized into networks that structure performance analysis. The networks, in combination with the IPI chart, enable variability analyses to be integrated efficiently into performance understanding.

Unit Four: Managing Uncertainty. Uncertainty and variability are unavoidable in building analysis. This unit introduces different ways to manage uncertainty and to communicate results to non-fire safety professionals.

Collectively, the four units address different aspects of fire safety analysis to provide a comprehensive treatment of a way to consistently evaluate the performance of any building.

In general, the chapters are "stand-alone" units that allow a reader to select topics that satisfy specific needs. Although there is a thematic structure, one need not move sequentially through intervening chapters. Rather, specific topics may be selected to augment information for component functions, operations, quantification, and analysis.

Part I

The Foundation

The primary objective of this book is to identify a framework to analyze the fire performance for any building.

The analytical framework is universal. It is not restricted by any geographical location, any jurisdiction that writes or enforces codes and standards, or any fire protection devices or actions that are intended to make the building perform better.

Although the framework is universal, quantification is local. Quantification is dependent on the building design, its location and all existing features that influence performance. The human element is also an important part of performance outcomes.

A performance analysis evaluates fire scenarios that link the fire, active and passive fire defenses, people, building architecture, and site conditions. Each component plays a role in the process. A performance analysis produces a clear understanding of what to expect during a building fire. This understanding becomes the basis of risk characterizations for people, property, operational continuity, exposed buildings and enterprises, and the environment.

Fire performance is dynamic because it is based on constantly changing fire conditions combined with phasing in and phasing out of different components. Time is the factor that links all components. The Interactive Performance Information (IPI) chart is introduced to record information and help to relate the constantly changing status of each component.

The framework logic is based on established procedures of other disciplines. These include event logic diagrams, logic networks, fault trees, and event trees. All parts of the fire safety framework have been adapted from techniques that have existed for half a century. Quantification is deterministic and based on the state of the art of fire science and engineering. All parts are interrelated by the IPI chart to examine component performance at an instant of time (static) and incorporate the results into a time-related (dynamic) relationship.

Unit One describes fundamental concepts for a methodical, consistent performance analysis. Although definitions are in general use within the fire safety community, a few important concepts are defined more carefully to establish a precise understanding. The concepts are described to enable one to understand the framework structure. Additional tools are introduced to organize performance evaluations. These include space–barrier networks to select and evaluate multi-room scenarios.

Fire Performance Analysis for Buildings, Second Edition. Robert W. Fitzgerald and Brian J. Meacham.
© 2017 John Wiley & Sons Ltd. Published 2017 by John Wiley & Sons Ltd.

2

Preliminary Organization

2.1   Introduction

Fire performance may be evaluated for any existing building or proposed new structure whose plans have progressed to the detailed design phase. Therefore, buildings constructed at any time, under any (or no) regulations, and in any geographical location are appropriate candidates for a performance analysis.

The analytical framework organizes the building's architectural design into manageable parts that contribute to component analysis. The components may be combined to describe holistic performance. Viewing each building with the same systematic framework enhances the ability to compare performance and risk in a consistent manner.

This chapter introduces three topics to establish an initial base for performance thinking. The first, *Part One: Organizational Concepts*, gives a brief overview of the analytical process and introduces a few concepts and definitions that help with organization. The next, *Part Two: Barrier–Space Modules*, identifies the role of barriers and barrier–space paths for analytical consistency. The final topic, *Part Three: Traditional Fire Defenses*, describes active and passive fire defenses that have been a traditional part of building fire safety. Performance evaluations organize these practices into an analytical framework that enables one to evaluate fire safety with consistent, logical procedures. Organizationally, these topics initiate a methodical way of thinking about building performance and its analysis.

2.2   Overview of Evaluations

A performance evaluation starts with gathering knowledge about the building's function and architecture, its fire defenses, and any special features that influence the building's fire performance and risk characterizations. This information establishes an initial knowledge base for analysis.

Scenario analyses provide the means to understand a building's performance. A scenario analysis starts by selecting a room of origin. Then, a *diagnostic fire* is selected to describe the time-related combustion product generation for this room and the additional rooms through which the fire can propagate. The diagnostic fire is used to evaluate the active fire extinguishment for local fire department extinguishment and automatic sprinkler suppression.

Fire Performance Analysis for Buildings, Second Edition. Robert W. Fitzgerald and Brian J. Meacham.
© 2017 John Wiley & Sons Ltd. Published 2017 by John Wiley & Sons Ltd.

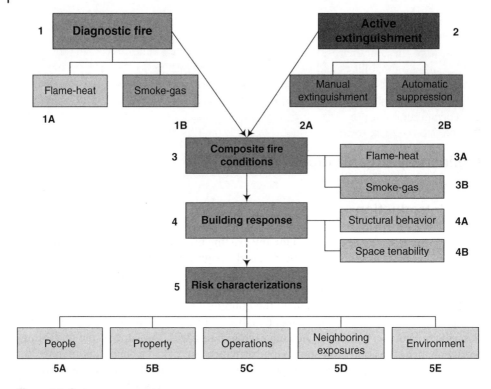

Figure 2.1 System components.

The *composite fire* combines the diagnostic fire with water application from fire suppression to describe changes in fire status from ignition to extinguishment. Knowledge of the building's composite fire conditions enables one to evaluate the structural response and estimate tenability conditions in important building spaces. Finally, the understanding that unfolds during a comprehensive performance analysis enables risks to be characterized for the exposed people, property, operational continuity, neighboring exposures, and the environment.

Time is the common factor that links all parts of the evaluation.

A brief overview of the procedure is based on the organization of Figure 2.1 and involves the following steps:

1) Select a room of origin and construct a diagnostic fire that includes the room of origin and additional rooms through which the fire can propagate. This fire provides a measure against which to evaluate fire defenses and characterize risks. A diagnostic fire describes the time and products of combustion (POCs) relationships for flame-heat and smoke-gas. The diagnostic fire assumes no extinguishment intervention.
2) Evaluate automatic sprinkler suppression, if a system has been installed.
3) Evaluate local fire department extinguishment.
4) Describe composite fire changes from ignition to extinguishment by combining the diagnostic fire and active extinguishment. Initially, the composite fire is the same as the diagnostic fire. This natural fire becomes disrupted after first water application

from a sprinkler system or the local fire department. Fire suppression actions continue until extinguishment has been completed.

5) Describe the building's response to the composite fire conditions. The composite fire affects tenability conditions in target spaces that house people or important contents on a permanent or temporary basis. This includes spaces that are used for occupant egress. Structural behavior is also part of a building's response.

6) The building's response for structural behavior and target space tenability are used to characterize risks to people, property, operational continuity, neighboring exposures, and the environment.

This process may be used for any scenario in any building.

PART ONE: ORGANIZATIONAL CONCEPTS

2.3 The Diagnostic Fire

A diagnostic fire establishes a measurement standard for a performance analysis. A diagnostic fire is the "load" that is tested against the "resistance" of fire defenses. The diagnostic fire is the natural fire that may be expected in a room of origin and propagation through a sequential cluster of additional rooms, assuming no extinguishment intervention.

A diagnostic fire identifies the development, properties, and movement of POCs. The main components are *flame-heat* and *smoke-gas*. Each moves through a building by different routes and at different speeds. Each has transfer mechanisms through different media. Each has a different, but important, influence on fire defenses and the exposed people, property, and mission.

2.4 Anatomy of a Representative Fire

Figure 2.2 shows a characteristic time vs. fire growth relationship. The process starts with an overheat condition that may produce smoldering or initial flaming. We define *ignition* as the initial self-sustained burning of the fuel. More specifically, we define ignition as the appearance of the first fragile flame on the fuel surface. For our purposes, overheat and smoldering conditions do not constitute ignition.

After ignition, the fire may continue to grow. When it reaches about 300 mm (12 in), i.e. nearly a small wastebasket size or knee height for an adult human, the fire begins to produce more predictable fire growth characteristics. The firepower for this condition is about 25 kW. This fire size defines *established burning* (EB).

After EB, the fire may continue to grow until *flashover* (FO) occurs. FO is a significant event in a room fire's history. It defines a condition in which all combustibles are surface burning. The fire growth from EB to FO describes an important segment in a building's performance.

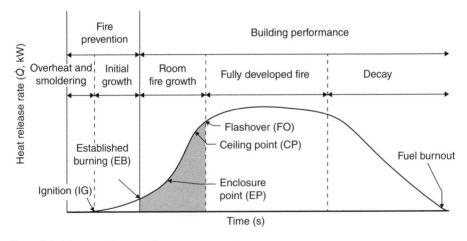

Figure 2.2 Characteristic room fire.

After FO, the fire continues to burn at a somewhat steady rate for an extended duration of time. Fuel quantity and ventilation are among the more important factors that influence duration of the fully developed phase.

Eventually, the fire consumes most of the fuel and begins to decay. This leads to eventual burnout with a residue of ash and some unburned fuel.

For our purposes, EB in the room of origin is the start of a building performance analysis. A fire of about 25 kW is a convenient and analytically credible fire size for this event. Global time measurements start at EB in the room of origin and all local component time durations are coordinated with this value.

2.5   Fire Prevention

Fire prevention is evaluated separately from a building's performance after the fire starts. Although fire prevention is important for a total package, we uncouple prevention from the building's performance after EB for three reasons:

1) Time durations and initial fire behavior are extremely variable before EB. However, fire behavior is relatively predictable for fires larger than EB. Confidence in evaluating fire defenses and risk comparisons is better, and outcomes are more consistent when the building analysis process starts at EB.
2) Effective prevention practices can mask significant shortcomings in a building's design. Risk management requires a clear knowledge of the manner in which a building will perform after the fire starts, regardless of the frequency of ignition and established burning.
3) Occupant extinguishment can be an important part of a complete fire safety program. However, we do not include occupant extinguishment in a building's performance analysis. Rather, we incorporate this action as a part of the prevent EB analysis. Therefore, fire prevention includes both ignition prevention (the more common concept) and prevention of EB after ignition. Chapter 23 describes fire prevention applications.

2.6   Fire Scenarios

A performance analysis methodically examines dynamic interactions of the fire, fire defenses, and the building. Often, the systematic examination of a well-thought-out scenario provides most of the information needed to understand the building's fire behavior and to characterize risks. This information can be augmented with a few strategically selected additional scenarios or "what if" conditions that may require only partial additional analysis. The aggregate knowledge gives a reasonably complete understanding of how the building will work during a fire and the type and severity of risks that may be expected.

A building fire scenario uses the diagnostic fire as an initial measure against which all other fire features and risks are examined. The main components for a complete building performance are:

• the diagnostic fire with separate flame-heat and smoke-gas descriptors;
• the fire and barrier effectiveness;

- the fire and fire detection;
- the fire and automatic sprinkler suppression;
- the fire and fire department extinguishment;
- the fire and structural frame behavior;
- the fire and smoke tenability in the building spaces for which risk must be characterized;
- the fire, smoke, and risk to people, property, operational continuity, exposed neighboring enterprises, and the environment.

PART TWO: BARRIERS, SPACES, AND CONNECTIVITY

2.7 Spaces and Barriers

A space is a volume enclosed by barriers. A barrier is any surface that can delay or stop flame-heat or smoke-gas movement in a building. All buildings are assemblies of spaces and barriers whose main architectural function is to make the building work.

Barriers are important for normal building functions as well as fire safety performance. Horizontal and vertical barriers separate spaces to define rooms and provide privacy, security, and noise control. Barrier openings provide routes for physical, visual, and informational communication using devices such as doors, windows, piping, and electrical conduits. They sometimes provide or hide routes to convey services such as electricity, water, waste disposal, heat, and air. Barriers channel occupant movement in and out of the building and between functional spaces. Barriers may or may not support structural loads.

Holes in barriers are essential for routine building functions. Most barrier holes are constructed intentionally because a building cannot work without them. These holes are important to a building's fire performance because they can permit combustion products to move from one space to an adjacent space.

Barriers serve many functions in a building's day-to-day operations. These functions dominate the long-term use, maintenance, and status of barriers. They also influence a building's fire performance.

2.8 Barriers and Fire

Because the combustion process needs oxygen, at least one barrier opening is necessary for a room fire to reach FO. An opening to the outside of the building can simultaneously feed oxygen to the fire and vent the heat and gases. When the opening is to the building's inside, the same functions occur. However, inside venting propagates the fire to other building spaces.

Barriers are important to fire fighting. Barriers can channel smoke and heat toward or away from fire fighters. They can prevent fire propagation or they may block fire fighters and water application. Barriers can help to establish a defense line for fire control. Barriers sometimes help fire fighting; sometimes they hinder effective fire fighting. Regardless of their usefulness, a building's existing barriers are a part of fire suppression and building performance evaluations.

A door or window, whether open or closed, is a part of the barrier. Any penetrations or openings, whether protected or unprotected, are parts of the barrier. We evaluate all barriers from a field performance viewpoint that incorporates any construction features, status of openings, fire endurance, or combustibility that may exist.

Many fire protection professionals think of barriers primarily in terms of fire resistance ratings that are provided by the standard fire endurance test (ASTM E-119 or ISO 834). However, the rated endurance time rarely has an influence on a building's fire performance. Instead, the holes that allow fire to propagate into adjacent rooms have a much greater influence. A performance evaluation considers all features that may affect fire propagation or suppression.

Time is the quantity that is common to all parts of fire performance for buildings. The barrier's ability to prevent movement of combustion products deteriorates during the time that it is exposed to a fully developed fire. This is particularly important with combustible barriers and wood doors. Changes of status during the fire are a part of a routine performance analysis.

2.9   Barrier Performance

The function of a barrier during a fire is to delay or prevent combustion products moving from the exposed side to the unexposed side. Building analyses evaluate barrier success for the duration of a fire. At any given time, a barrier can have only one of three possible states:

1) A small, hot-spot ignition source can exist on the unexposed side.
2) A large, massive ignition condition can exist on the unexposed side.
3) The barrier is successful and no ignition source can exist on the unexposed side.

These states are mutually exclusive. Only one state will be defined to exist at any instant of time. For example, if both hot spot and massive ignitions exist simultaneously at different locations, the massive condition will dominate fire extension into the adjacent room, and the analysis will consider only that state.

A small, hot-spot ignition may result from too much heat transmission through a barrier. It can also result from surface cracks that allow small flames to penetrate to the unexposed surface of the barrier. A hot-spot failure can also occur because of an inadequately protected through construction. For analysis purposes, a hot-spot failure is assumed to cause ignition and EB in the adjacent room. If fuels were present, fire growth would be "normal" although at an accelerated rate. If fuels are not near, ignition may be delayed.

A massive failure occurs when the barrier has a large opening. These failures may result from an open door, a broken window, partial or full barrier collapse, or when a hot-spot failure enlarges into a massive failure, as would occur with a combustible barrier or deteriorating gypsum panels.

A massive failure produces a large influx of fire gases into the adjacent space. These fire gases will cause an almost certain FO in the adjacent room within a minute or two. This FO will occur whatever the fuel content in the space. The equivalent of FO conditions can exist in spaces sterile of fuel, such as corridors, when a massive opening allows the POC from a flashed-over room to enter into the space.

These failure modes affect ignition, fire spread, and the time for FO in the adjacent room. A rough rule of thumb suggests that a hot-spot failure occurs with an opening of less than 6.25×10^{-2} m^2 (100 in^2). An opening of greater than 25×10^{-2} m^2 (400 in^2) normally identifies massive failure.

2.10   Space–Barrier Connectivity

A dynamic performance analysis examines a variety of movements through the building. The movement may involve flame and heat, or smoke and gas. The movement may relate to occupants, fire fighters, or hose lines moving to destinations in the building.

Estimating time durations for movement along specific paths is fundamental to a dynamic analysis.

Identification of path connectivity is fast and easy. Although the task may seem trivial in many situations, some buildings exist where the process may require creativity. Because the thought process for identifying paths of interest through a building is of such fundamental importance to performance analysis, we pause to discuss some useful concepts.

It is essential to focus clearly on the specific building features being analyzed and to communicate the outcomes. Normally, the space–barrier path identification is almost self-evident. For example, in Figure 2.3(a) one may wish to examine potential flame-heat movement from Room 1 to Room 2 to Room 5. Figure 2.3(b) describes the sequence for this movement. Figure 2.3(c) illustrates the space–barrier sequence for a different route that may be used during egress. In all cases, one merely traces the elements of a sequential space–barrier path of interest. Barrier status and the other factors that affect performance will be evaluated as the analysis unfolds.

We recognize that multiple potential paths usually lead from most spaces. However, only a single space–barrier path is examined for any specific investigation. Chapter 3 describes techniques for identifying multiple paths in complicated buildings and for isolating specific paths for analysis. For now, the goal is merely to present the concept that any specific path may be identified by a chain of interconnected spaces and barriers.

The space–barrier path descriptions need not be exclusively rooms and walls, floors, or ceilings. An analysis can examine any potential sequential chain of spaces and barriers regardless of architectural classification. For example, an engineer may wish to examine a particular route of interest during a fire reconstruction. A focus of attention may involve propagation in a concealed space, such as between studs of a wall. In this situation a fire can move from a room through a gypsum wallboard and then through the space between studs to another location. The initial space is a fire room. The first barrier is the gypsum wallboard surface, and the next space is the volume between the studs and the wallboards. Fire stopping might be the next barrier location along the propagation path. Thus, the space–barrier concept may be used to focus on any sequential path of movement to examine its significance for the problem being studied.

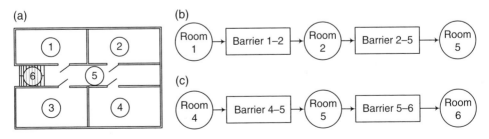

Figure 2.3 Space–barrier path.

2.11 Virtual Barriers

Barriers are normally envisioned as physical surfaces. However, virtual (i.e. imaginary) barriers provide another way to analyze complex arrangements. This allows one to tailor performance analyses for different types of buildings with a consistent procedure that expands the concept to a wide variety of architectural conditions.

A *virtual barrier* is a fictitious barrier that has no resistance to flame-heat or smoke-gas movement. We might call it a zero-strength barrier because POCs and people can move through it unimpeded. However, virtual barriers can divide large open volumes into smaller spaces to permit more detailed analysis. The (virtual) space–barrier segmentation enables one to gain a better understanding of time, combustion product transfer, people movement, and suppression effectiveness. Also, virtual barriers can be useful for conceptualizing alternative design solutions and developing risk management programs.

Uncompartmented large open spaces, such as "big box" retail stores, warehouses, manufacturing operations, parking garages, exhibition halls, restaurants, or theaters, can complicate some details of a performance analysis. However, when we divide these large open spaces into smaller segments (i.e. "rooms") with virtual barriers, we are able get a better understanding of fire spread, time, movement, and suppression effectiveness. Scenario analyses can clarify the importance of fire origins in different spaces.

Figure 2.4 shows a large, open volume building. Often, "natural" segmentation may exist, such as aisles or functional manufacturing operations. Sometimes division is more useful along natural architectural features. Regardless of the basis of segmentation, these features help to position virtual barrier locations for establishing spatial subdivisions.

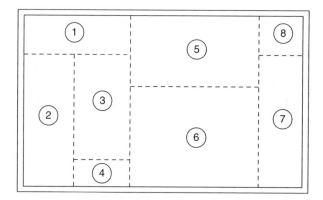

Figure 2.4 Virtual barrier separation.

2.12 Virtual Barrier Applications

The space–barrier representation can relate physical sequences and their interactions with fire defenses. However, one need not be confined to conventional rooms, walls, and floors. We will illustrate the concept with two examples where virtual barriers can help to structure analytical thinking.

Example 2.1   Figure 2.5 shows a partial section of a building. Assume that we wish to investigate the ease or time duration for fire to move from Room 1 to Room 2. We can examine Path A and Path B to compare the results.

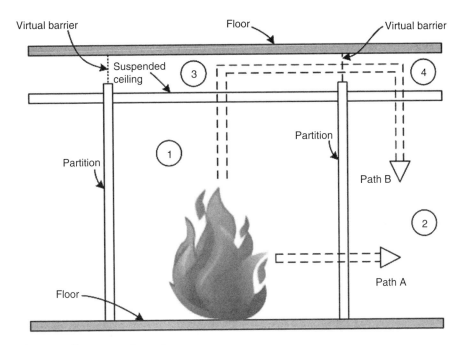

Figure 2.5 Fire propagation path.

Solution:
We define Path A as fire movement from Room 1, through Barrier B_{1-2}, to Room 2. Schematically this is described as follows:

Path B examines the movement from Room 1, through the suspended ceiling, into the plenum space, and drop down through the next ceiling into Room 2. Here, the wall partitions do not extend fully to the underside of the next floor, thus making a large plenum space over many rooms.

A fire that penetrates the Room 1 ceiling will fill the entire plenum. Although POCs could drop down into many rooms, assume that we wish to examine only Path B.

Virtual barriers may be "installed" above the suspended ceiling at the locations of the wall partitions. Thus, we have a virtual room above each actual room. This allows us to describe Path B as:

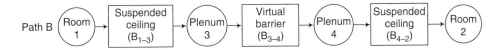

The schematic descriptions provide a clear picture of the thought process for each examination. The virtual barriers provide a convenient way to "package" the analysis into consistent space–barrier modules.

Discussion:
Because virtual barriers are "zero strength" and do not impede movement of flame-heat or smoke-gas, the evaluation effort is not increased. However, the flexibility to recognize potential alternative designs is often enhanced.

Let us discuss this idea with a brief illustration. Assume that the suspended ceiling is a perforated mesh to allow air communication between the plenum and the room. Assume further that the rooms are below-deck staterooms in a cruise ship. A ship is merely a "building" that can change its location easily.

The plenum returns air from the staterooms to a central heating, ventilating, air conditioning (HVAC) source. If a fire were to occur in a stateroom, the return air, containing smoke and gases could drop down into the corridors and other staterooms. This scenario actually happened, causing multiple life loss.

Although this book focuses on analysis, potential alternative solutions may come to mind when the time duration and magnitude of potential problems become evident. In this example, one may:

a) Replace the virtual barriers with physical barriers to impede smoke (and air) movement. Although this solution may be impractical because of design constraints for the HVAC system, it may be considered separately or as part of another alternative.
b) Replace the open plenum space with ducts to transfer stateroom air. The ducts could have dampers into staterooms.
c) Provide for rapid local (room) fire extinguishment with an automatic sprinkler system.

Perhaps additional alternative solutions can be considered as one examines the complete situation. Often, ship design has greater flexibility for solutions than a tenant operated building on land. Impediments for alternative solutions can sometimes be less constraining. However, the purpose here is to create awareness that real or virtual space-barrier concepts can channel thinking and facilitate ideas for finding a better solution.

Example 2.2   Fire can propagate from one floor to the floor above either internally through building components or externally up the façade. Figures 2.6(a) and (b) show a partial section of a multi-story building. Curtain walls are used for exterior wall construction.

A spandrel is the band of exterior wall at floor levels that extends from the head of a window on one floor to the sill of the window on the next floor. Some spandrel distances are small because the curtain walls may use glass panels from floor to ceiling. Other curtain walls may have panels that combine glass windows set into a frame. These curtain walls provide a spandrel distance height that is based on architectural and construction decisions. Regardless of the architectural treatment, one may examine this mode of fire propagation with virtual barriers.

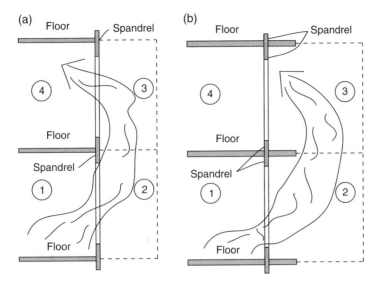

Figure 2.6 Virtual barrier separation. (a) Smooth exterior façade. (b) Exterior façade with eyebrows.

Figure 2.6 segments the path into four rooms using virtual barriers. The paths may be described as:

Discussion:
When a performance analysis indicates that the potential for exterior propagation is unacceptable, different solutions may be considered. An initial set of alternatives could include:

a) Construct an "eyebrow" at each floor level that extends beyond the exterior wall, as shown in Figure 2.6(b).
b) Extend the spandrel height to make smaller windows.
c) Construct balconies at each floor.
d) Install an automatic sprinkler system to extinguish the fire before FO.
e) Install a barrier sprinkler system to prevent flames from breaking glass and entering into the next room.

A fire safety evaluation would provide cost and effectiveness information for each of these solutions. Perhaps additional choices may become evident during a performance analysis. The architect can understand fire propagation implications with each design alternative. The information becomes another part of complex architectural decision-making.

2.13   Space–Barrier Discussion

Connectivity networks that describe space–barrier paths are powerful tools to organize fire analysis, systematize evaluations, and communicate with others. Fortunately, they are easy to formulate. Modern computational techniques such as adjacency (connectivity) matrix analysis can make the process even faster and more accurate for multiple paths in three dimensions from any room of origin.

Barriers may be physical or virtual. Spaces may be selected to provide rigor and understanding from an engineering analysis. Although identification of appropriate barrier-space chains has an obvious application for multi-room fire scenario identification, the concept is valuable for applications such as

- Identify potential paths for flame-heat fire propagation.
- Identify paths of specific interest for smoke-gas movement and tenability.
- Identify potential paths for occupant egress.
- Establish a relationship between the diagnostic fire and target spaces that affect risk characterizations.
- Identify operational sequences for fire department search and rescue.
- Identify attack routes for fire department extinguishment operations.
- Examine possible paths of fire propagation during a fire reconstruction.

Virtual barriers can simplify certain types of problems because the space–barrier module provides a consistent structure for analysis. For example, virtual barriers are useful to develop time movements for long corridors or large spaces. They are particularly valuable to relate time to smoke-gas movement and people movement. They can also structure three-dimensional analyses for open spaces, such as atria. These versatile tools enhance communication with others.

PART THREE: FIRE DEFENSES

2.14   Fire Defenses

Traditional fire defenses are essential for protecting buildings. The selection and location of fire defenses, as well as the quality of installation and maintenance are important to a building's performance. Evaluations involve understanding individual (micro) behavior within an interactive (macro) behavior of the complete building system.

Active and passive fire defenses affect a building's fire performance. Here, we inventory and briefly describe these fire defenses. Some terminology may be slightly different from common usage in order to establish an unambiguous representation of the performance function.

An *active fire defense* is a device or action that must receive a stimulus to act in a real or perceived fire condition.

Traditional active fire defenses include:

1) Automatic detection and alarm system
   a) Fire detection
   b) Alarm actions
   c) Occupant alerting (notification)
   d) Fire department notification
   e) Equipment activation
2) The automatic sprinkler system
3) Fire department operations
   a) Fire suppression
   b) Search and rescue
   c) Property protection and salvage
4) Trained building fire brigade
5) Special hazard or spot protection using automatic systems
   a) Carbon dioxide systems
   b) Dry chemical systems
   c) Foam agent systems
   d) Water-spray systems
   e) Halon replacement systems
6) Special features
   a) Smoke management
   b) Automatic elevator recall
   c) Automatic closing of selected doors or ducts
   d) Emergency lighting and signage
   e) Communication
   f) Emergency power
7) Occupant activities
   a) Detection, occupant alerting, and fire department notification
   b) Fire extinguishment
   c) Property protection
   d) Information receiving or transmitting
   e) Help others
   f) Escape.

A *passive fire defense* is a building component that remains fixed whether or not a fire emergency exists.

Traditional passive fire elements include:

1) Insulation of structural elements to delay or prevent failure.
2) Barriers to prevent extension of the flame-heat or smoke-gas from one space to another.
3) Opening protectives in barriers, such as doors or dampers, to inhibit the movement of flame, heat, smoke, or gases into the adjacent space.
4) The egress system.
5) Areas of refuge.
6) Fire attack routes.

2.15   Active Fire Defenses

The inventory of active fire defenses described above identifies the usual fire defense methods that affect performance. We will briefly describe their functions to provide a context for a more complete discussion of operations and functions in Unit Two. Unit Three will revisit these topics to structure performance evaluations.

2.15.1   Fire Detection and Alarm

Fire detectors are small devices that sense the intrusion of combustion products. Detectors can be designed to sense heat, flame, smoke, or other POCs. Performance effectiveness depends on their location relative to the fire, movement of the combustion products, and detector sensitivity. Fire size at detection is the performance measure for detectors. Fire size at detection may vary from smoldering to relatively large, depending on the fire characteristics, the detection instrument, and its location in the building.

Several different actions may occur after detection. The most common outcome after detection is to sound an alarm or other signal intended to alert occupants of a fire emergency in the building. This is called a local alarm system. A second type of activity transmits a signal to a location within or outside the building to summon an emergency response. A third type of operation occurs when the alarm triggers activation of other fire defenses such as recalling elevators, closing doors, or triggering automatic fire suppression or smoke control systems.

Alarm systems have a role in fire department notification in one of four ways. The first involves a *proprietary, supervisory station* system that transmits a trouble supervisory or alarm signal to a central location. The location may be within the building being protected or in another building controlled by the property owner. Proprietary supervising station fire alarm systems are common in large buildings or with owners of multiple buildings. Owners of these properties normally have a security force to monitor their own security, HVAC, fire alarm systems, or other building operation systems.

Another way to notify a fire department involves transmission of a trouble supervisory or alarm signal to a *central station*. A central station is a facility staffed 24 hours a day and operated by a commercial company that provides trouble supervisory and alarm monitoring and service for buildings. A central station will retransmit an alarm to the fire service communication center as well as provide repair services for the equipment.

A third method transmits a trouble supervisory or alarm signal to a *remote servicing station*. A remote servicing station is normally an alarm-monitoring facility located at the community fire service communication center. When located at a communication center, the attending personnel often retransmit the trouble and supervisory signals to another location that will respond to those signals. When the fire department does not wish to directly monitor the fire alarm systems in its jurisdiction, it may designate a private company to provide the service.

An *auxiliary fire alarm system* is a fourth type of fire department notification. An auxiliary system has direct connections from the building's fire alarm system to the local fire service communication center. This type of service provides the fastest notification to the fire department, although it is not available in all communities.

It is useful to distinguish the different functions that occur after detection by terms associated with specific performance activities. For example, we designate evaluation functions as *alerting* occupants, *notifying* the fire department, or *releasing* other fire protection devices. Each requires a different analysis. A few of these terms may differ slightly from building codes, fire standards, or the vernacular of the trade. However, we attempt to provide unambiguous descriptions of functional performance to structure specific activities for specific components.

2.15.2 The Automatic Sprinkler System

An automatic sprinkler system consists of a water supply, control valves, a piping distribution system, sprinklers, and a monitoring system for trouble and alerting alarms. The objective of the sprinkler system is to extinguish or control a fire within a size that will not be a threat to people, property, or operational continuity. Sprinkler systems have a broad range of design alternatives that allow one to tailor the hardware to meet the needs of a variety of hazards and environmental conditions.

A performance analysis examines an existing or proposed system with regard to its reliability (i.e. will water discharge from a fused sprinkler?) and its design effectiveness (i.e. will the sprinkler system control the fire within specified boundaries?). The reliability is constant for any system. The design effectiveness is dependent on the diagnostic fire characteristics, the fire suppression area, and location within the building.

2.15.3 Fire Department Operations

The local fire department is an important active fire defense system for a building. Many buildings rely exclusively on a combination of barrier effectiveness to delay fire propagation and the fire department to extinguish a hostile fire and aid occupants. Therefore, an understanding of the building's ability to help or hinder fire ground operations is an important part of performance analysis.

A local fire department has three major roles on the fire ground. One is to extinguish the fire, the second is occupant search and rescue, and the third is property protection and salvage.

Community fire emergency resources can vary broadly within any geographic region. It is important to know the staffing, equipment, and water availability for the community. It is equally important to understand the community emergency center operating procedures. In addition, the building location, site features, and building architectural plan all affect fire ground operations. Although the complexity may appear great, a systematic analysis enables a clear picture to emerge.

Life safety is the most important concern of the fire fighters. If it appears that lives may be at risk, a fire chief may redirect available fire fighters to search the rooms and rescue any trapped occupants. This activity may require many of the available fire fighters.

Another fire ground activity is property protection and salvage. Most commonly, this involves protecting exposed zones within the fire building or neighboring exposed buildings from fire extension. Although salvage after a fire was a routine operation in earlier times, this activity is not commonly done today.

The main function of a fire department at a building fire is to act as a change agent. The department takes a dangerous situation and changes the risk by actions that seem most appropriate with the available information and time duration into the incident.

2.15.4 Building Fire Brigade

Some buildings have a trained building fire brigade that can provide a more rapid initial fire attack. This early fire attack is augmented by actions of the local fire department when they arrive. Building fire brigades are common in larger industrial facilities and rare in most other occupancies.

The capability of building fire brigades varies enormously. Some building fire brigades are mini-fire departments with proper equipment and training. Other building fire brigades provide an inappropriate sense of security because they devote few resources to staffing, training, experience, and equipment. This contributes to ineffective fire fighting. When a building fire brigade is available for suppression operations, a performance analysis evaluates their expected capability and effectiveness.

2.15.5 Special Hazard Automatic Suppression Systems

Some buildings house materials that are easily ignitable and fast-burning. Flammable liquids, some gases, and certain chemicals fall into this category. Usually the building's functional operations use these materials in normal operations. For example, dip tanks, machining processes, spray booths, storage tanks for flammable liquids, and deep fat fryers for cooking may present special hazards in manufacturing or commercial buildings. Often these hazards are located in smaller areas within larger operational spaces. Automatic protection for these locations is called spot protection or specific hazard protection.

Automatic extinguishing systems are tailored to the hazard to provide quick extinguishing action and keep operational downtime to a minimum. Performance weakness occurs when the hazard and the means of extinguishing a fire are not matched properly. The most common types of special hazard extinguishing systems are carbon dioxide (CO_2), dry chemical, foam, and water-spray systems.

Special hazard systems supplement the built-in fire protection and are not a substitute for the automatic sprinkler system. The goal is to prevent a fire or explosion from extending beyond the specific area being protected. A major factor in the evaluation of special hazard systems involves the hazard and its characteristics. Assessment of the extinguishing agent capability and its design effectiveness to suppress this hazard requires a good knowledge of the hazard, extinguishing agent, and storage and delivery system.

2.15.6 Special Features

Buildings can provide many special features that help in fire suppression, egress, and emergency operations. Changing an air-handling system into an emergency smoke management mode is becoming more common in building design. The smoke management system may attempt to exhaust smoke from fire floors and pressurize other rooms, such as stairwells and corridors, preventing smoke from migrating into those spaces.

A design may integrate a variety of other features that enhance life safety features. For example, emergency lighting and signage can help occupants and fire fighters to move through the building with a better sense of direction and confidence. Communication systems for alerting occupants and providing emergency information are available. Many buildings have automatic elevator recall so that the fire department can control the use of elevators to reduce damage and life loss. Detection instruments can trigger the closing of selected doors held open by magnetic devices and the shutting of dampers in air ducts. Electrical reliability is enhanced when emergency electrical systems come online when the building's normal power supply is interrupted.

A broad spectrum of equipment and building operations is available to improve a building's performance and risk characterizations. In some cases, they are very effective and valuable. Other features may not provide much value for the construction costs. Nevertheless, their effectiveness becomes a part of a performance analysis.

2.15.7 Occupant Activities

The occupant often provides important contributions to some active fire defenses. While escape is the advice most often given when a fire occurs, the occupant may frequently try to extinguish the fire or protect other people and property.

Humans are sensitive when it comes to recognizing and discriminating a fire's combustion products. Occupants often become a critical link in the detection and notification process. The human interface is the most common means of notifying the fire department of a fire. In addition, occupants may alert other occupants to the existence of a fire emergency and also assist in their evacuation.

In fire ground operations, building occupants often provide valuable information as to the fire's location, potential hazards, and the status of others remaining in the building. When an individual can give detailed information about the existence and location of critical business operations, the fire ground officer may be able to provide strategic protection for those important spaces exposed to the fire.

Occupants often extinguish small fires early in the fire growth process. An occupant can be effective in extinguishing small smoldering and early (i.e. small) flaming fires. Fire extinguishers help an occupant to put out small fires.

Normally, occupant activities are not used in a performance analysis for fires larger than established burning. Nevertheless, active occupant participation is often valuable to save lives, reduce damage, and mitigate risk. Sometimes occupants do not participate. When they are an intentional part of a performance analysis, variability analyses using "what if" conditions evaluate outcomes.

2.16   Passive Fire Defenses

Passive fire defenses remain in position whether or not a fire exists. Often, passive fire protection features may provide additional time until active defenses can become effective. They also provide movement routes for occupants and emergency forces. We will briefly describe their functions to establish a context for building analysis.

2.16.1   Structural Fire Protection

All building materials deteriorate under the high temperatures produced by a fire. Besides high temperatures, the duration of heat energy application causes progressive deterioration to the point of collapse. Passive fire protection attempts to delay collapse of the structural system beyond the point at which the fire has burned out.

Structural steel, reinforced concrete, timber, and prestressed concrete are the most common structural materials used in construction. Different types of protective coverings delay the heat energy transfer from a fire to the structural element. This helps the structural frame to resist excessive deformation or premature collapse. The type of insulation protection depends on the type of the structural framing system.

Unprotected (i.e. exposed) structural steel loses strength rapidly in a fire. Protective coverings delay the heat energy transfer from the fire gases to the structural steel. Mineral spray-on materials or gypsum board coverings usually protect steel beams and girders. Concrete encasement was once a common method of providing structural fire protection, but it is not as common today. Intumescent coatings may have value for specialized applications.

A common way to provide protection for steel joist construction uses a suspended membrane ceiling. A membrane ceiling delays excessive heat accumulation in the space between the ceiling and the floor above. Membrane systems are constructed to allow thermal expansion of the ceiling framing without collapse. Lighting fixtures and air diffusers must be compatible with the ceiling design and construction. The entire system of structural framing, suspended ceiling, and lighting and air diffusers is described as a floor–ceiling assembly or a roof–ceiling assembly.

Reinforced concrete and prestressed concrete use a thickness of concrete cover to protect the steel reinforcing bars.

Timber can lose strength upon heating, and the material also augments the fire when it burns. When wood burns, a char forms to provide some insulation for continued pyrolysis. This may delay collapse for a time. Encasing the wood with gypsum boards is another common way to protect wood members.

The ability to calculate with confidence the expected behavior of any element of a structural frame has expanded greatly in recent years. More is known about structural behavior than any other part of the complete building–fire system.

2.16.2   Barriers

We define a barrier as any surface that will delay or prevent POCs from propagating into an adjacent space. This definition is broader than traditional descriptions because we are interested in the full range of barrier behavior rather than only the fire endurance duration from a standard test. Barrier performance identifies the time delay for flame-heat or smoke-gas movement into an adjacent space.

Typical barriers from a code viewpoint are floor–ceiling and roof–ceiling assemblies, interior partitions that separate building fire areas, firewalls, and enclosed corridors and shafts. Firewalls are specially constructed barriers intended to prevent any extension of fire from one side of a building to the other. Designers attempt to ensure that a firewall remains stable even if a fire destroys the structural framing on one side of the wall.

Code-defined partitions and firewalls specify a fire resistance rating based on the ASTME 119 or ISO 834 standard fire endurance test. The test classifies construction systems according to their relative ability to prevent fire propagation from one space to another. Classifications are based on a standardized fire and expressed as a time in hours or minutes. Many in the construction business are under the misconception that the hourly rating represents the time a barrier assembly will last in a building fire. The assembly may fail in a considerably shorter time or it may be successful for a longer time. Construction and field conditions influence actual barrier performance.

Although the standard fire test does provide some useful information, it is inadequate for a performance analysis. The field conditions, particularly with attention to openings and their status during a fire, are critical to fire behavior. Performance evaluations examine the full range of barrier conditions for the duration of the fire.

2.16.3 Opening Protectives

Many barriers have openings to allow the passage of people, light, heat, air, plumbing, electricity, or other building services from one room to another. The fire resistance changes whenever a barrier is opened or penetrated. Opening protectives, such as doors, windows, shutters, and dampers, restore some fire resistance to barriers. Also, fire stopping around pipes in through construction prevents fire propagation between adjacent spaces. Both their quality and their status during a fire influence performance.

When viewing an opening protective, the frame and its connection to the barrier become important to performance. From a code viewpoint, fire endurance ratings are accompanied by an associated endurance for the protectives.

2.16.4 The Egress System

A building is expected to provide a safe path for occupants to leave. A means of egress is defined as a continuous and unobstructed path from any location in a building to a public way. A means of egress consists of three separate and distinct parts. The first is the *exit access*, which is the path that connects any building location with an exit. Typically, corridors become the exit access for most buildings. The second component is the *exit*. An exit is a specially protected space separated from other parts of the building by construction that can provide a temporary area of refuge for occupant movement to the exit discharge. Stairwells and exterior doors are often the building exits. An *exit discharge* is the third component that terminates the means of egress. The exit discharge leads to a public way, usually outside the building.

A means of egress is a designed path to the outside that incorporates additional safety features. While no building code identifies an acceptable risk, all codes give rules for constructing egress paths. Many buildings do not have all the features of a code-complying means of egress. These nonconforming routes are often described as "ways out." While a layman's perception is that any way out is an "exit," this is not the technical definition of the term.

2.16.5 Area of Refuge

A building often has areas that can provide refuge from a fire's POCs. Most often, these locations are accidental rather than designed. When areas of refuge are a conscious part of a building design they become part of a performance analysis. The building exit (the second component described in the egress system) is intended to be a temporary area of refuge to permit passage of occupants to the exit discharge.

Intentional (or advertized) areas of refuge are evaluated to determine if occupants will remain safe for the duration of the fire. This function and the components of an area of refuge are discussed in Unit Two. The corresponding analysis is structured in Unit Three.

2.16.6 Fire Attack Route

For most practical situations, a fire attack route is the inverse of the egress path. Fire fighters must move themselves and their equipment to the fire location. This often involves movement in a direction opposite to the occupants who may be leaving the building.

An interior fire attack is usually launched from the stairwells that are the exits. Stairwells may contain standpipes and provide some protection to fire fighters from heat and smoke. This enables fire fighters to pause, secure their self-contained breathing apparatus (SCBA), and prepare to advance hose lines for water application.

2.17 Closure

Building codes and associated standards are important to gain regulatory approval for a new design. Conventional regulatory approval is not the focus of this book. Rather, we wish to document a framework with which to analyze any existing building or proposed new building for fire performance.

The product of a performance evaluation is an enhanced understanding of the way the building will work in a fire. This understanding is used to characterize risks to exposed people, property, operational continuity, neighboring exposures, and the environment in a way that the traditional regulatory process cannot. Performance understanding will enable one to make better decisions regarding risk management or other fire safety engineering projects.

This chapter introduces a few concepts that are necessary to structure an evaluation. The concept that a building analysis starts at EB is important. EB is the demarcation that separates fire prevention considerations from the building's performance expectations. The separation of fire prevention from building behavior is important to understanding holistic performance. Prevention effectiveness is normally examined when we understand how the building will behave after the fire starts.

Time is the factor that links all parts of a performance analysis.

3

Tools of Analysis

3.1   Introduction

The primary goal of this book is to describe a framework for analyzing fire performance in buildings.

A framework would be sterile without ways to quantify the components. While it is useful to distinguish between operational mechanics and quantification, they are a matched pair. Although the framework structure can remain constant, capabilities in numerical measures improve as the state of the art of fire science and engineering improves.

The analytical framework is intended to address the following attributes:

- Easy to use.
- Have transparent logic.
- Possess technical rigor that can be documented.
- Identify with a fire safety engineering thought process.
- Move easily between component behavior and holistic performance.
- Incorporate dynamic actions and their interfaces among components.

The framework organizes critical events of a building fire into packages that guide component performance evaluations. Although the framework guides the analytical process, numerical measures establish component behavior for selected time durations and conditions. Component evaluations have the following attributes:

- Based on deterministic analysis.
- Use state-of-the-art fire science and technology.
- Incorporate the influence of other elements and conditions that affect the component being examined.

The framework uses a variety of logic diagrams and other analytical relationships to structure fire performance. Part One introduces logic diagrams and their use to guide the thought process. Part Two describes dynamic movement analysis through space–barrier connectivity networks. Part Three introduces a few additional tools to help guide the process.

Fire Performance Analysis for Buildings, Second Edition. Robert W. Fitzgerald and Brian J. Meacham.
© 2017 John Wiley & Sons Ltd. Published 2017 by John Wiley & Sons Ltd.

PART ONE: THE LOGIC

3.2   The Framework Logic

Fire and buildings form a complex, dynamic system. Nevertheless, the system can be decomposed into manageable parts that may be evaluated as independent components. Then, the components can be recombined to provide performance understanding.

The framework adapts techniques from other disciplines to provide logical coherence. Their integration provides the basis for constructing an Interactive Performance Information (IPI) chart. An IPI chart orders time relationships of different components and records important information for evaluations and risk characterizations. The IPI chart is introduced in this chapter and is discussed in greater detail in Chapter 4.

The techniques that structure the framework and its associated IPI chart are:

a) *Fault trees* and *success trees* use deductive logic to structure component performance. These logic diagrams provide a "big picture" association of the major events and their relationships.
b) *Network logic diagrams* recast fault trees and success trees to incorporate sequence, dependence, and conditionality into evaluations.
c) *Event trees* incorporate the dynamics of time with status changes in component elements.
d) The *IPI chart* combines these logical structures into a form that provides both visual and informational descriptors. The IPI chart allows one to observe changes in any component over time as well as to examine the status of any or all components at any instant of time.

These techniques structure the analytical framework. Evaluations use a combination of quantitative and qualitative assessments. Variation ("what if") analyses enable one to augment performance understanding by examining differences in condition. The IPI chart is a major contributor to understanding interactive performance.

3.3   The Major Parts

Fire performance is dynamic because the building, the fire defenses, and human activities respond in different ways that affect, and are affected by, continually changing fire conditions. In addition, many fire defenses and human actions phase in or phase out of operation as the situation changes. Time links all parts of a complete building fire safety system.

Although every fire is different, the process for analyzing building performance and characterizing the risks is consistent, logical, and orderly. Figure 3.1 shows a static representation of the major components for a holistic view of building performance and risk.

An evaluation starts with a *diagnostic fire* scenario. The diagnostic fire describes the fire growth and propagation assuming no extinguishment intervention. Because barrier openings often produce multi-room fires, a diagnostic fire includes the room of origin and the additional cluster of rooms that can be involved due to openings or weakness in

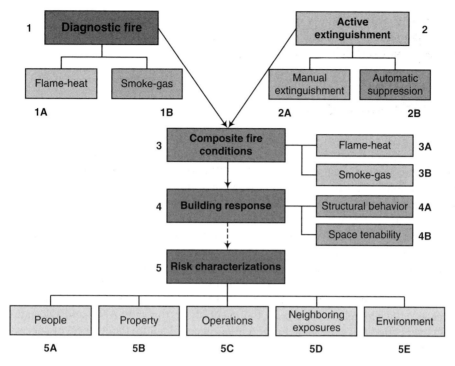

Figure 3.1 System components.

surrounding barriers. This natural fire describes the time and characteristics of the flame-heat and smoke-gas movement through the affected barrier–space modules.

Active extinguishment can occur only two ways: automatic sprinkler suppression and local fire department extinguishment. Only the local fire department can extinguish fires in any building that is not protected by an automatic sprinkler system. Although each extinguishment method is evaluated independently, the two methods can be integrated when appropriate.

The *composite fire* is identical to the natural (diagnostic) fire until initial water application disturbs the process. After the first suppression water is applied, the heat and smoke continue with a turbulent, changing rate until extinguishment is complete. Composite fire conditions continue to affect the structural behavior and the smoke tenability until the fire is put out.

The building response uses composite fire conditions to evaluate structural frame behavior and the tenability in target spaces. Target spaces may be identified because of their importance to risk characterizations. For example, spaces through which people attempt to leave a building or rooms that defend-in-place occupants must occupy are usually designated as target spaces. The hostile environment can also place at risk fire fighters using personal protective equipment. The composite fire can also threaten property (building, information, and contents), operational continuity, nearby exposed structures or enterprises, and the natural physical environment.

During a fire, building spaces may provide adequate protection for a period of time, but not necessarily continuously. Because fire performance is dynamic, we characterize space tenability with 'windows of time' that describe a space's ability to protect the exposed. Therefore, risk is associated with time durations during which the building is unable to maintain tenability in important spaces.

3.4 Event Logic Diagrams

Logic diagrams in the form of fault trees and success trees enable one to recognize performance relationships. These tools have been available for half a century. For example, the NFPA Fire Safety Concepts Tree (NFPA 550) uses an event logic diagram to guide design selections. We use event logic diagrams to identify logical relationships for analytical components.

An event is an occurrence. For example, a fire reaches the size of 800 kW; or a sprinkler actuates; or the local fire department is notified of a fire; or an individual safely reaches the exit discharge. These descriptions illustrate specific occurrences that we call events. Precise event descriptions are necessary to establish logical relationships.

Logic gates connect events. An AND gate (symbol: $\odot$) indicates that *all* events below that gate are necessary for the event above the gate to occur. An OR gate (symbol: $\oplus$) signifies an *either/or* relationship with the event above the gate. AND and OR gates connect events that trace the logic for potential causes of failure (fault trees) or success (success trees).

Fault Tree   A fault tree identifies a specific failure condition and traces (deductively) the root causes of failure. Although fault tree construction is more of an art than a science, it does require a thorough understanding of the system being studied and the role of each component within the system. One identifies all of the factors that contribute to or could cause operational failure of the system. These factors are ordered and grouped into distinct failure events which are synthesized into a logical framework. After initial completion, the tree logic is reviewed and tested. The main tests for a fault tree are: will it work, and is it logical?

Figure 3.2(a) shows the first levels of a fault tree for active extinguishment in a building fire. Note that the events are written in terms of failure, and all events are related to AND or OR gates.

Event descriptions use consistent symbolic representations. Here, a letter without a bar represents success of an event (e.g. M = manual extinguishment successfully occurs). A bar "−" over the symbol indicates "not" (e.g. $\bar{M}$ = manual extinguishment does not occur).

One can 'read' a fault tree, such as that of Figure 3.2(a), as follows:

Given: *A specific fire size and time from established burning (EB).*

1) *Failure of active extinguishment (top event) occurs when:*
    The fire is not suppressed by an automatic sprinkler system ($\bar{A}$)
    AND
    The fire is not manually extinguished by the fire department ($\bar{M}$).
2) *The fire is not suppressed by an automatic sprinkler system when:*
    A sprinkler system is not present ($\bar{AP}$)
    OR

(a)

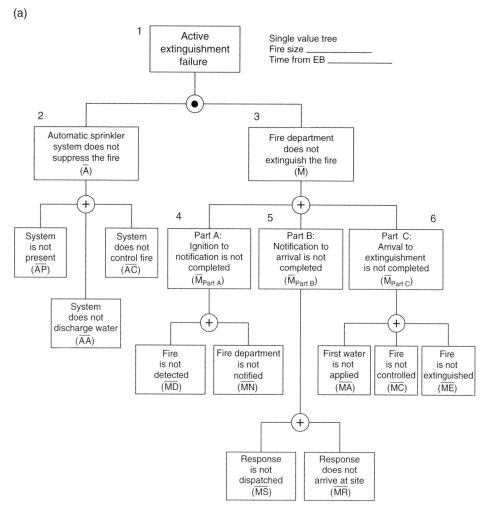

Figure 3.2 (a) Fault tree.

The sprinkler system does not discharge water (i.e. apply agent) from a fused
    sprinkler $(\overline{AA})$
OR
The sprinkler system does not control the fire before it extends to a larger size $(\overline{AC})$.
3) *The fire is not extinguished by the local fire department when:*
Part A (ignition to notification) is not completed $(\overline{M}_{Part\,A})$
OR
Part B (notification to arrival at the site) is not completed $(\overline{M}_{Part\,B})$
OR
Part C (arrival to extinguishment) is not completed $(\overline{M}_{Part\,C})$.
4) *Part A (ignition to fire department notification) is not completed $(\overline{M}_{Part\,A})$ when:*
The fire is not detected $(\overline{MD})$
OR
The fire department is not notified $(\overline{MN})$.

(b)

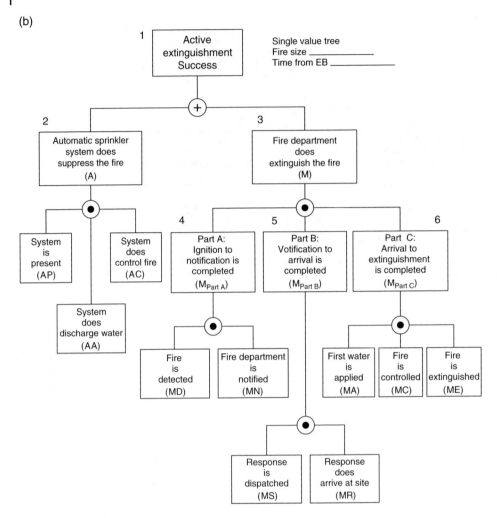

Figure 3.2 (Cont'd) (b) Success tree.

5) *Part B (notification to arrival at site) is not completed ($\overline{M}_{Part\ B}$) when:*
   The response companies are not dispatched ($\overline{MS}$)
   OR
   The response companies do not arrive at the site ($\overline{MR}$).
6) *Part C (arrival at site to extinguishment) is not completed ($\overline{M}_{Part\ C}$) when:*
   The fire department does not apply first water ($\overline{MA}$)
   OR
   The fire department cannot control the fire ($\overline{MC}$)
   OR
   The fire department cannot extinguish the fire ($\overline{ME}$).

A fault tree organizes the contributing events into a logical framework that enables one to deductively trace the causes of a failure event. Fault trees provide insight into the specific factors that can cause failure during a performance analysis.

Success Tree   A success tree views the same process in terms of success rather than failure. A fault tree can be converted into a success tree with only two modifications. The first changes failure event descriptions into terms of success. The second converts the logical OR gates to AND gates, and vice versa. Therefore, a success tree is an inverse of a fault tree. Figure 3.2(b) shows the success tree that corresponds to the fault tree for active extinguishment.

   This success tree may be 'read' as follows:

Given: *A specific fire size and time from EB.*

1) *Active extinguishment success (top event) occurs when:*
   The fire is controlled by an automatic sprinkler system (A)
   OR
   The fire is manually extinguished by the fire department (M).
2) *The fire is suppressed by an automatic sprinkler system when:*
   A sprinkler system is present (AP)
   AND
   The sprinkler system does discharge water from a fused sprinkler (AA)
   AND
   The sprinkler system does control the fire within the specified area (AC).
3) *The fire is extinguished by the local fire department when:*
   Part A (ignition to fire department notification, $M_{Part\ A}$) is completed
   AND
   Part B (notification to arrival at site, $M_{Part\ B}$)is completed
   AND
   Part C (Arrival to Extinguishment, $M_{Part\ C}$) is completed.
4) *Part A: Ignition to fire department notification ($M_{Part\ A}$) is completed when:*
   The fire is detected (MD)
   AND
   The fire service communication center receives notification of the fire and the building's address (MN).
5) *Part B: Notification to arrival ($M_{Part\ B}$) is completed when:*
   The fire service communication center dispatches the responding fire companies (MS)
   AND
   The assigned companies arrive at the site (MR).
6) *Part C: Arrival to extinguishment ($M_{Part\ C}$) is completed when:*
   The fire department applies first water to the fire (MA)
   AND
   The fire department controls the fire (MC)
   AND
   The fire department extinguishes the fire (ME).

A success tree shows a static description of all critical events that can provide successful performance. Although at certain times during the scenario some events will have been successfully completed while others have not yet started, the logic identifies the events that must be completed for success.

3.5 Event Logic Observations

Fault trees and success trees identify logical relationships among events. Although the event logic decomposition process can show component hierarchy and relationships, one must be aware of the following characteristics:

- The causal relationships do not incorporate time. Therefore, a tree may be viewed as a single frame of a continuous video in which the measures of each event continuously change. We call these *single value* trees because the success or failure modes have value or actuate at a specific instant of time. Thus, an event that may not be operational or have value at one instant of time may be successful at another instant of time. Time coordination is discussed later in this chapter.
- Only the events that contribute to success or failure are shown in the logical decomposed sequence. Sequencing and coordination of events are not always clear.
- Sequential, dependent, independent, conditional, and exclusive events are rarely identified. One must clearly understand event associations and their sequencing to evaluate outcomes. For example, in Figure 3.2(b), Parts A, B, and C of fire department suppression are sequential and conditional on successful completion of the previous event. Also, the sequence in which their sub-events are evaluated becomes important because conditionality among other events can affect quantification. For example, the diagnostic fire size at any particular time will affect all suppression event evaluations. On the other hand, in evaluating automatic suppression (A), the initial agent application (AA) and design effectiveness (AC) are independent components. The sequence in which these two events are examined does not matter, although both must act for a successful operation. We identify event relationships in Units Two and Three.
- Decomposition is carried to the level that is needed. Often, an engineering analysis will use additional levels of detail to understand performance. Fewer levels may be needed to communicate outcomes to other professionals. For example, the decomposition levels of Figure 3.2 illustrate the process. Evaluations could extend the trees to additional levels of detail to enable one to analyze performance with greater confidence.
- The mathematical logic of event tree construction allows one to calculate outcomes. At one time, probability had a role in the development of these trees and was the basis of quantification. We now use the trees to guide deterministic evaluations. Because site-specific building analyses have many dependent and conditional circumstances that are affected by unique situations, we find deterministic evaluations to be more informative.
- Logic diagrams can be constructed to describe either success or failure. However, both outcomes cannot be shown on the same diagram. Figure 3.2(a) shows a failure diagram. Figure 3.2(b) shows the corresponding success diagram. The only differences between these diagrams involves a change of event descriptions and the transference of AND and OR gates.

The choice between a success tree and a fault tree relates to application and needs of evaluation. We find fault trees to be useful for initially developing deductive relationships. However, success trees are better suited to understand overall performance.

3.6   Logic Networks

The event logic diagrams described in the previous section are useful to envision the "big picture" of what is important in a performance evaluation. Unfortunately, these diagrams are not useful in showing an analytical thought process. Nor can they conveniently relate events such as the evaluation of fire suppression for a sequence of rooms. To overcome this difficulty, we recast fault and success trees into a network hierarchy that portrays the same logic and also guides the analysis.

A network is a semi-graphical display of nodes and branches (lines) that portray a process. When the networks conform to mathematical logic, numerical measures of performance can be calculated, if desired.

Networks have two common forms. One shows activities on nodes (AONs). The other uses activities on branches (sometimes called arrows) (AOAs). One can easily convert from AONs to AOAs and vice versa. However, we will use only the AON networks because they are more suitable for fire performance applications.

We will illustrate one form of network construction and application by converting the fault and success trees of Figures 3.2 into an equivalent set of bimodal logic networks. Each logic network has two boundary conditions that define end events. The starting event is the status of a previous event or a clearly defined condition. The ending events define the outcome that is being evaluated. Figure 3.3 illustrates the form. Here, the

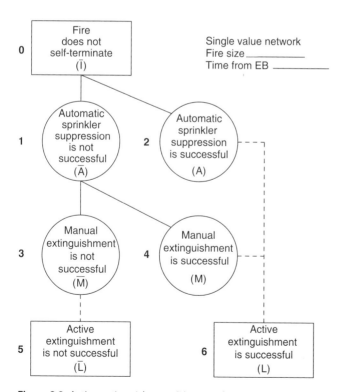

Figure 3.3 Active extinguishment (L) network.

starting point is a fire that continues to burn. The terminal events (L) and ($\bar{L}$) are the possible outcomes of active extinguishment for automatic sprinkler suppression and local fire department manual extinguishment.

We may 'read' Figure 3.3 as follows:

Initial condition: A specific fire size and time from EB

Path 0–1–3–5:
  GIVEN: A fire that does not self terminate, but continues to burn ($\bar{I}$)
  AND
  The fire is not suppressed by an automatic sprinkler system ($\bar{A}$)
  AND
  The fire is not extinguished by the local fire department ($\bar{M}$)
  THEN: Active extinguishment is not successful ($\bar{L}$). Terminal event 6 signifies that active extinguishment is successful. The outcome for this event can be traced through two paths (as follows).
Path 0–2–6:
  GIVEN: A fire that does not self-terminate, but continues to burn ($\bar{I}$)
  AND
  The fire is suppressed by an automatic sprinkler system (A).
  THEN: Active extinguishment is successful (L).
  OR
Path 0–1–4–6:
  GIVEN: A fire that does not self-terminate, but continues to burn ($\bar{I}$)
  AND
  The fire is not suppressed by an automatic sprinkler system ($\bar{A}$)
  AND
  The fire is extinguished by the local fire department (M).
  THEN: Active extinguishment is successful (L).

Therefore, active extinguishment can be successful when:

The fire is suppressed by an automatic sprinkler system (A)
OR
The fire is not suppressed by a sprinkler system but is extinguished by the local fire department (M).

The dashed lines are transfer (dummy) connections to provide network continuity.

Figure 3.3 recasts the top level of the fault and success tree diagrams of Figure 3.2 for active extinguishment into an equivalent network diagram. Network hierarchy guides an explicit analytical process.

AND gates connect any continuous chain of events in logic networks. Thus, the ending event of "Active extinguishment is not successful ($\bar{L}$)" shows the top level of the fault tree of Figure 3.2(a). On the other hand, OR gates connect two or more chain-ending events that lead to a common terminal event. The ending event of "Active extinguishment is successful (L)" shows the top level of the success tree of Figure 3.2(b).

3.7 Decomposing Logic Networks

A major attribute of network diagrams is the ability to guide the thought process of a logical performance evaluation. Although a separate network is needed to diagram each successive level of event logic trees, this does not pose any difficulty or added analytical time because networks are nested to form a coherent package. Network packages provide a consistent, clear thought process that can be decomposed to any desired level.

We will show a few levels of Figures 3.2 to illustrate the process.

Automatic Sprinkler System   The network of Figure 3.4 describes the next level of logic diagrams for automatic sprinkler suppression.

Figure 3.4 may be read as follows:

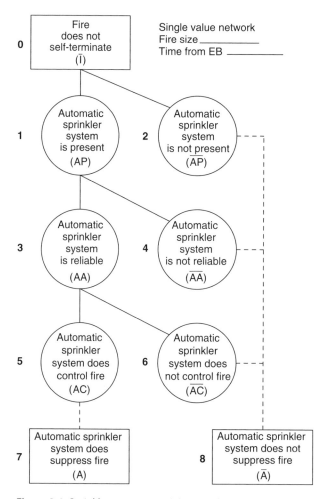

Figure 3.4 Sprinkler suppression (A) network.

Initial condition: A specific fire size and time from EB.

Event 7 notes that the automatic sprinkler system does suppress the fire (A) when:

Path 0–1–3–5–7.
  GIVEN: A fire that does not self-terminate, but continues to burn (Ī)
  AND
  The sprinkler system is present (AP)
  AND
  The sprinkler system does discharge water from a fused sprinkler (AA)
  AND
  The sprinkler controls the fire within the specified size (AC)
  THEN: The fire will be suppressed by the sprinkler system (A).

Event 8 notes that the automatic sprinkler system will not suppress the fire (Ā) when:

Path 0–2–8: The sprinkler system is not installed ($\overline{AP}$)
OR
Path 0–1–4–8: The sprinkler system is present (AP), but water will not discharge from a fused sprinkler ($\overline{AA}$)
OR
Path 0–1–3–6–8: The sprinkler system is present (AP) and water discharges (AA), but the system is unable to control the fire within a specified size ($\overline{AC}$).

The network uses exactly the same logic as that of the fault and success trees of Figures 3.2 (a) and (b). However, the thought process traces the sequence more explicitly.

Manual Extinguishment   Figures 3.5(a)–(d) show the networks that describe a manual extinguishment evaluation. Briefly, the network of Figure 3.5(a) may be read as follows:

Initial condition: A specific fire size and time from EB.

GIVEN: A fire that does not self-terminate, but continues to burn (Ī).

Path 0–1–3–5–7: The fire will be successfully extinguished (M) when Part A: Ignition to notification ($M_{Part\ A}$) is completed; AND Part B: Notification to arrival ($M_{Part\ B}$) is completed; AND Part C: Arrival to extinguishment ($M_{Part\ C}$) is completed.

The local fire department will not successfully extinguish the fire ($\overline{M}$) when:

Path 0–2–8: The fire department is not notified of the fire ($\overline{M}_{Part\ A}$)
OR
Path 0–1–4–8: The fire department has been notified ($M_{Part\ A}$) AND is unable to arrive at the site ($\overline{M}_{Part\ B}$)
OR
Path 0–1–3–6–8: The fire department has been notified ($M_{Part\ A}$); AND arrives at the site ($\overline{M}_{Part\ B}$); AND is unable to extinguish the fire ($\overline{M}_{Part\ C}$).

Figures 3.5 (b)–(d) structure the remaining analytical decomposition. Figure 3.5(b) examines the building management's responsibility described in Part A (Ignition to notification) as follows:

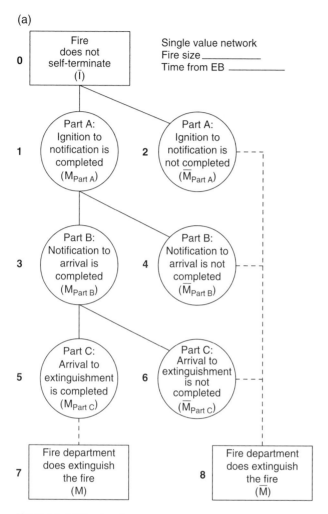

Figure 3.5 (a) M network.

Initial condition: A specific fire size and time from EB.

GIVEN: A fire that does not self-terminate, but continues to burn ($\bar{\text{I}}$).

Path 0–1–3–5: The fire is detected (MD) AND the local fire department has been notified (MN)
THEN: Part A (Ignition to notification) will be successfully completed ($M_{\text{Part A}}$).

Part A will not be successfully completed($\bar{M}_{\text{Part A}}$) when:

Path 0–2: The fire is not detected ($\overline{\text{MD}}$)
OR
Path 0–1–4–6: The fire has been detected (MD) AND the local fire department has not been notified ($\overline{\text{MN}}$)
THEN: Part A will not be completed ($\bar{M}_{\text{Part A}}$).

(b)

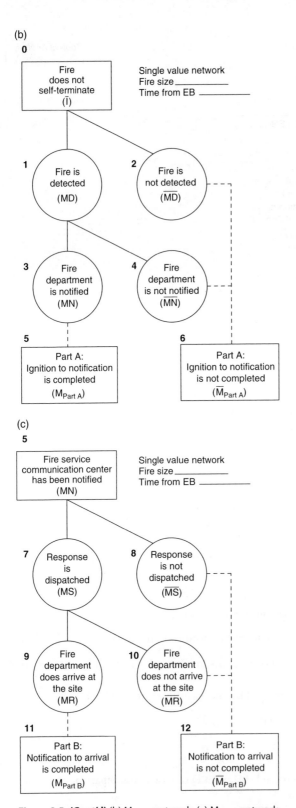

(c)

Figure 3.5 (Cont'd) (b) $M_{Part\,A}$ network. (c) $M_{Part\,B}$ network.

(d)

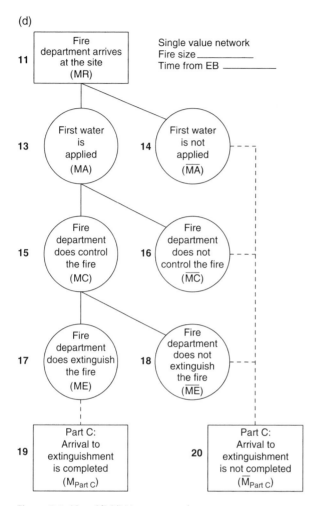

Figure 3.5 (Cont'd) (d) $M_{Part\,C}$ network.

Figure 3.5(c) examines the local community management's responsibility described in Part B (Notification to arrival) as follows.

Initial condition: A specific fire size and time from EB.

GIVEN: The fire service communication center has been notified of the fire and its location (MN).

Path 5–7–9–11: The responding companies have been dispatched (MS) AND they arrive at the building site (MR)
THEN: Part B (Notification to arrival) will be successfully completed ($M_{Part\,B}$).

Part B will not be successfully completed ($\overline{M}_{Part\,B}$), when:

Path 5–8: The fire service communication center has not dispatched the responding fire companies ($\overline{MS}$)
OR

Path 5–7–10–12: The responding companies have been dispatched (MS) AND they have not yet arrived at the site ($\overline{\text{MR}}$)

THEN Part B will not be completed ($\overline{\text{M}}_{\text{Part B}}$).

Figure 3.5(d) structures the evaluation of Part C (Arrival to extinguishment) as follows.

Initial condition: A specific fire size and time from EB.

GIVEN: The first-in apparatus has arrived at the site (MR).

Path 11–13–15–17–19: First water has been applied to the fire (MA) AND the fire is controlled (MC) AND the fire is extinguished (ME)

THEN: Part C (Arrival to extinguishment) will be successfully completed ($\text{M}_{\text{Part C}}$).

Part C (Arrival to extinguishment) will not be successfully completed ($\overline{\text{M}}_{\text{Part C}}$) when:

Path 11–14–20: First water has not been applied ($\overline{\text{MA}}$)

OR

Path 11–13–16–20: First water has been applied (MA) AND the fire has not been controlled ($\overline{\text{MC}}$)

OR

Path 11–13–15–18–20: First water is applied (MA) AND the fire is controlled (MC) AND the fire has not been extinguished ($\overline{\text{ME}}$)

THEN Part C will not be completed ($\overline{\text{M}}_{\text{Part C}}$).

3.8 Network Diagram Observations

Bimodal network diagrams recast event logic trees to provide easier analysis. Although fault and success trees can portray a complete package of events on a single diagram, networks guide performance evaluations more explicitly. The mathematical logic is the same for both.

Network diagrams have the following characteristics:

- Networks provide a routine, consistent process to evaluate performance. Sequencing and coordination of events in networks are usually clear. Exclusive or independent events can be noted when they occur.
- Networks do not incorporate time into the causal relationships. Networks may be considered as a single frame of a continuous video. The video concept emerges because the event status changes during the sequential time periods. This limitation becomes manageable with the IPI charts discussed in Section 3.12.
- This chapter only describes the thought process for constructing networks. Units Two and Three discuss component operations and event quantification.
- The networks display both success and failure on the same diagram. This enables one to mix and match event quantification information using the most convenient techniques.
- Decomposition is carried to the level that is needed to create a good understanding of performance. The structure enables one to bridge knowledge gaps when state-of-the-art quantification resources are inadequate.

3.9 Single Value Networks

We describe the evaluation networks discussed above as single value networks (SVNs) to emphasize that each event depicts a state of behavior at a single instant of time. Because all events are dynamic and change over time, one selects appropriate time intervals to evaluate performance.

A major strength of SVN networks is the ability to examine and depict a status at a specific instant of time. A SVN analysis "stops the world" to examine in detail the interactive performance of system parts at a specific moment in time.

3.10 Time Relationships Using Event Trees

Time coordination is essential to understand fire performance. Conventional risk analysis procedures use event trees to describe time-related behavior. We adapt these techniques to incorporate time into the performance framework.

An event tree is a logic diagram that starts with a defined initial event and establishes a forward (inductive) logic to organize future events. A traditional risk analysis typically uses event trees to describe a scenario involving dissimilar events. We adapt that concept to describe the changes in each event that occur over time. This enables one to identify component status changes over time durations.

Figure 3.6 illustrates this concept to describe fire department extinguishment. The evaluation at each node answers the question: "Will the fire department extinguish the fire *n* minutes after EB?" This enables one to use SVN analyses to evaluate the dynamic event changes over time.

The event tree depicts changes in status of an event for a continuum of time intervals.

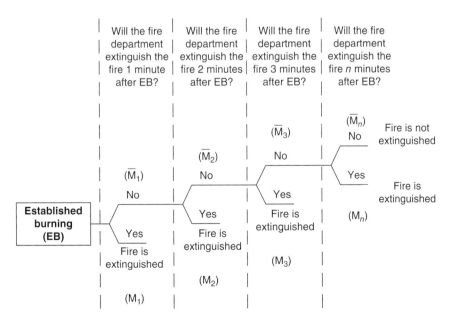

Figure 3.6 Event tree (M).

3.11   Continuous Value Networks

Time coordination within and among components is essential to understand dynamic fire performance. Continuous value networks (CVNs) order the process.

Any conventional event tree can be recast into an equivalent continuous value network. Figure 3.7 shows the CVN that is equivalent to the event tree of Figure 3.6. The symbols identify whether the local fire department has extinguished the fire (M) or has not extinguished it ($\overline{M}$). The subscripts identify the time after EB.

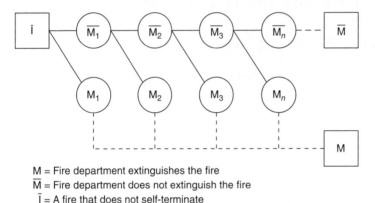

M = Fire department extinguishes the fire
$\overline{M}$ = Fire department does not extinguish the fire
$\overline{I}$ = A fire that does not self-terminate

Figure 3.7 Event tree (M).

A CVN is analogous to a continuous video. One can stop the CVN at any time to examine its status with the corresponding SVNs at that instant. After sufficient information is gained from the instantaneous SVN examination, we continue with the CVN to the next event or time of interest.

The CVNs and SVNs enable one to coordinate time durations with diagnostic fire environment growth and status changes in other fire defenses. We show a CVN as a horizontal network to emphasize that the events display a continuous passage of time. Conversely, a SVN is shown as a vertical network to highlight the instantaneous status of an examination.

3.12   The IPI Chart

A building's fire safety system involves components that affect and are affected by other components. Although the basic components and their functions can be described as a general anatomy, their interactions in the time duration of the fire make each building fire unique. Performance is significantly influenced by the way in which components work together and the times during which they interact.

The IPI chart provides a way to show actions and organize performance data. An IPI chart integrates the basic tools of:

1) Success trees that organize the major components for an evaluation. Performance events define the organizational structure of the IPI chart.
2) CVNs that describe time-related changes in status from ignition to extinguishment for each scenario component. These are shown by the horizontal rows of the IPI chart.

	Section	Symbol	Event	Time 5	10
Building performance / Fire response	1	$\bar{I}$	**Diagnostic fire**		
	A	$\bar{I}_F$	**Flame-heat**		
	B	$\bar{I}_S$	**Smoke-gas**		
	2		**Active extinguishment**		
	A	M	**Fire department extinguishment**		
	a	$M_{Part A}$	Part A: Ignition to notification		
		MD	Detect fire		
		MN	Notify fire department		
	b	$M_{Part B}$	Part B: Notification to arrival		
		MS	Dispatch responders		
		MR	Resonders arrive		
	c	$M_{Part C}$	Part C: Arrival to extinguishment		
		MA	Apply first water		
		MC	Control fire		
		ME	Extinguish fire		
	B	A	**Automatic sprinkler suppression**		
	a	AA	Sprinkler system is reliable		
	b	AC	Sprinkler system controls fire		
	3	L	**Composite fire**		
	A	L_F	**Flame-heat**		
	B	L_S	**Smoke-gas**		
Building response	4	R	**Building response**		
	A	St	**Structural frame**		
	B	TS	**Target space tenability**		
Risk characteristics	5	L	**Risk characteristics**		
	A	LS	**People**		
	B	RP	**Property**		
	C	RO	**Operational continuity**		
	D	RE	**Exposure protection**		
	E	RN	**Environment**		

(Working IPI template)

Figure 3.8 Working Interactive Performance Information chart.

3) SVNs that examine details of a component's performance at a specific instant of time or at a critical event in a building fire. Vertical columns show each component's sub-events as an integrated package for evaluation. Vertical columns enable one to examine the status of any or all components at any specified time.

Figure 3.8 shows a basic IPI chart. The general organization is based on relationships of Figure 3.1. The horizontal rows describe CVN status changes from EB to extinguishment for components that affect performance. The vertical columns show SVN information at any specific instant of time. Because IPI charts are so important to organizing, evaluating, and communicating fire safety performance, Chapter 4 discusses additional features.

Although the organizational form of the IPI chart is constant, additional decomposition levels for events may be incorporated. Appendix A provides a complete array of IPI charts that have been decomposed to levels for complete analysis. These IPI charts enable one to examine component parts to a depth appropriate for the needs.

One may use spreadsheet capabilities to examine individual components separately or in combination. Different ("what if") scenarios may be examined as separate evaluations and then compared easily. IPI charts are versatile tools that enable one to examine a wide variety of conditions with a reasonable expenditure of professional time.

3.13   Coding

Every building is in some ways different from every other building. Every building fire scenario is different from every other scenario. Nevertheless, the organizational structure with which to analyze fire safety is methodical and consistent.

There are many different components that relate to building fire safety. Also, different scenarios could provide additional insights into reasonable variations that may be expected for the building. Sometimes it is difficult to keep scenarios in order to avoid unintentional mix-ups. We reduce the opportunity for inadvertent errors by coding component packages with exclusive colors and symbols.

The major components of the IPI chart have been given color codes. These are shown in the centerfold. For example, all fire department components are shades of green, to distinguish successive levels of decomposition. Therefore, when using several spreadsheets involving variations in different components or scenarios relating to the fire department, the opportunity for error involving other components is reduced. The pattern is repeated with different color codes that are consistent for each component. The centerfold illustrates the consistency of the color code.

Each component is also given a unique symbol. Component decomposition has a logical progression. For example, each major component is given a base symbol in capital letters such as Ī for the fire, M for manual suppression, A for automatic suppression, LS for life safety, etc. The next level of decomposition normally uses two (or sometimes three) letters to designate the event. For example, the second level for A (automatic sprinklers) is AA (agent application) and AC (automatic sprinkler control). The next level of decomposition for these events uses three letters. Thus, AC is decomposed to fac (sprinklers fuse), dac (water density is sufficient), cac (water is continuous), and wac (water distribution is unobstructed).

The number of letters indicates the level of decomposition and the last letters identify the main component from which the event is based. The pattern enables one to use consistent shorthand notation for calculation or organizational purposes. Each event within the nesting has a unique symbol designation to maintain organization of complex relationships in fire safety analysis.

PART TWO: SPACE–BARRIER CONNECTIVITY

3.14   Introduction

Fire scenario evaluations focus on specific locations and clusters of rooms in a building. Sometimes those locations relate to the fire and its suppression. Sometimes they involve people movement toward or away from the fire. Sometimes they examine products of combustion (POC) movement that may threaten remote spaces containing people or objects at risk. The examination of any movement through a building uses barrier–space modular networks to coordinate scenario evaluations with architectural form and function.

Many barrier–space networks that guide performance analyses can be constructed by observation, as described in Sections 2.7 to 2.12. However, room paths in complicated buildings or buildings having multiple movement paths can sometimes be difficult to recognize. For example, smoke movement may involve unique combinations of barrier–space modules to examine tenability at remote locations, such as the stack effect contaminating remote spaces of a high-rise building.

Barrier–space networks structure a thought process that enhances numerical analysis and communication. These networks are used for a wide variety of different performance applications.

The fire, POCs, and people move in space–barrier–space–barrier–... sequences. We simplify the process with two modules:

- The room of origin module.
- A barrier–space module for each space beyond the room of origin.

We use a unique number to indicate each barrier–space module. For example, the schematic descriptor of Path B in Figure 2.5 is shown as barrier–space module numbers in Figure 3.9.

Modules enable one to trace movements, For example, Figure 3.9 shows movement from Module 1 to Module 3 to Module 4 to Module 2. Shared barriers between modules are a part of both designated modules. Thus, Barrier 3–4 is a part of both Module 3 and Module 4.

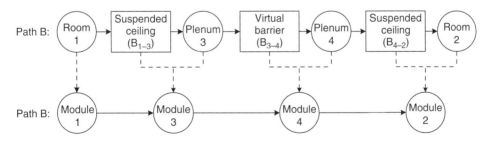

Figure 3.9 Space–barrier modules.

3.15   Room Connectivity

We illustrate modular network construction with a few examples. Initially, we will use a simple small building to describe the concept. Then, we will discuss other ways to structure more complicated buildings.

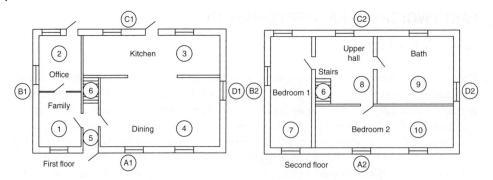

Figure 3.10 *(Note: No Caption)*

Example 3.1 Figure 3.10 shows floor plans for a small, two-story house. Construct general barrier–space modular networks of room-to-room interconnectivity for paths that can allow people or POC movement. Assume that the solid barriers will successfully prevent any POC movement into the adjacent space.

Although the barrier between the kitchen and dining room is open, we still call this incomplete separation a "barrier." For consistency, outside "rooms" are surrounded by virtual barriers and designated with an A (for the address side), and then clockwise with B, C, and D. Each floor can also have a unique designation.

Solution:
Figure 3.11(a) shows a general barrier–space diagram for room interconnectivity. Any type of movement can be traced through this general interconnectivity diagram. This network assumes that doors are open. Connectivity through solid (i.e. unpenetrated)

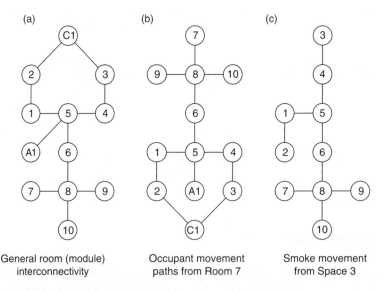

Figure 3.11 Room (module) connectivity.

barriers is not shown in this example because neither airborne POCs nor people move through these barriers.

A network may be recast to portray an analysis more conveniently. For example, Figure 3.11(b) shows occupant movement originating in Room 7. Although exit to the building outside (A) is most obvious, we have shown all possible paths. Modules 9 and 10 are included to show potential movement to a dead end. If an occupant moved to a dead end, the path must be retraced to leave the building. Figure 3.11(c) traces smoke movement from a fire in Room 3.

Discussion:

This simple example illustrates general network construction and ways to guide or communicate the thought process. Engineering analyses can quantify time and tenability conditions. Modules 5, 6, and 8 have the most commonality, and one might consider them as "target rooms" for special attention in risk characterizations.

Appropriate "what if" situations can identify alternatives that change risk characterizations. For example, if the door between Module 4 and the lower hall, Module 5, were closed, tenability time would be extended in other parts of the building.

Networks guide an analytical thought process. Sometimes clusters of rooms do not enter into the analysis and may be omitted. However, a clear, complete, and accurate representation of the rooms that focus on the component being evaluated is essential to a technical analysis.

3.16   Building Interconnectivity

Coordination of dynamic movements through a building is central to fire performance analysis. The fire can propagate from the room of origin through a cluster of rooms. Smoke travels on a different medium and can spread throughout the building via barrier–space modules. Occupants must traverse these modules to leave the building. Fire fighters travel through architectural segments to attack the fire or rescue occupants. Barrier–space modules define all architectural segments through which any type of movement occurs.

Defining an appropriate modular sequence is necessary for all component evaluations. Network selection is based on the component to be evaluated and the level of detail appropriate to the analysis.

It is possible to use a computer program to convert architectural plans to show all barrier–space modular interconnectivity in any building or ship. This type of program can show both penetrated and unpenetrated barrier paths to enable one to examine unimpeded movement as well as the effect of time-related barrier failures. Any room may be designated as a room of origin, and any or all paths from that room can be identified.

When performance evaluations are incorporated, outcomes can be sorted to identify special features, such as the most vulnerable locations for fire origins. Although this computer program has existed for many years, we only wish to make the reader aware that information technology can use the concepts of this book for a broad set of innovative applications, if desired.

3.17 Segmenting Buildings

Modular networks structure all scenario evaluations. One must select enough spaces to give a good understanding of building performance. Selecting too many rooms is inefficient, and the additional rooms provide little added value for the increased professional time expenditure. On the other hand, too few rooms give an incomplete picture of building behavior. Also, poorly selected room clusters may not examine features that can be critical to performance. Although selection of spaces of origin and connectivity paths through a building does not usually pose difficulty for an experienced fire safety engineer, users new to the concept may need some thought and experimentation.

The model building described in Appendix B is the basis for most of the examples in this book. This model building is used to organize a variety of performance application examples. It has many features that are routinely encountered in fires.

Normally, careful selection of a single diagnostic fire scenario becomes the basis for obtaining most of the information needed to understand performance. This means that one may examine a part of the building with care while maintaining awareness that the remaining parts of the building merely contribute to the total understanding. Thus, one may select a room of fire origin and the cluster of rooms through which the diagnostic fire can propagate without considering complete building burnout. Similarly, one may examine the critical rooms through which smoke or people can move with little attention to other rooms. Rooms that are used as fire fighter attack routes may be examined as separate segments within the complete building.

We handle these different needs by segmenting the building. The important concepts are described with a series of examples using the model building plans.

3.18 Summary

Spaces and barriers define the architecture of all buildings. We note that two modules can define the architecture of all buildings. One is a space of origin and the other is a space surrounded by barriers. A performance thought process examines the routes of the components along a modular chain. The arrangement of modules traces the thought process associated with the component analysis. Thus, a modular chain for fire spread is different from that of smoke movement; egress movement is different from fire fighter attack movement.

All parts of a performance fire analysis involve movement. For example, fire through a cluster of rooms, smoke transport through the building by air movement, occupant egress, and fire fighter operations all involve movement through barriers and rooms over a period of time. Coordination of component actions compares conditions where people or things (including POCs) occupy the same space at the same time. Barrier–space modules form chains that enable one to examine movement over time to compare space occupancy situations.

The integration of time and dynamic component changes may be recorded on the horizontal rows of the IPI chart. The comparison of the status of components at any given time may be made on the same IPI chart.

PART THREE: ADDITIONAL TOOLS

3.19 Networks and Charts

The fault/success trees, SVNs, CVNs, and the integrative IPI chart provide a logical framework for analysis. They are interrelated and have a correspondence with mathematical relationships that enable one to calculate outcomes with appropriate numerical input. However, our main purpose for this set of tools is to guide analytical thinking with state-of-the-art fire science and engineering.

Other types of networks and charts that do not provide the mathematical rigor of logic networks can provide guidance for fire analysis. Sometimes these tools may organize one's thinking to focus on specific or alternative applications. At other times they may ensure that important conditions for the building have been considered. They often help to identify appropriate scenarios for risk characterizations or "what if" analyses.

Although reading networks and charts is almost self-intuitive, we illustrate a few representative examples to illustrate their general use. They will be useful in the component descriptions of Units Two and Three.

3.20 Organizational Charts

Figure 3.12 shows an organizational chart that identifies different building status conditions associated with human detection of a fire. Each condition would produce different time durations, and, therefore, different risk characterizations. One would select those building status conditions that are appropriate for the performance being evaluated. Often, one may want to think about different conditions for the same diagnostic fire scenario and then focus on an appropriate scenario for risk characterizations.

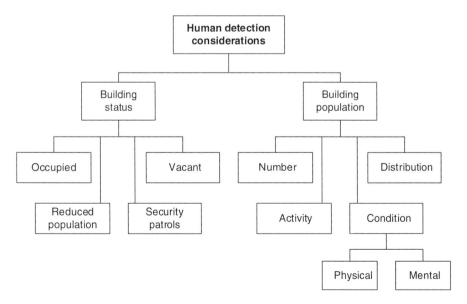

Figure 3.12 Organization chart: human detection.

Another type of organizational chart is illustrated in Figure 3.13. This chart identifies common building features that affect instrument fire detection. An individual may identify the conditions that can affect quantitative analysis outcomes from computer models or spreadsheets. When conditions are present that are not incorporated into the mathematical models, qualitative analysis helps to adjust results to reflect actual conditions.

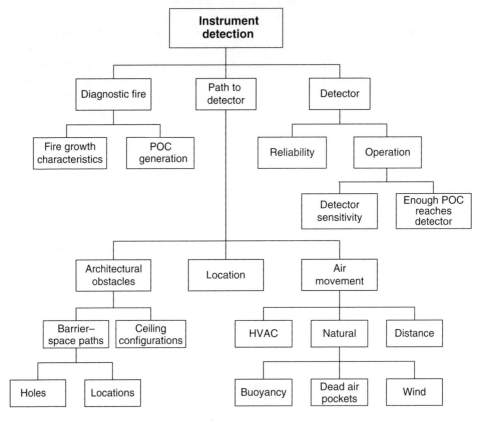

Figure 3.13 Organization chart: instrument detection.

3.21 Organizational Networks

Figure 3.14 illustrates a different organizational network. This network shows all possible outcomes for life safety in a fire. If one wants to study the egress from a specific room, only Path 0–3–6–17 is of considered. For defend in place, such as a hospital or nursing home, only Path 0–2–6–17 is of interest. However, life safety for a designed area of refuge within the building would examine Path 0–1–4–8–14–17. This type of network helps to identify specific scenario evaluations.

Although organizational networks and charts do not have the mathematical structure associated with logic trees and networks, they provide an organized set of factors that

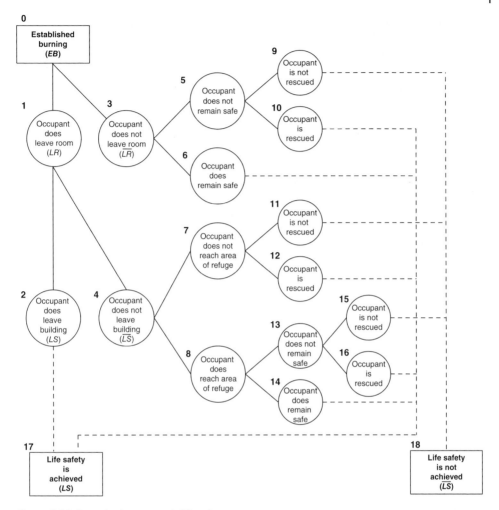

Figure 3.14 Organization network: life safety.

affect performance analysis. One only selects the factors that are appropriate for the scenario of interest.

3.22 Closure

Performance based fire safety analysis and design for buildings is far more complex than other disciplines that are associated with planning, design, and operation. In addition to continually changing fire conditions, different components enter or leave the process at various times, depending on the specific building. Fortunately, the challenge of dynamic changes and integration is made manageable by an analytical framework that:

- Identifies the performance-based components.
- Decomposes the larger components into smaller constituent parts.

- Uses CVNs to relate changes in the components and constituent parts. This is analogous to a video of the entire process that is filtered to allow each component to be viewed in isolation.
- Converts traditional event logic fault trees and success trees into equivalent analytical networks. These SVNs trace an analytical thought process that enables one to make a static performance analysis for all contributing factors at any instant of time. The networks allow one to conceptually "stop the world" at any critical event or instant of time to examine all factors that contribute to individual performance and integrate them into the evaluation.
- Shows performance results on an IPI chart. This time- and performance-based spreadsheet shows the start and stop of all components. It portrays a visual representation of phasing of all components that affect the fire and suppression, the building response, and risk to people and other exposures.
- Records performance information in the cells of the IPI chart. Examination of interrelationships provides a clear picture of dynamic interactions of the components. The spreadsheet attributes of the IPI chart enable much of the analytical and observational data to be available for evaluations.
- Structures analyses with barrier–space networks that organize scenarios and evaluations.
- Uses the state of the art of fire science and engineering as the basis for measures of performance.
- Augments evaluations with supplemental information provided by organizational charts and networks. A combination of the analytical networks, organizational charts and networks, and deterministic measures enables one to "connect the dots" to describe performance continuity with confidence.

Time is the single element that is common to all parts of a dynamic fire performance analysis.

Event trees, fault trees, success trees, and network diagrams are interrelated and coordinated in the IPI chart. Barrier–space networks organize scenarios. The aggregate of block diagrams, networks, logic diagrams, and the IPI chart all provide a structure on which to base analytical decisions. A performance analysis uses all information that can be accessed during the time available to acquire it.

4

An Introduction to the Interactive Performance Information Chart

4.1 Introduction

The primary goal of this book is to describe an analytical framework and procedures for evaluating fire performance in buildings.

The Interactive Performance Information (IPI) chart is the most important single tool for evaluating fire performance of buildings. The IPI chart organizes information, stores data, enables one to understand dynamic interactive behavior, and aids communication.

This chapter describes the organization and introduces a few functions of the IPI chart in performance analysis and risk management.

4.2 The Basic Template

To illustrate the IPI concept, imagine Superman hovering above a building that has just had a fire ignition. Assume that Superman records the entire operation with a magic camera that can "see" through the building and filter each part of the process as a separate activity. Thus, at the end of the scenario one can view the holistic process and also examine a filtered video of each part in isolation. In addition, the video can be stopped at any instant of time to examine the status of each component and recognize how events or actions affect and are affected by other activities.

The IPI chart is our Superman video. It allows us to examine both micro and macro performance. It enables us to look at any component in isolation and recognize when the actions of another part causes its behavior to change. We can "stop the world" at significant events during the process to examine any status or conditions where components interact. Although the complete process is complex, the IPI chart enables one to examine individual components or the "big picture" at any intermediate time. Thus, we can understand what is happening and recognize the effectiveness of individual performance on holistic outcomes.

Figure 4.1 identifies the major components of the complete fire safety system. Because these components are a part of all comprehensive fire safety performance and risk evaluations, they structure the IPI template.

Figure 4.2 shows the basic template organization for the IPI chart. The main components for this template are fixed for all buildings and geographic locations. Although the template fixes the positions of the major components, we can insert additional rows to

Fire Performance Analysis for Buildings, Second Edition. Robert W. Fitzgerald and Brian J. Meacham.
© 2017 John Wiley & Sons Ltd. Published 2017 by John Wiley & Sons Ltd.

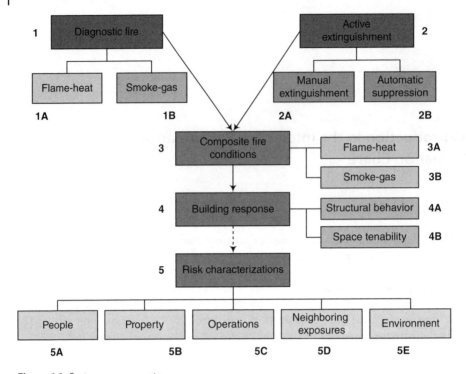

Figure 4.1 System components.

				Working IPI template																		
												Time											
		Section	Symbol	Event					5					10						15			
Building performance	Fire response	**1**	$\bar{I}$	**Diagnostic fire**																			
		A	$\bar{I}_F$	Flame-heat																			
		B	$\bar{I}_S$	Smoke-gas																			
		2		**Active extinguishment**																			
		A	M	Fire department extinguishment																			
		B	A	Automatic sprinkler suppression																			
		3	**L**	**Composite fire**																			
		A	L_F	Flame-heat																			
		B	L_s	Smoke-gas																			
	Building response	**4**	**R**	**Building response**																			
		A	St	Structural frame																			
		B	TS	Target space tenability																			
Risk characteristics		**5**	**L**	**Risk characteristics**																			
		A	LS	People																			
		B	RP	Property																			
		C	RO	Operational continuity																			
		D	RE	Exposure protection																			
		E	RN	Environment																			

Figure 4.2 Interactive Performance Information chart organization.

accommodate unique needs for any specific building. The additional rows decompose components into their constituent parts and focus on specific room fire conditions that are important to performance.

4.3 The Working Template

The working template is shown in Figure 4.3. This template decomposes the basic components to show the major events for a holistic building evaluation. The decomposition is based on fault/success trees and the associated networks introduced in Chapter 3. These are combined with event trees to incorporate time into the system. Collectively, the IPI integrates all of these concepts into the framework that can examine specific building conditions during critical events or time durations.

The main value of an IPI chart is enhanced understanding of performance evaluations. The IPI chart is a tool that allows one to work interactively among the components during an analysis.

Although calculated numerical measures are an obvious content, cells can be used for a wide variety of additional information. For example, supplemental sheets may be used

Section	Symbol	Event	Time (5 / 10 / 15)
1	$\bar{I}$	**Diagnostic fire**	
A	$\bar{I}_F$	**Flame-heat**	
B	$\bar{I}_S$	**Smoke-gas**	
2		**Active extinguishment**	
A	**M**	**Fire department extinguishment**	
a	$M_{part\ A}$	Part A: Ignition to notification	
	MD	Detect fire	
	MN	Notify fire department	
b	$M_{part\ B}$	Part B: Notification to arrival	
	MS	Dispatch resonders	
	MR	Resonders arrive	
c	$M_{part\ c}$	Part C: Arrival to extinguishment	
	MA	Apply first water	
	MC	Control fire	
	ME	Extinguish fire	
B	**A**	**Automatic sprinkler suppression**	
a	AA	Sprinkler system is reliable	
b	AC	Sprinkler system controls fire	
3	**L**	**Composite fire**	
A	L_F	**Flame-heat**	
B	L_s	**Smoke-gas**	
4	**R**	**Building response**	
A	St	**Structural frame**	
B	TS	**Target space tenability**	
5	**L**	**Risk characteristics**	
A	LS	**People**	
a	LS_E	Egress per-evacuation delay	
b	LT	Egress travel: Tenability criteria	
c	LS_P	Defend in place: Tenability criteria	
B	RP	**Property**	
C	RO	**Operational continuity**	
D	RE	**Exposure protection**	
E	RN	**Environment**	

(Left side row groupings: Building performance → Fire response (sections 1–2) and Building response (sections 3–4); Risk characteristics (section 5).)

Figure 4.3 Interactive Performance Information chart: working template.

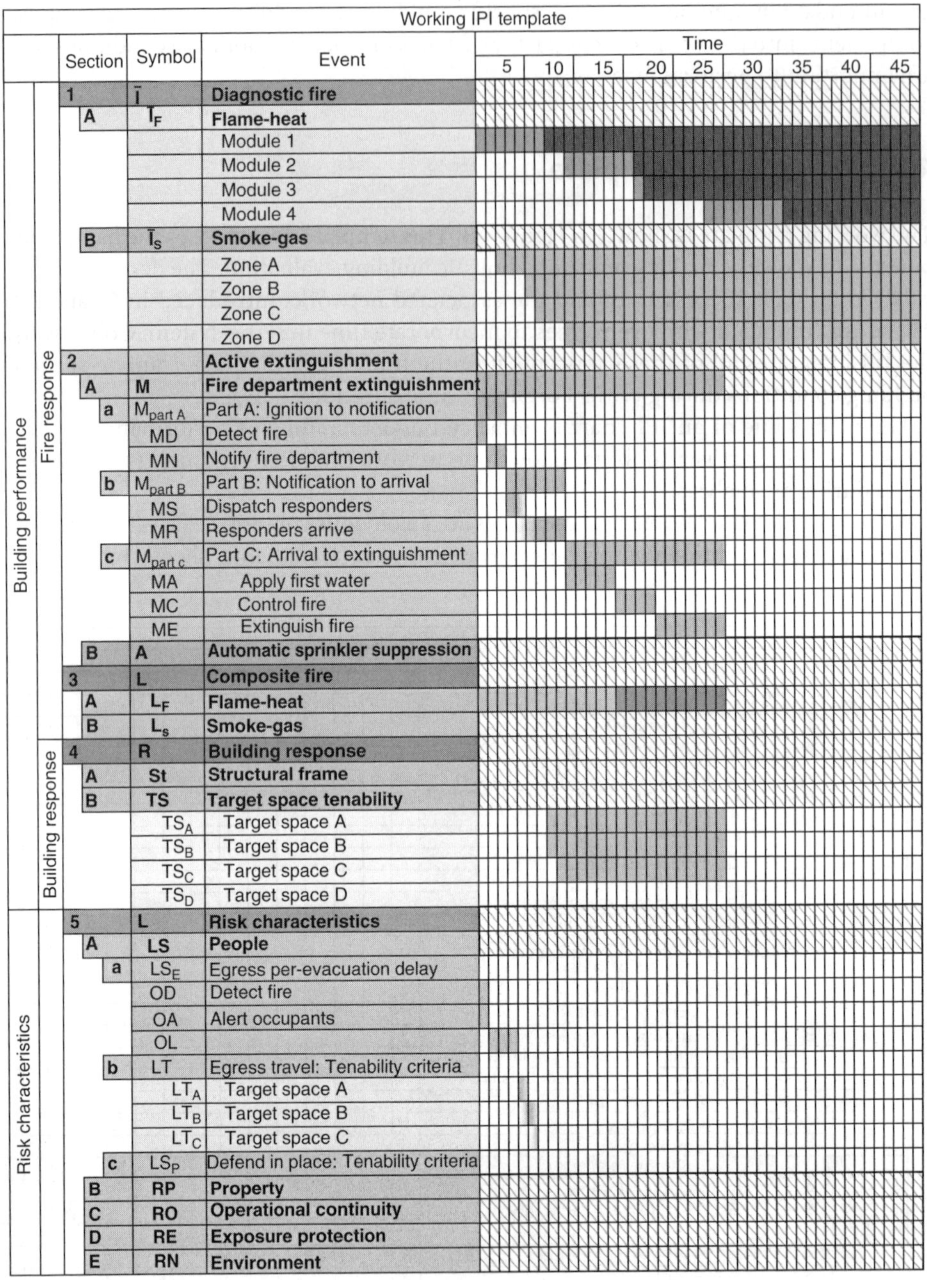

Figure 4.4 Interactive Performance Information: Building A.

The table "Working IPI template" contains the following rows (Section / Symbol / Event) with a Time axis marked 5, 10, 15, 20, 25, 30, 35, 40, 45:

- Building performance — Fire response
  - 1 / Ī / Diagnostic fire
    - A / Ī_F / Flame-heat
      - Module 1
      - Module 2
      - Module 3
      - Module 4
    - B / Ī_s / Smoke-gas
      - Zone A
      - Zone B
      - Zone C
      - Zone D
  - 2 / / Active extinguishment
    - A / M / Fire department extinguishment
      - a / M_part A / Part A: Ignition to notification
        - MD / Detect fire
        - MN / Notify fire department
      - b / M_part B / Part B: Notification to arrival
        - MS / Dispatch responders
        - MR / Responders arrive
      - c / M_part c / Part C: Arrival to extinguishment
        - MA / Apply first water
        - MC / Control fire
        - ME / Extinguish fire
    - B / A / Automatic sprinkler suppression
  - 3 / L / Composite fire
    - A / L_F / Flame-heat
    - B / L_s / Smoke-gas
- Building response
  - 4 / R / Building response
    - A / St / Structural frame
    - B / TS / Target space tenability
      - TS_A / Target space A
      - TS_B / Target space B
      - TS_C / Target space C
      - TS_D / Target space D
- Risk characteristics
  - 5 / L / Risk characteristics
    - A / LS / People
      - a / LS_E / Egress per-evacuation delay
        - OD / Detect fire
        - OA / Alert occupants
        - OL /
      - b / LT / Egress travel: Tenability criteria
        - LT_A / Target space A
        - LT_B / Target space B
        - LT_C / Target space C
      - c / LS_P / Defend in place: Tenability criteria
    - B / RP / Property
    - C / RO / Operational continuity
    - D / RE / Exposure protection
    - E / RN / Environment

Figure 4.4 Interactive Performance Information: Building A.

for photographs of field conditions, rationale of selections, experimental data. Additional sheets containing "what if" conditions may be used to compare variation analyses with base evaluations. Cells can identify critical events such as flashover, detection, first water application, or onset of untenable conditions in target spaces. Unit Four describes ways to manage uncertainty and the cells can record probability estimates of performance. These can be used to construct graphs and tables. The scope of IPI chart capability with information technology is enormous.

4.4   Reading IPI Charts

Using an IPI chart is straightforward. One adapts spreadsheet technology to a structured format. However, one can also get an intuitive sense of building performance from visual observations of IPI patterns. "Reading" IPI charts can give an awareness of the strengths and weaknesses of a building.

The IPI can serve as a rapid screening tool for comparing different buildings and selecting priorities to sequence changes for risk management. We illustrate IPI comparisons with visual information for two buildings. Buildings A and B have the same size, construction, architecture, and occupancy. However, they have different features that affect fire performance and risk. Assume that you have no plans or additional information about these buildings, but you are asked to "read" the IPI charts. Even this limited information gives an insight into what one may expect during a fire.

Example 4.1   Figure 4.4 shows the results of a performance analysis for Building A. Although numerical measures and other information are normally shown in the cells, only solid bars are used to indicate phases of component activity. Describe what one may expect of this building for the fire scenario depicted.

Solution:
Although numerical measures are not shown, the solid blocks provide clues to performance.

Section 1: Diagnostic fire
Section 1A: Flame-heat movement

- The fire in Module 1 (room of origin) starts at established burning (EB) time = 0 and flashover (FO) occurs in 7 minutes.
- The barrier between Modules 1 and 2 fails at 9 minutes after EB.
- The fire grows to flashover in Module 2 in an additional 7 minutes. The time for Module 2 to reach FO is 16 minutes after EB.
- The barrier between Modules 2 and 3 fails at the same time as FO in Module 2.
- Module 3 experiences FO at 1 minute after failure. Thus, Module 3 experiences FO at 17 minutes from EB.
- The barrier between Modules 3 and 4 has a $\overline{T}$ failure at 6 minutes after FO in Room 3 and Module 4 experiences FO at 8 minutes after that failure. Thus, FO occurs in Module 4 at 31 minutes after EB.

Section 1B: Smoke-gas movement

- Zone A (room of origin) is untenable at 2 minutes after EB.
- Zone B (corridor from Module 1) is untenable at 5 minutes after EB.
- Access corridor Zone C is untenable at 6 minutes after EB.
- Zone D (Exit) is untenable at 9 minutes after EB.

Section 2: Active extinguishment
Section 2B: Automatic sprinkler suppression There is no automatic sprinkler suppression in the building because this section is unfilled.
Section 2A: Manual extinguishment

- *Part A: Ignition to notification*
  - Section MD shows that detection (MD) occurs 1 minute after EB.
  - Section MN shows that notification of the fire department (MN) occurs 2 minutes after detection. Thus, $M_{Part A}$ (fire department notification) occurs 3 minutes after EB.
- *Part B: Notification to arrival*
  - Section MS shows the time duration for alarm processing is 1.5 minutes. MS occurs 4.5 minutes after EB.
  - Section MR shows that the time duration between MS and first-in arrival (MR) is 4.5 minutes. Thus, $M_{Part B}$ MN to MR is 6 minutes, and first-in arrival (MR) occurs 9 minutes from EB.
  - The diagnostic fire condition at first arrival is:
    - Module 1 (room of origin) reached FO 3 minutes earlier.
    - The barrier between Modules 1 and 2 has just experienced a hot-spot ($\overline{T}$) failure into Module 2, causing EB in a second room.
- *Part C: Arrival to extinguishment*
  - Section MA shows that first water can be applied (MA) at 5 minutes after arrival. Thus, first water is estimated at 14 minutes after EB.
  - The diagnostic fire at initial MA shows the room of origin has flashed over and Room 2 has not yet reached FO, but the room fire size is large.
  - Section MC shows that the fire is expected to be controlled (MC) 4 minutes after first water application. Thus, one estimates fire control at 18 minutes after EB.
  - The diagnostic fire size at MC shows Modules 1 and 2 to be fully involved and Module 3 has also reached FO. However, first water should be applied before Module 2 has reached FO. Initial water application completely changes fire characteristics and the diagnostic fire is no longer valid. The timing is sufficiently close that a few minutes one way or another could result in very different outcomes. Fire ground operations are discussed more completely in Unit Two.
  - Section ME shows that extinguishment (ME) is expected to be completed 7 minutes after control (MC).
  - Section $M_{Part C}$ shows that *Part C: Arrival to extinguishment* can be completed about 16 minutes after arrival.

Combining Parts A, B, and C provides an estimate of extinguishment (ME) to occur 25 minutes after EB.

Section 3: Composite fire The composite fire is the same as the diagnostic fire until first water application. Then, turbulence ensues and fire conditions become very uncertain until extinguishment. Nevertheless, one can get a sense of behavior during the turbulent period.

Section 4: Building response
Section 4A: Structural performance This row is blank, indicating that structural behavior is not significant. It would be possible to describe deflection or signs of distress in the cells, if useful.

Section 4B: Target room tenability The lack of floor plans makes this section difficult to envision. In this example, the building describes only for life safety by egress. Target modules (designated A, B, C, etc.) form a continuous path from the rooms of occupant origin to the exit discharge. Thus, the target spaces are different from the diagnostic fire spaces of Section 1B. The bar charting fill indicates the time duration during which the space is *not* tenable.

Section 4B shows the time durations during which the spaces are tenable for occupant egress:

- Target space A is tenable from EB to 7 minutes and after 25 minutes.
- Target space B is tenable from EB to 7 minutes and after 25 minutes.
- Target space C is tenable from EB to 8 minutes and after 25 minutes.
- Target space D is not shown here because it does not influence the occupant egress being evaluated.

Section 5: Risk characterizations The following observations may be made:

- Fire detection (OD) is the same as Section 2A at 1 minute after EB.
- Alert occupant (OA) occurs at the same time.
- Occupant pre-movement time is estimated at 3 minutes. Thus, the occupant leaves the room 4 minutes after EB.
- Travel time through Target space A is 0.5 minutes.
- Travel time through Target space B is 1 minute.
- Travel time through Target space C is 0.2 minutes.

Comparisons of the time that target spaces remain tenable (Section 4B) with the time that representative occupants are expected to be in those target spaces (Section 5) indicates that these occupants will not experience untenable conditions during egress.

Example 4.2 Building B is in the same fire response district as Building A. Although Building B is similar in size, construction, and use, it has a somewhat different architectural plan and fire safety features. Describe what one may expect of this building from the IPI chart shown in Figure 4.5.

Solution:
Section 1. Diagnostic fire The diagnostic fire depicted in Section 1 is the same as Building A.

Section 2: Active extinguishment
Section 2B: Automatic sprinkler suppression There is no automatic sprinkler suppression in the building because this section is unfilled.
Section 2A: Manual extinguishment

- *Part A: Ignition to notification*
  - Detection (MD) occurs 3 minutes after EB and fire department notification (MN) occurs 4 minutes after detection. This building completes $M_{Part\ A}$ at 7 minutes after EB.
- *Part B: Notification to arrival*
  - The alarm handling time and first alarm response time are the same as in Example 4.2 This places $M_{Part\ B}$ at 7 + 6 = 13 minutes after EB.

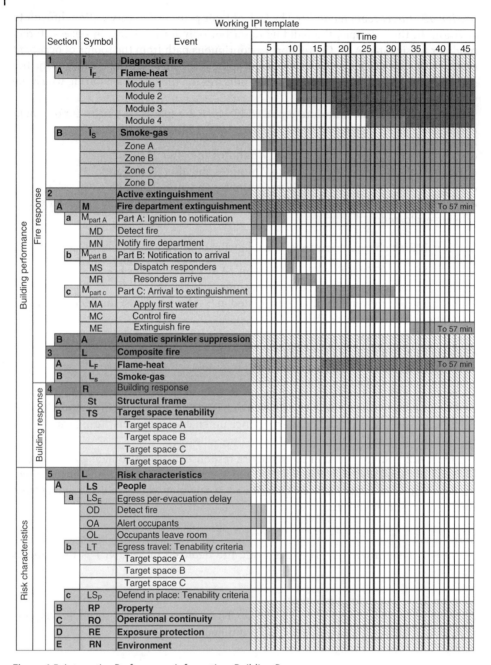

Figure 4.5 Interactive Performance Information: Building B.

- *Part C: Arrival to extinguishment*
  - At first fire department arrival, the room of origin is fully involved and the second module has substantial fire development. These conditions make theoretical first

water application more difficult. The estimated time for theoretical MA grows to 7 minutes making a total of 20 minutes after EB. The diagnostic fire now involves three rooms – a very different set of conditions than exists at first water application in Building A.

- Section MC indicates the fire is expected to be under control after an additional 12 minutes. This makes a time of 20 + 12 = 32 minutes after EB. Four rooms are likely to be involved. The fire ground becomes a very different situation that that of Building A. In addition, the diagnostic fire should be extended to determine time durations when additional rooms may become involved.
- The IPI chart indicates that extinguishment (ME) is expected about 25 minutes after control. This indicates that extinguishment can be expected about an hour after EB.

Section 3: Composite fire The composite fire is the same as the diagnostic fire until first water application. Then, turbulence ensues and fire conditions become very uncertain until extinguishment. Nevertheless, one can get a sense of behavior during the turbulent period.

Section 4: Building response
Section 4A: Structural performance This row is blank indicating that structural behavior is not significant. It would be possible to describe deflection or signs of distress in the cells, if useful.
Section 4B: Target room tenability The lack of floor plans makes this section difficult to envision. In this example, the building is being examined only for life safety by egress. Target modules (designated A, B, C, etc.) form a continuous path from the rooms of occupant origin to the exit discharge. Thus, the target spaces are different from the diagnostic fire spaces of Section 1b. The bar charting fill indicates the time duration during which the space is *not* tenable.

Section 4B shows the time durations during which the spaces are tenable for occupant egress:

- Target space A is tenable from EB to 7 minutes and after an hour.
- Target space B is tenable from EB to 7 minutes and after an hour.
- Target space C is tenable from EB to 8 minutes and after an hour.
- Target space D is not shown at this time.

Section 5: Risk characterizations The following observations may be made:

- Fire detection (OD) is the same as Section 3A at 3 minutes after EB.
- Alert occupant (OA) occurs at the same time.
- Occupant pre-movement time is estimated at 3 minutes. Thus, the occupant leaves the room 6 minutes after EB.
- Travel time through Target space A is 0.5 minutes.
- Travel time through Target space B is 1 minute.
- Travel time through Target space C is 0.2 minutes.

Comparisons of the time that target spaces remain tenable (Section 4B) with the time the representative occupants are expected to be in those target spaces (Section 5) indicates that these occupants will experience untenable conditions during egress through Target space B. Also, tenability and occupant movement are very close. A faster fire or

POC movement through the building, or slower occupant reaction or movement will create life safety concerns.

4.5 Building Comparisons

The IPI chart can be used to compare different buildings or different design considerations of a single building. Numerical values, digital images, or other information sources enable one to associate the performance or one design with another. This enables one to develop effective risk management programs more easily.

Even without Initial building selections that order performance expectations may be done quickly by comparing IPI charts. Because the working template uses a consistent format one may visually compare buildings (or design alternatives) with quick visual comparisons. Figure 4.6 illustrates the concept with Buildings A and B. One may easily recognize that Building A has features that distinguish it from Building B. Thus, Building B would be a better candidate for upgrades if one could allocate additional resources.

4.6 IPI Enhancements

The main function of an IPI chart is to organize and display information. This information creates a performance understanding to guide rational decisions for improving or managing risks associated with an unwanted fire. The cells of an IPI chart may be used in a variety of ways, such as:

- Shade the phase in and phase out of components for rapid visualization of actions.
- Store data that are used for component evaluation.
- Store data concerning the status of component performance.
- Link with other computer programs to establish quantification for conditions relating to component performance.
- Link with other computer programs to identify information such as room geometry, building architectural plans, or features that affect fire performance of the building.
- Create variability analyses relating "what if" alternatives to assess the effect of differences in condition.
- Deal with uncertainty.
- Construct time-dependent graphs or charts to portray performance and behavior.
- Become the file for report documentation.

Individual components behave dynamically. In addition, one component can affect or be affected by the actions of another. The value of the IPI becomes enriched as one examines quantitative and qualitative details of performance to augment understanding. Conceptually, completed IPI charts may be viewed as unique "fingerprints" of a building's performance. This enables one to distinguish among the many issues and evaluate the effects of changes more easily.

4.7   Summary

The IPI chart integrates attributes of Gantt charts used in project management and computer-generated spreadsheets with the logic of systems analysis and risk management. The organizational structure is specially adapted to evaluate the fire safety performance of buildings. Thus, the IPI chart enables one to address any unique application with a consistent framework and logic.

The hierarchy of the IPI chart identifies the major components of the analytical framework of fire safety. The cells relate component behavior and time. They provide a location to calculate, store, describe, and portray the data that define the performance of a building.

The working IPI template may be adapted for specific needs of any evaluation. For example, rows may be added to show each barrier–space module as a separate unit with information and quantification that is associated with that module. Worksheets may be isolated to concentrate on specific quantification and applications of analysis. Hiding non-essential information can reduce distractions. Unhiding the information needed for component interaction aids an evaluation. Thus, one may work with uncluttered IPI sheets during specific examinations and then select essential information to communicate results.

The versatility of IPI chart applications is particularly suited for performance evaluations. The templates provide a consistency for analysis and an organization for information relevant to the building.

Importing and storing information from other technology sources can assist in documentation of decisions. For example, digital photographs of building or site conditions may be incorporated into the documentation. Separate quantification of the time and flow for water supply sources to come on line becomes part of manual suppression analysis. Adjacency matrices can identify potential propagation paths for fire spread or egress paths for occupant escape. A wide variety of information technology applications can be imported into the documentation and incorporated into an analysis.

The IPI chart is fundamental to understanding the fire performance of buildings. It is the device that links all analytical parts with building performance.

(a)

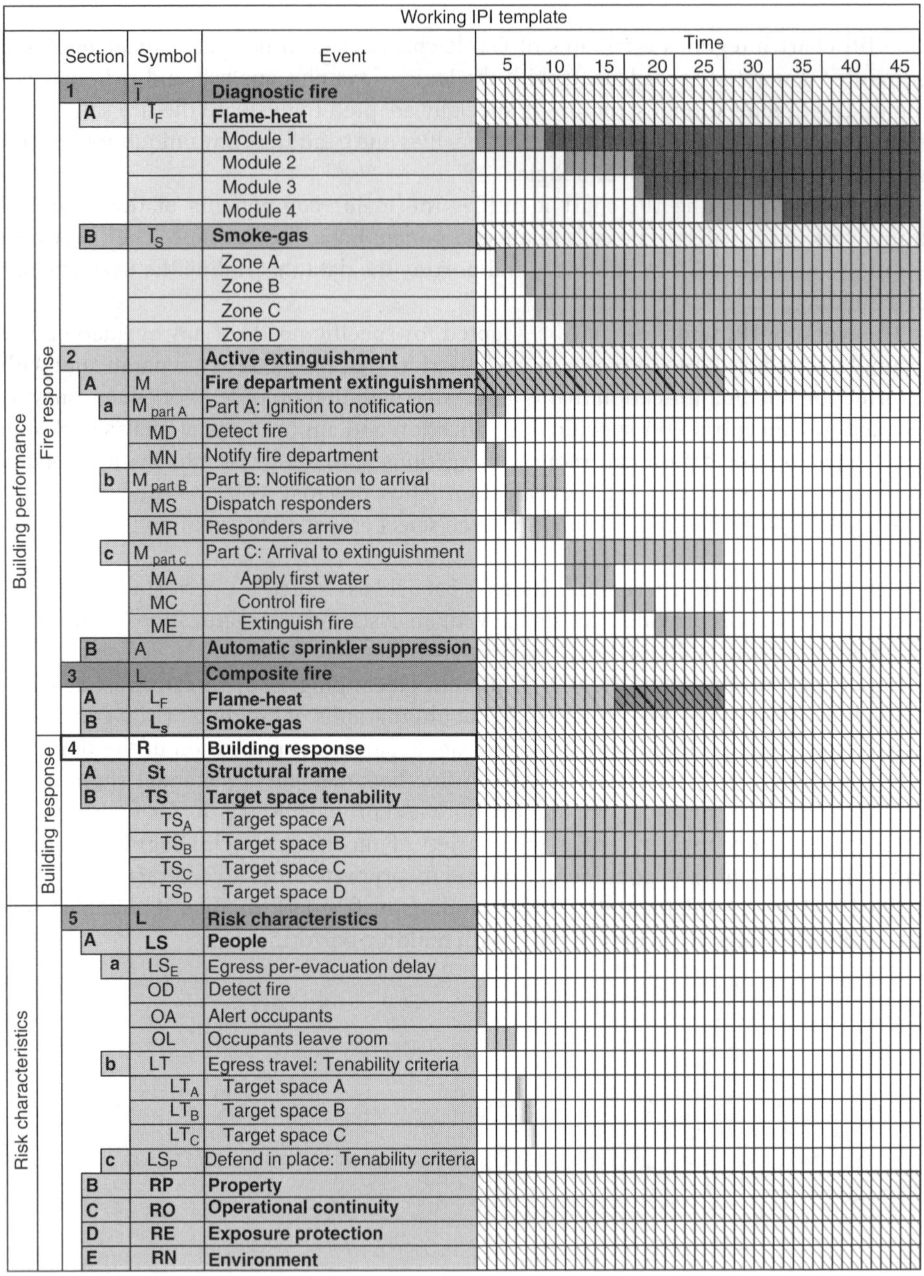

	Section	Symbol	Event	Time								
Working IPI template				5	10	15	20	25	30	35	40	45
Building performance → **Fire response**	1	Ī	**Diagnostic fire**									
	A	T_F	**Flame-heat**									
			Module 1									
			Module 2									
			Module 3									
			Module 4									
	B	T_S	**Smoke-gas**									
			Zone A									
			Zone B									
			Zone C									
			Zone D									
	2		**Active extinguishment**									
	A	M	**Fire department extinguishment**									
	a	$M_{part A}$	Part A: Ignition to notification									
		MD	Detect fire									
		MN	Notify fire department									
	b	$M_{part B}$	Part B: Notification to arrival									
		MS	Dispatch responders									
		MR	Responders arrive									
	c	$M_{part c}$	Part C: Arrival to extinguishment									
		MA	Apply first water									
		MC	Control fire									
		ME	Extinguish fire									
	B	A	**Automatic sprinkler suppression**									
	3	L	**Composite fire**									
	A	L_F	**Flame-heat**									
	B	L_s	**Smoke-gas**									
Building response	4	R	**Building response**									
	A	St	**Structural frame**									
	B	TS	**Target space tenability**									
		TS_A	Target space A									
		TS_B	Target space B									
		TS_C	Target space C									
		TS_D	Target space D									
Risk characteristics	5	L	**Risk characteristics**									
	A	LS	**People**									
	a	LS_E	Egress per-evacuation delay									
		OD	Detect fire									
		OA	Alert occupants									
		OL	Occupants leave room									
	b	LT	Egress travel: Tenability criteria									
		LT_A	Target space A									
		LT_B	Target space B									
		LT_C	Target space C									
	c	LS_P	Defend in place: Tenability criteria									
	B	RP	**Property**									
	C	RO	**Operational continuity**									
	D	RE	**Exposure protection**									
	E	RN	**Environment**									

Figure 4.6 Building comparisons.

(b)

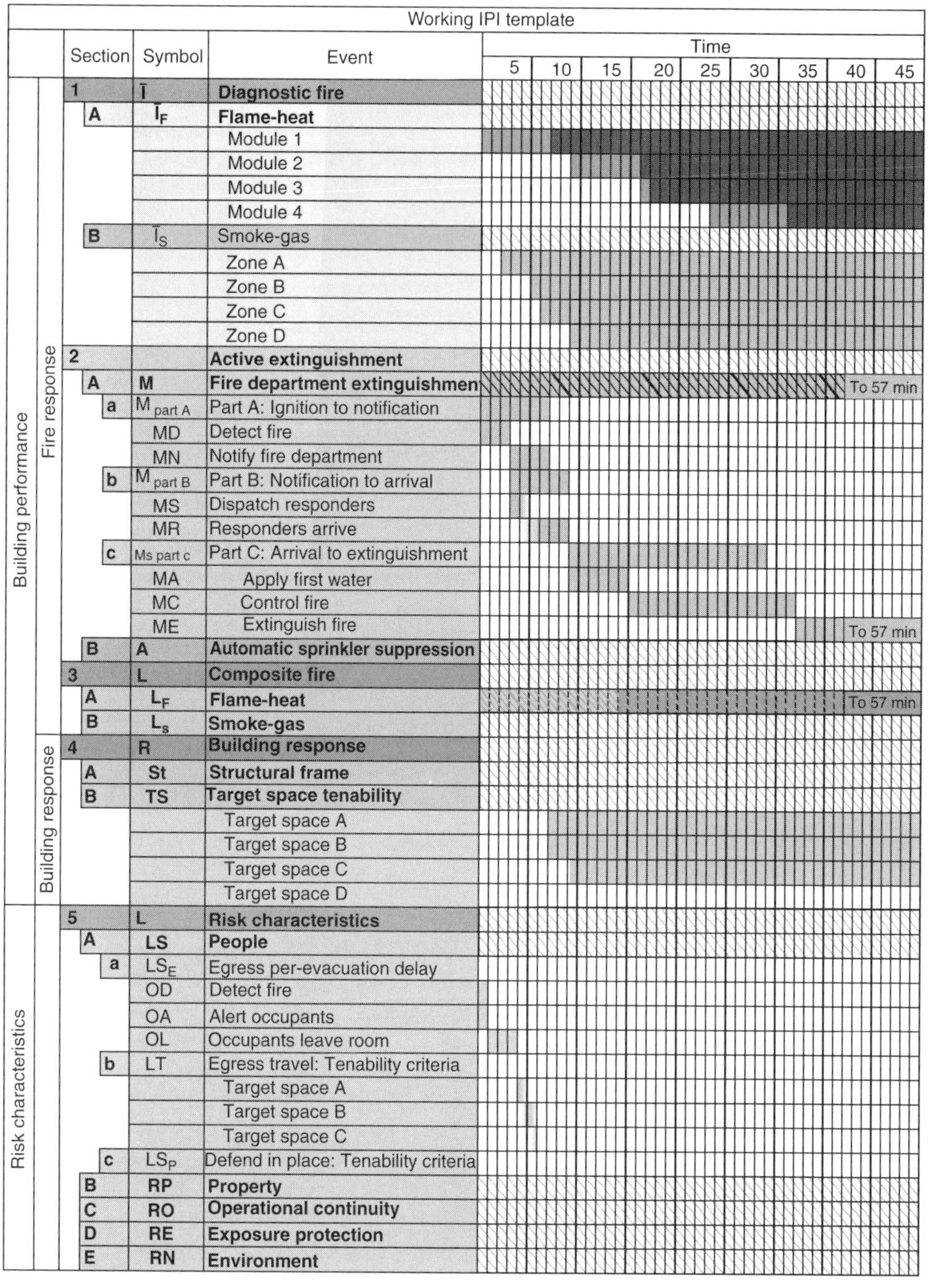

	Section	Symbol	Event	Time
				5 · 10 · 15 · 20 · 25 · 30 · 35 · 40 · 45
1		$\bar{I}$	**Diagnostic fire**	
	A	$\bar{I}_F$	Flame-heat	
			Module 1	
			Module 2	
			Module 3	
			Module 4	
	B	$\bar{I}_S$	Smoke-gas	
			Zone A	
			Zone B	
			Zone C	
			Zone D	
2			**Active extinguishment**	
	A	M	Fire department extinguishment	To 57 min
	a	$M_{part A}$	Part A: Ignition to notification	
		MD	Detect fire	
		MN	Notify fire department	
	b	$M_{part B}$	Part B: Notification to arrival	
		MS	Dispatch responders	
		MR	Responders arrive	
	c	$Ms_{part c}$	Part C: Arrival to extinguishment	
		MA	Apply first water	
		MC	Control fire	
		ME	Extinguish fire	To 57 min
	B	A	**Automatic sprinkler suppression**	
3		L	**Composite fire**	
	A	L_F	Flame-heat	To 57 min
	B	L_s	Smoke-gas	
4		R	**Building response**	
	A	St	**Structural frame**	
	B	TS	**Target space tenability**	
			Target space A	
			Target space B	
			Target space C	
			Target space D	
5		L	**Risk characteristics**	
	A	LS	**People**	
	a	LS_E	Egress per-evacuation delay	
		OD	Detect fire	
		OA	Alert occupants	
		OL	Occupants leave room	
	b	LT	Egress travel: Tenability criteria	
			Target space A	
			Target space B	
			Target space C	
	c	LS_P	Defend in place: Tenability criteria	
	B	RP	**Property**	
	C	RO	**Operational continuity**	
	D	RE	**Exposure protection**	
	E	RN	**Environment**	

Row groupings (left vertical labels): Building performance → Fire response (Sections 1–3); Building response (Section 4); Risk characteristics (Section 5).

Working IPI template

Figure 4.6 (Cont'd)

5

Quantification

5.1 Performance Evaluations

Fire safety engineering practice involves a broad spectrum of applications ranging from routine regulatory-based design to sophisticated risk management where a solution is based on performance expectations. Some management decisions ask the question: "How can I do better with my available resources?" Another may examine technical decisions such as: "Are the fire defenses appropriate for my objectives and how can I improve their performance quality."

A performance evaluation creates an understanding of what to expect during a building fire and the associated risk characterizations to people, property, operational continuity, neighbors, and the environment.

Evaluation procedures integrate two distinct parts:

1) An analytical framework to provide a systematic, methodical organization to analyze individual components and integrate all parts into a description.
2) Quantification to provide numerical measures of performance.

The primary goal of this book is to identify a framework for analyzing fire performance in buildings. However, a framework is sterile without ways to quantify the critical events. One cannot function without the other.

Fire safety is an emerging engineering discipline. Consequently, performance quantification is uneven. Some components such as structural analysis and detector actuation are relatively well developed. Room fire models can provide accurate representations of behavior within their limits of theory and input knowledge. On the other hand, certain aspects of manual fire extinguishment and automatic fire suppression are inadequate for comprehensive numerical analysis, although enough knowledge exists to develop a performance evaluation.

Quantification uses any information or calculation tool that is relevant and seems appropriate to obtain the necessary numerical measures. Thus, any sources such as computer programs, experimental data, calculated values, observational facts, and failure analyses become resources for quantification. Quantification may be viewed as a set of tools. An engineer is expected to select the appropriate tool for the need. The framework provides a platform for the quantitative measures of performance.

Fire Performance Analysis for Buildings, Second Edition. Robert W. Fitzgerald and Brian J. Meacham.
© 2017 John Wiley & Sons Ltd. Published 2017 by John Wiley & Sons Ltd.

5.2   Information Accessibility

Among the greatest assets of the information age are the storage and rapid retrieval of information. Although personal knowledge and experience are important, the Internet provides almost instant access for data and relevant information. In addition, social networking enables geographically separated individuals to communicate and exchange information or opinions in a real-time context. Thus, information is relatively broad in scope and rapidly accessible. Circumstances connecting field observations, library research, and office deliberations become blended. This enables one to make faster and better decisions.

A building evaluation uses all information that is accessible within the available time frame to provide quantitative measures for performance events. Many computer models can provide insights into complex interactions. These models are continually improving in scope and accuracy. Experimental data provide information that enhances calculations. Digital recordings can give visual descriptions of room arrangements, building features that influence component performance, and conditions that affect interactions. The comprehensive array of information and information sources is a major resource to be utilized in building evaluations.

The analytical framework provides structure for using information to understand building performance. The Interactive Performance Information (IPI) chart provides an organization to store and access relevant information. The IPI chart also uses the framework to enable one to analyze components and coordinate with other parts of the system. The IPI chart helps to communicate individual component or holistic behavior information to differing audiences.

5.3   Quantification

The goal of a performance analysis is to understand expected building behavior and its associated risk characterizations. Building evaluations for specific fire scenarios are the foundation for acquiring this understanding.

A scenario evaluation uses three types of analysis. A *quantitative analysis* calculates numerical outcomes. However, fire safety has not yet evolved to provide reliable quantification for the range of conditions that we routinely encounter in buildings. Therefore, a *qualitative analysis* augments calculated outputs to provide a sense of proportion for expected behavior.

Computer models and other fire safety data are growing at an exponential rate. Extrapolating the development of information and its accessibility during the past decade into the next decade shows that quantification will continue to grow. In addition to computer analysis, any resource that can provide numerical measures becomes part of a quantitative analysis.

Although quantitative analyses provide a valuable insight into complex relationships of fire and fire defenses, all quantitative analyses have shortcomings. Sometimes approximations and simplifications do not accurately represent actual conditions. Sometimes the theory is inadequate. Not all components have been modeled to provide numerical measures. Nevertheless, quantitative analysis is essential to performance-based engineering.

The quality of an evaluation becomes greatly enhanced when quantitative analyses are augmented by qualitative analyses. Qualitative analysis integrates function, theory, and field conditions to estimate the behavior. A qualitative analysis can incorporate many features that affect quantitative outcomes but are unable to be modeled at this time.

Quantitative analyses and qualitative analyses complement one another and are used interactively in evaluations. A quantitative analysis is the main source for performance measures. Yet, one must ensure that a calculated value is reasonable and incorporates the important features that affect performance

In other situations a qualitative analysis is the dominant evaluative tool, and quantitative resources may be used to enhance or validate certain key approximations. Also, qualitative analyses are normally used to select initial scenarios that become the basis for a performance analysis.

Both quantitative and qualitative analyses are sensitive to changes in condition. For example, the status of a door being open or closed may significantly affect performance. Alternative construction materials for common fuel packages may have significantly different burning characteristics that affect the time-related outcomes that, in turn, affect the performance of other components. Often, "what if" questions become evident during the decision-making function of scenario identification and one wants to examine the differences that may occur.

A *variability analysis* is the third type of analysis that provides a basis for ascertaining if possible changes will significantly affect performance outcomes or will have a relatively benign influence. A variability analysis identifies important questions that could affect quantitative or qualitative analysis outcomes. It establishes a basis for a "performance window" to better understand the design features that affect fire safety.

5.4 Performance Estimates

Most engineering estimates are based on calculations. For example, if computer models calculate that flashover will occur 18 minutes after established burning or that the fire size at the time of detector actuation is 50 kW, these values usually have credibility. The term "estimate" is not commonly associated with this type of numerical outcome. Nevertheless, those values are, in fact, estimates. Although one may have confidence in their accuracy, they are still estimates.

An estimate is defined in Webster's Dictionary as:

- to form an approximate judgment or opinion regarding the amount, size, weight, etc., of;
- to determine roughly the size, extent, or nature of;
- an approximate value or judgment of something as value, time, weight, etc.;
- a judgment or opinion of ... the qualities of something.

The quality of an estimate depends on the amount of information that is used, the skill in translating that knowledge into performance understanding and the time available to evaluate relevant details and conditions. The confidence in an estimate reflects the belief that the information accurately represents the conditions to be expected.

Because fire safety is an emerging engineering discipline, most of the calculations represent idealized conditions. The mathematical models may not adequately represent all factors that affect outcomes. Also, the complexity of interactions and the inadequate information for some parts of performance evaluations leave gaps in knowledge. Quantitative imprecision occurs even for conditions in which we have some confidence.

These shortcomings in the technology of state-of-the-art fire science and engineering do not cause problems for performance analyses. Enough is known about most components that one can "fill in the gaps" with some degree of confidence. This is particularly the case when the analytical framework bounds the gaps of knowledge with conditions that force one to deal with specific status situations. Thus, the framework combines state-of-the-art technology with knowledge of component operations and capabilities to give greater confidence in predicting performance outcomes.

We use the term *estimate* to quantify an outcome regardless of the manner in which the values were determined. Thus, computer outputs, calculation results, and observational opinions are all described as "estimates" in performance evaluations.

Estimating involves an integration of technology, experience, judgment, and creativity. Information and knowledge form the basis of judgments and decisions. There is never enough information and knowledge of details to eliminate uncertainty. Consequently, uncertainty is an integral part of all estimates.

5.5 Uncertainty in Performance Estimates

Concepts of classical safety analysis may be adapted to organize uncertainty information associated with fire performance evaluations. Variability and uncertainty may be grouped into three categories:

- Stability of conditions
- Variability in physical parameters
- Professional practices.

Conditional stability is associated with changes in the situation being evaluated. Stability factors reflect the influence of common arrangement changes to predicted performance. For example, how will rearranging the room contents affect the fire growth potential for the room? How will the repositioning of movable room partitions influence fire propagation or smoke movement? Conceptually, conditional stability relates geometrical and dimensional variability to evaluation scenarios.

Physical parameter variability combines two aspects. The first involves variability associated with the physical performance of different materials. For example, burning characteristics of a cotton batting upholstered chair will be different from the characteristics of a functionally similar chair constructed with foam plastics. Similarly, operating and extinguishing characteristics of a common sprinkler will differ from those of an early suppression fast response sprinkler. In other words, from a fire performance viewpoint, all chairs and sprinklers are not created equal.

Behavioral stability is another type of physical variability. For example, values of thermal conductivity in many materials vary with temperature and testing procedures. Similarly, the coefficient of linear expansion for materials may change with temperature. This type of uncertainty can affect calculated outcomes that predict expected performance.

A third uncertainty category relates to the formulation and use of the deterministic relationships and calculation procedures used in *professional practice*. For example, what are the limits of room dimensions with regard to flashover correlation accuracy? When do limits of validity begin to affect outcomes significantly? Computer programs have default values and limits of applicability incorporated into their codes. The professional practice category identifies and addresses uncertainty and inaccuracies associated with calculation procedures.

This discussion has not included the topic of statistics. Statistics and classical probabilistic analysis have a role in the complete fire safety system. Biases inherent in judgmental decisions are also present in statistical studies. Unfortunately, this is not always evident. However, because a building scenario analysis evaluates a singular, case-specific installation, statistical performance is normally not used with these evaluations.

Uncertainty is a natural part of all engineering work. Fire safety has more uncertainty than mature engineering disciplines. Nevertheless, uncertainty is not to be feared. It must be handled in a rational manner by understanding the details that influence performance.

5.6 Philosophical Reflections

The terms *science* and *engineering* are often used as though they were a single entity. To be sure, one moves back and forth between information and application in solving technical problems. Although these disciplines are fellow travelers, sometimes their roles can be blurred.

We make a distinction that engineers solve problems – with constraints. Scientists create information. Engineering is the basis for performance decisions, including how much is enough. Science provides information to quantify outcomes.

The understanding that forms the basis for engineering decisions uses a mix of observation, technology, and judgment. During the infancy phase of an engineering discipline, judgment is a dominant influence. As technology improves, judgment is supplanted by analytical capabilities that become established with experience and testing. However, judgment is never eliminated and numerical measures are not ignored. Judgmental components become integrated with the evolving technological information base.

Engineers work in a very imperfect world where they must make decisions to solve problems in a timely manner. Where possible, knowledge from available scientific sources is used. When knowledge is inadequate, the gaps must be bridged by other sources and engineering judgment. Koen defines an engineering method as "the strategy for causing the best change in a poorly understood or uncertain situation within the available resources." Available resources include knowledge, information, equipment and procedures, confidence, money, and time. Rarely are any of them sufficient, and "best" is not faultless. Thus, while answers may not be perfect, they are based on the best information available at the time.

5.7 Closure

The goal of an engineering evaluation is to understand a building's fire performance and tell its story to others.

Assessment of the fire and its interaction with fire defenses is essential to understand the way a building will work in a fire. Assessments use the full range of knowledge and information within the available time and budget.

We are in the Age of Information. The ability to collect, retrieve, and use information is extraordinary. Computer models enable one to evaluate fire components that were beyond the scope of anyone a generation ago. One can record actual field conditions with a wide variety of digital and video recording devises, and instantly transmit the information to another location for consultation or storage. Much supplemental information is available quickly and comprehensively on portable devices.

Information and calculations alone do not provide a solution. The human provides the link between the problem to be solved and the best way to solve the problem. The IPI chart is the organizational storehouse for information involving fire performance. The IPI chart is also a systematic, interactive way to manage the information and acquire performance understanding. The role of this book is to describe a way of thinking that enables one to understand how a building will perform during a fire.

Part II

The Parts

Fire safety is an evolving engineering discipline.

Engineering disciplines advance from infancy through adolescence to maturity. Some parts of fire safety engineering have reached early maturity while other parts are now making the transition from infancy to adolescence. The state of the art is uneven. Nevertheless, general knowledge has progressed to the point where one can provide a credible evaluation of fire performance for buildings.

Unit One describes tools with which to structure a cohesive, disciplined analytical framework. The framework is founded on established techniques from other disciplines and has evolved to the point where confidence can be placed in its application.

Although the analytical framework structures the process, one must quantify outcomes with deterministic evaluations for the components. Thus, engineering evaluations must integrate a framework for thinking and component quantification to understand and compare alternatives. One is sterile without the other.

Unit Two describes the way each major component of the process works and procedures to quantify important events. Two elements are common to each part of an evaluation: (1) time, and (2) the fire and its products of combustion from ignition to extinguishment. The organization is as follows:

1) *Diagnostic fire.* Chapter 10 describes the philosophy, logic, and Interactive Performance Information (IPI) chart descriptions for the diagnostic fire. Performance selections are based on fire fundamentals, computer analyses, and observational estimates. Chapters 6 and 7 discuss these topics. Multi-room fires are normally due to barrier construction conditions rather than fire endurance ratings. Chapter 8 describes modes of barrier failure and multi-room fire propagation for diagnostic fire scenarios. Chapter 9 discusses smoke-gas propagation as a constituent of diagnostic fire scenario selection.
2) *Active extinguishment.* Fire department extinguishment and automatic sprinkler suppression are the only ways to put out a hostile fire. Understanding the interaction of fire extinguishment, the fire, and the building is central to any performance analysis.

    Fire department extinguishment combines actions that are under the control of three different organizations. First, detection and fire department notification is the responsibility of building management. Second, the local community is responsible for fire department response time and its composition. Finally, the fire ground commander is responsible for fire extinguishment. Collectively, time durations for these

Fire Performance Analysis for Buildings, Second Edition. Robert W. Fitzgerald and Brian J. Meacham.
© 2017 John Wiley & Sons Ltd. Published 2017 by John Wiley & Sons Ltd.

three sequential parts, combined with diagnostic fire conditions, become the basis of manual extinguishment analysis.

Chapters 11 and 12 describe detection and alarm. Chapter 13 discusses the community fire response. Descriptions of community fire capabilities are provided as background for readers who may have limited experience with these organizations. In addition to the analytical events, the discussion provides an overview of manual extinguishment.

Fire suppression starts with estimating the fire size for which initial (first) water can be applied. Based on that information, fire control and extinguishment are evaluated. The fire, the building, and local fire fighting resources all interact until the fire is put out. Chapters 14 and 15 describe essential events and the basis for analysis. These topics anticipate that readers have limited or no fire ground experience. Chapter 16 discusses sprinkler system suppression for fire control.

3) *Composite fire conditions.* The interaction of the diagnostic fire with fire department extinguishment or automatic suppression produces composite fire conditions. Eventually, extinguishment occurs. Chapter 17 briefly discusses this process.

4) *Structural behavior.* The structural frame and barriers resist the heat energy produced by the composite fire conditions. Modern structural calculations are changing the significance of traditional code requirements for fire endurance ratings. Nevertheless, it is appropriate to understand historic code requirements in the context of modern structural engineering capabilities. Structural behavior is the component in which structural engineers and fire safety engineers share responsibility. Each has a distinctive role in structural performance. Chapters 18 and 19 provide a perspective on these topics.

5) *Space tenability.* The movement of smoke and gases into target rooms and other building spaces away from the fire has an important impact on the safety of human life as well as property preservation and operational continuity. Chapter 20 describes smoke movement and tenability considerations for spaces beyond those rooms involved in flame movement.

6) *Risk characterizations.* Risk assessment for humans, property, operational continuity, and other conditions is a usual function of fire performance analysis. Chapters 21 and 22 discuss concepts for risk analysis of these topics.

7) *Prevent established burning.* Fire prevention is an important part of risk management. Although fire prevention is separated from building behavior given established burning, we do not overlook this important aspect of fire performance for buildings. Also, occupant actions can have a major effect on fire safety. Chapter 23 incorporates occupant fire extinguishment and special hazards fire suppression as parts of the prevent established burning evaluation.

6

The Room Fire

6.1 Introduction

Some fires are fast; others slow. Some fires exhibit high temperatures; others burn much "cooler." Some fires self-terminate before becoming a real threat; others grow to monster proportions, posing major dangers to a building and its occupants. The type of fire that can occur and its characteristics are important to the way a building will respond to a hostile fire.

Regulatory applications do not require knowledge of fire dynamics because understanding natural fire behavior is not necessary to use prescriptive codes and most fire standards. On the other hand, knowledge of how fires behave is essential for performance analysis and risk characterization.

This book describes a framework for analyzing the fire performance of buildings. Because a diagnostic fire is so important to fire defense analyses and risk characterizations, this chapter reviews a few important concepts and definitions. *Part One: Room Fire Concepts* describes important definitions for a systematic building analysis and room fire behavior. *Part Two: Room Fire Descriptors* discusses some useful experimental and theoretical relationships that influence room fire development.

These concepts and relationships are the basis of quantitative fire analysis. They also are the foundation for the qualitative evaluations described in Chapter 7. Quantitative and qualitative techniques may be viewed as a "matched pair" of tools to understand fire behavior in a room.

The way to discuss a topic as complex as room fires presents a dilemma in trying to answer the question, "How much is enough?" Although room fires are an important part of a holistic analysis, this is not a book on fire dynamics. Rather, we provide a cohesive way of thinking about quantitative and qualitative fire scenarios to supplement contemporary theory and numerical modeling.

Fire Performance Analysis for Buildings, Second Edition. Robert W. Fitzgerald and Brian J. Meacham.
© 2017 John Wiley & Sons Ltd. Published 2017 by John Wiley & Sons Ltd.

PART ONE: ROOM FIRE CONCEPTS

6.2 Fire

Fire is a chemical reaction involving a mixture of combustible gases and oxygen. Heat is necessary to initiate the process, and the reaction produces exothermic heat of combustion. Let's examine this process a little more.

From a combustion viewpoint, a fuel is anything that will burn in a fire. When a fuel is heated, pyrolysis produces volatile gases. When these gases mix with oxygen from air within a specific range of proportions, a flammable mixture is produced. Flaming occurs when this mixture reaches a high enough temperature to initiate a chemical reaction. The reaction releases a variety of combustion products, including light, heat, water vapor, gaseous species, and soot particulates.

6.3 The Role of Heat: Ignition

Heat sources exist in many different forms because all buildings have operations or occupants that use heat in some manner. Conditions in which the heat is too great and out of control can cause an unwanted fire situation.

Two types of heat sources can produce flaming ignition. A *piloted* ignition occurs when an open flame or a spark causes the gas and air mixture to flame. An *auto-ignition* (or *self-ignition*) can occur without a pilot source when the gas and air temperatures are high enough to produce flames. Although either of these sources may cause an ignition, both have roles in the subsequent fire growth.

Ignitability describes the ease of ignition for a material. In a laboratory, ignitability is measured as the time to raise the surface temperature of a material to its piloted ignition temperature. The pyrolysis rate depends on the heat flux that is applied to the surface of a material. Laboratory tests can provide a sense of proportion for the relative ease with which different materials can ignite. Figure 6.1 illustrates the concept of time of heat flux application and material ignitability.

Figure 6.1 Ignitability comparisons.

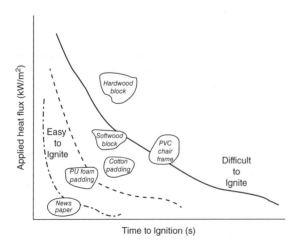

Table 6.1 Ignition temperature comparison.

Material	Piloted ignition temperatures [°C (°F)]	Auto-ignition temperatures [°C (°F)]
Paper		233 (451)
Cotton furniture covering	260 (500)	400 (752)
Thermoplastic solid (rigid PVC)	390 (734)	450 (842)
Thermoplastic foam (flexible PVC)	330 (626)	385 (725)
Thermosetting plastic solid (polyester)	400 (752)	500 (932)
Soft wood (red wood)	204 (399)	350 (662)
Hard wood (red oak)	275 (527)	

Temperatures necessary to ignite the air and gas mixture are lower for piloted ignitions than for auto-ignitions. Table 6.1 illustrates the kinds of differences one may expect for a few common materials.

6.4 The Role of Heat: Heat of Combustion and Heat Release Rate

Different materials release different amounts of heat in a fire. The heat of combustion is the quantity of heat that can be theoretically released during a fire.

In the early 1900s the fire load was defined as the total heat of combustion in a room. At that time the total room heat content was normalized to describe the fire load as pounds of cellulosic fuel per square foot of floor area (psf) (kg/m^2). In the 1970s the expression was changed to psf (Btu/ft^2 or kJ/m^2) of bounding surface area because that value correlated better with fire testing. Although an observational recognition of fuel quantity is useful, this approach is archaic because computer models provide more accurate analyses.

The heat release rate (HRR), $\dot{Q}$, is a property of much greater significance. For example, the ordinate of a room fire descriptor, such as that shown in Figure 6.3 (later), is now the HRR rather than ceiling temperature (T), which was used in the past. Methods of testing individual fuel objects and groups of fuels provide data to estimate room behavior much more accurately. For now we want to create awareness that although many factors can influence room fire development, the HRR ($\dot{Q}$) is the most significant property in describing a fire.

6.5 The Role of Heat: Heat Transfer

Heat flows from higher temperatures to lower temperatures. The transfer of heat influences the speed and certainty of growing fires and fire propagation. Heat transfer conditions influence the complete range of fire behavior.

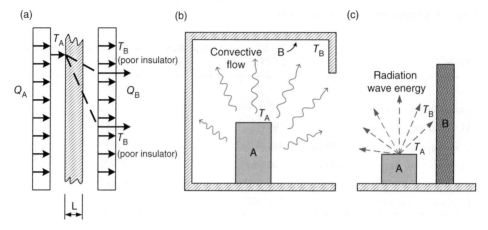

Figure 6.2 Heat transfer modes.

There are three mechanisms that transfer heat from hotter to colder locations. Let us briefly review some heat transfer concepts that influence fire growth.

Conduction moves heat in a solid from one region to another. We illustrate the process by examining the movement of heat energy from face A to Face B in Figure 6.2 (a). The heat transfer by conduction can be expressed as

$$\dot{q}''_{cond} = k\frac{\Delta T}{L} \tag{6.1}$$

where:

> $\dot{q}''_{cond}$ = conduction heat transfer per unit of surface area (W/m²);
> k = material conductivity (W/mK);
> ΔT = temperature difference between Face A and Face B (K);
> L = distance between Face A and Face B (m).

Conduction heat transfer is directly proportional to the conductivity and temperature differences, and inversely proportional to the distance. The heat leaving Face B (Q_B) is equal to the heat entering Face A (Q_A) minus heat losses during the movement through length, L. This may be expressed as

$$Q_B = Q_A - (\text{losses from A to B}) \tag{6.2}$$

Losses are due to molecular collisions that cause the temperature to rise inside the body transferring the heat. The losses may be calculated for any distance into the material and normalized to calculate a heat flow. The rate of conduction heat transfer is influenced by the thermal conductivity (k), density (ρ), and specific heat (C_p). Although we normally perceive conduction through solids, it also occurs in air and other gases.

One can calculate conduction heat transfer in an object to relate heat, time, and materials. Poor insulating materials (e.g. metals) have a high thermal conductivity and low losses in the material. Good insulators (e.g. wood, rockwool, foam) have a low thermal conductivity that causes less heat to transfer through the material.

Convection transfers heat through a circulating fluid of gases or liquids. The circulating fluid is the means for transferring the heat. However, the complete transfer process from one solid object to another is a combination of conduction and convection. For example, Figure 6.2(b) illustrates heat transfer from a hot object (A) to a cooler object (B) by convection. Convective heat transfer can be expressed as:

$$\dot{q}''_{conv} = h\Delta T \tag{6.3}$$

where:

$\dot{q}''_{conv}$ = convection heat transfer per surface area (W/m^2);
h = convective heat transfer coefficient (W/m^2 K);
ΔT = temperature difference between face A and a the circulating fluid (K).

Convection heat transfer combines the convective heat transfer coefficient and the temperature difference. In Figure 6.2(b), conduction transfers heat from the hot surface of object A to the air. The (warmer) buoyant air circulates and moves the heat to distant objects. When the hot air comes into contact with another solid object (B), heat is transferred to that object by conduction. The heat transfer coefficient enables one to calculate the quantity of heat that is transferred from the air to surface B, as well as the temperature at the surface of B.

Additionally, some of the heat is transferred from hot air to cooler air with which it comes into contact. This changes air temperatures and sets up convective movements.

Radiation is electromagnetic wave energy that travels through space without an intervening medium such as a solid or fluid, as illustrated in Figure 6.2(c). When radiant energy reaches another object, the energy may be absorbed or reflected.

Absorptivity is the fraction of radiant heat that is absorbed by the body. Black bodies have an absorptivity of 1.0 and absorb all of the incident heat. Low absorptivity surfaces (e.g. metal surfaces) have an absorptivity of closer to 0.0 and reflect most of the incident radiation.

The Stefan–Boltzmann law calculates radiant heat emitted from a source. The heat transfer is proportional to the fourth power of the absolute temperature (of the emitting surface), as described in equation (6.4).Thus, hotter fires emit more radiation than cooler fires:

$$\dot{q}''_{rad} = \varepsilon\sigma T^4 \tag{6.4}$$

where:

$\dot{q}''_{rad}$ = radiant emission per surface area (W/m^2);
ε = surface emissivity;
σ = Stefan–Boltzmann constant (56.7×10^{-12} kW / m^{2}K^4);
T = absolute temperature (K).

Although the quantity of radiant heat is not diminished with distance or by passing through most gases, water vapor in the air will absorb some of the heat. Thus, a fraction of the emitted heat is lost before it reaches the target object. Distance becomes a factor in this situation.

Radiation is an important cause of ignition, fire growth, and propagation. When radiation impinges on a fuel, the fuel absorbs the heat energy, increases in temperature, and releases more volatile gases. Auto-ignition can occur when the target fuel is distant from the burning object, and piloted ignition can occur if flames are near.

In a larger fire, approximately 30–50% of the heat energy is released in the form of radiation. About the same amount of convective energy is released. The remainder is incomplete combustion in the form of soot, carbon monoxide, or other products.

6.6   Realms of Fire Growth

Time is the common element that relates all parts of a building's fire performance. Time is related to conditions that affect fire growth and the generation of products of combustion (POCs). In order to examine the different parts of the process we compartmentalize fire conditions into realms of fire growth. Each realm has a set of factors that dominate behavior within that segment of a room fire.

A diagnostic fire relates time and the POCs. The flame-heat part of combustion is the central focus of fire growth descriptors. Although it is often perceived that room fire development is a single smooth process from established burning (EB) to flashover (FO), this is not accurate. Although a single mathematical relationship is sometimes used to describe fire growth, it is easier to understand the complete process by examining details within segmented growth regions.

Fire makes transitions from a fire free status (FFS) to ignition and EB, to FO, and into a fully developed fire. We call the important segments *realms of fire growth*. Each realm has somewhat different characteristics and factors that influence fire behavior within the segment. Enough critical factors must exist for a fire to continue burning and transfer into the next realm.

A representative room fire profile is shown in Figure 6.3. Although each fully developed room fire may follow this general process, the time durations and the certainty of fire growth can differ greatly because of differences in fuels, ventilation, and the room enclosure.

We will attempt to establish a base of understanding by discussing the process of Figure 6.3 to describe the characteristics of a room fire and the factors that affect its behavior. This integration of fire characteristics and factors associated with realm behavior provides an insight into room fire scenarios.

Realm 1 (pre-burning) describes the transition from a FFS to ignition (IG). Realm 2 (initial burning) examines fire growth from IG to EB. Although we consider these realms as a part of fire prevention, they are important to the complete picture of fire development.

The room fire growth from EB to FO is segmented into three realms. Realm 3 (vigorous burning) extends from EB to the enclosure point (EP). This realm represents free burning to the condition when the room enclosure begins to affect fire development. Realm 4 (interactive burning) describes fire growth from EP to the ceiling point (CP). This important event requires certain room–fuel–ventilation conditions to be present for flames to reach the ceiling. Realm 5 (remote burning) is the final segment of room fire growth. This realm describes fire growth from the CP to FO. FO is a very significant event in room fire behavior and building performance evaluations.

Realm 6 (full room involvement) is the final segment in the history of a room fire. This realm describes burning conditions from FO until most of the fuel is consumed and the fire decays and burns out.

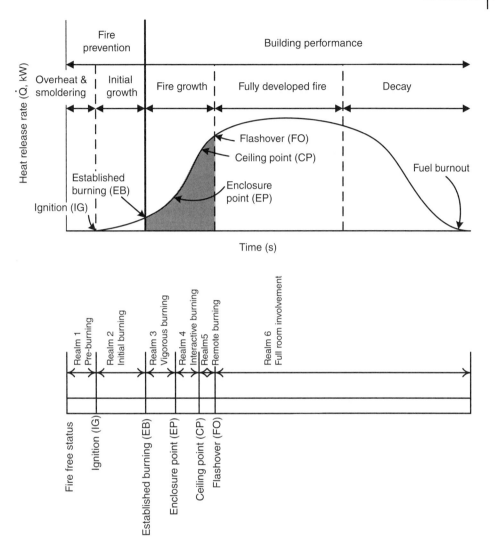

Figure 6.3 Realms of fire growth.

The realms of fire growth described in Figure 6.3 represent the complete process of fire development in small rooms. Conditions and characteristics involving fire size, fuels, ventilation, and room enclosure dominate the fire behavior and influence the time and transition into the next realm.

6.7 Fire Development: Fire Free Status to EB

Realms 1 and 2 describe the initial segments of the fire growth process.

- *Realm 1: Pre-burning.* When an abnormal heat increase or an external heat source is on or near a fuel, the temperature rise causes increased volatilization. The fuel may discolor or smoldering may occur in some materials. Even though smoldering may involve self-sustained combustion of the fuel, we do not define this condition as ignition. We define ignition as the appearance of the first small, fragile flame of the burning process.

- *Realm 2: Initial burning.* During the period immediately after ignition, the fire may go through a period of uncertain development. The flame may go back and forth between flaming and smoldering. It may go out. Or it may gain in strength and grow to a size sufficient for one to predict its future behavior with better confidence. We call this size "established burning." A firepower intensity of about 25 kW is a useful definition for EB. For mental estimates, EB may be visualized as a flame of about 300 mm (12 in). When the fire reaches EB it becomes established or stabilized, and its future course of action is more predictable.

The time duration to move through Realms 1 and 2 depends on the fuel and heat. On one side of the spectrum, a flammable liquid and spark will cause a rapid, almost instantaneous fire. On the other side, a condition may exist in which a conduit heats a fuel over a long period of time. When the combination of temperature and flammable gas–air mixture is right, smoldering or flames may break out. It is possible for smoldering to have a long incubation before going out or bursting into flames.

6.8 Room Fires

The fire growth from EB to FO is the most significant interval from a building performance viewpoint. This period from EB to FO gives occupants, fire fighters, and suppression methods a window of time to act. Fire characteristics during this segment affect fire detection, automatic sprinkler operations, smoke production, fire venting, and some building emergency operations.

After FO occurs, the heat, smoke, and general conditions change quickly and significantly. Differences before and after FO are similar to a step function in their effects on performance. Let us pause briefly to describe a room fire before FO.

Before discussing fire development in a room, first consider an unconfined fire, as shown in Figure 6.4(a). As the fire grows, air is drawn into the combustion reaction and a fire plume forms. This fire plume is a column of hot gases and smoke that rises due to the buoyancy of the fluid. As the column of hot air and fire gases rises,

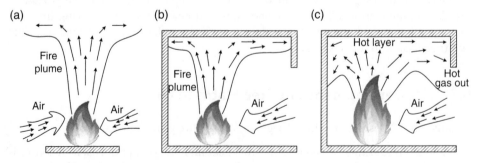

Figure 6.4 Room fire: enclosure effects.

additional air is entrained and the hot air and gases begin to cool. When enough heat has been lost to the entrained air, the gases lose their buoyancy and diffuse into the atmosphere.

Now consider a similar fire within a room enclosure. During its early stages, the fire will behave much the same as if it were free burning in the open air. Room openings supply air, and the fire plume develops and grows. However, as the fire continues to grow, the room enclosure causes fire behavior to change. The ceiling interrupts the fire plume, forming a hot gas layer as in Figure 6.4(b). As the hot gas layer descends below the top of the opening, fire gases that flow out are replaced by new air that flows into the room, as in Figure 6.4(c). This new air provides fresh oxygen to the fire, allowing it to burn more vigorously and increasing the quantity of heat and fire gases.

The bounding surfaces of the ceiling and walls influence the fire growth characteristics and speed. For example, as the hot gases of the fire plume impinge on the ceiling, a shallow layer of relatively hot gases directly beneath the ceiling begins to radiate out. This rapidly moving layer of hot gases that is driven by buoyancy of the hot plume is called the ceiling jet. As the ceiling jet moves away from the impingement location, it begins to lose heat to the ceiling and the surrounding air. This causes the layer to increase in thickness and decrease in speed. Eventually, as the fire continues to grow, the ceiling jet merges with the other gases to form a thicker, more uniform layer of hot fire gases. A strong ceiling jet can influence the operating speed of sprinklers and detectors.

As the hot layer of gases and smoke mushroom out from the ceiling, the walls interrupt the flow. This causes directional changes and creates turbulent flow. The hot gases radiate heat to the other combustibles that have not yet ignited, causing them to volatilize more rapidly. When the volatile gases of the fuels in the room reach their ignition temperature, and enough oxygen is present, the fire spreads very rapidly. This phenomenon is called "flashover."

6.9   Feedback

Heat causes volatile gases to be released from a fuel. When the volatile gas and air mixture reaches an ignition temperature, flames that define the chemical reaction we call fire appear. This chemical reaction produces exothermic heat. This exothermic heat has a major role in fire behavior after ignition.

After ignition, conductive heat within the fuel near the flame causes some pyrolization. Convective heat transferred from heated air in contact with unburned fuel releases additional gases. However, radiative heat impinging on the fuel causes most of the volatile gas release. Understanding the roles of heat transfer in releasing volatile gases helps one to recognize fire growth.

Feedback occurs when heat from one surface is reflected to other objects. The heat from feedback pyrolizes unburned fuel, causing increased fire growth. To illustrate the concept of heat transfer in fires, consider two pieces of wood placed close together, but not touching, as shown in Figure 6.5(a).

A single paper match can ignite the wood when we take advantage of heat transfer and fuel burning characteristics.

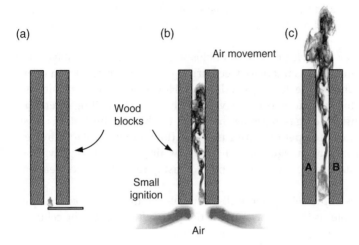

Figure 6.5 Feedback.

After the first small ignition, several simultaneous actions take place. First, a small amount of conductive heat warms the wood near the flames. Some of the heat is transferred by conduction into the wood, and gases pyrolize near the open flame.

Secondly, air is drawn by the chimney effect between the wood pieces, as shown in Figure 6.5(b). The flames heat this air and some of the convective heat is transferred to each piece of wood, pre-heating the wood to accelerate the volatilization process.

Third, the flames radiate heat in all directions. Some radiated heat from the flames on Block A impinges on Block B. This pyrolizes more of the gases of Block B. Simultaneously, the flames from Block B feed back radiative heat to Block A, causing additional pyrolization in A.

As the gases from the unburned fuel are volitalized by the different forms of heat transfer, the fire grows by continued piloted ignition. The arrangement of fuel and the effectiveness of heat transfer modes all contribute to a growing fire. It is more difficult for a single fuel block to produce a fire with the single match source because too much of the heat of combustion is transferred (lost) to the air and radiation feedback does not occur.

In a room fire, any hot surface radiates heat to the other objects. When an object radiates heat to another object, and heat is radiated back, the process is called feedback. Feedback radiation augments pyrolysis in the initial burning material, thus causing the fire to grow more rapidly. Some fire growth can be augmented by continuous piloted ignitions, as in the example above. In room fires the radiative heat transfer (with some convective heat transfer) can raise temperatures to auto-ignition levels, and fuels can burst into flames without an open flame. All of these mechanisms play a role in fire growth within a room.

6.10　Flashover

Flashover is a significant event in room fires. It marks the rapid transition between a growing room fire and a fire in which all room fuels are burning. Smoke, heat, and general fire conditions are very different before and after the FO event. Sometimes knowledge of the time from EB to FO is enough to characterize risks. Often, additional information is needed. Nevertheless, an understanding of the phenomenon of FO is important for fire safety engineering evaluations.

Essentially, a growing room fire continues to generate heat. This heat from the fire is transferred somewhat by convection and particularly by radiation to the bounding surfaces of the room and to other unburned fuels in the space. Feedback is a major influence for the increasingly rapid rate of production of volatile gases of the unburned fuel.

The mechanism of FO may also be described as a very rapid increase in burning within a room within a very short time. Often, the transition from a room fire to FO can occur in seconds. Before FO occurs, discrete objects in a room are burning. After FO, all objects in a room are burning.

6.11   Fully Developed Fire

After FO occurs, the fire will burn for an extended time until the fuel is nearly consumed or until it is extinguished. This is usually called a *fully developed fire* or the *post-FO fire*. A fire in this realm burns at a somewhat uniform rate for an extended time until the combustible contents are consumed and the fire decays and burns out. The time to burnout may be several minutes or several hours depending on the amount of fuel in the space and the ventilation conditions.

A fully developed fire produces the greatest heat release for the longest duration. This causes significant heat transfer to the barriers and the structural frame. Most structural and barrier fire endurance analyses are based on fully developed fire conditions.

Fully developed fires also produce enormous quantities of smoke. When this smoke is released inside the building rather than outside, smoke contamination can be widespread.

6.12   The Role of Ventilation

Air that supplies oxygen for combustion is important at all stages of room fire development. Within Realms 1, 2, and 3, enough air is available in the room without the need for external sources of ventilation. However, as the fire grows into Realms 4, 5, and 6, the location and size of room openings that supply replacement air and remove fire gases and smoke have an important effect on fire behavior and the POCs.

Flashover cannot take place unless a large enough opening provides the necessary air supply and fire gas removal. When a room is closed (except for leakage around openings) FO cannot occur. Nevertheless, this condition often maintains fire gas temperatures that are high enough to support combustion. However, the room is oxygen-starved, resulting in a lean mixture that creates a dangerous condition. If a barrier were to suddenly open (e.g. door opening), air enters the room quickly and mixes with the unburned, but hot, fire gases. This creates a combustible mixture that produces a very rapid fire, called a backdraft. A backdraft is a rapid deflagration (explosion) that causes much damage and danger to any humans in the vicinity.

Barriers normally provide ventilation through open doors and windows. Sometimes a window will break when the fire reaches Realms 4 or 5 causing a rapid influx of air that changes the time and certainty of fire growth.

The question becomes, how much is enough ventilation? Generally, a door that is open about 300 mm (12 in) is considered the smallest opening size that can produce FO. Even this opening size will produce an oxygen-deprived fire that burns slowly with relatively low temperatures. As the ventilation area increases, the fire gas–air mixture becomes enriched with oxygen, and the fire burns more efficiently. When the ventilation area is large, the available oxygen produces a very hot, fast fire that approaches free-burn conditions.

The time to FO is related to the ventilation area and its locations. Computer models and mathematical relationships can estimate fire growth time or conditions for which FO can be expected. Ventilation and its relation to the HRR are prominent factors in the calculations.

Ventilation has a particularly important effect on the fully developed fires of Realm 6. Burning characteristics are often viewed in terms of fuel-controlled burning and ventilation-controlled burning:

- *Fuel-controlled burning.* When a large amount of ventilation is available, fuel-controlled burning takes place. The rate of heat release of the fire is limited only by the fuel surface available for combustion. Fuel-controlled burning occurs in spaces having large ventilation openings that supply large amounts of fresh air. Temperatures are higher than for ventilation-controlled burning, and much heat energy is lost through the openings. The transfer of unburned fire gases outside the room causes burning when the gases mix with the available oxygen.
- *Ventilation-controlled burning:* When the ventilation area is restricted, reducing the supply of fresh air, the fire's HRR is limited by the air supply available for combustion. This ventilation-controlled burning causes fuels to burn more slowly, more inefficiently, at lower temperatures, and longer.

6.13 The Role of Barriers

Barriers provide openings for fire ventilation. However, barriers also influence fire growth by their geometry. For example, the walls of long, narrow rooms such as hallways or mobile homes enhance burning because radiation feedback increases temperatures of room contents faster than in more spacious configurations. This increases pyrolysis and creates faster fires. The proximity of flames to the walls also affects flame heights.

When the room has very high ceilings, such as atria, a CP cannot occur for flames to mushroom across the surface. Feedback radiation from the ceiling is eliminated, producing conditions closer to free burning. On the other hand, "ordinary" rooms with ceiling heights of 2.4–3 m (7.9–9.8 ft) provide conditions that permit flames to move across the surface and create remote burning.

Many buildings have rooms whose ceiling height is intermediate between "ordinary" and "high." For example, slower fire growth can occur in auditoria, assembly rooms, and dining rooms because the ceiling effect is less dominant. Even "ordinary" rooms in some buildings may be near to 4 or 5 m (13–16 ft.). These higher ceilings make reaching FO longer and more difficult.

In addition to the geometrical location of barriers, construction materials affect fire growth. Barriers may provide good insulation to the room. These barriers act like an oven to retain heat, causing higher temperatures and a faster time to FO. On the other hand, some barriers are poorly insulated. This allows heat to escape from the room more easily and reduce temperatures and cause a longer time to FO.

6.14   The Fire Development Process: EB to FO

Figure 6.3 describes fire growth and development and their relationships with realm segments. The growing fire from EB to FO requires attention because it is so critical to performance analyses. This part is segmented into three realms:

- *Realm 3: Vigorous burning.* Starting at EB, a fire behaves as a free-burn up to the EP when the room barriers begin to influence its behavior and character. A flame height of 1.5–2 m (4.5–6 ft) is about the fire size at which an ordinary room's geometry begins to affect fire development.
- *Realm 4: Interactive burning.* This realm extends from the EP to the CP, which is the event when flames first touch the ceiling. Within this realm, fuel objects near, but physically separated from, the initial fire source begin to ignite. Ignition is usually an auto-ignition caused by radiation heat transfer. However, more importantly, when the flames have enough substance to reach the ceiling, the likelihood of the fire continuing to grow to reach FO is very high. The CP is an important fire condition (size) on the path to FO.
- *Realm 5: Remote burning.* This realm describes the segment between flames initially touching the ceiling (CP) and FO. The ceiling jet and other ceiling gases radiate heat to remote objects causing increased fuel pyrolization. Also, walls augment remote object heating. Increased volatile gas production combined with high temperatures enable auto-ignitions to cause FO.

6.15   The Fire Development Process: FO to Burnout

Collectively, Realms 3, 4, and 5 describe the fire growth in a room from EB to FO. After FO, the fully developed fire burns until the fuel is consumed and only ash and unburned pieces remain. Figure 6.3 describes this as *Realm 6 (fully developed fire)*. This duration of this post-FO burning is influenced by the amount of fuel, the ventilation conditions, and the barrier insulation characteristics.

6.16   Summary

Room fires are complicated processes. Many interrelated parts mix and match to determine the progressive outcomes after ignition and EB. The particular relationships of fuel, building container, and ventilation combine to define the time-dependent fire conditions that will occur. The goal of *Part One: Room Fire Concepts* is to "tell the story" of general room fire development in such a way that one can understand the important parts and their roles in the process.

Fire science has quantified many of the concepts. The component roles and their interactions are intended to give a "big picture" holistic description of this complex process. *Part Two: Room Fire Descriptors* identifies experimental and theoretical information that will provide a sense of proportion for fire estimates and data for computer modeling and numerical relationships.

PART TWO: ROOM FIRE DESCRIPTORS

6.17 Introduction

The topic of room fire behavior has been the subject of intense research since the 1960s. A great deal is now known about the subject. Much more remains to be discovered. Nevertheless, fire science, combined with procedures from other disciplines, provides numerical measures that can convey a better understanding of room fire behavior.

Here, we re-examine room fire behavior with a focus on quantification for practical building conditions. We will not derive relationships from basic principles. The reader either knows this material as a part of background knowledge or can refer to other sources to obtain the necessary knowledge. Instead, this chapter identifies information to make better performance estimates.

In addition to describing numerical measures, several additional concepts are presented to offer a better understanding of diagnostic fires. These concepts and definitions enable one to be more consistent in structuring applications.

6.18 Fuels

A fuel is anything that will burn in a fire. The physical and chemical characteristics of fuels and their locations are important to how the fire will burn and the time for significant events to occur.

Fuels may be classified in several different ways. One way to look at fuels is according to their function in a building. For example, fuels may be *building fuels* or *contents fuels*. A building fuel is part of the permanent structure. Combustible parts of the structural frame, service fuels, and non-structural elements, such as non-bearing partitions and interior finish, fall into that category.

Contents fuels relate to the use and occupancy of the building. Contents fuels may be furnishings and decorations (e.g. draperies, curtains). Contents may be the goods sold in commercial stores or materials stored in warehouses. Contents may also be described as materials associated with a commercial or manufacturing operation. Occupant-related goods (ORGS) describe another type of contents fuel that is involved with general functioning of individuals in the building. ORGS may be books, papers, clothing, food or any objects that a human would use for functional living in the space.

Another way to look at fuels is according to their physical state. All fuels are derived from animal, vegetable, or petroleum-chemical sources. Fuels may be solid, liquid, or gaseous. Each is controlled and regulated in a different way, and each can produce a distinctive type and speed of fire development. Table 6.2 provides several examples of each.

Although solid cellulosic fuels comprise most building and contents fuels, petroleum-chemical (plastics) fuels are close behind. Liquid fuels and gases are more common for commercial and manufacturing uses, although they may be present in residential occupancies for cooking and heating.

Table 6.2 Forms of fuel.

	Solid	Liquid	Gas
Animal	Leather Fur Meat Grease	Fat	
Vegetable (Cellulosic)	Wood Paper Cotton	Spirits Wine Beer Cooking oils	
Petroleum chemical	Plastics Chemical- compounds	Oil Gasoline Kerosene Chemical- compounds	Natural gas Propane

The form of materials, their fuel properties, and their thickness and surface roughness are other ways to describe fuels. For example, the fire behavior of thin paper sheets or thin wood shavings is different from the same materials tightly packed in bundles or blocks. The surface area to mass ratio provides a measure to distinguish between these conditions. This ratio is also useful to recognize that some rough-textured surfaces (e.g. shag carpeting) have a high surface area to mass ratio when one includes the area of the strands in the surface area component. Although the heat of combustion may be the same, the form of the fuel produces a different burning behavior for materials having a high surface area to mass ratio than for the same materials boxed or bundled into smaller packages. Thus, the form of fuel will have a different effect on fire growth in Realms 1–3 than for the same material in Realms 4–6.

For most building applications, fuels may be viewed as cellulosic or petroleum-chemical. Cellulosic fuels are plant products such as wood, cotton, and animal fats. Petroleum-chemical products are broadly classified as plastics, petroleum products, and natural gases.

Plastics can be classified as thermoplastic plastics and thermosetting plastics. Thermoplastic products soften when heated. In addition to burning *in situ*, these materials can melt and form pool fires that produce additional burning. Thermosetting plastics are often used as a matrix for fiberglass to form a composite material. These plastics do not soften when heated, although they can burn vigorously when ignited.

Fire retardant treatments can raise ignition temperatures, thus impeding fire growth in Realms 1–3. However, after they ignite, the products usually burn vigorously. Consequently, the effectiveness of fire retardant treatments in Realms 4–6 is usually discounted. Both thermoplastic and thermosetting plastics can have many forms and conditions, including foam and solid. At some temperature (often widely variable), almost all petroleum-chemical materials will burn in a fire.

6.19 Fuel Packages and Fuel Groups

A *fuel package* is an arrangement of combustibles through which a fire can propagate by radiation. A *fuel group* is an arrangement of combustibles through which a fire can spread by direct flame contact alone. The separation distance between items distinguishes a fuel group from a fuel package. Fuel groups are a subset of fuel packages.

The fuel package concept combines heat release values for multiple fuel items. Equation (6.5) calculates the maximum separation distance to ignite a second item.

$$
R_{\mathrm{SD}} = \left(\frac{Q_r}{4\pi q_r^{''}} \right)^{0.5}
\tag{6.5}
$$

where R_{SD} is the separation distance, Q_r is the radiant HRR, and $q_r^{''}$ is the intensity of thermal radiation.

When the second item's separation distance from a burning object is less than that needed to cause ignition, the second item is considered to be part of the fuel package. This requires calculation or guidelines for distance estimates with the irradiance of second fuel items. Fuel groups are easier to recognize when using qualitative flame spread estimates.

When fuel items are separated, the fuel package concept requires radiation heat transfer calculations to estimate if fuel items are a part of the package. A fuel group is defined when direct flame contact is the mechanism for fire spread. Figure 6.6 illustrates these definitions.

These distinctions may appear trivial and perhaps unnecessary. However, an awareness of details is useful in interpreting computer model results used to construct diagnostic fires. They have an even greater role in fire investigations. Qualitative comparisons for real conditions are associated with quantitative conclusions to recognize the significance of possible outcome variability. Sometimes differences are inconsequential. At other times they become important. Nevertheless, rapid, visual recognition of relationships between quantitative and qualitative expectations gives an insight into performance forecasting.

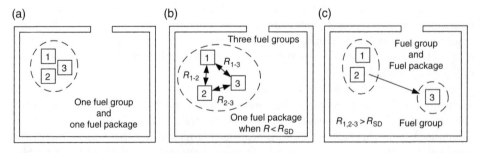

Figure 6.6 Fuel package-fuel group definitions.

Up to this point, no mention was made of the interior finish. If the finish on the walls and ceiling were non-combustible (e.g. gypsum board), the fuel group and fuel package designations would be as described above. On the other hand, if the walls were combustible (e.g. Luan mahogany) and a fuel item was adjacent to the wall, the interior finish becomes part of the fuel group or fuel package. If the item were moved away from the wall, the interior wall finish becomes a separate (additional) fuel package. If the ceiling were combustible and flames could cause ignition, the ceiling would be included as a part of each group or package.

Thus, both room contents and interior finish are parts of the room fire. Recognition of the potential for flame spread through fuel items is essential to understand room fires. Chapter 7 discusses this concept with regard to defining model rooms for computer analysis or observational categorization.

6.20 Heat Release Rate

The HRR, given the symbol $\dot{Q}$, is the most important property in diagnostic fire quantification. The HRR describes heat energy released from burning fuel per unit of time. In building enclosures, $\dot{Q}$ is a roughly exponential function from ignition to FO. After FO, $\dot{Q}$ is relatively uniform until the fire begins to decay.

Figure 6.3 illustrates a representative time vs. $\dot{Q}$ relationship for a room fire. Here, we will discuss quantification concepts of $\dot{Q}$ for individual fuel groups or packages while not becoming entangled in interactive details.

Figure 6.7 divides a room fire's history into two segments:

- Segment O–A represents a growing fire, ignition to FO.
- Segment A–B–C characterizes a fully developed fire that continues burning until fuels are consumed and the fire decays and goes out. Segment B–C denotes the decay phase.

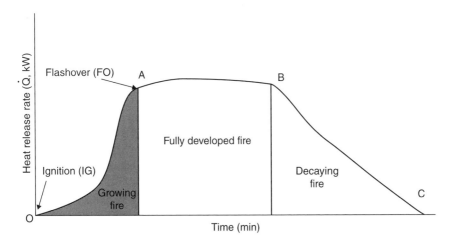

Figure 6.7 Room fire segmentation.

The quantification of $\dot{Q}$ for these segments also enables one to estimate temperatures, fire size, and smoke production at any instant or cumulatively over any time period. In addition, FO estimates and detector and sprinkler actuation is related to $\dot{Q}$. Thus, an ability to determine $\dot{Q}$ provides a wealth of valuable knowledge about the fire effects in enclosures.

All of these quantities may be incorporated into an Interactive Performance Information (IPI) chart. This enables one to compare conditions at selected instants of time along the continuum of the fire scenario.

6.20.1 Determining the HRR

Quantification in all engineering disciplines integrates theory and experiment. The HRR is a prime example of this type of interaction. A diagnostic fire must provide a reasonable estimate of the $\dot{Q} - t$ relationship for the room of origin and the additional cluster of rooms that can become involved.

The HRR $\dot{Q}$ may be calculated from equation (6.6) as

$$\dot{Q} = \dot{m}\Delta H_c \tag{6.6}$$

where

$\dot{Q}$ = rate of heat release (kW);
$\dot{m}$ = mass loss rate (kg/s);
ΔH_c = heat of combustion(kJ/kg).

Although equation (6.6) is simple in concept, its application to fuel items within a room is complex. Quantification must be determined by laboratory experiment. The heat of combustion is not constant, and the influence of construction on the item being measured adds complexity. The usual practice is to measure the HRR rather than to calculate $\dot{Q}$ values from equation (6.3). This is normally done by a furniture calorimeter, a cone calorimeter, or a room fire test.

The furniture calorimeter was initially developed as a laboratory test in the 1980s to measure the HRR from full-size items of furniture. Because of the expense and time in obtaining furniture calorimeter test results, the cone calorimeter was developed to provide a small-scale bench test to determine the HRR for components. The goal has been to calculate an equivalent HRR for a full-scale item with appropriate combination of bench test results.

We will briefly describe the furniture calorimeter and the cone calorimeter tests with some discussion of using the test results for computer modeling and establishing a diagnostic fire.

6.20.2 Furniture Calorimeter

The apparatus in Figure 6.8 shows the concept of furniture calorimeter testing. A furniture item is placed on a load cell to measure the continuous mass loss, $\dot{m}$, as the item burns. A hood collects and measures the oxygen concentration, light transmission in the exhaust duct, and effluent flow rates. The HRR and smoke production rates are calculated from these measurements.

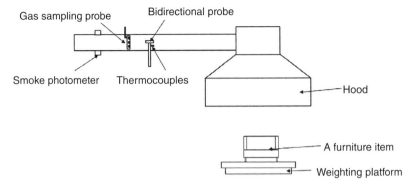

Figure 6.8 Furniture calorimeter.

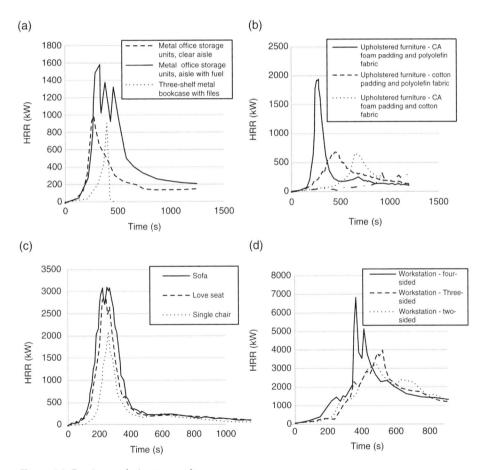

Figure 6.9 Furniture calorimeter results.

The room's contents will influence computer fire modeling and diagnostic fire estimates. Figure 6.9 show the types of results that can be obtained from a furniture calorimeter for representative contents fuel items. One can see that room contents and fuel interactions will influence the diagnostic fire, variability of outcomes, and

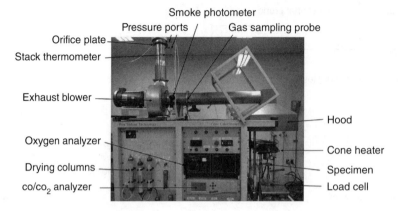

Figure 6.10 Cone calorimeter.

associated performance forecasts. Uncertainty and variability between modeling decisions and possible future contents are discussed in *Unit Four: Managing Uncertainty*.

6.20.3 Cone Calorimeter

The cone calorimeter (see Figure 6.10) is a small-scale bench apparatus that tests materials to determine their HRR, extinction rate (visibility relationship), soot, and selected fire gases.

The objective of a cone calorimeter test is to obtain information that can be combined in a predictive model to calculate the burning behavior for the composite item. The goal is to obtain results that can replicate the outcomes of the more expensive furniture calorimeter test.

Much progress has been made in developing analytical models using cone calorimeter results to predict the HRR for fuel items and fuel packages. While it remains a work-in-progress, expectations are high that procedures to predict HRR rates for fuel packages will be established.

6.20.4 Pause for Discussion

The HRR ($\dot{Q}$) is the dominant measure of room fire behavior, the diagnostic fire, and many of its combustion products. Although much progress has been made, the task is far from complete. Nevertheless, currently available information can be used with computer models and other estimates to provide a reasonable base for engineering quantification of fire behavior. This section discusses a few factors that should be recognized in using test data.

Enclosure Effects
Laboratory tests must use a consistent protocol in order to compare different fuels and relate the results to behavior in an enclosed space. Thus, specimen preparation, ignition type and location, and other variables are standardized to compare results between laboratories as well as between fuel items.

Ventilation and feedback are important laboratory test conditions. The furniture calorimeter test is essentially a free burn. Ventilation is not restricted, and feedback from room surfaces does not occur. Considering fire growth realms, the results should be accurate for Realms 1–3. As the barrier effects of the enclosures begin to contain smoke and heat, feedback begins to cause $\dot{Q}$ to increase above that shown in an open burning furniture calorimeter test.

Ventilation Effects
The laboratory tests have free burn conditions that provide adequate ventilation through the entire test. A room fire may have variable room ventilation conditions that range from small to very large. Restricted oxygen supply will affect $\dot{Q}$. This effect has an influence in Realms 4 and 5 for a pre-FO fire.

Ventilation has a very strong effect for post-FO fires. $\dot{Q}$ is significantly affected by both the size and location of openings. The influence is quantified as $A_v \sqrt{H_o}$ and discussed in Section 6.23.

Fuel Package Effects The laboratory test follows protocol for ignition source, intensity, and location. Because piloted ignitions are used, auto-ignition sources must be estimated. The fuel package may have different configurations and ignition locations or types than are used in the test protocol. Protected areas from feedback can affect the $\dot{Q} - t$ relationship. Unusual fuel package organization can affect this relationship. Sometimes combustion efficiency (χ) and surface area (m^2) are used to modify the burning rate

6.21 Fire Size Measures

Fire size is a primary basis for identifying realms that segment room fire characteristics and factors that influence fire behavior within the realms. Depending on the application, the fire size may be described in several ways. Certainly flame heights are useful to envision conditions. Estimating flame height becomes a base for understanding. Numerical measures associated with flame heights include the HRR and temperature of the flame body. Older measures that use ceiling temperature measured by thermocouples provide additional data.

Figure 6.11 relates free burn fire size (i.e. flame height) with associated physical measurements. This information is useful for associating observational estimates with

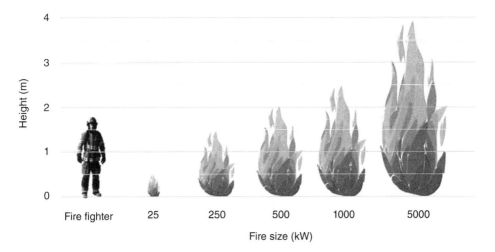

Figure 6.11 Fire size measures.

Table 6.3 Heat release rate comparison.

Free burn flame height	Free burn $\dot{Q}$	Representative Item
0.5 m	25 kW	Soft case, waste basket
1.5 m	250 kW	Small bookshelf
2 m	500 kW	Wooden desk
2.5 m	1000 kW	Small display kiosk, office chair, metal office storage unit
4 m	5000 kW	Four-sided Office workstation

behavior. Table 6.3 shows some maximum $\dot{Q}$ values for fuels and fuel packages to provide a sense of perspective to the relationships.

6.22 Overview of Factors that Affect Room Fire Behavior

Different factors dominate the fire behavior within each realm of fire growth. When enough factors are present, the fire can develop and move into the next realm. When too many factors are weak or missing, the fire can burn for a time and decay without moving into the next realm.

It is difficult to be more specific regarding factors or numerical quantities because of the synergistic nature of the components. Table 6.4 expands important factors that were introduced in Part One. All of these factors affect and are affected in some way by $\dot{Q}$.

6.22.1 Realm 1: Pre-burning

The transition from a FFS to ignition examines fuel and heat interaction. The response of the fuel to a heat source may be described broadly as its ignitability. Fuels susceptible to easy ignition may be described as "kindling fuels." This description includes both the form (high surface area to mass ratio) and the fuel's propensity to volitalize combustible gases.

Heat application has a broader scope. Ignition heat involves the source (e.g. electrical, mechanical, or chemical) and its delivery agent, the magnitude and form of heat flux (i.e. localized source or larger application area), the fuel surface area receiving the heat flux, and the distance from the heat source to the fuel.

6.22.2 Realm 2: Initial burning

Initial burning is the period during which the first small, fragile flames attempt to generate some substance and strength in order to grow to EB. During this phase the ignition source may remain in place, thus contributing to increased pyrolysis and a faster fire development. On the other hand, the ignition source may be removed, and continued development to EB depends on the heat transfer from the initial flames to the air or to the fuel.

The ignitability and thickness (surface to mass ratio) remain important. However, other factors contribute to the fire growth during this phase. Continuity is important here.

Table 6.4 Room fire growth factors.

Controlling factor	Realm 1 (FFS–IG)	Realm 2 (IG–EB)	Realm 3 (EB–EP)	Realm 4 (EP–CP)	Realm 5 (CP–FO)	Realm 6
Fuel	Surface area receiving heat flux Duration of heat application Ignitability	Continuity Ignitability Surface roughness kpc of fuel Thickness Surface area	Continuity and feedback kpc of fuel Surface area Ignitability Type of fuel Quantity	Fuel group arrangement Feedback Surface area Tallness Quantity Type of fuel	Quantity Arrangement	Quantity
Room enclosure			Proximity of flames to walls Wall finish combustibility Room insulation	Proximity of flames to walls Ceiling combustibility Ceiling height Wall finish combustibility Room insulation	Room insulation Ceiling height Aspect ratio	Room insulation Ceiling height
Ventilation				Size and location of openings HVAC operation	Size and location of openings HVAC operation	Size and location of openings HVAC operation
Terminal condition	IG	EB	EP	CP	FO	Decay and self-termination

FFS, fire free status; IG, ignition; EB, established burning; EP, enclosure point; CP, ceiling point; FO, flashover.

(a)

Heat

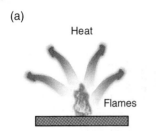

Flames

(b)

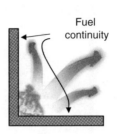

Fuel
continuity

Figure 6.12 Fuel continuity.

Continuity is positioning the fuel at or above the flames, as illustrated in Figure 6.12. A fire on a flat surface cannot propagate because much of the heat generated is lost to the air and feedback is too weak to pyrolize volatile gases. However, when the fuel is in a position for feedback to generate enough volatiles, the fire can propagate. For example, a carpet laid flat on the floor will self-terminate because feedback is insufficient to cause fire growth. On the other hand, that same carpet on a vertical wall provides conditions for easier fire spread because of its fuel continuity.

Surface roughness influences early fire development. Smooth surfaces do not propagate flames as easy as rough surfaces. Rough surfaces have a larger surface area to mass when all of the 'nooks and crannies' of the surface are considered.

The thermal inertia ($k\rho c$) is another fuel characteristic that has an important influence on fire growth. The thermal inertia is a measure of the time for a fuel surface to pyrolize gases when subjected to a radiative or convective heat flux. Fuels with a low thermal inertia, such as foam plastics, are good insulators. It is difficult for heat to move through these materials by conductive heat transfer. On the other hand, fuels with a higher thermal inertia, such as wood, are comparatively poor insulators because heat can be conducted through the material more quickly.

Thermal inertia is the main reason why foam plastics are such "bad actors" with regard to fire growth than "not as bad actors" such as wood and cotton. Relatively speaking, the heat cannot move easily through materials of low thermal inertia. Low thermal inertial fuels gasify at the surface very quickly, rather than transfer the heat into the depth of the material. Thus, excessive volatile gases form at the surface and are quickly released to form the combustible gas–air mixture. This is the primary reason why foam plastics provide a more severe and fast fire spread problem than do cellulosic materials such as wood or cotton.

6.22.3   Realm 3: Vigorous Burning

Vigorous burning describes the domain in which an EB grows within the fuel package and develops a strength and stamina of its own. The fire size grows from knee high (25 kW) to nearly the height of a person (400 kW) during this realm.

This realm is characterized by fire growth within the initial fuel package. When the fire reaches the size of a person (1.5–2 m), it changes in character. This change is caused by the enclosure surfaces beginning to influence fire growth because of feedback.

Ignitability, continuity, thermal inertia, and feedback remain factors that affect fire behavior in this realm. However, their influences decrease as the fire grows to the enclosure point. The quantity of fuel in the fuel package, as well as its size and arrangement, influences the growth in Realm 3. The location of the fuel package also begins to influence outcomes.

The flame height begins to be affected by its proximity to walls during Realm 3. As a rule of thumb, the effect of walls on fire size is as follows:

- Fuel package distant from walls (i.e. "free burning"): $\dot{Q} = \dot{Q}_{base}$.
- Fuel package is adjacent to wall: $\dot{Q} = 2\dot{Q}_{base}$.
- Fuel package in a corner: $\dot{Q} = 4\dot{Q}_{base}$.

These conditions are represented in Figure 6.13. The increase in flame height (and in $\dot{Q}$) is based on two factors. Air entrainment is the main factor that affects fire size. When the flames are distant from barriers, air can entrain in the fire plume through a 360° surface area. When the flames are against the wall, the flame surface area is reduced by half. Thus, in order to get the same air entrainment, $\dot{Q}$ and the flames double in size. When the fuel package is in a corner, the flames must increase by a factor of 4 to maintain the same surface area to entrain air.

The second factor that influences the flame height is the feedback radiation from the walls to the fuel package. Although this is less dominant than surface area for air entrainment, it remains a contributing factor.

As the fire grows in Realm 3, some factors begin to lose importance while others begin to become more meaningful. Locations of fuel packages within a room begin to influence the ease and speed of fire growth.

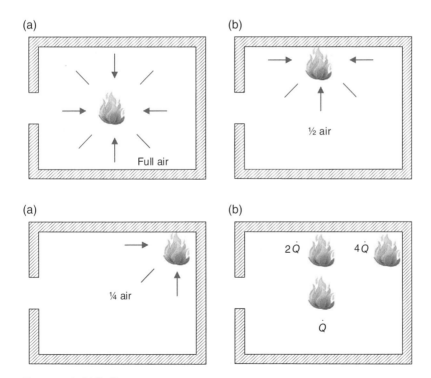

Figure 6.13 Wall effect.

6.22.4 Realm 4: Interactive Burning

Interactive burning is associated with flame heights between the EP and the CP. Here, the ceiling fire event should not be just a "glancing blow," such as with burning curtains that may briefly touch the ceiling and fall down. Rather, it indicates that the ceiling–flame contact is more substantive. When enough conditions exist for a substantive impact of flames to the ceiling, FO is very likely to occur as the fire continues into the next realm. If conditions and factors make it difficult for flames to reach the ceiling, the fire will be slower and may go out before FO occurs.

Fuel packages near the walls continue to have greater flame heights because of barrier feedback and the needs for oxygen to supply the flammable mixture. When rooms have higher than "normal" ceiling heights, fire growth is more difficult and requires increased time. Flashover may not even be achieved for rooms with high ceilings, such as atria.

The size and location of openings that supply air to the fire gain significance in this realm. An oxygen-starved fire burns slower and cooler, thus slowing down the process. If the HVAC system continues to operate in normal mode during the fire, both supply air and exhaust air will create faster and more certain burning conditions.

The fuels as well as their sizes and locations are important to continued fire growth. Some fuels that have a very high $\dot{Q}$ can impact the ceiling and cause FO independently. Thus, $\dot{Q}$ output is a major interest in this realm, and conditions often favor ignition of additional fuel items. The term interactive burning signifies the potential for extending the fire to additional fuel items. Coalescing of flames from two or more fuel groups causes increased flame heights.

The question "How much is enough?" applies to distances between fuel items. Heat from the environment increases temperatures and pyrolysis in the unburned fuels. This causes the fire to extend to additional fuel items. During the interactive realm, nearby fuel items can be ignited by auto-ignition created by radiation from the burning item. Local flying brands may also cause piloted ignitions in this realm.

Although convective heat transfer contributes to the unburned fuel heating, radiation is the primary cause. Figure 6.14 shows a burning fuel item, A, and a target item, B.

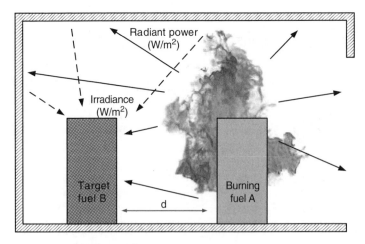

Figure 6.14 Separation effect.

Irradiance is the normalized incident power (W/m^2) on the surface of the target fuel. Radiance is the normalized power (W/m^2) emitted from the burning fuel. Irradiance is a function of the flame size and the distance between the flames and the target fuel package.

The maximum distance, d, between fuel packages at which ignition of the target fuel can occur may be estimated by testing the ignitability of the target fuel (Section 6.3) and the irradiance that will be received from radiant emissions of the burning fuel. Brabauskas[1] examined this question and published the results shown in Figure 6.15. Categories were determined from ignitibility tests. Items such as curtains or drapes are especially easy to ignite. Items such as wood blocks are difficult to ignite.

The tests were done in isolation. In a room enclosure, additional heat sources such as convective heat transfer and feedback from the ceiling jet and walls and ceiling will enable the second item to ignite faster.

Table 6.4 introduces several new factors into the process. In the fuel category, the surface area of secondary fuel items exposed to irradiation and the location of fuel packages relative to the burning items affect the ease of fire spread.

Tallness of fuels occurs in two ways. One is a physical tallness of fuel items, such as bookcases and combustible interior finish. The second is a flame height that is influenced by the HRR. Fuels having a low thermal inertia produce high (and fast) flame heights, because the combustible gases are released in greater quantity and speed. This produces fires with much greater flame heights, sometimes causing FO with a single fuel package.

Ventilation openings and their locations have a much greater influence on burning in Realm 4. Although smaller fires develop with existing air in the room, fires in Realm 4 require additional oxygen to grow. Consequently, examination of ventilation sources and amounts are important for this phase.

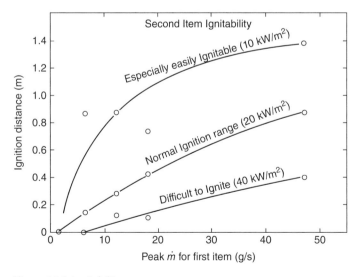

Figure 6.15 Ignitability.

---

1 Vytenis Babrauskas, Will the second item ignite? NBSIR81-2271, 1981.

The room enclosure begins to have a greater influence on fire growth in Realm 4. Room insulation begins to affect the burning rate. Well-insulated rooms act like an oven to retain heat. This increases temperatures and thermal feedback. The aspect ratio (length/width) influences fire growth. For example, long narrow rooms provide greater feedback from walls than do rooms with an aspect ratio closer to 1.0.

6.22.5 Realm 5: Remote Burning

Beyond the CP, a fire mushrooms along the ceiling and radiates heat energy to other fuels, causing an increased rate of volatilization. Some fuels remote from the initial ignition may experience auto-ignition, and additional fires may start in the room. Frequently, other items may be near the auto-ignition point, and fuels remote from the initial burning ignite almost simultaneously. This phenomenon is called FO.

Table 6.4 notes that the factors that influence Realm 4 remain important to achieving FO (Realm 5). The quantity of fuel in the space assumes greater significance in providing enough fuels to have a significant fire in the room.

6.23 Flashover

Flashover is a very significant event. The phenomenon is easy to recognize when it happens. The process is also easy to describe qualitatively. However, a diagnostic fire must determine if conditions exist to cause FO. Fire science has developed equations and models to predict FO. Simple algebraic equations are often useful to get a sense of proportion to determine whether the conditions exist that can lead to FO. These equations have been developed from correlations of fire tests combined with theoretical relationships of the combustion process. The most common FO correlations are as follows:

$$\text{Babrauskas}: \quad \dot{Q}_{\text{fo}} = 750\, A_{\text{v}} \sqrt{h_{\text{v}}} \tag{6.7}$$

$$\text{Thomas}: \quad \dot{Q}_{\text{fo}} = 7.8\, A_{\text{T}} + 378\, A_{\text{v}} \sqrt{h_{\text{v}}} \tag{6.8}$$

$$\text{MQH}: \quad \dot{Q}_{\text{fo}} = 610 \left(h_{\text{k}} A_{\text{T}} A_{\text{v}} \sqrt{h_{\text{v}}} \right)^{1/2} \tag{6.9}$$

where

$\dot{Q}_{\text{fo}}$ = HRR needed to produce FO (kW);
A_{v} = ventilation area (m^2);
A_{T} = total room surface area (m^2);
h_{v} = average height of ventilation openings (m);
h_{k} = effective heat transfer coefficient (kW/m^{2}K).

The MQH (McCaffrey, Quintiere, Harkleroad) correlation assumes FO to occur at 500°C while Thomas uses 600°C. Therefore, the estimate for FO will be more conservative with the MQH equation than for the Thomas correlation.

Figure 6.16 Typical single-person office in model building.

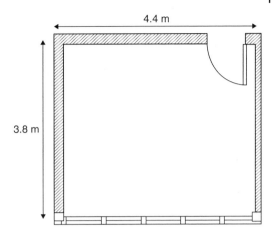

4.4 m

3.8 m

Although these equations are valid only for small rooms, they provide a sense of proportion for the required HRR needed to cause FO. The HRR may be obtained by the furniture calorimeter. If test results are not available, one may estimate $\dot{Q}$ from bench scale cone calorimeter tests and mathematical calculations. If neither of these is satisfactory, one must estimate outcomes from available data.

Example 6.1 Figure 6.16 shows a typical single-person office in the model building. The room geometry and ventilation area are as follows:

Room dimensions: 4.4 m × 3.8 m × 2.44 m (14.5 ft × 12.5 ft × 8.0 ft)
Openings: Door: 2.04 m × 0.91 m (6.7 ft × 3.0 ft)
              Windows: 1.22 m × 0.91 m (4 ft × 3.0 ft)
              Window head: 1.83 m (6 ft) above floor

Calculate the HRR needed to cause FO for the room of Figure 6.16 for the following conditions:

a.   Door open, no windows open
b.   Door open, two windows open.

Solution:
(a) Values for A_T, A_v, and h_v are:

$$A_T = 2\left(4.4\times3.8\right)+2\left(4.4\times2.44\right)+2\left(3.8\times2.44\right)-\left(2.04\times0.91\right)=71.6\,\text{m}^2$$

$A_v = (2.04 \times 0.91) = 1.86\,\text{m}^2$
$h_v = 2.04$ m
h_k = effective heat transfer coefficient for the boundary surface inside the compartment.

Values change with time up to the thermal penetration. Here, we assume that the exposure time is greater than the thermal penetration time and use $h_k = 0.03\ \text{kW/m}^2\text{K}$.

The required HRR to cause FO is calculated as:

Babrauskas: $\dot{Q}_{fo} = 750(1.86)\sqrt{2.04} = 1990\,\text{kW}$

Thomas: $\dot{Q}_{fo} = 7.8(71.6) + 378(1.86)\sqrt{2.04} = 1560\,\text{kW}$

MQH: $\dot{Q}_{fo} = 610\left[\left(0.03 \times 71.6 \times 1.86 \times \sqrt{2.04}\right)\right]^{1/2} = 1460\,\text{kW}$

We note that a single item of furniture can cause FO when the $\dot{Q}$ is similar to the larger values shown in Figure 6.9. On the other hand, if the item has a $\dot{Q}$ lower than the value to cause FO, item combinations in a fuel group or fuel package may be sufficient.

Solution:
(b) Values for A_T, A_v, and h_v are:

$$A_T = 2(4.4 \times 3.8) + 2(4.2 \times 2.44) + 2(3.8 \times 2.44) - (2.04 \times 0.91) - 2(1.22 \times 0.91) = 69.4\,\text{m}^2$$
$$A_v = (2.04)(0.91) + 2(1.22 \times 0.91) = 4.1\,\text{m}^2$$
$$h_v = \left[A_{door}h_{door} + 2\left(A_{window}h_{window}\right)\right]/A_v = \left[1.86 \times 2.04 + 2(1.22 \times 0.91)\right]/4.1 = 1.47\,\text{m}$$

where A_{door} and A_{window} are the opening areas of a door and a window, respectively, and h_{door} and h_{window} are the opening heights of a door and a window, respectively ($h_k = 0.03$ kW/m^2).

Babrauskas: $\dot{Q}_{fo} = 750(4.1)\sqrt{1.47} = 3730\,\text{kW}$

Thomas: $\dot{Q}_{fo} = 7.8(69.4) + 378(4.1)\sqrt{1.47} = 2420\,\text{kW}$

MQH: $\dot{Q}_{fo} = 610\left[\left(0.03 \times 69.4 \times 4.1 \times \sqrt{1.47}\right)\right]^{1/2} = 1960\,\text{kW}$

We note that the larger openings permit a greater heat loss, requiring a larger $\dot{Q}$ to cause FO. The HRR values of individual items and fuel groups or packages are used to estimate if FO will occur.

Discussion:
These calculations show a range of HRRs causing FO from 1460 to 3640 kW, and this is due to variations in room openings alone. The door opening is easy to identify. It can be more difficult to identify the status of windows.

Windows often break just before or just after FO occurs. The type of glazing can be important to their behavior. For example, a common glazing is ordinary window glass. This glass breaks easily because of temperature differentials. Glazing in high-rise buildings is very thick and does not break easily. Additionally, some glazing may be double- or triple-pane. This glazing is becoming popular because it provides better insulation for temperature control and heating/cooling savings.

Thick glass and double and triple glazing withstand temperature differentials that routinely cause ordinary glass to break in a fire. Often, the local fire department intentionally breaks windows to vent smoke and heat for firefighting. Thick glass and multiple pane glass are difficult to break, even with an ax. This affects firefighting operations both for intentional venting of the fire and because backdraft conditions can become more likely.

This example used a representative small office space in the model building. If the required FO potential were calculated for larger rooms in the same building, correlation accuracy is reduced. Certainly, the calculations would not be valid for the open office of the top floor. Judgment must be exercised in interpreting results from calculations.

The questions arise for a diagnostic fire: "Will FO occur?" and "What should I use for my analysis?" Although uncertainty and variability are apparent in these calculation methods, the application of determining the $\dot{Q}_{fo}$ required to produce FO and the $\dot{Q}$ provided by the room contents provide a sense of proportion for the fire growth hazard of the room. Some amount of judgment is necessary to select appropriate values for the diagnostic fire.

6.24 αt^2 Fires

The three realms of fire growth from EB to full room involvement involve different factors and conditions, as noted in Table 6.4. Although the realms may be distinct, transitions from one realm to the next are often assumed to be smooth. This allows one to characterize the segment from EB to FO by a power law, $\dot{Q}=\alpha(t-t_i)^p$, where t_i is a reference time from ignition. The term α is a fire intensity coefficient that reflects the influence of fuel on the speed of fire growth.

For many materials in buildings, $p = 2$. If one ignores the incipient stage of the fire (before EB), these fires can be characterized by the equation $\dot{Q}=\alpha t^2$ where t is measured from EB. This relationship is often termed a "t^2 fire."

The relationship $\dot{Q}=\alpha t^2$ provides a useful way to estimate the time to FO. Although this power law is valid only for fire growth to the CP, that condition is reasonably close to FO for many small rooms.

The detection industry uses specific values of α to define ultrafast, fast, medium, and slow fire types.[2] Experiments with either single items or multiple item packages provide experimental values for the fire intensity coefficient, α. Figure 6.17 shows representative time relationships for αt^2 fires.

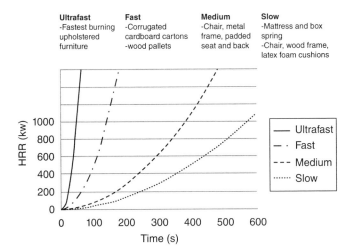

Figure 6.17 αt^2 fires. *Source*: U.S. National Institute of Standards and Technology (NIST), NBSIR 88-3695, NBSIR82-2604.

2 NFPA 72 National Fire Alarm And Signaling Code, 2016.

6.25   Realm 6: Fully Developed Fire

Realm 6 of a natural room fire produces a lengthy period of burning. In this realm the combustibles are essentially consumed and the fire decays to ash and a few unburned products. The fire conditions during Realm 6 are influenced predominantly by the quantity of fuel, the room insulation, and the ventilation conditions. Although Chapter 18 (Codes, Standards, and Practices) describes traditional ways of calculating post-FO fire durations, most of those methods are no longer necessary with available computer models.

A fully developed (post-FO) fire produces enormous amounts of smoke and heat. Room openings provide both the air supply and the venting of heat and smoke. The rate of burning affects these quantities. The rate of burning is influenced by the amount of air for combustion. Fires may be *fuel-controlled* or *ventilation-controlled*. A fuel-controlled fire occurs when a large amount of air is available, such as when the room has large ventilation openings. The rate of heat release is limited by the fuel surface available for combustion. A ventilation-controlled fire occurs when openings are small and the air supply is restricted. This condition causes fuel to burn longer, more inefficiently, and cooler than fuel-controlled burning.

The insulative qualities of the room also influence the time–temperature conditions. Well-insulated (thermally thick) rooms do not lose much heat to the outside. The room acts more like an oven and the fire burns hotter. On the other hand, barriers in poorly insulated (thermally thin) rooms transmit more heat to the outside. This reduces the heat energy of the room and causes the fire to burn somewhat cooler.

Figure 6.18 depicts representative variations that can occur with different ventilation conditions.[3]

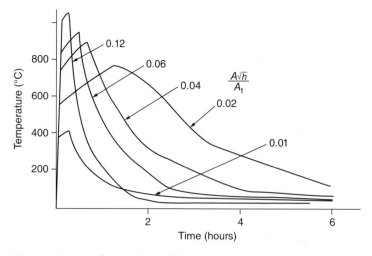

Figure 6.18 Room fire ventilation effects.

3 S.E. Magnusson, S. Thelandersson, Temperature-Time Curves of Complete Process of Fire Development, Lund Institute of Technology, Lund, Sweden, 1970.

6.26   Limits of Applicability

Be careful! One must be ever vigilant when using computer models and relationships validated by "small" rooms. For example, theory and practices for the phenomena and calculations of FO are valid only for "small rooms." The room size dimensions to define a small room are not completely numerical, but rather relate to the mechanics of reaching FO. As a general guide, one can assume that "ordinary" rooms of most buildings may be considered "small." When the geometric dimensions begin to increase, numerical calculations lose accuracy.

Certainly, large spaces such as "big box" stores, warehouses, and manufacturing rooms, particularly with high ceilings, are recognized as large rooms in which correlations and FO models are not valid. Atria are also in this category because the high ceilings prevent FO. However, what criteria define limits of applicability? An intimate understanding of the theory and programmed default values, correlation limitations, input accuracy, and validation comparisons is necessary. In the absence of this detailed knowledge, judgment must be used.

We encounter many conditions where one must evaluate calculated results for applicability. For example, computer results for open offices having ordinary ceiling heights and large areas for operations must be scrutinized. Although the lower ceiling produces a ceiling jet and ceiling feedback radiation, the lack of wall enclosures would change FO relationships. Restaurants, open mercantile stores, and auditoria have spaces where calculation theory that is influenced by wall and ceiling boundaries would also be questionable. Nevertheless, many calculation results can be adapted to these types of building spaces with engineering judgment based on fundamental fire concepts.

6.27   Large Rooms: Full Room Involvement

Fire propagates in a large space by progressive flame movement from fuel group to fuel group, rather than by the rapid FO that is characteristic in smaller rooms. Some fuel and room geometry configurations can produce this almost simultaneously. Although these phenomena may theoretically describe a boundary between large and small rooms, we shall consider full "room" involvement to represent the large, intensive burning of a part of the entire space. Essentially, the "room" can be defined with the virtual barriers described in Chapter 2. Even though the entire large space may not be involved, the virtual barriers enable one to analyze segments of the building with the same rational procedures as that of compartmented buildings.

High ceilings, large volumes, and spaced fuel groups characterize large rooms. Spreadover, rather than FO, is the usual method of flame propagation. Radiation heat transfer is the principal mechanism of fire spread between fuel groups. In these situations, an analysis starts with an ignition and EB in an initial fuel item. The conditions for Realms 1 and 2 of Table 6.4 remain valid for this early fire growth. However, as the fire grows larger, some of the small room realm conditions disappear.

Because radiation heat transfer is the primary cause of fire spread between fuel items, the fire size, ignitability of the adjacent fuel item, and the separation distance between fuels dominate fire propagation estimates.

The characteristic spreadover mechanism is as follows:

- A fire starts and grows in a fuel item.
- The burning fuel radiates heat.
- The heat flux pyrolyzes fuels in adjacent fuel items.
- Auto-ignition temperatures of adjacent fuels are reached. Sometimes flying brands can cause piloted ignitions.
- Fire in an adjacent fuel item grows.
- Heat radiates to additional fuel groups and the process continues.

The most significant factors that affect the process are:

- Fuel group sizes and distribution.
- Size of the radiating item generating the heat flux.
- Spacing between the emitting heat source and the target surface.
- Heat flux received by the target fuel.
- Ignitability of the target fuel and the critical heat flux (irradiation) in the target fuel needed to cause ignition.
- Duration of heat flux received by the target fuel item.

The fire conditions and proximity of adjacent fuel packages may cause successive ignitions. As the fire continues its spread to other fuel packages, conditions worsen. While hot layers do form at the ceiling and back radiation does pyrolyze unburned fuels, the principal mechanism of fire growth is radiation-induced fire propagation throughout the arrangement of fuel packages. The time duration for fire spread is an important quantity.

6.28   Fire Safety Engineering in the Information Age

Fire safety performance is dynamic and multi-dimensional. The analytical framework and IPI chart provide a structure with which to organize and understand performance. Although all parts of the holistic system are dynamic, understanding fire behavior is central to relating its interactions to the other "moving parts" of the system.

Fire growth within enclosures is enormously complex. Computational models and analytical methods are essential to provide quantitative measures for performance-based analysis and design. Appropriate use of these tools to support building fire performance analysis relies heavily on:

- understanding the problem;
- selecting an appropriate tool for the job;
- recognizing what can be reasonably expected from the tool;
- selecting and using valid and appropriate input data;
- operating the tool correctly;
- exercising reasonable engineering judgment in interpreting and using the output.

Several excellent resources can assist with selection of appropriate tools. *The SFPE Engineering Guideline on Substantiating a Fire Model for a Particular Application* guides the reader through defining the problem, selecting a candidate model, and validation and verification of output. This document also discusses how to identify and treat uncertainty, as well as analyzing sensitivity. The document also discusses important fire-related phenomena such as gas layer, ceiling jet temperature and velocity, flame heights, heat flux, and response of a target to heat flux for algebraic, zone (lumped mass), and field (computational fluid dynamics) models.

The *SFPE Engineering Guide on Room of Origin Fire Hazards* describes deterministic fire effects modeling. This document discusses types of available tools, what they can predict, and the type of input data that is needed. Figure 6.19 reproduces figure 4 and

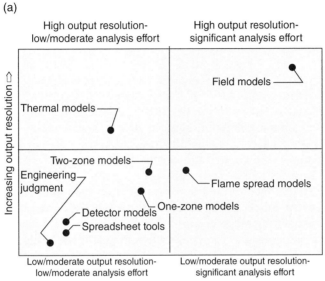

(a)

Input	Output	Applicability
The model input should consist of: • The effective enclosure dimensions • Opening dimensions • Mechanical ventilation flow rates and locations within the fire enclosure • Fuel characteristics • Fire size (transient or constant) • Boundary material thermal characteristics Other inputs include: • Details on the combustion chemistry • The limiting oxygen index • Fuel composition • Fraction of energy released as radiation Models that can predict detector actuation will require details on the detector spacing and actuation characteristics. Specifications necessary to obtain output at the correct location may be necessary, such as the location and orientation of a target.	The model output will include: • Time–temperature profile within the compartment • Composition of various compartment gases • Fire size (post-flashover models) • Boundary temperatures • Heat flux at various boundaries • Heat flux at a target • Burning inside and outside the compartment • Unexposed boundary temperature • Ignition of combustible objects • Whether or not flashover would occur	Two-zone models have broad applicability for predicting the pre-flashover environment in one or more spaces. Care is required to ensure the model assumptions remain valid for the geometry considered. Very large spaces, highly irregular spaces, or spaces with large aspect ratios are examples of geometries that could challenge some of the model development assumptions.

Figure 6.19 Ventilation effect. (a) Relative comparison of the input detail and analysis effort with the output resolution among common fire hazards analysis tools. (b) Two-zone input, output, and applicability. *Source*: Reproduced with permission of SFPE.

table 4 from this document. This information describes relationships between analysis effort and output resolution, types of inputs, outputs, and applicability of two-zone models that were used for this example.

One of the challenges of engineering is to understand the significance of holistic responses to component variability that can result from reasonable changes in condition. Computer models can be accurate for specific conditions that are defined for the calculations. However, they describe outcomes for a specific set of input information. Uses, conditions, and materials change over the years. The engineer must examine how holistic building performance can change with the different component conditions. How will performance differences change the risks to people, property, and operational continuity?

This is the same issue that the structural engineering profession faced a century ago in codifying appropriate live loads with which to design the supporting structure. Although that problem was easier to solve than comparable fire selections, nearly 20 years of technical and philosophical debate were needed before consensus live load values were selected for building codes. Design loads for snow, wind, and earthquake are more complicated and needed most of the 20th century to achieve consensus. Lack of consensus values did not stop applications. Nor did achieving consensus reduce the need for additional research on the topics. To solve the contemporary needs of practice, engineers combined available information, holistic behavior, and engineering judgment.

The fire safety engineering profession is attempting to identify appropriate "design" fire conditions for engineering practice. Until consensus fire conditions become codified, the fire safety engineer must use judgment to select diagnostic or design fire conditions that are appropriate for the building. Computer analyses can enhance understanding and provide a rationale for diagnostic fire selections.

Computer models provide rapid, cost-effective ways to understand the effect of enclosure and contents changes on a fire. "What if" variation studies enable one to describe a range of plausible deviations for a room or cluster of rooms. A window of reasonable fire condition differences, used with IPI charts to gain an understanding of holistic effects, enables one to identify significant features that affect the building's performance. This provides sensitivity to associated risks and enables one to clarify weaknesses in fire defenses or risk management programs.

A wide range of tools is available to enhance understanding of fires. Figure 6.20 identifies several categories. One can estimate fire conditions with simple algebraic relationships. Knowledge is expanded with spreadsheet calculations, two-zone models, and computational fluid dynamics models. The *SFPE Handbook of Fire Protection Engineering* describes these methods and the Internet provides other resources.

Numerous sources exist for model input data. Internet searches can identify test reports, experimental information, and other useful publications. However, the data must be appropriate to the problem and conditions. For example, the type of computer workstation or the materials used for upholstered furniture can have an enormous effect on fire growth and POCs. Differences can be an order of magnitude or more. Therefore, data that do not reflect contemporary materials or realistic conditions may produce misleading results.

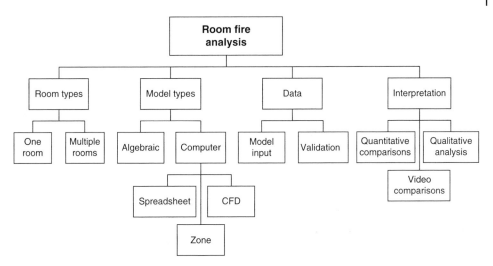

Figure 6.20 Organization chart: room fire estimate.

Fuel package input data must closely match the situation being modeled. Fortunately, the Internet provides access to a wealth of fire performance data. For example, search engines, such as the Fire ReSearch Engine (http://www.iafss.org/fire-research-engine/) look specifically at universities, research institutions and laboratories with a focus on fire. It taps into specific information resources, such as the National Institute of Standards and Technology (NIST) fire on the web materials, and the University of Maryland burning item database (http://www.firebid.umd.edu/). These are useful sources of input data for fire performance analyses.

Videos are another useful tool for interpreting fire performance analyses. These can also include instructions and examples of how to use computational tools, such as fire effects models and evacuation models. They may also include experiments or actual scenes of burning items, rooms, and buildings. Online learning also provides web-based educational opportunities

Today, information technology has a variety of forms to collect digital data, store it, and make it accessible. For example, YouTube, a web-based video channel, has resources on building fires, fire experiments, use of models and much more. The Underwriters Laboratories fire fighter safety site (https://www.youtube.com/user/ULfirefightersafety) provides videos of experiments, tactical responses, and lectures on fire dynamics. Web searches can provide video tutorials on use of models, performance of fire protection products and more.

In this information age we have a gift of easy accessibility to knowledge and information. The challenge becomes how to use this gift wisely. Computer fire modeling is an accepted tool that contributes to understanding complex fire applications. This book recognizes their importance, but does not describe the basic principles of fire dynamics and computer modeling. Chapters 6–9 describe concepts that augment digital-based information to interpret output results better. Often, these concepts are used before selecting the digital models. Qualitative evaluations enable one to be more efficient at organizing the study (i.e. the problem), selecting appropriate data, and interpreting the accuracy of calculated results.

6.29   Closure

Room fire growth is complicated. Much theoretical and experimental research has been conducted over the past half century. The computer has made many of the interactive calculations tractable. This combination has advanced understanding of how room fires burn. While the process is not yet completed, the theory, calculations, and experimentation can augment human judgment and observations substantially. Although a definitive picture has not yet emerged, available knowledge and information, combined with an organized thought process, provide a good understanding of what to expect in room fires.

This chapter has attempted to paint a picture of room fires. We combine descriptive portrayals with an abridged summary of relevant theory and calculations. The goal is to organize the process. The synergistic descriptors identify the way in which fire fundamentals fit into the process.

A combination of observation, data, and comparison is important when using computer models to simulate behavior. There is a wide variety of computer models, calculations, and information storage. All give a piece of the picture. The engineer must integrate all pieces to gain a good picture of what will happen. This requires one to relate a mental model of the room with theoretical and computational assumptions and limitations to understand expected behavior.

7

The Room Fire: Qualitative Analysis

7.1   The Role of Qualitative Analysis

Mathematical relationships and computer models are essential for modern fire safety engineering. Complex interactions that change dynamically require computer capabilities to portray the associations. Without these computational tools, our understanding of fire, fire defenses, and risk would be inadequate. Nevertheless, one must recognize that present capabilities are just a way station on the path to computational maturity. Contemporary interpretation and confidence in results constantly improve in the normal evolution of all engineering disciplines.

At this time in the evolution of fire safety engineering it is also important to have rapid, approximate qualitative procedures to augment computational results. Quantitative procedures and qualitative analyses provide skills that can be used independently or in combination because they have the same theoretical base. Their synergism enables one to understand room fires better and portray them more appropriately.

Comparing qualitative and quantitative expectations provides several advantages. Certainly, a major benefit is the ability to examine computational outcomes to ensure that they represent reasonable expectations. Other applications involve fire investigation reconstruction. Rapid observational techniques that predict fire behavior provide an insight into likely scenario causes as well as guidance for quantitative analyses.

Preliminary qualitative estimates can save analytical time and provide a more meaningful quantitative analysis. Much professional time can be saved in complex buildings by recognizing important room fire conditions.

The ability to describe likely outcomes in layperson's language is a useful skill for a fire safety engineer. Communication of differences in building performance for "what if" alternatives is a valuable asset for risk management because it provides a more complete picture regarding the importance of certain routine decisions. For example, what if an occupant wishes to cover the walls of a room with combustible interior finish rather than non-combustible linings? Or, what if an office manager selects modern office furnishings, such as fabric over fiberglass batting with a hardwood core, rather than traditional wood or steel workstations? What if furniture uses cotton padding rather than expanded polyurethane foam? Alternatives that affect holistic building performance can be broad. Their influences may be significant or

Fire Performance Analysis for Buildings, Second Edition. Robert W. Fitzgerald and Brian J. Meacham.
© 2017 John Wiley & Sons Ltd. Published 2017 by John Wiley & Sons Ltd.

minimal. The ability to give a realistic sense of proportion to others within a short time frame becomes a valuable skill.

7.2 Qualitative Estimates for Room Fires

There are two different types of qualitative observational assessments for room fire behavior. One is a "top-down" assessment that involves rapid screening techniques to order the relative hazards of different room interior designs. This process is useful to get an initial perspective of the range of problems that exist. The other is a "bottom-up" analysis that examines scenario behavior for a specific ignition location. Because both procedures are rapid, one often moves back and forth between the two to gain a better perspective.

Top-down concepts are valuable for diagnostic fire selection and preliminary risk analysis. A *model room* is defined to be a conceptualization of a room and its contents with regard to fire behavior. Top-down estimates portray the model room's general fire growth potential (FGP) rather than the outcome for a specific fire growth scenario. Only a few FGP categories for top-down estimates are needed to rank room hazards.

A "bottom-up" analysis is useful to examine specific fire scenarios. For example, several different ignition locations may be identified in a room to envision their FGP. Rapid approximations of the FGP at different locations can give a "feel" for the general outcomes that may be expected, as well as a better sense for interpreting computer simulation results.

A bottom-up scenario analysis is particularly useful for conducting fire reconstruction analyses. Given a possible ignition location, the realms of fire growth described in Chapter 6 provide a basis to evaluate the expected outcomes.

This chapter will discuss both specific ("bottom-up") fire scenarios and general ("top-down") FGP classifications. Each has a role in fire safety applications. The techniques can be applied rapidly and consistently to gain an insight into building performance.

Rapid visual estimates are particularly useful when combined with the capabilities of technology. For example, one can make video or digital pictures of conditions and relate them to model rooms, computer results, and corporate information sources. Sorting procedures involving a small number of categories can improve estimating consistency.

PART ONE: BOTTOM-UP ESTIMATES

7.3 Bottom-up Scenario Estimates

Qualitative observational estimates based on the theory and concepts of Chapter 6 can provide a first-order, bottom-up analysis to organize thinking and predict outcomes. The realms of fire growth shown in Table 7.1 provide the framework for thinking. Scenario outcome estimates assess all relevant factors within each realm to estimate whether enough conditions exist to cause transition into the next realm.

For any time period and within any realm, a fire may

- continue to grow and move into the next realm; or
- burn for a time and eventually self-terminate within the realm because not enough conditions exist to cause transition into the next realm.

A scenario estimate may start at the fire free status (FFS) or at ignition (IG). Alternatively, and more usual for building analysis, it may start at established burning (EB). This discussion will start at the FFS to describe the process more completely. For now, we separate the fire growth process into two segments: FFS to EB; and EB to flashover (FO). The *room fire growth* from EB to FO (Realms 3, 4, and 5) is the significant part of a diagnostic fire, and this segment is the focus of room classifications described in Part Two. *Fire prevention* (Realms 1 and 2) uses the segment FFS to EB. Realm 6 uses the total heat energy to evaluate barriers and the structural frame.

7.3.1 Realm 1: FFS to IG

The scenario analysis for Realm 1 examines the fire prevention question of whether a fire can occur with the fuel and heat conditions of the situation. The basis for this decision is:

- **Given: *FFS and overheat condition***
- **Question: *Will IG occur?***

The estimate examines the situation to determine if enough factors are present to cause IG. The availability of air (ventilation) and the room container attributes are rarely important to outcomes in this realm. The basis for this estimate considers: the type, magnitude, location, and duration of heat application; the surface area receiving heat flux; and the ignitability of the target fuel.

Figure 7.1 organizes common factors that are important to estimating the outcome of this event. Time estimates are not included in this realm from FFS to IG. Although the time duration may be of interest for fire prevention purposes, the critical event for this estimate is ignition, IG.

7.3.2 Realm 2: IG to EB

The scenario for Realm 2 examines whether conditions exist that enable a fire to reach EB. This decision is based on:

- **Given:**
  - *IG*
  - *Time _____ minutes from IG*
- **Question: *Will EB occur?***

Table 7.1 Room fire growth potential factors.

Controlling factor	Realm 1 (FFS–IG)	Realm 2 (IG–EB)	Realm 3 (EB–EP)	Realm 4 (EP–CP)	Realm 5 (CP–FO)	Realm 6
Fuel	Surface area receiving heat flux Duration of heat application Ignitability	Continuity Ignitability Surface roughness kpc of fuel Thickness Surface area	Continuity and feedback kpc of fuel Surface area Ignitability Type of fuel Quantity	Fuel group arrangement Feedback Surface area Tallness Quantity Type of fuel	Quantity Arrangement	Quantity
Room enclosure			Proximity of flames to walls Wall finish combustibility Room insulation	Proximity of flames to walls Ceiling combustibility Ceiling height Wall finish combustibility Room insulation	Room insulation Ceiling height Aspect ratio	Room insulation Ceiling height
Ventilation				Size and location of openings HVAC operation	Size and location of openings HVAC operation	Size and location of openings HVAC operation
Terminal condition	IG	EB	EP	CP	FO	Decay and self-termination

FFS, fire free status; IG, ignition; EB, established burning; EP, enclosure point; CP, ceiling point; FO, flashover.

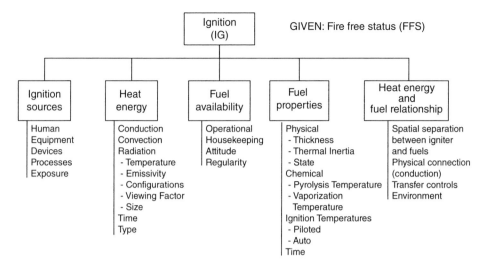

Figure 7.1 Factors: ignition.

This evaluation examines the conditions that affect fire growth in this realm as well as the time duration between IG and EB. If the time to grow from IG to EB is not important, one need address only the question of "Will EB occur?" On the other hand, if the time between IG and EB is important to an analysis, the time duration enables one to estimate this value.

The estimate is based on whether enough factors are available to enable a small ignition to grow to EB. Room air is adequate for fires of this size. Room size and insulation are rarely important in this realm. The major factors that influence this realm are: fuel continuity; fuel ignitability; fuel surface roughness; thermal inertia ($k\rho c$) of the target fuel; fuel thickness (i.e. fuel area to mass ratio); and surface area of the fuel. The time estimation, t_n, may provide useful information for building analysis or fire prevention analysis.

Figure 7.2 organizes these factors for usual scenarios.

7.3.3 Realm 3: EB to Enclosure Point (EP)

This segment is the start of a building performance analysis. Examination of the conditions that lead up to EB need not be a part of this analysis because performance before EB is a part of fire prevention. The decision for this segment of fire growth is based on:

- **Given:**
  - *EB*
  - *Time _____ minutes from EB*
- **Question:** *Will the fire grow to the EP?*

Sufficient room air is normally available to provide a free burn fire within this realm. Because the enclosure point describes the fire size at which the room barriers begin to affect fire growth, the proximity of fuel package distance to barriers has some effect near the boundary of this realm.

Fuel conditions remain the same as Realm 2 with an addition of the quantity of fuel to allow the fire to continue to grow. Figure 7.3 organizes the factors that influence a Realm 3 evaluation.

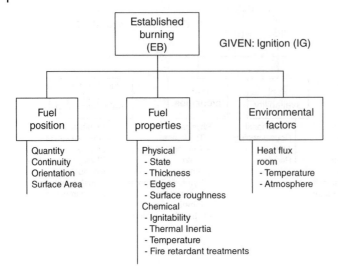

Figure 7.2 Factors: established burning.

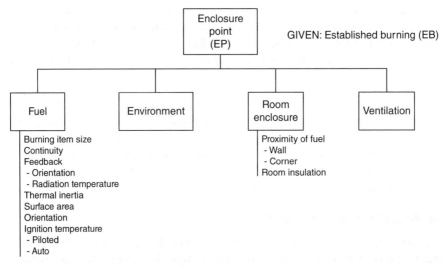

Figure 7.3 Factors: enclosure point.

7.3.4   Realm 4: EP to Ceiling Point (CP)

Realm 4 is very significant to room fire behavior. When conditions are not sufficient for a fire to grow to the CP, FO seldom occurs. When the flames have sufficient strength to impact on the ceiling, continued development to FO is very likely. The decision for flames to reach the ceiling is based on answering the question:

- **Given:**
  - *A fire the size of EP*
  - *Time _____ minutes from EB*
- **Question:** *Will the fire grow to the CP?*

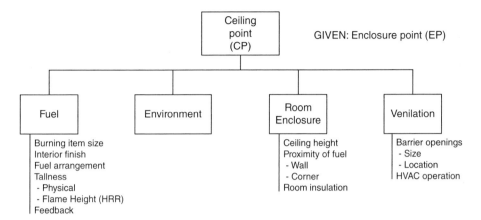

Figure 7.4 Factors: ceiling point.

Figure 7.4 identifies the factors that influence this event.

Ventilation conditions in this realm begin to affect the fire. Ventilation has a greater influence as the fire grows. Barrier openings must augment supply air from outside the room and vent smoke and fire gases. The evaluation for this realm examines the size and location of barrier openings as well as possible heating, ventilation, and air conditioning (HVAC) operations to determine ventilation conditions.

Both interior openings (e.g. door openings) and the status of windows contribute to this event. Ordinary single-pane window glazing often breaks in this realm or when the fire grows into Realm 5. However, double- and triple-pane glass or special tempered glazing may not break. When glazing does not break, the combustion process slows. In addition, the potential for backdraft conditions may be present if no other ventilation (except leakage) is available.

The ceiling height begins to assume greater importance. Low ceilings enhance fires reaching the CP, and FO can occur more easily and quickly. High and moderate ceiling heights cause slower growth and the fire does not extend into the next realm as easily.

Fuel group materials, sizes, and their arrangements begin to have a greater influence on fire growth. Fuel item spacing affects the irradiance and the ease of multiple item involvement. Also, the ceiling effects radiate heat to unburned fuels, causing more pyrolysis and easier fire extension. Although physical tallness of furniture items contributes to high flames, fuels having a high heat release rate (HRR) can also have high flame heights. Therefore, fuel materials assume greater importance. For example, in Figure 6.9(d), furniture calorimeter values of $\dot{Q}$ for an office workstation constructed of old-style wood furniture can be about 1 MW. On the other hand, a modern workstation having four-sided acoustic panels can reach over 6 MW.

An interior designer may view this to be an unimportant alternative. A fire safety engineer would consider this a major change in fuel conditions that affects the diagnostic fire, automatic sprinkler performance, fire department operations, and available time for the egress system to remain tenable.

Variability evaluations can describe the differences and communicate with others the effect of seemingly insignificant changes of interior design. The fire risk for an office building with open wood workstations is very different from the risk with modern enclosed acoustic workstations. Neither prescriptive codes nor sprinkler standards reflect the performance differences in the speed and intensity of fires with these different materials.

7.3.5   Realm 5: CP to FO

Flashover is a very significant event in a room fire and holistic building performance. The decision as to whether FO will occur answers the question:

- **Given:**
  - *A fire the size of CP*
  - *Time _____ minutes from EB.*
- **Question:** *Will the fire grow to FO?*

If flames can reach and mushroom across the ceiling, FO is almost certain and usually occurs quickly. Many factors that are important to sustain and expand fire growth in earlier realms lose importance during Realm 5. Ventilation remains important. The room's aspect ratio is also a consideration, and the quantity of fuel and its arrangement assume a greater importance. Figure 7.5 organizes the factors that influence Realm 5 behavior.

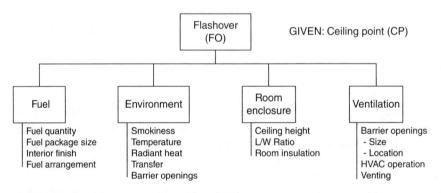

Figure 7.5 Factors: flashover.

7.4   Time and the Fire Growth Potential

Time is the element that links all parts of the fire safety system together. Fire performance and associated risks are very different when we create conditions that produce fast fires rather than slow fires. Some room conditions can produce flashover in as little as two or three minutes after EB. Other rooms may not reach FO for more than 20 minutes after EB – if they reach FO at all.

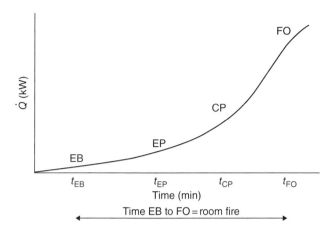

Figure 7.6 Room fire transitions.

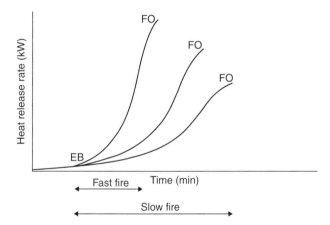

Figure 7.7 Flashover (FO) speeds.

The *fire growth potential* (FGP) describes the ease for room conditions to reach flashover. Normally, fast fires are linked with a high FGP; slow fires are usually associated with a low FGP. Thus, FGP created by the room's interior design and the time to reach flashover are related.

The room fire growth for any specific scenario and the time from EB to FO may be estimated with fire fundamentals, concepts of realm transitions, and HRR. Although the realms of fire growth are defined by specific criteria, the actual transition between realms is more blended, as represented by Figure 7.6. Thus, we recognize that each realm boundary is more of a narrow band, rather than a specific position.

Conceptually, a FGP represents the ease and speed of fire growth for a room. However, one may recognize that each different ignition location produces a different scenario behavior and has a different potential for reaching FO. Figure 7.7 illustrates hypothetical outcomes for different ignition locations. What is not shown is that some scenarios will certainly result in flashover. Other scenarios do not have conditions that allow transition to a higher realm and self-terminate before ever reaching FO conditions.

Example 7.1 Estimate the FGP outcomes for an ignition at Location A in the room of Figure 7.8(a). Use a bottom-up analysis.

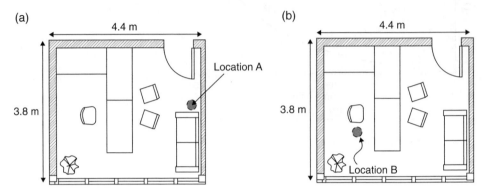

Figure 7.8 Office ignition locations.

Dimension: 4.4 m × 3.8 m × 3 m (14.4 ft × 12.5 ft × 9.8 ft).
Interior finish: Mahogany wood panels.
Contents: Office chairs, wood desk, small sofa with foam padding and cotton fabric, waste basket.
Kindling fuel (and IG location): Waste basket (Location A).
Ventilation: Door 2.04 m × 0.91 m (6.7 ft × 3.0 ft).

Solution:
Table 7.2 shows the rationale for FGP decisions in the realms. Transition factors are identified from Table 7.1. Time durations are estimates based on factors of Table 7.1 and furniture calorimeter data from Section 6.20.2.

An ignition at Location A for this scenario is expected to produce a FO at about 7 minutes after EB.

Example 7.2 Estimate the FGP outcomes for an ignition at location B in the room of Example 7.1 (Figure 7.8b). Use a bottom-up analysis.

Solution:
Table 7.3 shows the rationale for FGP decisions in the realms. Transition factors are identified from Table 7.1. Time durations are estimates based on factors of Table 7.1 and estimated furniture calorimeter data.

An ignition at location B for this scenario would be expected to self-terminate within Realm 4. FO is not expected for an ignition at Location B.

7.5 FGP Adjustments

Rooms with high ceilings, large volumes, and, often, large fuel package sizes must be viewed differently from smaller rooms. Differences are more than just a matter of size. When FO is the mechanism that causes all fuels in a space to ignite nearly

Table 7.2 Example 7.1: fire growth potential transition rationale.

Realm	Behavior in realm	Applicable realm transition factors	Rationale
Given EB at Location A			
3 (EB to EP)	Flames from EB in waste basket can spread to the surface of the sofa	Fuel continuity; fuel $k\rho c$; ignitability; surface area; close spacing; quantity Time estimate: 2 min	Cotton cover and polyurethane foam of the couch has low thermal inertia and it allows flame spread Relatively large surface of the couch provides enough fuel to reach enclosure point
4 (EP to CP)	The flame can reach to the ceiling	Fuel type; continuity with interior finish; proximity of fuel to wall Estimated time: 3 min	Furniture calorimeter information of Section 6.20.2 indicates the HRR can reach 600 kW or much more. Contact with combustible interior finish shows that any type of small sofa will easily reach the CP. Most sofas will reach the CP independently. Some may have more difficulty Proximity of fuel to wall increases flame height Ventilation is adequate Flames can easily reach 3 m ceiling height
5 (CP to FO)	Flames touch the ceiling with a robustness that causes FO in the compartment	Fuel quantity and arrangement Time estimate: 2 min	Flames impact the ceiling with robustness because of the HRR of the fuel combined with the interior finish Other fuel items in the room will pyrolize and ignite from auto ignition – combined with the continuity of combustible interior finish

EB, established burning; EP, enclosure point; CP, ceiling point; FO, flashover; HRR, heat release rate.

simultaneously, the realm concept is valid. When the room is larger and fire grows by spreading from one fuel package to the next, qualitative adjustments must reflect differences in conditions. A range of in-between sizes and conditions can blur differences between FO and spread-over.

Full involvement in large rooms starts with a fire in an initial fuel item. The fire propagates when fire size and proximity of adjacent fuel items cause radiation-induced ignition in the adjacent fuel. The fire continues spreading to other fuel packages until conditions reach a level where the equivalent of full involvement occurs in a part of the building, often defined by virtual barriers. While hot layers do form at the ceiling, and back radiation does contribute to unburned fuel pyrolysis, the principal mechanism of fire growth and spread is radiation-induced fire propagation through the fuel arrangement. Moderate-sized rooms may have an initial fire growth by spread-over until room dimensions and ventilation characteristics create FO conditions.

Table 7.3 Example 7.2: fire growth potential transition rationale.

Realm	Behavior in realm	Transition to next realm	Rationale
Given EB at Location B			
3 (EB to EP)	The established burning inside the waste basket will have difficulty spreading to the chair or to the desk	Fuel item spacing; poor feedback; second item, fuel ignitability; fuel quantity Time: 8 min (if it does occur)	The spacing of the EB source from the second item, combined with the relatively low radiation energy because of the small waste basket flame, makes multiple item fires difficult The materials of the chair and desk become very important. If they have low ignitability (e.g. wood), fire spread becomes very difficult; weak feedback conditions make ignitions difficult. Waste basket has little substantive fuel
4 (EP to CP)	Fire does self-terminate before reaching CP	Fuel orientation; continuity; ignition of second item target fuels Growth to CP is not expected	Fuels have little continuity to aid spread of small flames Second item target fuel orientation and spacing makes fire growth difficult
5 (CP to FO)	NA	NA	NA

EB, established burning; EP, enclosure point; CP, ceiling point; FO, flashover; NA, not applicable.

The characteristics of a spread-over mechanism are:

- A fire starts and grows in a fuel item.
- The burning fuel radiates heat across a spatial separation that pyrolyzes adjacent fuels.
- The separate adjacent fuel package is ignited, usually by auto-ignition.
- Fire in this adjacent fuel item grows.
- The coalesced fire radiates heat to additional fuel packages and the process continues.

Realm 1, 2, and 3 descriptors are similar for both spread-over and FO. However, as the fire continues to grow, conditions begin to change. The dimensions, fuel packages, and enclosure effects affect the fire differently. The FO mechanism is not present and ceiling effect is not as dominant. The factors that affect fire growth estimations are:

- The size of the radiating item generating the heat flux.
- Spacing between the emitting heat source and the target surface.
- The heat flux received by the target fuel.
- The critical heat flux (irradiation) to cause ignition in the target fuel.
- The duration of heat flux.

7.6 Estimating Spread-over Scenarios

Estimating fire growth in Fuel Package 1 from FFS to IG, IG to EB, and EB to the equivalent of EP is the same as for small rooms. These events and their time durations involve the same factors. However, as the fire continues to grow, evaluation questions must be adapted to reflect fire spread-over conditions:

- **Given:** *A fire the size of EB in Fuel Package 1.*
- **Question:** *Will the fire self-terminate within Fuel Package 1 ($\overline{ST_1}$) before igniting Fuel Package 2 (ST_2)?*

$\downarrow$

- **Given:**
  - *A fire in Fuel Package 1 that does not self-terminate ($\overline{ST_1}$)*
  - *Time _____ minutes from EB.*
- **Question:** *Will the fire ignite Fuel Package 2 ($\overline{ST_2}$)?*

$\downarrow$

- **Given:**
  - *A fire in Fuel Package 1 ($\overline{ST_1}$)*
  - *Ignition of Fuel Package 2 (IG_2)*
  - *Time _____ minutes from EB.*
- **Question:** *Will the fire self-terminate within Fuel Package 2 ($\overline{ST_2}$) before igniting Fuel Package 3?*

$\downarrow$

- **Given:**
  - *Fires in Fuel Packages 1 and 2 that do not self-terminate ($\overline{ST_1}$) ($\overline{ST_2}$)*
  - *Time _____ minutes from EB.*
- **Question:** *Will the fire ignite Fuel Package 3 (IG_3)?*

$\downarrow$

- **Given:**
  - *Fires in Fuel Packages 1 and 2 ($\overline{ST_1}$) ($\overline{ST_2}$)*
  - *Ignition of Fuel Package 3 (IG_3)*
  - *Time _____ minutes from EB.*
- **Question:** *Will the fire self-terminate within Fuel Package 3 ($\overline{ST_3}$) before igniting Fuel Package 4?*

$\downarrow$

- *Continue the process of questioning ST and IG until the fire size is clearly established.*

Figures 7.9 and 7.10 organize the primary factors that influence FGP estimates.

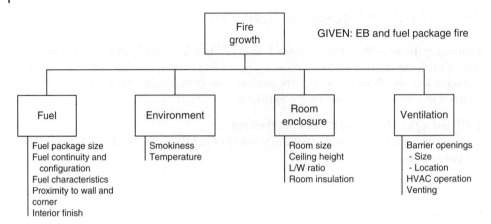

Figure 7.9 Factors: fire growth.

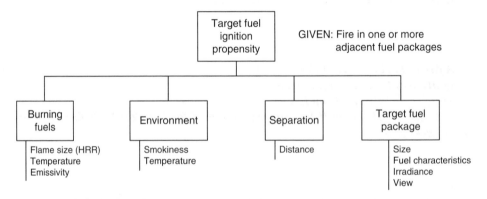

Figure 7.10 Factors: ignition.

The process of assessing the path of successive ignitions and fire growth continues through as many fuel groups as necessary to gain an understanding of the building and to describe the fire. Beyond the first fuel group, the fire may simultaneously expose several adjacent fuel groups to radiation heat. These, in turn, continue the potential spread in two or three dimensions. The analytical thought process methodically examines the time duration for fire propagation and HRR among fuel groups. Networks are useful to relate time and sequence of fuel group involvement.

PART TWO: TOP-DOWN ESTIMATES

7.7   Qualitative Room Classifications

An initial "walk-through" building assessment provides an opportunity to under-stand the way the building works and gives a sense of the effect of a fire on its opera-tions. A walk-through "scoping" of the building and its risk sensitivities can incorporate many simultaneous tasks. In addition to an initial screening of FGP for the rooms, the walk-through will identify barrier conditions, egress issues, detec-tion and fire suppression features, and anything that affects holistic performance. This rapid "top-down" qualitative assessment provides information for organizing fire scenarios.

Among their many uses, top-down qualitative FGP classifications order room fire hazards. Assessments are based on the fire fundamentals of Chapter 6 and use concepts of qualitative bottom-up estimates described earlier in this chapter. Chapter 8 incorpo-rates barrier failure conditions for multi-room fire scenarios.

The classification process recognizes that although different bottom-up fire growth scenarios may exist in any room, classifying the room's FGP as a single hazard is a useful way to describe hazards. A qualitative top-down FGP classification describes the rela-tive composite hazard of a room and a time to reach FO.

Rapid visual estimates are particularly useful when combined with the capabilities of information technology. For example, one can make video or digital pictures of conditions and relate them to model rooms and computer results. Sorting proce-dures for the small number of FGP categories can provide consistency between observations and computer results. A sense of proportion for room fire behavior improves confidence in relating computer-based outcomes to observational judgment.

7.8   FGP Comparisons

Many professions struggle to communicate relative severity among a spectrum of hazards. For example, wind engineers use the Fujita Scale (F0–F5) to categorize tornado damage. Emergency management categorizes hurricanes by combining wind speed with potential damage descriptions (Cat 1–Cat 5). Construction planners label flood plain conditions by 100-year (or 50 or 25) flood descriptors, even though this label has relatively little to do with time frequencies.

Each of these descriptors allows individuals from all walks of life to comprehend specific severity zones within the range of possible conditions. All of these descriptors integrate mathematics and science with observation and judgment. All have significant technical limitations in their organization. Uncertainty is inherent in the groupings. Category identification involves observationally based decisions and uses arbitrary boundaries to differentiate the fuzzy region between groups. Yet, each method concep-tualizes conditions and communicates effectively with other professionals and the public.

Fire growth hazard classifications for rooms can categorize ranges of behavior in ways similar to the other technical professions. Fire safety applications can group the relative fire severity for rooms with descriptors that integrate technology and observations

Room A	Room B	Room C

Figure 7.11 Room fire growth potential comparisons.

within a range of conditions. While there may be some uncertainty about selections, and adjacent boundaries may be indistinct (similar to the realms of fire growth), fire growth hazard groups provide value for fire safety applications and communication.

To introduce the concept, consider the three office spaces shown in Figure 7.11(a)–(c). These rooms have the same occupancy, size, and construction characteristics. We recognize that Room C has a much greater FGP than Room A. Room B has an FGP between those of Rooms A and C.

Although this example is trivial, because differences involve only the quantity of fuel and its arrangement, the simplicity demonstrates that different rooms of the same occupancy can have very different FGP conditions. Group identification becomes more difficult when we compare the FGP for rooms having variable geometry and size, different fuel materials and interior finish, or differences in construction characteristics. Nevertheless, a "top-down" analysis that can classify FGP severity is of value for organizing performance analysis and communicating with others.

7.9   Interior Design and Model Rooms

A room's *interior design* is defined for fire growth assessments by interaction of the following:

- Room contents and their arrangement
- Interior finish
- Size and shape of the room
- Barrier construction
- Barrier openings.

A *model room* is the specific interior design on which a diagnostic fire is based. Because a model room is clearly identified, one is able to envision natural fire behavior. The model room also describes conditions for computer fire modeling.

Combustible room contents provide the primary fuel for a fire. The room contents, fuel groups, fuel group locations, and material characteristics that affect fire behavior are all included in model room information.

Interior finish is the surface covering of the floors, walls, and ceiling. Surface covering characteristics can have an important influence on the speed and ease of fire growth. Combustible wall or ceiling coverings, such as wood paneling, combustible fiberboard, or carpeting can contribute significantly to fire growth and FO. Non-combustible interior finish reduces the room's FGP.

The location, orientation, and method of attachment of these surface-covering fuels can be important to fire behavior. For example, carpeting on the floor rarely contributes to fire development from EB to FO. On the other hand, that same carpeting on a vertical surface can contribute significantly to conditions that encourage and accelerate FO. Thin wall or ceiling coverings, such as paint, wallpaper, and the paper on gypsum wallboard have insignificant influence on surface flame spread when applied directly to substrates such as plaster, concrete block, and reinforced concrete.

The *room geometry* can influence fire development. Rooms with low ceilings interrupt the fire plume earlier, causing the flames to mushroom across the ceiling. This mushrooming effect radiates heat and pyrolyzes combustibles remote from the seat of the fire. Long, narrow rooms feed the heat back from walls more easily to cause faster, easier fire growth than in rooms with an aspect ratio closer to 1. The insulation qualities of the barriers affect fire development. Well-insulated rooms retain heat better than poorly insulated rooms, increasing the ease of reaching FO.

Ventilation allows air exchange through the size, height, and location of room openings. When openings are small, inefficient combustion occurs, resulting in lower fire temperatures. Although some of this comparatively lower heated gas is released from the room, most of the heat remains in the room, creating the potential for a backdraft explosion. On the other hand, when the openings are large, more efficient combustion occurs, producing higher fire temperatures and a more rapid burnout. Large openings cause large amounts of hot gases to leave the room. Time durations for fire growth are different between poorly ventilated and well-ventilated rooms.

Observational estimates handle ventilation in two ways. Initially, we assume that sufficient ventilation is available to allow efficient combustion to occur. Thus, classifications are not based on the chance opening status of doors and windows, but rather on the room's interior design characteristics. The objective is to classify the room's interior design for FGP. After classifying the room's interior design, we can incorporate ventilation conditions, if necessary. It is important to recognize that room FGP classifications are based only on the interior design and not on ventilation effects.

7.10 FGP Classification Groups

The relative FGP of a room's interior design should convey understanding. The assessment is based on technical principles and observational comprehension, and one must be able to identify a category very quickly.

A FGP answers two main questions:

1) What is the relative ease for the room's interior design to reach FO, given adequate ventilation?
2) If FO occurs, what is a functional time between EB and FO?

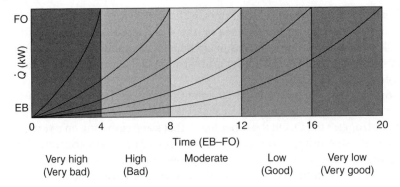

Figure 7.12 Room classification groups.

	Alpha	Default time (min)	Estimated time (min)	Characteristics
Very high	0.06	1–4	_____	Almost certain FO; much room "clutter;" ignitable combustibles; combustible wall and/or ceiling finish; high HRR materials
High	0.03	4–8	_____	FO relatively likely; high fuel surface to bounding surface ratio; easy for fuels to produce flames touching the ceiling.
Moderate	0.003	8–12	_____	"Comfortable" room layout; fire growth can reach ceiling with moderate ease; material HRR moderate with a few high items.
Low	0.002	12–16	_____	A few fuel packages that can produce CP conditions; non-combustible interior finish; generally low HRR fuels.
Very low	0.001	16–20	_____	Very difficult to drive the room to FO; non-combustible interior finish; spaced combustible fuel packages.

Figure 7.13 Room classification characteristics.

We use five categories to group the composite "top-down" description of a room's FGP. It has been observed that rooms producing fast fires have a high FGP. Conversely, rooms with slowly developing fires traverse from one realm to another with difficulty and have a low FGP.

The at^2 concept is a generally accepted descriptor for fire growth from EB to about 1 MW. This usually covers the region from EB to FO. Additionally, FO for fires in most "normal" sized rooms rarely exceeds 20 minutes from EB. Five equal segments provide a practical and convenient grouping for time intervals, as shown in Figure 7.12.

Although we recognize that at^2 fires may not be accurate for fires in the low and very low categories, the concept remains useful and practical.

Figure 7.13 offers an alternative way of showing these groups with default values for expected time durations to reach FO within the group. The last column provides an opportunity for an individual to estimate the time to FO for observed conditions.

7.11   Selecting FGP Groups

Fire growth potential group identification is based on a composite unit rather than on a specific scenario. Neither potential ignition sources and locations nor ventilation conditions are considered in a room classification. The objective is to classify a room's interior design features.

Precision in selecting a classification category is not as important as recognition of the synergistic factors that affect a room's fire development. This is not to imply that inattention to detail or big differences in classification groups are unimportant. On the contrary, one of the attributes of this procedure is improving cognitive skills in relating fire science and engineering knowledge to performance applications.

7.11.1   Evaluation Guidelines

Holistic observations provide a good "feel" for the ease with which the room's interior design can support combustion. The following organization guides the thought process for category selection.

Room Enclosure Geometry

What is the room size? Large rooms are more difficult to drive to FO than small rooms.

What is the ceiling height? Low ceilings interrupt the flames and cause them to mushroom across the ceiling. The ceiling's back radiation causes unburned fuels to pyrolize and create conditions that enhance FO.

What is the aspect ratio? Narrow rooms create feedback conditions that help fuels to burn faster.

Interior Finish

Do the walls and ceiling have combustible interior finish? Note that the criterion for combustibility is, "Will it burn?" It is not necessary to know the ASTM E-84 classification.

Does the ceiling have combustible finish? Hot fire gases are interrupted and collect at the ceiling. The ceiling jet and hot gases enhance pyrolysis of combustible ceilings. Combustible ceilings allow auto-ignition or more rapid piloted ignition to speed fire development in the room's fuel packages.

How many walls are covered with combustible wall covering? What is the substrate? Combustible wall coverings can ignite and burn, often rapidly. However, a more significant role is the fuel package size increase when wall coverings are in contact with combustible contents. Also, the substrate has an effect on the speed and ease of propagation. When a small pocket of air exists between the combustible wall covering and the substrate, flame spread is greatly enhanced. When the wall covering is directly attached to an energy absorbing material (e.g. concrete, gypsum), flame propagation becomes more difficult. When the covering is directly attached to a material with a low thermal inertia (e.g. foam plastic insulation, wood), flame propagation increases, although not as much as with an air void.

Fuel Groups

What are the fuel groups? Fuel groups allow fire to move through the items by direct flame contact alone. Identification of the number and size of discrete fuel groups gives a good "feel" for the ease or difficulty for fire propagation. Very large fuel groups or fuels with a high HRR can drive a room to FO independently.

What is the separation and distribution of fuel groups? Well-separated fuel items make small fires difficult to grow because fire spread is due to radiation. When fire sizes are within Realm 3, a distance of 0.5 m (20 in) makes radiant fire propagation difficult for "light" materials. The distance is reduced to 0.25 m (10 in) for "heavy" materials.

What is the location of fuel groups relative to the walls and corners? Fuel groups that have contents fuels in contact with combustible interior finish combine to create very large fuel packages. They may even encompass the entire room. Fires near non-combustible walls or corners will have higher flame heights than if they have greater distance separation.

Contents Materials

What are the room content materials? Combustion characteristics of the room's fuel groups influence the ease and speed of fire development. Cellulosic materials like wood and paper behave differently from foam plastic materials. Fire size also has an effect on the speed and ease of propagation. The thermal inertia can greatly influence the flame height and speed of development. Contents that are encapsulated in non-combustible containers (e.g. file cabinets and safes) do not contribute to fire growth.

Kindling Fuels

What and where are the kindling fuels in the room? Kindling fuels are the light, often very thin combustibles that are present in most rooms. These fuels are easily ignited and allow rapid flame spread. While they sustain combustion for only short periods of time, they can become conduits for fire propagation. Drapes and curtains, loose or light paper products, and light furniture coverings are illustrations of kindling fuels. By this definition, open flammable liquid gases could also be viewed as kindling fuels. Certainly, kindling fuels can be moved easily. However, their presence gives a sense of relative speed of fire growth and development.

Organizational Chart

For the purpose of classifying rooms, Figure 7.14 organizes these factors that dominate the classification for the FGP of a room's interior design.

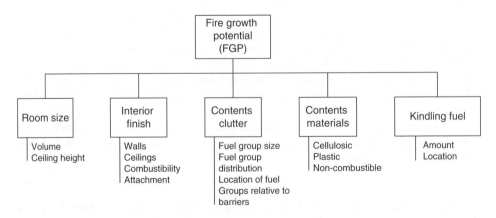

Figure 7.14 Factors: room classification estimates.

7.11.2 Classification Examples

A top-down holistic assessment gives a rapid, yet surprisingly good "feel" for the relative FGP of a room. Often, this level of classification is adequate for an initial screening to give an understanding of the types of conditions that can be anticipated in a more carefully constructed diagnostic fire.

This qualitative classification ranks a room's interior design according to its potential to reach FO. It is important to recognize that this assessment evaluates only a room's interior design. Ventilation conditions are not a part of the selection.

Assessments mentally integrate the factors of Figure 7.14 to arrive at one of the five classification groups of Figure 7.13. However, one dominant condition often gives a good indication of the room's FGP. This condition assesses the ease of flames reaching the CP. FO generally occurs when conditions enable flames to touch and move across the ceiling. Therefore, conditions involving large fuel groups, high HRR materials, combustible ceiling finish, and combustible wall coverings become candidates for high or very high fire growth hazard (FGH) classifications.

We illustrate techniques for making top-down assessments with the following examples. Figure 7.15 shows four common office layouts. Building code and standards organize business (office) use with a "one size fits all" grouping. Performance design is able to discriminate fire growth hazards. A top-down fire growth hazard classification can initially order relative hazards for ease and time to reach FO. This classification is used, with knowledge of the building's architecture and potential risks, to identify locations

(a) (b)

(c) (d)

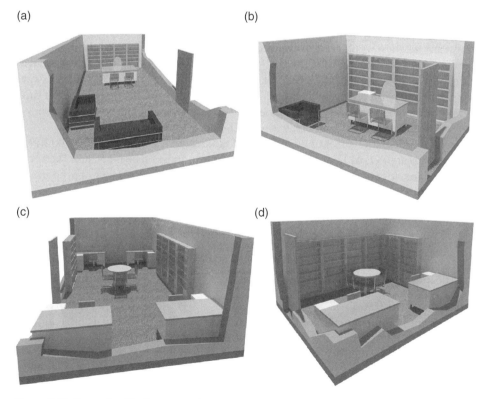

Figure 7.15 Room classification comparisons.

for appropriate rooms of origin. Although the process is very rapid, the insight into organizing a performance analysis becomes valuable.

Example 7.3   Figure 7.15(a) shows an interior design for an office, Room A. Table 7.4 describes its attributes. Select an FGP classification for this room.

Table 7.4 Room A description.

Room size	$W = 4.4$ m (14.5 ft), $L = 7.6$ m (25.1 ft), $H = 3$ m (9.9 ft)
Interior finish	Walls: painted gypsum plaster
	Ceiling: painted gypsum plaster
	Floor: hardwood
Contents	Older cellulosic-based (wood) office furniture
	Cotton batting padding in upholstered furniture
Arrangement	Approximately as shown
	Waste baskets and incidental occupant-related goods are not shown
Kindling fuels	Normal expectations for an executive office
Ventilation	Adequate for stoichiometric burning, although not a consideration in room interior design classification

Solution:
Classification: very low FGP.

Discussion:
The factors of Figure 7.14 provide the rationale for the factors of Table 7.5. A fire in almost any location would have a low likelihood of causing full room involvement (FRI). The combined fuel groupings of the chair and love seat provide the greatest potential to achieve FRI. Awareness of the HRRs of the furnishings (Section 6.20) indicates that flames can barely touch the ceiling. Even if FO did occur from this fuel group, the time duration would be very long. The judgment is that this room and its contents provide a very low likelihood of reaching FO.

Room FO calculations (Section 6.23) would indicate that approximately 3700 kW is needed to create FO. Although these calculations are questionable, because the room size is too large for equation validity, the number gives only a crude sense of proportion. Even with this estimate, the likelihood of achieving FO is very low.

Example 7.4   Change the attributes of Room A to those of Room B as shown in Table 7.6. Figure 7.15(b) shows the relative distribution of contents. Select a FGP for Room B.

Solution:
Classification: moderate FGP.

Discussion:
Although the room contents are the same as Room A, the floor area is 50% and the volume is 41%. This contributes to a greater sensation of contents crowding with a clutter ratio of 0.4, which is twice as great as Room A. The ceiling height is lower, making it easier for a fire in the couch and love seat to produce flames that can touch the ceiling. It remains unlikely that a fire in other fuel packages would cause flames to touch the ceiling or spread to other items.

Table 7.5 Room A classification rationale.

Room size	Floor area = 16.7 m^2 (181 ft^2), volume = 50.1 m^3 (180 ft^3), ceiling height = 3m (9.9 ft)
	The materials and sizes of the fuel packages will make it difficult for flames to touch the ceiling
Interior finish	Energy-absorbing; will not contribute to flame spread
Contents clutter	Wide separation of furniture except at couch and love seat; estimate about 12% of floor area covered by furniture; gives a feeling of spaciousness
	A general indication of space utilization is the ratio of surface area of room contents to the surface area of the room interior. This room has a value of 0.2. Individual observations seem to discriminate room space utilization (i.e. "clutter") into a few groups that can be ordered with the ratio of content surface area to the room's bounding surface area. Although this concept has not been studied carefully, preliminary testing seems to suggest this concept may be useful for fire growth hazard analysis
Contents material	Wood and cotton base of the upholstered furniture does not produce a high peak heat release rate. The major fire concern would be the coalescence of combined burning of the couch and love seat. Flames can perhaps reach the ceiling; nevertheless, the likelihood of this fuel package causing full room involvement is low. The likelihood of fire spread to other fuel packages is low
Kindling fuels	A few waste baskets, magazines, etc. and occupant-related goods are likely to be in the room. These should not contribute much to fire growth or spread after established burning (EB), although they could be involved in the ignition to EB part of the process. As the classification is based on a given EB, the kindling fuels will not be a factor in the classification selection

Table 7.6 Room B description.

Room size	L = 4.4 m (14.4 ft), W = 3.8 m (12.5 ft), H = 2.44 m (8 ft)
Interior finish	Walls: painted gypsum plaster
	Ceiling: painted gypsum plaster
	Floor: hardwood
Contents	Older cellulosic-based (wood, cotton) furniture
	Cotton batting padding in upholstered furniture
Arrangement	Approximately as shown; waste baskets and incidental occupant-related goods are not shown
Kindling fuels	Normal expectations for a living room in a home
Ventilation	Adequate for stoichiometric burning, although not a consideration in room interior design classification

A table such as Table 7.7 can be used to adjust factors from the previous room. In Table 7.7 each factor can raise (i.e. make worse) or lower (i.e. make better) the likelihood that FO will occur. Not all factors contribute equally to the fire hazard, so we recommend a four-star rating system to describe importance. A rating of * provides little change, whereas a rating of **** indicates a significant importance level. Identifying these changes enables one to blend differing influences to provide a rationale for adjusting FGP classifications.

Table 7.7 Classification adjustments: changing Room A to Room B.

Factor	Direction of change from Room A	Importance of change from Room A[a]	Discussion
Lower ceiling to 2.44 m (8 ft)	↑ (worse)	*	Easier for the flames to reach ceiling and the room to contain heat
Change floor area and volume	↑ (worse)	**	Based on smaller volume (easier to reach flashover) and closer spacing of the fuel packages, makes fire development more likely; incorporated into the room size change above
Interior finish Contents clutter	No change	–	Spacing is more conducive to fire development; this factor was incorporated into the room size change above
Contents materials	No change	–	
Kindling fuels	No change	–	Perhaps a slight increase in the likelihood of full room involvement because of room size reduction

[a] A rating system describes the importance of the factor in changing the classification from the comparison (model) room. A four-star (****) designation signifies a major influence while a one-star (*) designation indicates a minor effect.

The conditions that contribute to Room B reaching FO have increased and the time duration to FRI has decreased because of the room's dimensional changes. FO calculations (Section 6.23) are more valid for these sizes and indicate that about 2000 kW is needed to reach FO. HRR estimates show that the couch and chair fuel package is close to this value.

Example 7.5 Change the attributes of the model Room A to those of Room C in Figure 7.15(c), as shown in Table 7.8. Select an FGP classification.

Table 7.8 Room C description.

Room size	$W = 4.4$ m (14.5 ft), $L = 7.6$ m (25.1 ft), $H = 3$ m (9.9 ft)
Interior finish	Walls: painted gypsum plaster Ceiling: painted gypsum plaster Floor: hardwood.
Contents	Modern two-sided office workstations Other furniture is wood Occupant-related goods are substantially greater because of work being done
Arrangement	Approximately as shown; waste baskets and incidental occupant-related goods are not shown
Kindling fuels	Normal expectations for a working office space
Ventilation	Adequate for stoichiometric burning, although not a consideration in room interior design classification

Solution:
Classification: high FGP.

Discussion:
Although the room size is the same as that of Room A, its office function is different. In addition, the illustrative HRR for modern workstations is high (Section 6.20).

There is an uncertainty about whether to judge this room as "low" in the "high" classification or as "high" in the "moderate" classification. It seems more appropriate to place it somewhere near the boundary between these groups. Here we will select a single classification of high potential to reach FO.

Table 7.9 shows the logic for the fire growth hazard selection.

Table 7.9 Classification adjustments: changing Room A to Room C.

Factor	Direction of change from room A	Importance of change from room A[a]	Discussion
Ceiling height	No change	–	
Floor area and volume	No change	–	
Interior finish	No change	–	
Contents clutter	↑ (worse)	**	The room layout gives the sensation of greater clutter. This may be recognized as the clutter ratio changes from 0.2 in Room A to 0.45 in Room C
Contents materials	↑ (worse)	***	High heat release rate and increased flame heights from the workstations increase significantly the likelihood of flames touching the ceiling and reaching full room involvement. The room size and ceiling height mitigate the increase slightly
Kindling fuels	No change	–	

[a] A rating system describes the importance of the factor in changing the classification from the comparison (model) room. A four-star (****) designation signifies a major influence while a one-star (*) designation indicates a minor effect.

Example 7.6 Change the attributes of Room A to those of Room D, as shown in Table 7.10. Select an FGP classification.

Solution:
Classification: very high FGP.

Discussion:
This room combines the smaller volume with combustible interior finish and modern furniture, having expectations of a high HRR. Table 7.11 shows the factors to blend in deciding the classification.

The combination of a relatively small room, a high HRR of contents, and combustible interior finish all contribute to conditions in which FO will be easy to achieve. The

Table 7.10 Room D description.

Room size	$L = 4.4$ m (14.4 ft), $W = 3.8$ m (12.5 ft), $H = 2.44$ m (8 ft)
Interior finish	Walls: mahogany plywood cover over gypsum plaster Ceiling: painted gypsum plaster Floor: hardwood
Contents	Modern two-sided office workstations Other furniture is wood Occupant-related goods are substantially greater because of work being done
Arrangement	Approximately as shown; waste baskets and incidental occupant-related goods are not shown
Kindling fuels	Normal expectations for a working office
Ventilation	Adequate for stoichiometric burning, although not a consideration in room interior design classification

Table 7.11 Classification adjustments: changing Room A to Room D.

Factor	Direction of change from room A	Importance of change from room A	Discussion
Lower ceiling to 2.44 m (8 ft)	↑ (worse)	**	Easier for the flames to reach ceiling; high heat release rate (HRR) of furnishings can reach the ceiling
Change floor area and volume	↑ (worse)	**	Based on smaller volume (easier to reach flashover) and closer spacing of the fuel packages, makes flashover more likely
Interior finish	↑ (worse)	***	The continuity provided by the combustible interior finish greatly increases the size of the fuel packages. It contributes to increased HRR and flame spread
Contents clutter	↑ (worse)	**	Fire group spacing is conducive to flame coalescence. The clutter ratio is 0.55
Contents materials	↑ (worse)	***	High HRR and increased flame heights from the upholstered materials increase significantly the likelihood of full room involvement
Kindling fuels	No change	–	

[a] A rating system describes the importance of the factor in changing the classification from the comparison (model) room. A four-star (****) designation signifies a major influence while a one-star (*) designation indicates a minor effect.

importance of each as a separate consideration in Table 7.11 is difficult to distinguish because of their interactions. However, one can recognize that each factor provides an additional contribution to the potential for FO.

7.12   Discussion

Although many possible room interior designs and styles can be created, most room types use a relatively small group of common features. Dimensions, arrangements, content materials, and furnishings are quite manageable from a fire analysis viewpoint. The dimensional and use standards used by architectural and building and interior design professionals can provide a basis for model room development by the fire protection community.

Architects, interior designers, and developers often use standardized information for residential and commercial buildings. One could construct a catalog of sample model rooms showing dimensions and arrangements. Computer analyses provide FO time durations for scenarios. If consensus FGP categories were established for specific model rooms, differences in condition would become easy to identify.

Creation of model rooms is fairly easy because commerce and building design are relatively uniform and modular. For example, one can organize most commercial office spaces into four groups:

- Single-person office
- Two-person office
- Four-person office
- Open office landscape.

Dimensions for these offices usually vary only within narrow limits. In addition, furnishings tend to be relatively stable and manufacturers produce generally consistent dimensions. Thus, model office rooms and FGP classifications are relatively "easy" to provide "consensus" definitions. The process is the same for other occupancies and room uses.

Fuel amount and arrangement can be defined with a space utilization factor ("clutter factor"). For example, as part of a student project on residential fire growth hazards, a team of students examined perceptions of fuel package sizes with occupant descriptions of room contents. This unscientific, although well conceived, *ad hoc* study found that individuals with both non-fire and fire backgrounds identified fuel package clutter in rooms as "comfortable," "cluttered," and "uncluttered."

The results were very consistent. However, the more interesting conclusion noted that the ratio of the surface area of contents to the surface area of the room was a constant for each category across many types of individuals. Although the ratio was consistent for any specific room type, the numerical value was different for each type of room. For example, living rooms, bedrooms, offices, kitchens, etc. each had a different ratio. The numerical relationship was constant over a wide range of individuals. Perhaps numerical relationships that relate to observational variations for FGP groups may be a useful future psychological research topic. In general, it has been noted that differences in observational categories are relatively minor, except when perceptions (e.g. fuel type) are very different.

The factors shown in Figure 7.14 show an organization suitable for information technology applications. For example, digital pictures, dimensional values, and material identification for the contents may be used to construct a catalog of rooms and their classifications. These can be correlated with computer modeling and calculation results to provide consistency.

7.13 Closure

The integration of observation, judgment, and technology to classify risks involving tornadoes, hurricanes, and floods is useful for engineering applications and valuable to communicate with the public and professionals of other disciplines. FGP is organized in a similar manner to enable room conditions to be compared and selections to focus on important building conditions. In addition, the room classifications help to identify appropriate diagnostic fires.

A room's FGP classification is the starting point in evaluating how a building will behave in a fire. The objective of classifications is to rank interior designs according to their FGP, not to calculate outcomes of specific fire scenarios. For many applications, this rapid assessment is sufficient. For others, it becomes a basis for a more careful additional analysis.

Classifying room interior designs uses judgment to integrate recognized field conditions with principles of fire dynamics. While calculations can give confidence in describing certain expectations, one should embrace observational skills, particularly in evaluating outcomes of computer analyses. This thought process is useful to describe likely outcomes to others, as well as to augment computer output for more detailed investigations.

The basis for FGP classifications is useful to identify variation analyses. Certain contents or interior finish changes can affect the room fire significantly, but may have little effect on building risk characterizations. Other changes may have a major impact on a building's performance.

The fire is the focus of a complete evaluation. Identification of reasonable alternatives or variations that can influence outcomes significantly is a role that a professional assumes in managing risks, by discriminating important details from the less important features.

This skill is useful for communicating effects to professionals in other fields. For example, would reasonable rearrangements of fuel package affect the outcome? How would changes in the contents affect fire behavior? Many possible alternatives may be quickly evaluated by mentally estimating fire behavior to get a sense of proportion regarding the importance. This filtering procedure is rapid and effective in establishing particular "what if" conditions that are important to bracketing behavioral variation.

8

Beyond the Room of Origin

8.1 Introduction

Although a single room fire can cause risks, discomfort, and some damage to property and buildings, the one-room fire seldom poses a significant building threat. Even though we strive to contain fires to a single room, the building often fails to do so.

The fire function of a barrier is to delay or prevent movement of products of combustion (POCs) from one space into an adjacent space. A barrier's performance can change during the fire because heat energy causes deterioration and actions of other parts of the holistic fire safety system can cause physical modifications.

Traditional practices prescribe fire endurance ratings for barriers. These hourly ratings encourage robust rather than flimsy construction, which is beneficial. However, fire endurance ratings rarely prevent multi-room fire propagation in buildings. Initial room-to-room fire spread is almost always due to pre-existing holes or other conditions that allow fire propagation.

Fire safety engineers, structural engineers, and architects have different roles with regard to barriers. The fire safety engineer selects the multi-room diagnostic fire while the structural engineer describes structural deformations that would be caused by that fire. The architect creates conditions that can cause room-to-room fire spread.

The fire safety engineer is responsible for identifying barrier openings, weaknesses, and other conditions that allow fire to move into the next space. Thus, the status of opening protectives and recognition of other conditions that allow POCs to move from one space to another are within the scope of fire safety engineering. The structural engineer describes structural behavior based on diagnostic fire conditions provided by the fire safety engineer.

This chapter discusses barrier failure modes, qualitative hazard estimates, and diagnostic fire applications. Collectively, this information helps to organize quantitative procedures to select diagnostic scenarios.

8.2 The Inspection Plan

Selection of an appropriate room of origin for a diagnostic fire can be difficult in complicated buildings when barriers allow multi-room fire propagation. The process of evaluating diagnostic fire room clusters can be simplified by using a preliminary

Fire Performance Analysis for Buildings, Second Edition. Robert W. Fitzgerald and Brian J. Meacham.
© 2017 John Wiley & Sons Ltd. Published 2017 by John Wiley & Sons Ltd.

qualitative analysis and Interactive Performance Information (IPI)-based sorting procedures.

An old fire service technique for pre-incident fire inspections combines visual identification of avenues of fire spread with fire growth hazards (FGHs). The fire officer starts at the top floor and examines each room for ways in which fire can move into the room and the potential hazards that may be expected from room contents. The inspection progresses down through the building, identifying conditions that may cause difficulties or danger during fire fighting operations.

This chapter describes procedures for fire safety engineers to adapt this fire service technique to select appropriate room clusters for diagnostic fires. The procedure integrates three interactive parts:

Part One: Barrier Effectiveness describes barrier functions and failure modes that contribute to multi-room fires.

Part Two: Barrier–Space Modules organize units for a rapid qualitative sorting of FGH conditions. Comparative FGH categories are based on room fire fundamentals described in Chapters 6 and 7. Barrier failure modes are incorporated to qualitatively approximate a time to flashover (FO). The technique provides a systematic method for preliminary identification of a room's FGH potential.

Part Three: Building Applications combine barrier failure modes and modular FGH classifications with IPI charts for preliminary diagnostic fire identification. Techniques are described to illustrate networking enhancements that use quantitative analysis.

Collectively, the techniques provide a rapid, systematic procedure for field inspections or conceptual performance design.

PART ONE: BARRIER EFFECTIVENESS

Barrier fire endurance ratings have been the basis for organizing prescriptive code requirements for nearly a century and are a major focus of the building code and fire community. Although fire endurance testing has a role in fire safety, it rarely provides significant information for the important time periods of a performance analysis. This does not imply that flimsy construction is appropriate or unimportant. It does mean that barriers require more comprehensive attention to understand their influence on building performance.

Barrier effectiveness describes the composite barrier's ability to delay or prevent an ignition into an adjacent space. Composite barrier construction is the focus of barrier performance.

The composite barrier incorporates all of the openings, features, and construction methods inherent in a functioning building. A clear recognition of these influences on holistic barrier behavior enables performance to be evaluated more realistically. Although *Part One: Barrier Effectiveness* discusses barrier failure in isolation, the type and location of the failure affect fire behavior beyond the room of origin.

Knowledge and experience of fully developed fires and their relationship to barrier testing have a role in prescriptive applications. Chapter 18 discusses this topic more explicitly. Also, barrier performance cannot be divorced from the structural frame behavior that is discussed in Chapter 19. Collectively, the barrier fire endurance, barrier openings, and structural support provide a complete picture of barrier fire performance.

8.3   Barrier Functions in Buildings

Horizontal and vertical barriers serve many functions in a building. They separate spaces to define rooms as well as to provide privacy, security, and noise control. Barriers provide routes for physical, visual, and informational communication with devices such as doors, windows, piping, and electrical conduits. They often provide and hide routes to convey services such as electricity, water, waste disposal, heat, and air through a building. Barriers sometimes support structural loads.

Barriers can also delay or prevent fire and smoke movement from one space to the adjacent space.

Barriers serve many functions other than stopping or delaying smoke and fire propagation. A building's day-to-day operations dominate long-term barrier use, maintenance, and status. These regular building functions have an important influence in evaluations of a building's fire performance.

8.4   Barrier Fire Functions

The main roles of barriers in fires are to:

- Protect people from combustion products during building evacuation or while they remain in the building.
- Shield property or downtime conditions from combustion products.

- Influence fire damage and smoke tenability by channeling the movement of combustion products along certain paths.
- Influence the transport of combustion products to detectors.
- Influence sprinkler control or extinguishment effectiveness.
- Influence fire department manual extinguishment operations and decisions.
- Influence fire department search and rescue operations.
- Affect the time of multi-room fire propagation.

The influence of barriers during a fire may be helpful or detrimental. For example, when barriers contain the steam generated from a sprinkler discharge, fire extinguishment is faster and more certain because oxygen depletion augments the water's extinguishment role. On the other hand, when a barrier location obstructs water discharge, sprinkler system effectiveness is reduced.

Barriers may help or hinder fire department operations. For example, barriers can block a hose-stream from reaching the seat of the fire, or they can break up a solid stream into droplets to improve suppression effectiveness. An attack route that requires forcible entry or directional changes will delay finding the fire and stretching attack hose lines. A room with few openings can contain steam generation and assist in extinguishment. On the other hand, that room may create backdraft conditions that can increase damage and risks to fire-fighter safety. Barriers can help fire fighters establish defensive positions to stop or control a fire. A barrier's performance value depends on the building architecture, the fire, fire ground operational needs, and barrier effectiveness.

Performance analysis provides a way to understand how architectural layouts, barrier construction, and barrier penetrations affect damage and risk. Although we will remain aware of multiple barrier functions during a fire, this chapter will address only fire propagation and its movement through barriers.

8.5 Concepts for Barrier Evaluations

A barrier is any surface that will delay or prevent combustion products moving from one space to another. Any barrier that exists is incorporated into the analysis. Therefore, barriers may be:

- Weak or strong
- Complete or incomplete
- Penetrated or unpenetrated
- Combustible or non-combustible
- Load-bearing or non-load-bearing
- Fire-rated or non-fire-rated.

A door or window, whether open or closed, is a part of the barrier. Any penetrations or openings, whether protected or unprotected from flame, heat, and smoke, are parts of the barrier. We evaluate all barriers from a field performance viewpoint, whatever their construction, fire resistance rating, or combustibility.

In addition to real physical barriers, we use *virtual barriers* to organize performance analyses in open spaces. A virtual barrier is a fictitious barrier that has no resistance to flame-heat or smoke-gas movement. Its function is to segment open spaces to describe fire/behavior better. Virtual barriers are also valuable in evaluating automatic and manual fire suppression conditions for large open spaces.

A diagnostic fire incorporates the influence of local barrier behavior on three-dimensional fire propagation. The global time clock starts with established burning (EB) in the room of origin and incorporates flame-heat movement involving barrier failure and FO in sequential rooms. The local time clock for a barrier normally starts when the involved room flashes over, although an analysis can adapt to situations where a fire moves through a barrier before FO occurs.

We define three states of barrier performance by the ignition potential on the unexposed side:

1) A small, hotspot ignition (designated $\bar{T}$) can occur on the unexposed side.
2) A large, massive ignition (designated $\bar{D}$) can occur on the unexposed side.
3) The barrier is successful (designated B) and no ignition can occur on the unexposed side.

These states are mutually exclusive, and only one state can be present at any time. Thus, if both hotspot and massive ignitions exist simultaneously, the massive condition will dominate the fire propagation and be considered the only state. Figure 8.1 shows the organization.

A barrier's status can change with time. The continuum of a barrier's behavior during a fire is analogous to a video. The barrier status at any instant in time is analogous to a

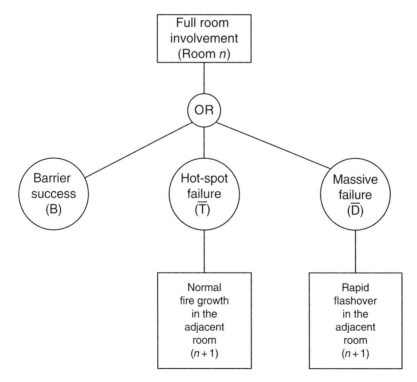

Figure 8.1 Barrier failure effects.

photograph. The video (and associated photographs) changes with each increment of time and heat energy release.

A small, hot-spot ($\overline{T}$) ignition may be caused by excessive heat transmission through a barrier. It can also result from surface cracking that allows small flames to appear at the unexposed surface of the barrier. Hot-spot failures are often caused by through construction penetrations that have not been protected adequately. When fuels are present at the unexposed side, a hot-spot failure causes an ignition and EB. This EB produces a predictable fire growth in the adjacent room.

A massive ($\overline{D}$) failure occurs when a large opening appears in the barrier. Massive failures may result from an open door, a large broken window, or a partial or full barrier collapse. Massive failures can also develop from hot-spot failures that enlarge over time, such as with combustible materials or gypsum board that has lost hydration.

A $\overline{D}$ failure produces a large influx of fire gases into the adjacent space. This will cause an almost certain FO within a minute or two, regardless of the fuel content in the space.

The barrier failure mode may be associated with barrier opening size. A rough rule of thumb suggests that a hot-spot ($\overline{T}$) failure occurs with an opening area of less than 6.5×10^{-2} m^2 (100 in^2). An opening of greater than 26×10^{-2} m^2 (400 in^2) produces a massive ($\overline{D}$) failure.

8.6 Barrier Failure Modes

Performance analyses apply both to existing buildings and to proposed new buildings. Potential barrier failure modes are easier to recognize in existing buildings. They must be anticipated or consciously controlled in proposed new buildings. Variability ("what if") analyses are useful for recognizing situations where attention to construction details is important.

A fire propagates from one room to an adjacent room through barriers. It is possible for fire to spread from building to building due to flying brands or because inadequate separations permit radiant heat ignitions. Although a complete performance analysis would consider this mode of spread, building-to-building fire spread is not described in this book.

The following general modes of barrier failure describe potential avenues of fire spread from one room to another. Some modes are common while others relate to specific construction practices.

1) Large Openings
a) Doors The most common cause of room-to-room fire propagation is through large openings that may or may not be covered with a door. An open door has no resistance to a massive ignition in the adjacent room.

A closed combustible door is described in Mode 6 below.

b) Entryway For our purposes, an entryway is a large barrier opening that does not have a protective covering. An entryway is very common in many building occupancies to allow passage through several interconnected rooms when a door may be architecturally inappropriate.

c) Pass Through A pass through is a smaller opening through which objects can be moved from one room to another.

d) Interior Windows   Windows that allow visual communication between rooms are common in many buildings. When the covering fails, a massive ignition can occur in the adjacent space. The type of protective covering can be plain glass that fails within a minute or two of FO. On the other hand, the covering may be of polycarbonate plastic that may withstand the heat from a fully involved room for long periods of time before melting and contributing fuel to the fire. Because time durations are important to building performance during the early stages of a fire, the type of window covering can influence multi-room fire estimates. In addition, double- and triple-pane glazing as well as polycarbonate glazing have an impact resistance that makes firefighting ventilation difficult.

e) Grilles   Grilles are common in many buildings to facilitate air circulation. They may be simple louver-covered openings of thin metal or wood, or they may be woven metal coverings for HVAC (heating, ventilation, and air conditioning) ducts.

Figure 8.2 illustrates these barrier openings.

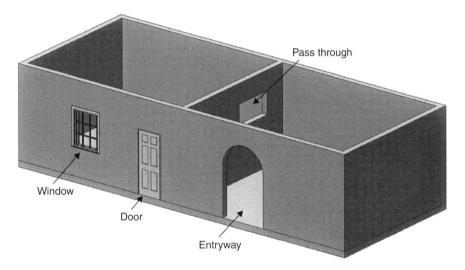

Figure 8.2 Barrier openings.

2) Small Openings

a) Through Construction   Electrical or plumbing lines in a building commonly move through small holes in the walls and ceilings. These openings are commonly called "poke throughs" or "through construction." Although building codes require protection of these small holes to prevent fire propagation, one often finds holes that are unprotected.

b) Small Holes or Cracks   Although these openings should not be present, they often exist, particularly in older buildings. Frequently, but not always, these openings are covered by combustible wall linings or suspended ceilings. This type of opening causes a weakness that compromises fire endurance. In addition, cracks around construction elements such as panels or slabs can permit premature hot-spot failures.

c) Small Coverings   Holes and smaller ducts often have coverings to prevent dirt and dust from entering into the opening or conduit. These coverings are similar to larger grilles, but small enough to cause a hot-spot, rather than a massive, failure.

d) Heat Conduction   Metal pipes and plates can transfer heat through barriers. Over time, a fire on one side can cause enough conductive heat transfer to ignite kindling fuels on the other side of a barrier.

e) Conductive Heat Transmission   Full room involvement causes barriers to heat during the post-FO conditions. When kindling fuels are present on the unexposed side, and the fire duration is long enough, ignition can occur. This is one of the failure criteria of the ASTM E-119 and ISO 834 fire endurance tests discussed in Chapter 18. Although thermal transmission is more common with floor–ceiling assemblies, it may occur on the unexposed side of any barrier.

Figure 8.3 illustrates these failure types.

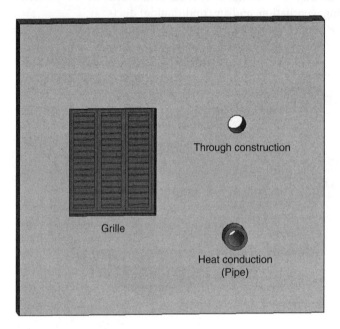

Figure 8.3 Barrier openings.

3) Exterior Propagation

a) Spandrel   Exterior fire propagation can occur when a non-combustible spandrel distance is insufficient to prevent flames from lapping from floor to floor through exterior windows. Figure 8.4(a) illustrates this condition.

b) Eyebrows or balconies   The installation of eyebrows or balconies between floors provides an architectural treatment that is often attractive and, if wide enough, can prevent floor-to-floor exterior propagation. Figure 8.4(b) illustrates this condition. Exterior balconies are also useful to interrupt floor-to-floor propagation.

c) Combustible Façade   Exterior floor-to-floor propagation can also occur when a combustible exterior covering is installed, as illustrated in Figure 8.4(c). A combustible façade allows fire to spread much faster because no delay for sequential room involvement is necessary. Radiation through windows can ignite near-by kindling materials.

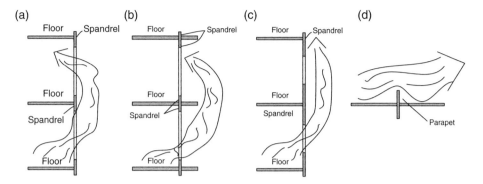

Figure 8.4 Exterior fire propagation.

d) Exterior Windows   Differences in exterior window glazing can affect fire propagation. One way for fire to propagate through an exterior window is by radiation that can
ignite kindling fuels such as curtains or drapes. When the glazing is ordinary glass,
flames and heat can easily break the glass. When the window is constructed of double
or triple layers or of thick plate glass, breakage is more difficult, although radiation
transfer still occurs.

e) Roofs   Sometimes combustible roofs or roofs that have become fully involved are
separated by parapet walls. When the parapet is not high enough, fire can propagate to
the adjacent roof surface, as shown in Figure 8.4(d).

4) Spaces Inside Barriers
a) Interstitial Space   It is common to have a space between a suspended ceiling and a floor
above. When the vertical barriers extend completely to the floors above, this space
becomes an extension of the room. However, walls often extend only a short distance
above the suspended ceiling. This can create a very large space for fire to extend and drop
down into other rooms. Figure 8.5(a) illustrates this condition.

b) Floor Voids   Certain functions, such as computer rooms, have voids under the floor.
These voids may provide air circulation or electrical wiring trays to service computers
or other equipment that requires easy access for servicing. Often, these raised floors
have holes or may be removable, enabling a fire to drop down and spread to additional
spaces. Figure 8.5(b) illustrates this construction

c) Inside Barrier Construction   Sometimes a ceiling is attached directly to an overhead
structural frame. When the frame is of joist construction, such as open web joists or
wood joists, any compromise of the ceiling will allow fire to move through the void
spaces, as illustrated in Figure 8.5(c).

d) Inside Walls   When a fire breaches the covering of a stud wall, flames can move
through the space between the studs. If platform construction has been used, the fire
will be interrupted at the next floor level. However, if balloon construction is used, the
fire will have uninterrupted movement between the studs for the entire wall height.

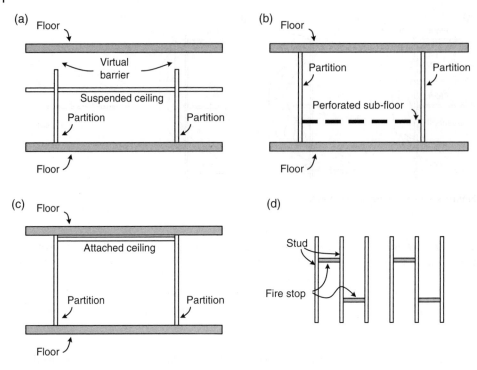

Figure 8.5 Barriers.

Installation of fire stops prevents rapid, extensive flame spread. Figure 8.5(d) illustrates these conditions.

e) Inside Ducts   Most modern buildings have ducts to transport air or remove unwanted gases. Fire in a duct can be transported to remote locations in the building. This condition can be avoided by installing dampers as barriers to interrupt movement of flames or hot gases.

5) Shafts
The most common vertical shafts in buildings are elevators and stairwells. However, a building examination often reveals other vertical or horizontal conduits through which fire can spread.

a) Corridors, Stairwells, and Elevators   These building elements are so common for people movement that the transport of flame-heat and smoke-gas products can be overlooked.

b) Dumb Waiters   Dumb waiters are small freight elevators that carry small objects between floors. Although they are often associated with food movement in restaurants, dumb waiters are used in commercial and industrial buildings to transport small items efficiently and quickly. It is easy to overlook dumb waiters as a conduit that will propagate fire.

c) Chases   A chase is a vertical or horizontal conduit for plumbing (pipe chase) or electrical wiring (electrical chase). It is common for chases to extend vertically or horizontally through several rooms. Chases are usually protected when they pass through barriers to make fire propagation difficult. However, inspections often disclose unprotected openings through barriers, particularly in hidden spaces or after building renovations have been completed. Sometimes these unprotected openings are important; sometimes they are not.

Figure 8.6 illustrates these features.

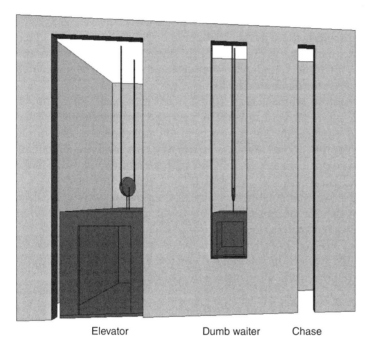

Elevator   Dumb waiter   Chase

Figure 8.6 Vertical shafts.

6) Combustible Barriers

Wall panels provide a comfortable ambiance for residents or many business environments. They are often an inexpensive choice for building renovations. Combustible wood panels are common in spaces where building regulations do not prohibit them. They are often discovered even where prohibited.

a) Panels One can often observe combustible wood panels placed directly over wood studs or over gypsum panels that may have holes or extensive areas removed. Regardless of the construction methods, when combustible barriers are recognized, their behavior becomes an important part of a performance analysis.

A combustible panel contributes to the FGH as a part of the room's interior finish. In addition, the panels can burn through and allow fire to propagate into the adjacent space. Typically, a flat panel will initially form a small hot-spot hole. As the fire continues, this hole expands to the size of a massive failure. Thus, combustible barrier performance goes through transitions as time progresses.

b) Doors A wide variety of wood doors and frames are available for buildings. When these barriers are closed and have integrity, they prevent free extension of a fire into an adjacent room. Hollow core doors may burn through in about 10 minutes. Solid core doors last longer, perhaps 25–30 minutes. Both exhibit the transitional behavior of progressing from a small, hot-spot failure to a large, massive failure.

7) Virtual Barriers

A virtual barrier is a fictitious barrier that has no resistance to flame-heat or smoke-gas movement. A virtual barrier provides a way to analyze certain types of spaces using consistent modular techniques. Sections 2.11 and 2.12 discuss virtual barriers.

Building corridors, particularly those having changes in direction, can be segmented with virtual barriers. This technique can simplify fire, smoke, or egress analysis and improve communication with others. Undivided stairwells and shafts can be segmented to describe time and distance for flame-heat and smoke-gas movement.

Virtual barriers allow one to segment large rooms, such as manufacturing floors, warehouse storage, and "big box" retail stores. Fire spread through different "virtual rooms" can be described using regular barrier–space modular techniques. The virtual rooms can be used with the IPI chart to analyze a variety of time-sensitive components such as fire department extinguishment, automatic suppression, egress, and smoke movement.

Fire and smoke move along corridors or up stairs as time progresses. Segmenting corridors or stairwells at specific locations allows one to relate time-contamination to the floor plan. This aids related analyses for egress or fire department suppression.

Hot smoke in a vertical shaft cools as heat is transferred to the walls. As the smoke loses its buoyancy, it begins to affect the stack effect. Virtual barriers provide a convenient way to relate smoke removal with floor levels in the building.

8.7 Barrier Failures and Building Performance

One role of the fire safety engineer is to develop the diagnostic fire for multi-rooms. Holes in barriers are the primary cause of fire propagation between rooms. The status of barriers may be recognized in existing buildings relatively easily to enable the diagnostic fire to reflect appropriate situations for a performance analysis.

Potential failure modes for proposed new buildings may be handled through scenario comparison using variability analysis. The status of interior doors or other openings may be incorporated into appropriate "what if" scenario evaluations. When the status of specific large openings causes important diagnostic fire abnormalities, appropriate solutions can be recommended. Similarly, when conditions that can create severe multi-room fire potential become apparent, attention may be given to design specifications and construction inspections.

Many barriers will be successful and will stop any POC movement into the adjacent space. Others will fail because of certain construction features. A barrier failure may or may not be important, depending on other conditions and what is in the adjacent room. A holistic analysis can provide answers about the significance of any barrier failure. IPI charts are useful to document the influence of time on the changes.

PART TWO: BARRIER–SPACE MODULES

8.8   Introduction

A diagnostic fire describes the relationship between time and combustion products for the room of origin and the fire propagation through barriers into additional rooms. The diagnostic fire assumes no extinguishment intervention.

Although a fire will attack barriers before FO, the greatest heat energy impact occurs after a room is fully involved. At any given time, a barrier can exhibit only one of three mutually exclusive states:

1) Barrier success (B) allows no ignition source on the unexposed side.
2) A small, hot-spot failure ($\overline{T}$) in which a localized ignition source appears on the unexposed side.
3) A large, massive failure ($\overline{D}$) where a large volume of heat and fire gases flow into the adjacent space.

The type of barrier failure affects the speed and certainty of fire spread in the adjacent room. When a hot-spot ignition occurs, heat and smoke from the ongoing fire somewhat accelerate fire growth in the next room. This creates a natural, although faster fire development in a newly ignited space. On the other hand, a massive barrier failure allows a large influx of fire gases into the adjacent room. This produces FO in a short time, usually within a minute or two for most rooms.

8.9   Barrier–Space Modules

An architectural space is a volume surrounded by barriers, either real or virtual. Regardless of the building's use or complexity, all architectural plans may be described by a combination of barrier–space units or modules. The modules may have different sizes, shapes, functions, and interior designs. However, their assemblage defines the building. Regardless of architectural complexity, a picture of multi-room fires emerges when one portrays the connectivity with a modular network.

During a fire, any room is either a room of origin, or a room beyond the room of origin. The modular concept simplifies the organization of multi-room diagnostic fires. Although current attention will focus on flame-heat movement, the concept can be adapted for other types of analysis such as smoke movement, people movement (egress), and manual fire suppression. Modular networks in combination with the IPI chart allow one to compare different sets of time-dependent alternatives.

Any space enclosed by real or virtual barriers is considered a modular unit. The architectural space plan defines the specific arrangement of modular units for a building. Figure 8.7 illustrates the concept. Barriers that separate adjacent rooms are a part of each module.

8.10   Massive Barrier Failure ($\overline{D}$)

A massive ($\overline{D}$) failure occurs when a large opening exists in a barrier. The barrier failure may be instantaneous, such as would occur through an open door. A massive failure may be delayed, such as when a window breaks during the fire or a barrier shows initial

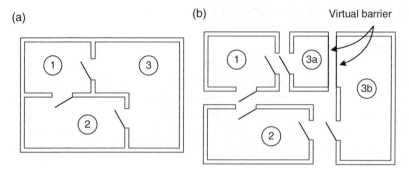

Figure 8.7 Barrier–space modules.

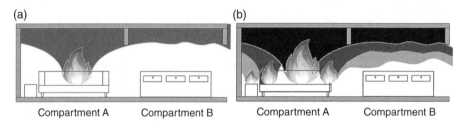

Compartment A   Compartment B   Compartment A   Compartment B

Figure 8.8 Massive failure effects.

fire endurance before eventually collapsing. A massive failure can be progressive, as when a combustible barrier burns through. In this situation, a small ($\bar{T}$) failure initially occurs and grows progressively into a massive ($\bar{D}$) failure.

Although realms of fire growth remain valid for fire development in the next module, the speed of growth may be so fast that the transitions become obscured. Figure 8.8 illustrates the mechanics of fire propagation through a large opening.

Figure 8.8(a) shows a pre-FO fire in Compartment A. The adjacent Compartment B supplies air for combustion through the $\bar{D}$ barrier opening. Smoke and hot gases flow from Compartment A into Compartment B. Ignition in Compartment B does not yet occur because the flames must spread across the ceiling of Compartment A as a precursor to FO. After FO occurs in Compartment A, the excess fire gases of Compartment A (seeking oxygen to form the combustible mixture) extend into Compartment B.

Although it is possible for conditions to exist in which fire can propagate into Compartment B before FO of Compartment A, that situation is not common. Therefore, we make a simplifying assumption that ignition and EB will not occur in an adjacent room until FO occurs in the preceding room. This simplification enables time relationships to be ordered more easily.

Figure 8.8(b) shows fire spread from Compartment A into Compartment B after FO. The flow of smoke and hot gases from Compartment A into Compartment B before FO does pre-condition the contents of Compartment B. Radiation through the opening combined with increased temperatures of the barriers and contents of Compartment B from the smoke and hot gases cause very rapid FO conditions. This normally happens within 1 or 2 minutes after FO in Compartment A.

Significant differences in room sizes may influence the spread. For example, when a very small room flashes over and a $\bar{D}$ failure allows flame movement into a much larger room, the flame front may be perceived as a large ignition source. It would produce a somewhat "normal," although more rapid, fire growth into the next room. In the opposite condition of flames moving from a large room into a small room, FO is certain to occur almost instantaneously. Most atypical conditions are easily recognized, and fire growth and time durations may be estimated using fire fundamentals.

Normally, FO will occur in Compartment B regardless of contents. For example, fire gases from a flashed over room into a sterile corridor still produce enough flaming to be depicted as a FO condition. Flame movement distance and time may be conveniently described in sterile rooms with virtual barrier segmentation.

8.11 Hot-spot Barrier Failure ($\overline{T}$)

A hot-spot ($\overline{T}$) failure occurs when a barrier permits fire propagation by surface radiation to ignite kindling fuels near the surface. However, the more common $\overline{T}$ failures originate from small holes in the barriers. Figure 8.9 illustrates modular fire spread for $\overline{T}$ failure conditions.

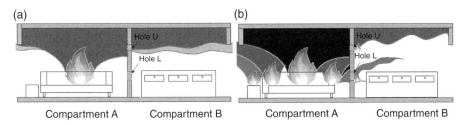

(a)   (b)

Hole U   Hole U
Hole L   Hole L

Compartment A   Compartment B   Compartment A   Compartment B

Figure 8.9 Hot-spot failure effects.

Although fire development in Compartment B progresses through the normal realms, as described in Chapters 6 and 7, smoke and higher temperatures accelerate the process. When easy-to-ignite fuels are located near a $\overline{T}$ barrier failure, ignition can occur. Therefore, all of the usual fire fundamentals apply to estimating fire growth, except that time durations are shorter and realm transitions are easier. When kindling fuels are not present near a $\overline{T}$ failure, no ignition can occur. Thus, fuel package contents and locations become more important for a $\overline{T}$ failure.

Although smoke and fire gases will pass through the small opening before FO, flames will not issue from a $\overline{T}$ location until after FO. The opening location influences ignitability and ease of spread. For example, an opening near the floor has less significance than an opening near the ceiling before FO (Figure 8.9a). After FO the conditions change and flames, heat, and smoke will move from one room to the next, as shown in Figure 8.9(b). We again make the simplifying assumption that local time for barrier attack starts at FO of Compartment A. Local time for an adjacent room starts at EB in that room.

8.12 The Role of Interior Finish

Although combustible interior finish is important to fire growth in the room of origin, it assumes a greater significance for fire propagation in rooms beyond. Combustible ceiling finishes are particularly important, because hot fire gases pre-heat the upper surface and ignition can occur very quickly. As depicted in Figure 8.9, combustible ceiling finishes can produce very fast FO conditions when $\overline{T}$ failures are located high in the room.

Combustible wall finishes also provide fuel locations and conditions that provide easier room fire development. Almost any $\overline{T}$ location will have conditions that can cause fire propagation over combustible surfaces.

8.13   Virtual Barriers

A virtual barrier has no resistance to smoke or flame movement. Therefore, a virtual barrier allows a $\overline{D}$ failure at all times. Nevertheless, virtual barriers are useful for describing time relationships involving fire spread or movement into adjacent compartments.

Virtual barriers were introduced in Section 2.11. The concept is useful for providing barrier–space boundaries to maintain analytical consistency and to simplify analysis.

8.14   Qualitative Diagnostic Fire Analysis: Room Classifications

A preliminary qualitative diagnostic fire analysis examines clusters of candidate rooms to select diagnostic fire locations and scenarios. Although a primary objective is to select an appropriate room of origin and multi-room diagnostic fire, several secondary goals may also be achieved. The most useful outcome of a qualitative analysis is acquiring a "feel" for fire behavior in the building. The observational relationships help to ensure that computer modeling is representative. Also, a qualitative examination discloses ways in which fire can propagate to additional spaces.

A qualitative analysis also provides an informational foundation for the quantitative analysis. Integrating qualitative and quantitative analyses helps to provide a greater understanding of fire calculations and produces more confident outcomes.

The qualitative analysis is rapid, yet able to identify details important for fire behavior. Descriptions use the terms "room" and "module" to be the same entity. Initially, the FGH classification for the room and the effectiveness of barriers are evaluated separately. Then, the two parts are combined to describe the fire growth time.

The FGH assessment is based on top-down evaluations described in Chapter 7. The process may be outlined as follows.

1) Identify a candidate cluster of rooms that seem significant to the holistic building analysis. Fire extension in three dimensions is common and must be considered.
2) Examine each room in the cluster to select a FGH classification using top-down classification estimates. The FGH classification evaluates the room interior design for ease and speed of achieving FO. *Initially, each room is evaluated as a room of origin.* Assume ventilation is adequate to support combustion because openings are not a factor in interior design FGH classifications.
3) Estimate a time from EB to FO for the room *as a room of origin*. The default times in Figure 8.10 provide a perspective for at^2 fires. The inspector selects a time duration that seems appropriate to the conditions.

	Alpha	Default time (min)	Estimated time (min)	Default time (min)	Estimated time (min)	Default time (min)	Estimated time (min)
Very high	0.06	1–4	___	1–3	___	0–1	___
High	0.06	4–8	___	3–6	___	0–2	___
Moderate	0.003	8–12	___	6–9	___	1–2	___
Low	0.002	12–16	___	9–12	___	1–3	___
Very low	0.001	16–20	___	13–16	___	1–4	___
		Room interior design as a room of origin		Room interior design as a subsequent room with a $\overline{T}$ failure		Room interior design as a subsequent room with a $\overline{D}$ failure	

Figure 8.10 Room classification selections.

4) Now consider the same room and FGH classification as if the room were a room beyond the room of origin. That is, a fully involved fire exists in an adjacent room and a $\overline{T}$ failure occurs in a barrier.
5) Fire development could be influenced by fuel availability near the failure location. When fuels are not present near a $\overline{T}$ failure location, ignition will not occur. When fuels are available, EB and fire growth can occur.
6) One has two alternatives for assessing fire propagation for an adjacent room. The first considers a specific fire scenario propagation from the barrier's $\overline{T}$ ignition. The second recognizes that fuel locations are easily changed and assumes that fuels will exist near the $\overline{T}$ location. In the absence of specific needs, we recommend that routine analyses assume kindling fuels exit near the $\overline{T}$ location and the time to FO be based on the room's interior design. Special situations that affect performance can be easily documented.

   Estimate the time duration from EB to FO for a fire caused by a $\overline{T}$ barrier failure. Default time estimates may be adjusted to reflect conditions observed by the inspector. In this evaluation, one need not actually have $\overline{T}$ failure conditions. The actual failure mode and its location will be recognized by barrier analysis in another part of the inspection.

   Finally, consider the same room and FGH classification as if the room had a $\overline{D}$ failure. That is, a fully involved fire exists in an adjacent room and a $\overline{D}$ failure occurs in a barrier.
7) Estimate the time duration from EB to FO for a fire entering the room from a $\overline{D}$ barrier. As with the $\overline{T}$ analysis, we do not need to actually have $\overline{D}$ conditions to make the assessment. The barrier analysis will incorporate that situation if it exists.

The qualitative FGH ranking of the room's interior design is based on concepts discussed in Chapter 7. A single FGH classification describes the room's interior design. The FGH does not change with ignition location, ventilation conditions, or whether the room is a room of origin or a subsequent room. Figure 8.10 provides a convenient means to identify the three values for time to FO according to a room's interior design.

8.15   Qualitative Diagnostic Fire Analysis: Barrier Contributions

After a relative FGH potential is estimated for each room in the cluster, the barriers are examined for performance. Concepts described in this chapter provide guidance for recognizing potential failure modes and their significance.

Trained observers can complete an inspection relatively quickly. Digital photographs can document a basis for estimates and be used for quantitative analysis at a later time. The IPI chart can store information and sort rooms to identify the influence of different rooms of origin on multi-room fires.

8.16   Qualitative Diagnostic Fire Analysis: Modules

The final part of the modular performance description combines the FGH for the room's interior design with barrier performance to estimate fire behavior beyond the room of origin. Because top-down FGP descriptions do not identify specific ignition locations, one assumes that fuels will be present to produce EB at the barrier failure location.

PART THREE: QUALITATIVE FIRE ANALYSIS

8.17   Introduction

A qualitative evaluation of room hazards combines top-down evaluations described in Chapter 7 with barrier failure modes of this chapter. Evaluations provide useful information for performance applications by identifying conditions that impact multi-room fires. Rapid "what if" variability analyses for different opening conditions help to identify important scenarios in complex buildings with a small investment of time.

Here, we describe the mechanics of organizing a complex analysis. Examples using a hypothetical room cluster describe the process and illustrate important concepts. The procedure is appropriate for multi-level conditions and any barrier failure mode. Sequential ignitions and multi-path routes involving three-dimensional arrays may be organized in the IPI chart.

8.18   The Process

An initial "size up" of the building identifies potential rooms as candidates for the diagnostic fire. Scenarios for life safety evaluations may be different from scenarios to examine property or operational continuity. Different rooms of origin can influence fire defense performance and risk characterizations.

A rapid qualitative analysis of room clusters is useful for selecting appropriate diagnostic fires. The process first identifies a candidate cluster of rooms and classifies the fire growth potential (FGP) for each room's interior design using top-down qualitative estimates. Then (or concurrently), barrier failure conditions are examined. The barrier–space modules combine the FGP classifications and barrier failure types to estimate relative times to FO. Networks combine the geometry and time estimates to describe fire propagation.

Examples 8.1–8.3 illustrate ways to analyze a cluster of rooms for different rooms of origin and barrier conditions. This initial screening provides an insight into selecting appropriate rooms of origin. The methodical process is suitable for rapid computer analysis.

Example 8.1   Figure 8.11 shows a cluster of five rooms. Identify room-to-room fire propagation for the cluster and show the results on an IPI chart. The scenario uses:

- Room 1 as the room of origin
- All doors are open.

An initial top-down qualitative evaluation using the groups of Figure 8.10 produced the FGP classifications described in Table 8.1. Specific interior designs are not shown, in order to focus on the mechanics of analysis rather than to justify FGP evaluations.

Table 8.2 combines the FGP and barrier effectiveness to describe qualitative barrier–space modular data. The size of barrier openings and their locations can alter time to FO. An inspector selects FO time durations based on observations and judgment of field conditions.

An examination of barrier construction features provides the information of Table 8.3. The time durations for $\bar{T}$ and $\bar{D}$ are inspector estimates. The unpenetrated barrier segments are not expected to allow fire propagation during the scenario.

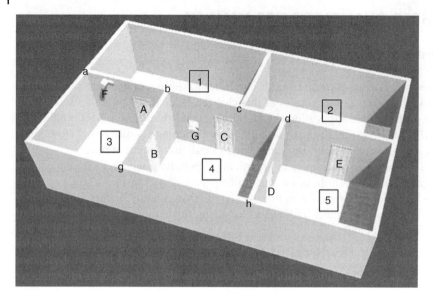

Figure 8.11 Example 8.1: room fire propagation.

Table 8.1 Example 8.1: fire growth potential room classifications.

Module	Fire growth potential classification
1	Moderate
2	High
3	Low
4	Moderate
5	Low

Table 8.2 Example 8.1: time to flashover (FO) estimates.

Room	Fire growth potential	Room of origin: time EB to FO (min)	Time EB to FO for the room if a $\overline{T}$ failure occurs (min)	Time EB to FO for the room if a $\overline{D}$ failure occurs (min)
1	Moderate	12	6	3
2	High	6	3	1
3	Low	18	8	2
4	Moderate	9	7	3
5	Low	13	5	1

EB, established burning.

Table 8.3 Example 8.1: barrier construction estimates.

Room separation	Barrier segment	Openings and penetrations	Construction	Status	Location	Time to $\overline{T}$	Time to $\overline{D}$
1–3	ab	Door (A)	Solid core wood	Closed	Normal	20 m	25 m
1–3	ab	Door (A)		Open		0 m	0 m
1–3	ab	Pipe, through construction (F)	Protected	Closed	At ceiling	60 m	80m
1–4	bc	Door (C)	Solid core wood	Closed	Normal	20 m	25 m
1–4	bc	Door (C)		Open		0 m	0 m
1–4	bc	Pass through (G)	(W) 24 in (610 mm) (H) 18 in (457 mm)	Open	Bottom 80 mm above floor	0 m	0 m
1–4	bc	Pass through (G)	(W) 24 in (610 mm) (H) 18 in (457 mm)	Wood Covering	Bottom 80 mm above floor	–	2 m
1–2	cf	None	NC	NA	NA	–	–
3–4	bg	Door (B)	Metal	Closed	Normal	30 m	50 m
3–4	bg	Door (B)		Open		0 m	0 m
4–2	cd	None	NC	NA	NA	–	–
4–5	dh	Door (D)	Hollow core wood	Closed	Normal	8 m	10 m
4–5	dh	Door (D)		Open		0 m	0 m
5–2	de	Door (E)	Hollow core	Closed	Normal	8 m	10 m
5–2	de	Door (E)		Open		0 m	0 m

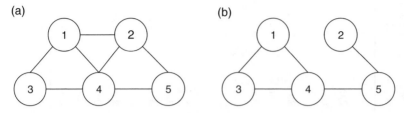

Figure 8.12 Example 8.1: module interconnectivity.

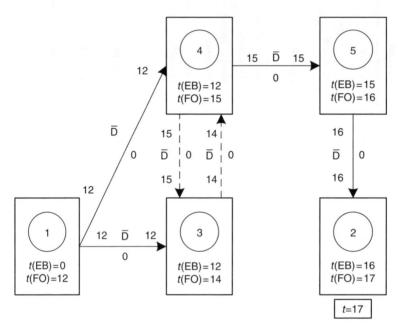

Figure 8.13 Example 8.1: interconnectivity analysis.

Solution Figure 8.12(a) shows the complete room-to-room connectivity network (see Section 3.14). However, the qualitative analysis indicates that the unpenetrated barriers between Rooms 1 and 2 and Rooms 4 and 2 will not allow fire to spread through them. Thus, the working network may be simplified in Figure 8.12(b) by eliminating these connectivities.

Room 1 is the room of origin. Figure 8.13 recasts the network to envision the module-to-module propagation easier.

Although an IPI chart can show the information without using Figure 8.13, a network is useful to show the thought process. The network nodes show the rooms (modules) and the global times for EB and FO. The branches show the type of barrier failure between modules and the global time for barrier failure.

The time at the start of any branch shows the global (cumulative) time of FO in the preceding room. The end number of a branch is the sum of this value plus the time delay for EB in the next room caused by the barrier. Although all failure modes in this example are caused by open doors and there is no barrier time delay, barriers frequently provide some resistance before a $\bar{\text{T}}$ or $\bar{\text{D}}$ failure occurs.

This example introduces multiple paths of propagation. Room 4 can reach EB either directly from Room 1 or through path 1–3–4. Similarly, Room 3 can reach EB either directly from Room 1 or through Path 1–4–3. In this case, all doors are open and these paths do not control the time to EB. We show dashed arrows to indicate non-controlling potential movement between rooms. However, propagation time durations and movement paths would be different if selected doors were closed.

Fire propagation time and sequencing of room involvement may be documented by tracing the sequential paths of the network. Here, the route is Room 1 (12 m)–Room 3 (14 m) and Room 4 (15 m). Then, Room 4 to Room 5 (16 m) to Room 2 (17 m).

The IPI chart shows the sequence of rooms and time durations. Seventeen minutes would elapse from ignition in Room 1 to having all five rooms fully involved.

Discussion The IPI chart in Figure 8.14 shows all of the information of Figure 8.13 in bar chart form rather than as network analysis. One could go directly from Figure 8.12 (b) to the IPI chart. The modular data of Tables 8.1–8.3 can be incorporated and the IPI chart may be programmed to provide the same results.

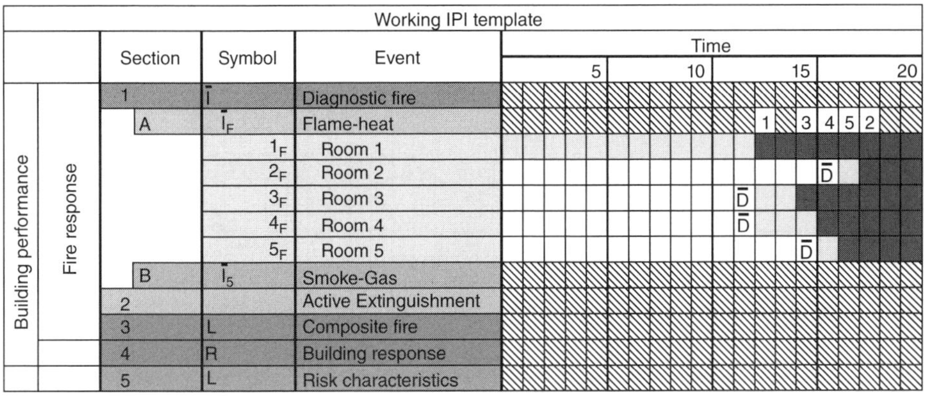

Figure 8.14 Example 8.1: Interactive Performance Information (IPI) representation.

This IPI chart portrays only the qualitative fire progression. Fire duration to burnout, barrier failure modes, and other relevant data may also be incorporated into a quantitative IPI chart.

Example 8.2 Solve Example 8.1 if the room of origin were Room 2 rather than Room 1.

Solution Figure 8.15 shows the network diagram that describes the fire propagation. Tracing the time durations the fire propagation path is 2–5–4–1 and 3. This time for the five-room involvement is 13 minutes. Figure 8.16 shows the corresponding IPI chart.

Discussion The IPI chart of Figure 8.16 shows that Rooms 3 and 1 experience EB at the same 10 minutes. However, Room 1 reaches FO in 13 minutes while Room reaches FO in 12 minutes. As with the previous example, the shorter times take precedence and simultaneous branches having longer times are disregarded.

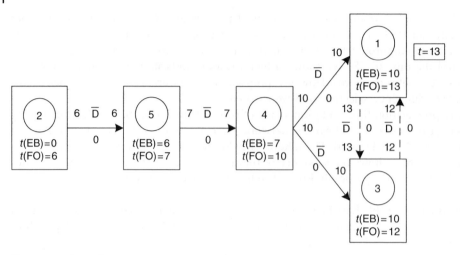

Figure 8.15 Example 8.2: interconnectivity analysis.

Working IPIT template									
		Section	Symbol	Event	Time				
						5	10	15	20
Building performance	Fire response	1	Ī	Diagnostic fire					
		A	Ī_F	Flame-heat			2 5	4 3 1	
		1_F	Room 1				D̄		
		2_F	Room 2						
		3_F	Room 3				D̄		
		4_F	Room 4			D̄			
		5_F	Room 5		D̄				
		B	Ī_5	Smoke-gas					
		2		Active Extinguishment					
		3	L	Composite fire					
		4	R	Building response					
		5	L	Risk characteristics					

Figure 8.16 Example 8.2: Interactive Performance Information (IPI) representation.

Example 8.3   Solve Example 8.1 if the room of origin were Room 2 and the status of barrier openings were as follows:

Rooms 2–5: Door E open
Rooms 5–4: Door D closed
Rooms 4–3: Door B closed
Rooms 4–1: Door C closed
Rooms 4–1: Pass through panel G closed
Rooms 3–1: Door A closed.

Solution   The status of barrier protectives significantly changes the time duration for fire propagation through barriers. This causes network to appear more complicated because a T̄ failure produces a different time to FO than a D̄ failure. Therefore, one must examine more alternatives. Although some calculations could be eliminated by observation because only the shortest time durations control, Figure 8.17 shows outcomes of all paths to illustrate the process. The nodes are segmented to show all values

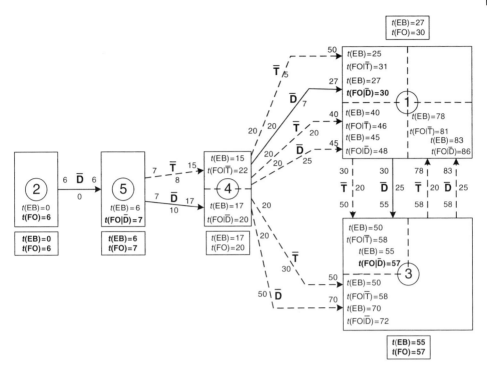

Figure 8.17 Example 8.3: interconnectivity analysis.

with governing time durations in bold. After examining the node time durations, we show solid branch lines that govern outcomes. Dashed lines do not control the results.

The network calculations are described as follows:

Room 2–5: Room 2 reaches FO in 6 minutes and moves through the open door to cause Room 5 FO 7 minutes after EB.

Room 5–4: The hollow core door separation allows a $\overline{T}$ failure at 8 minutes after Room 5 FO. This $\overline{T}$ failure grows into a $\overline{D}$ failure after an additional 2 minutes. Room 4 can reach FO from the $\overline{T}$ failure in $7 + 8 + 7 = 22$ minutes. FO can occur from a $\overline{D}$ failure in $7 + 10 + 3 = 20$ minutes. The smaller number governs and Room 4 is expected to reach FO in 20 minutes.

Room 4–3: The closed door has a $\overline{T} = 30$ minutes and a $\overline{D} = 50$ minutes. This allows two possible times to reach EB and FO:

From $\overline{T}$: Room 3, EB = 20 + 30 = 50 minutes, FO = 50 + 8 = 58 minutes

From $\overline{D}$: Room 3, EB = 20 + 50 = 70 minutes, FO = 70 + 2 = 72 minutes.

The smaller value of $t_{FO} = 58$ minutes would be selected. However, this value is temporary until we check the time to FO from all other paths. Here, we consider the path Room 1–3.

Room 4–1: Two barrier failure locations exist from Room 4 to Room 1. These produce possible times to FO as follows:

Pass through $\overline{T}$: Room 1, EB = 20 + 5 = 25 minutes, FO = 25 + 6 = 31 minutes

Pass through $\overline{D}$: Room 1, EB = 20 + 7 = 27 minutes, FO = 27 + 3 = 30 minutes

Door $\overline{T}$: Room 1, EB = 20 + 20 = 40 minutes, FO = 40 + 6 = 46 minutes

Door $\overline{D}$: Room 1, EB = 20 + 25 = 45 minutes, FO = 45 + 3 = 48 minutes.

The smallest of these values controls and t_{FO} = 30 minutes. We again consider this a temporary until we check all other paths. Here, we examine Path 3–1.

Room 3–1: The closed solid core wood door provides the following values:

Door $\bar{T}$: Room 1, EB = 58 + 20 = 78 minutes, FO = 78 + 3 = 81 minutes

Door $\bar{D}$: Room 1, EB = 58 + 25 = 83 minutes, FO = 83 + 3 = 86 minutes.

Room 1–3: The same solid core wood door provides the following values:

Door $\bar{T}$: Room 3, EB = 30 + 20 = 50 minutes, FO = 50 + 8 = 58 minutes

Door $\bar{D}$: Room 3, EB = 30 + 25 = 55 minutes, FO = 55 + 2 = 57 minutes.

Selecting the control values shows the following fire propagation:

Room 2: FO = 6 minutes

Room 5: EB = 6 minutes (from $\bar{D}$); FO = 7 minutes

Room 4: EB = 17 minutes (from $\bar{D}$); FO = 20 minutes

Room 1: EB = 27 minutes (from $\bar{D}$); FO = 30 minutes

Room 3: EB = 55 minutes (from $\bar{D}$); FO = 57 minutes.

Figure 8.18 shows an IPI chart with values described above. The IPI chart notes that the natural fire would propagate along path 2–5–4–1–3. Although Rooms 2 and 5 will be fully involved after 7 minutes, the barrier protectives begin to slow the fire. Room 4 reaches FO after 20 minutes because of the $\bar{D}$ failure.

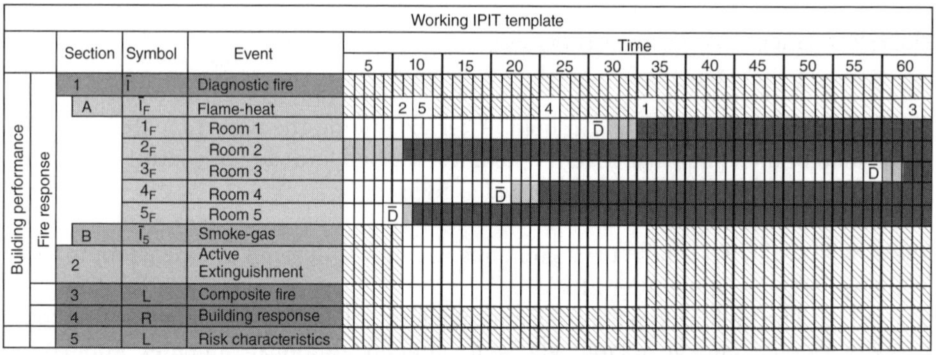

Figure 8.18 Example 8.3: Interactive Performance Information (IPI) representation.

Although the final rooms can reach FO through several paths of propagation, the fire will move from Room 4 to Room 1 after a global time of 30 minutes because of the $\bar{D}$ failure in the pass through. The fire then moves from Room 1 to Room 3 and reaches FO at global time 57 minutes. The propagation is due to a $\bar{D}$ failure in the closed door protective separating the rooms.

8.19 Discussion

Examples 8.1–8.3 illustrate a way to identify time sequencing of room fire propagation. A major value for the procedure is speed in ordering room clusters for diagnostic fire selection. Large buildings can have complicated functional plans. Modular rankings for individual rooms using 'top down' hazard classifications and barrier effectiveness identifies relative fire performance. Qualitative variability analysis identifies important conditions for detailed quantitative studies. This reduces attention for less significant situations.

The IPI chart can be programmed to select the controlling numbers and disregard time durations that are irrelevant and replace the need for networks. The qualitative filtering process both simplifies selection of appropriate rooms of origin and identifies the number of rooms that can become involved within important time periods of a building fire.

After one gains a sense of proportion for qualitative analysis, the process becomes routine for early size up of the building. Quantitative analysis can become more clearly identified and professional time requirements become minimal.

8.20   Information Technology Enhancements

Modern technology provides an array of opportunities to augment field inspection information. For example, co-workers at remote locations (e.g. corporate offices) can supplement voice communication with visual information. Panoramic views for room interior designs and barrier openings give an immediate opportunity to discuss improving estimates or alternative variability analysis considerations.

The Internet provides instant access to documented quantitative analyses or experimental data. A variety of information sources provide valuable opportunities to make better estimates for quantitative analysis and documentation. High-definition video and photographs are inexpensive and allow one to share visual information across the globe. Recent advancements in virtual reality and unmanned aerial vehicles can make inspections faster, more accurate, and safer. A variety of resources are available to improve important qualification estimates.

Preliminary building evaluations can identify those clusters of rooms that are particularly sensitive to performance and risk. Combining organizational resources helps to quickly identify those clusters of rooms that have the greatest influence on fire suppression or risk characterizations.

The IPI chart can serve a variety of roles in the process. It can help to identify particular rooms of origin that can produce multi-room fire conditions of importance to performance. The IPI chart can be programmed to order time and spaces to give perspective to performance. This provides a basis for detailed quantitative analyses. However, the most valuable aspect of IPI analysis is to relate time and different component events in a way that can be documented.

9

Smoke Analysis

9.1 Introduction

The flame-heat component dominates a building's performance because it defines the fire behavior and gives a clear picture of strengths and weaknesses for most parts of the fire safety system. However, the smoke-gas component has a major influence on the risk to people, property, and operational continuity. Visible smoke and invisible toxic gases have different effects on people and equipment. Because they are "fellow travelers" in the same medium, we will describe all airborne products of combustion as "smoke" for convenience.

Smoke is generated in a fire and transported by air movement. Forces that affect air movement cause smoke to move through building paths created by holes in barriers. Smoke obscures visibility in spaces through which people travel, and it affects property and electronic data storage and operations. Spaces can be protected from smoke penetration with one or a combination of barrier protection, ventilation practices, and mechanical air handling controls.

The following quotes describe some actual human observations of smoke environments in a burning building:

> "The space was as clear as outside. Then, the dense smoke dropped like a curtain. Within five seconds it was so black that I became disoriented."
>
> "It is difficult to understand the incredible blackness. You can't see your hand in front of your face. You must move by instinct and by following barriers."
>
> "I could see long distances through the smoky atmosphere. But, when I actually entered into the environment, the irritants were so stinging that I couldn't keep my eyes open."
>
> "We were just sitting in the [seventh floor] lobby talking. The lobby was clear. Then, smoke [from a second floor fire] billowed into the space so quickly that we could not find our rooms."
>
> "The smoke was so voluminous and dense that I couldn't see the building that was on fire."
>
> "The smoke in the space was dense. But, within seconds it cleared and the atmosphere was free."

Fire Performance Analysis for Buildings, Second Edition. Robert W. Fitzgerald and Brian J. Meacham.
© 2017 John Wiley & Sons Ltd. Published 2017 by John Wiley & Sons Ltd.

For practical purposes, smoke moves as air in the building. The smoke from a fire is normally dense, voluminous, and hot. We can gain a qualitative "feel" for smoke movement by applying physical principles that affect air movement with observations of building features that affect flow paths. Quantitative analysis using computer models provides a clearer picture of time relationships and environmental conditions. As with flame-heat movement, combining quantitative analyses with qualitative observations and estimates gives a more complete understanding of the dynamic smoke environment.

9.2   The Plan

We evaluate a building for smoke and its effect on people or functionality in two parts. The first part describes natural smoke movement through the building. Movement from the room of origin is based on physical principles that affect natural air flow and assumes no extinguishment intervention or mechanical force actions.

The second part examines tenability for target spaces distant from the diagnostic fire origin. A target space is a room that is important to life safety or building functionality. For example, a target space may be a segment of a continuous egress path; or it may be a room that houses important data or operations that are sensitive to corrosive gases or smoke; or it may be a defend-in-place room that houses people unable to travel. Thus, a target space is any building space away from the fire in which knowledge of the room's tenability is important to risk management.

This chapter discusses natural air movement from the diagnostic fire. Chapter 20 describes tenability and ways to protect spaces from smoke contamination. Chapter 32 shows the framework for performance analysis. Collectively, these chapters examine smoke generation, tenability, and the analytical thought process.

9.3   Smoke

We have all experienced friendly or hostile fires and can envision a rising smoke cloud. Smoke quantity, quality, characteristics, and behavior can vary enormously among fires.

Although both smoldering and flaming fires generate smoke, building evaluations use fires larger than established burning. It is possible for smaller fires to influence fire detection, and this situation is routinely included in a building evaluation. However, a diagnostic fire used for smoke analysis describes the rate and quantity of smoke generation, its characteristics of temperature, visibility, and toxicity, and the forces that cause natural movement.

Smoke obscuration is caused by soot, which is an unburned product of combustion. Some materials produce more soot than others, and inefficient combustion (e.g. poorly ventilated fires) produces more soot than does efficient combustion.

Smoke may be non-irritant where visibility reduction is due to light attenuation of the soot particles. Smoke may also be an irritant that causes the eyes to sting and tear. One may have a long ocular visibility in irritant smoke, but may be unable to keep one's eyes open because of the pain and discomfort. This reduces movement speed and causes anxiety. Sometimes a fire may produce both irritant smoke and low visibility, although the two characteristics do not necessarily accompany one another. Although irritant smoke is a common problem, we do not know how to measure and evaluate its conditions. Therefore, tenability criteria are usually based only on the visibility reduction due to smoke obscuration.

Entrained air in the fire plume becomes the medium in which smoke is transported. The rate of smoke production is nearly the same as the rate of air entrainment in the combustion process. Fuels, fire size, ventilation, and room characteristics are the most important influences on air entrainment. Room characteristics include room volume, ceiling height, and the size and location of the ventilation openings. The generation rate produces a volume of smoke that has characteristics of temperature, visibility, toxicity, and corrosion. Although these attributes can vary greatly in building fires, one can crudely estimate the volume of smoke and its characteristics.

A description of the time-related volume of smoke production is needed for building evaluations. The rate of air entrainment may be calculated for growing compartment fires as:

$$M = 0.188 P_f y^{3/2} \tag{9.1}$$

where:

   M = rate of air entrainment, i.e. smoke production (kg/s);
   P_f = fire perimeter (m);
   y = distance between the floor and the underside of the smoke layer below the ceiling (m).

The mass rate of smoke production can be converted into a volume rate by calculations using the ideal gas law at different temperatures.

One must be cautious when using equation (9.1) because the relationship is valid only in Realm 3 (see Section 6.6). Its reliability diminishes as the fire grows larger.

9.4   Buoyancy Forces

Smoky air that flows out from a room fire is hot. Hot air is a fluid that behaves in accordance with Boyle's law, $PV = RT$. In this expression the pressure, P, is not affected significantly because room sizes and openings in buildings do not vary greatly. The constant, R, is consistent for any specific gas. Therefore, a given air mass in building fires will change volume in proportion to its absolute temperature.

The absolute temperature of fire gases in a room fire is about three times the normal room temperature. Therefore, immediately upon flowing out of a room, the volume of hot smoke can increase as much as three times while keeping the same mass as cold (room temperature) smoky air.

We may envision smoky air as "magic" balloons called "smoke tracking units" (STUs) that have unusual characteristics. The original composition of a STU can change shape, temperature, mix with entrained air, and deposit soot as it moves away from the fire. Consequently, one may envision changes in the properties of a STU as it moves through a building. For example, a 0.025 m³ (0.9 ft³) STU volume at room temperature will expand to about 0.075 m^3 (2.6 ft^3) in a fire, while retaining the same amount of soot and the other constituents of smoke. This decreased density (buoyancy) allows colder units to push the hot STUs upward to the ceiling and then out through the room openings.

Several things happen to the STUs as they leave the room of origin and move through the building. They will lose some heat to the building surfaces and entrain (substitute) additional cool air into the units. These actions cause the smoke to cool, reducing the total volume while retaining the particulates and gases. Also, some soot will deposit on the surfaces along which the smoke flows. At a greater distance from the fire, the smoke may approach the temperature of ambient air. This cold smoke contains particulates that reduce visibility as well as toxic and corrosive fire gases while having the flow characteristics of the ambient air.

The transition from hot smoke to cold smoke is caused by the cooling effect of air entrainment and heat losses to boundary surfaces. Characteristics of hot smoke and cold smoke are different. Hot smoke is buoyant, causing an upward movement unless constrained by building surfaces. It is sometimes possible to recognize a stratified, clear plane of demarcation between the hot smoke layer and the cold air layer in a building space.

The buoyant pressures caused by heat from a fire cause substantial smoke flow. As air moves away from a fire, it cools relatively quickly, causing a loss in buoyancy and greater diffusion. The condition of a uniform smoke mixture from floor to ceiling is described as diffusion. Diffuse, cold smoke moves like the air in which it is mixed.

9.5 Natural Air Movement

All buildings are "holey" and have openings in and around barriers to allow air to move from one space to another. Large openings, such as open doors or grilles, transfer large quantities of smoke and air. Smaller openings, such as cracks and leakage areas between rooms, allow smaller, but not insignificant, movements of air and smoke. The basis for smoke analysis is: "Smoke will flow where air can go."

After smoke leaves a fire area across barrier openings it moves through the building and to the outside. In addition to buoyancy forces, natural air currents such as stack effect and wind influence movement.

Air moves through a building by forces caused by differences in pressures. The slight increase in pressure caused by higher temperatures is enough to cause smoke to migrate from a space of higher temperature into one of lower temperature. Boyle's law for two different spaces may be expressed as:

$$R = \frac{P_1 V_1}{T_1} = \frac{P_2 V_2}{T_2} \tag{9.2}$$

where:

P = fluid pressure (N/m^2);
V = volume (m^3);
T = temperature (K).

Equation (9.2) enables one to calculate the pressure differential across barriers when the temperatures and volumes are known.

The stack effect, sometimes called the "chimney effect" because it describes vertical air movement, also causes natural air movement in buildings. The stack effect is caused by differences in temperature between the inside and the outside of a building.

To illustrate the concept, assume initially that the air outside a building is cold relative to the inside. The cold air has higher density, which creates an outside pressure greater than that inside the building. A building normally has many openings (leakage) throughout its height. This leakage is both internal and through the outside building envelope. Also, doors are often open on lower floors. Cold air moves into the building through these low openings, causing the warmer inside air to be forced upward.

The inside air moves upward through elevator shafts, stairwells, other vertical shafts, and via floor-to-floor leakage. Because all buildings have leakage around openings and penetrations, the inside air is forced upward and out of the building through leakage and other openings that may be present. This upward air circulation is described as the *stack effect*.

Somewhere near the middle of the building is a neutral pressure plane (NPP). Its location depends on the building opening areas and their distribution. At this level, the pressure to move from the outside into the building is equal to the pressure to move from the inside to outside the building. Figure 9.1 shows the general pressure distribution and air movement caused by the stack effect. When outside air temperatures are higher than the air on the inside, the movement is inverted and a reverse stack effect is created.

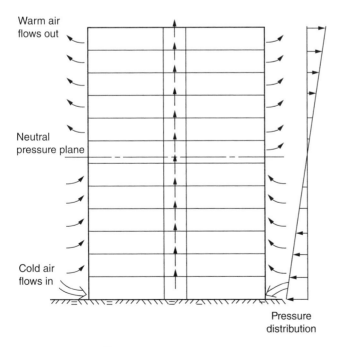

Figure 9.1 Stack effect.

The location of a fire with respect to the NPP influences the smoke movement. For example, assume that a fire occurs on a lower floor of a tall building. Buoyancy caused by high fire gas temperatures will cause upward pressures. As the hot gases cool due to air entrainment and heat loss to the boundaries, the buoyancy is reduced until it reaches ambient temperature. However, the stack effect also influences smoke movement to the upper floors.

The largest pressures occur near the top of the building. Thus, higher floors may be more smoke-logged than floors closer to the fire. Floors near the NPP will have some smoke migration, although the concentration will not be as great as at the top floors.

When a fire occurs above the NPP, some smoke migration exists above the fire due to buoyancy and leakage. This causes some greater smoke contamination at the top floors. Smoke will also leak out of a building when the inside pressures are greater than outside. Figure 9.2 illustrates these conditions.

9.6 Wind

When wind blows on a building, an array of positive and negative (suction) pressures are established. Positive pressures develop on the windward side. Negative pressures will occur on the adjacent and leeward sides. Figure 9.3(a) illustrates these pressures.

Wind pressures on a roof may be positive or negative depending on the slope. Air flow causes a negative pressure on flat roofs, as shown in Figure 9.3(b). Sloping roofs develop a positive pressure on the windward side and a negative pressure on the leeward side, as shown in Figure 9.3(c).

Wind pressures can complicate a smoke analysis. This is important when either automatic venting exists or intentional fire department ventilation occurs. For example, in a flat-roofed building as in Figure 9.3(b), wind flow across an open roof vent will create larger negative forces and help air removal from inside the building. On the other hand, if the roof is sloped, as in Figure 9.3(c), the side on which venting occurs will affect internal smoke movement. If the windward slope is vented, wind will force escaping smoke back into the building. If the leeward slope is vented, the negative pressure helps to remove the inside air faster.

A fire may vent itself by breaking a window. Smoke from a fire on the windward side will be driven toward the building interior. A fire on the leeward side lessens interior smoke conditions. We do not know the direction of the wind until the time of the fire. Therefore, variability analyses involving different scenarios become useful to understand how wind conditions can affect smoke movement.

9.7 Tenability Considerations

Tenability definitions are discussed more completely in Section 21.3. However, for our immediate needs, assume that tenability will be defined as the ability for an occupant to see a specified distance in smoke. To accommodate both hot smoke and cold smoke, a height from the floor level is often specified.

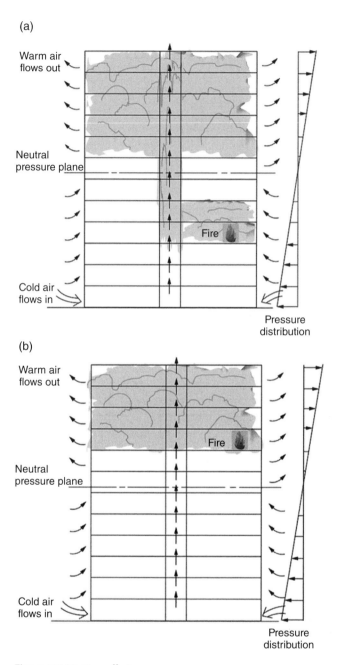

Figure 9.2 Venting effects.

Section 1B of the Interactive Performance Information (IPI) chart describes the smoke component of the diagnostic fire for selected rooms spreading out from the room of origin. Values in the IPI cells are usually expressed in terms of visibility distance above a predetermined height. A more complete discussion is available in Chapter 21.

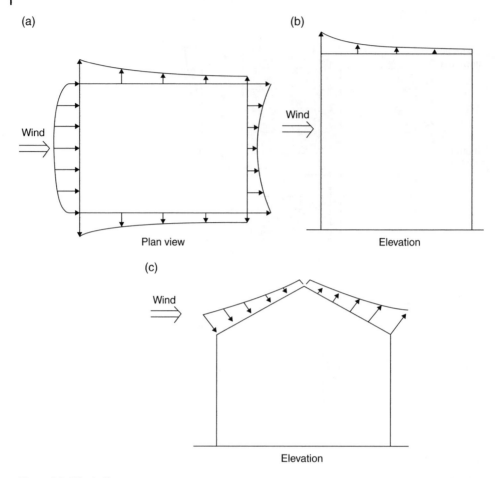

Figure 9.3 Wind effects.

9.8 Smoke Movement Analysis

Interactive Performance Information cell data support the type and needs of an analysis. Data for preliminary screening are less detailed than documentation needed for an important risk management study. Nevertheless, any data that seem relevant to a study can be included in various sheets of the IPI chart.

Section 1B of the IPI chart shows the time-related changes in smoke conditions for the diagnostic fire. The basis for each cell description for the smoke tenability conditions of a diagnostic fire is evaluated as:

- **Given:**
  - *Room of origin _____*
  - *Room of evaluation _____*
  - *Time _____ minutes from EB.*
- **Question: *Will room _____ be tenable from smoke propagation?***

Values may range from a simple "yes or no" to more detailed data such as smoke temperature, ocular density, smoke layer height, or toxicity conditions. Cells in related sheets may record qualitative or quantitative analysis data.

Smoke moves through the building faster and through more spaces than flame-heat. If smoke movement to all rooms were recorded, the IPI chart would become difficult to manage. Combining rooms into zones alleviates this problem by enabling the diagnostic fire to describe time and concentration for important locations. After gaining a comprehensive picture of this smoke movement, one can then examine specific target rooms (IPI Section 4B) in greater detail.

9.9   Smoke Movement Networks

Smoke movement is relatively easy to conceptualize, but difficult to analyze. A building fire is a complicated happening with many forces that influence smoke movement in three dimensions. The fire generates smoke and acts as a pump with driving forces of buoyancy and fire pressures of the heated air. Natural pressures, such as the stack effect and wind, influence smoke movement within a building. Openings during the fire may be caused by the fire itself, ventilation activities, leakage, and the opening and closing of doors. These actions occur during the fire, making the actual smoke movement a dynamic process. Nevertheless, a rapid qualitative evaluation can give a "sense" of the building's smoke behavior.

One value of a qualitative analysis relates to recognizing the influence of barrier openings on smoke movement. Another provides an insight into evaluating outcomes of quantitative computer modeling for important target spaces.

Example 9.1   Identify primary zones for smoke movement in the model building when Room 3013 is the room of origin.

Solution   Although this relatively small building has a large number of spaces, Figures 9.4(a) and (b) designate zones to identify important smoke movement paths.

Here, we segment corridors, lobbies, and stairwells into zones to get an overall perspective on smoke movement. Although the status of doors (open or closed) is important for estimating movement and tenability time, initial zone selection is based on building geometry. Major movement paths and fire defense considerations, such as occupant egress or fire attack routes, influence zone selection. Virtual barriers are useful to segment some zones.

Figure 9.5 shows a network that connects the zones of Figure 9.4. The network provides both visual connections and a way to organize time and tenability estimates.

Example 9.2   Zones provide an initial perspective of general smoke movement. While these zones may be used for quantitative computer results, they are too large for qualitative smoke estimates based on physical principles. Decompose the zones of Example 9.1 into smaller segments for making better estimates or relating tenability conditions to individual target spaces.

(a)

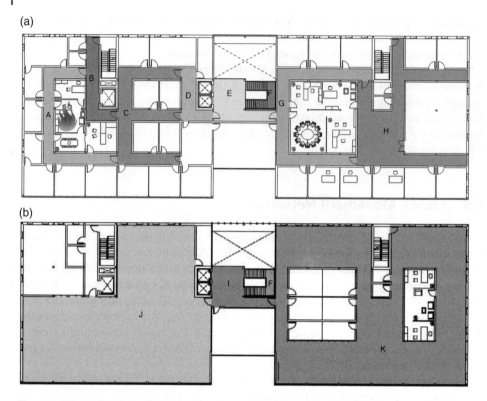

(b)

Figure 9.4 Example 9.1: smoke zoning.

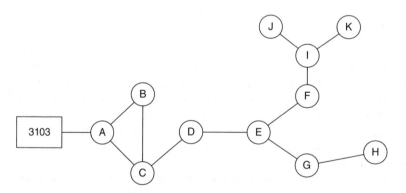

Figure 9.5 Example 9.1: smoke zone network.

Solution Figures 9.6(a) and (b) decompose the zones described in Example 9.1. Note the identification of virtual barriers to segment spaces more clearly. Although most virtual barrier locations may have logical positions, some flexibility exists in placing the virtual barriers.

Each segment is given a consecutive number that relates to the zones as noted in the network of Figure 9.7.

(a)

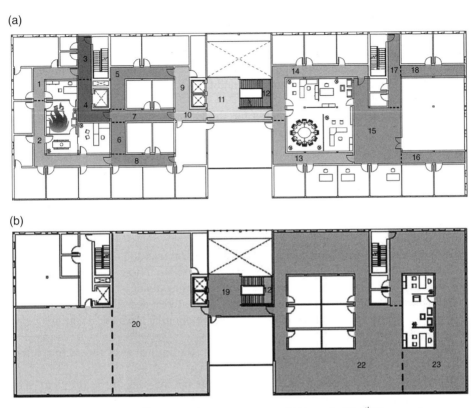

(b)

Figure 9.6 (a) Example 9.2: 3rd Floor smoke zone segments. (b) Example 9.2: 4th Floor smoke zone segments.

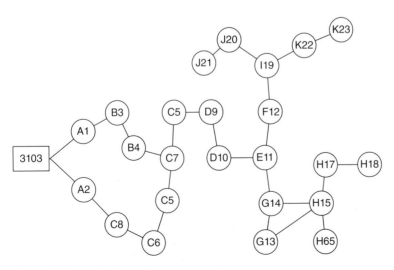

Figure 9.7 Example 9.2: smoke zone segments.

9.10   Qualitative Smoke Movement Analysis

Qualitative smoke movement analysis provides an understanding of the roles of the building's geometry, the diagnostic fire, and the forces that affect smoke movement. This knowledge enables one to understand conditions for better interpretation of quantitative smoke model outputs.

After networks have been constructed for smoke zones and important rooms, one may use these networks to estimate smoke movement. The process involves the following parts:

1) Describe the smoke generation rate from the diagnostic fire. Ventilation areas dominate smoke volume and temperature.
2) Construct networks to identify space-to-space smoke movement. The IPI chart organizes the time-step estimates for smoke to enter and leave each space of the network.
3) Estimate losses for smoke movement through small cracks and openings into spaces that are not included in the network modular analysis. These spaces may include adjacent rooms, vertical shafts (e.g. elevator), and the outside.
4) Estimate smoke movement through the spaces of interest described by the network. Initial smoke movement is estimated by relative opening sizes and locations using incremental time steps.
5) Adjust speed and direction of movement from the influence of forces that affect natural air movement.
6) Select module smoke contamination descriptors for the cells of the IPI chart. The descriptors may be verbal (e.g. low, high, OK, etc.) or numerical (e.g. visibility = 4 m, 6 m, 1 m, etc.).

A qualitative evaluation is fast, inexpensive, and gives a sense of what one may expect for the scenario. It also gives some insight into output evaluation from computer model analyses. Combining quantitative and qualitative assessments gives a better understanding of performance.

The influence of mechanical air-handling equipment is not a part of the natural air movement for a diagnostic fire. These effects are included in the target room analysis (Section 4B of the IPI chart). Chapter 20 discusses air-handling effects for target room analysis.

9.11   Quantitative Analysis

A computer analysis for the natural smoke movement provides a practical way to develop the smoke component of a diagnostic fire. Computer models represent complex combustion phenomena with a set of mathematical equations. Fire and smoke are combined in the analyses.

Algebraic, zone, and field models can represent fire phenomena. Knowledge of the complex interactions can provide better information with which to make engineering estimates. Examples 9.3 and 9.4 illustrate the type of information produced by computational fluid dynamics (CFD) field models.

Example 9.3 A fire occurs in Room 3013 of the model building. Develop a diagnostic smoke scenario for a fire of Example 9.1. Assume that all doors on the third and fourth floors are open. Show the results on an IPI chart.

Solution This quantitative analysis is based on the field model, Fire Dynamic Simulator, version 6. Figure 9.8(a) shows a furniture calorimeter heat release rate for the fire used in the simulation. Figure 9.8(b) shows a smoke generation curve for this fire.

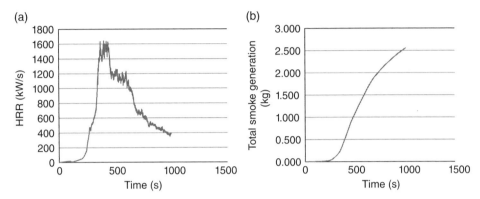

Figure 9.8 Example 9.3: fire-smoke description. (a) Furniture calorimeter heat release rate (HRR) for the fire used in the simulation. (b) Smoke generation curve for this fire.

Figures 9.9(a)–(d) show the results of this simulation for visibility distance of 3 meters and time durations of 2, 4, 8, and 16 minutes. It is possible to track smoke spread for any optical density (visibility) and any height below the ceiling. Additional outcomes can be modeled, depending on the needs of the analysis.

Figure 9.10 shows an IPI chart for the results of this computer simulation for the third floor zones described in Section 9.9. The numbers in the IPI cells indicate visibility levels.

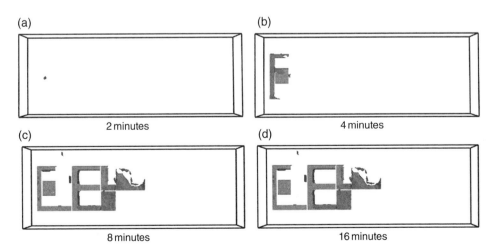

Figure 9.9 Example 9.3: smoke spread.

Building performance	Fire response	Section		Event	Working IPI template — Time (min)			
					1–5	6–10	11–15	16–20
		1	Ĩ	**Diagnostic fire**				
		A	ĨF	Flame-heat				
			1F	Module 1				
			2F	Module 2				
			3F	Module 3				
			4F	Module 4				
		B	ĨS	**Smoke-gas**				
			AS	Zone A	3 1			
			BS	Zone B	3			
			CS	Zone C	3			
			DS	Zone D	3	1		
			ES	Zone E		3		
			FS	Zone F		3		
			GS	Zone G		3		
		2		Active extinguishment				
		3	L	Composite fire				
		4	R	Building response				
		5	L	Risk characteristics				

Figure 9.10 Example 9.3: Interactive Performance Information (IPI) representation.

Example 9.4 Solve Example 9.3 with the door between Spaces D10 and E11 closed.

Solution Figure 9.11 shows the results for the same fire and conditions, except that the door to the third floor lobby is closed.

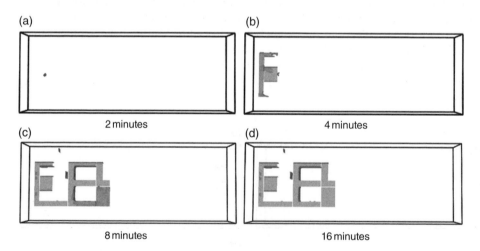

(a) 2 minutes   (b) 4 minutes   (c) 8 minutes   (d) 16 minutes

Figure 9.11 Example 9.4: smoke spread.

9.12 Discussion

The diagnostic fire smoke analysis identifies the smoke tenability of rooms that start at the room of origin and spread in a three-dimensional pattern to other parts of the building. Smoke movement is affected by many interrelated factors such as buoyancy, stack effect, venting, and wind. Although the actual building conditions may not be known for the actual fire, the diagnostic fire scenario will establish reasonable scenarios with which to bracket behavioral conditions and gain an understanding of the important conditions that affect performance.

For any given scenario, the quantitative and qualitative analyses work together to give a better understanding of likely outcomes. The IPI chart provides a way to sort the conditions and important scenarios. Interrelating the approaches enables one to develop diagnostic fires with greater confidence and understanding.

Computer analyses are able to incorporate complex relationships that cannot be done with hand calculations. Nevertheless, even these outcomes have limitations. For example, the fire used for the CFD model described above was based on laboratory furniture calorimeter data. The room enclosure changes fire characteristics (Chapter 6) and smoke conditions. Quantitative modeling results can be improved with qualitative adjustments to reflect these conditions.

A variety of other variations can exist. For example, doors may be open or closed; leakage into adjacent rooms may be ignored; leakage to the outside may be ignored; building conditions such as stack effect may not be included; and wind due to intentional or accidental ventilation may not be incorporated. Variations such as modeling simplifications, default selection, or common variations such as those above can change a building's performance.

Computer analyses are valuable to provide a sense of proportion for selected scenarios. However, professional practice must compare the value of information with the cost of its acquisition. Combining quantitative information with qualitative judgments enables one to improve the efficiency and products of each.

The integration of qualitative judgment with quantitative analysis enables potential "what if" variability analyses to be sorted with regard to feasibility and importance. An initial set of conditions can be examined for possible variations, such as the effect of opening or closing doors and venting, to rank conditions for building and risk analysis scenarios.

Section 1B of the IPI chart describes the diagnostic smoke conditions that are used for the fire scenario for risk characterizations and fire department suppression conditions. Closely associated with the fire spread is the evaluation of target room conditions for IPI Section 4B. The target room analysis included the influence of building design features such as HVAC actions and barrier effectiveness.

10

The Diagnostic Fire

10.1 Diagnostic Fires

A diagnostic fire is the natural development of a fire in a room of origin and its propagation through sequential rooms when no extinguishment intervention occurs.

A diagnostic fire describes time-related changes in the flame-heat component and the smoke-gas component from ignition to first water application by an automatic sprinkler system or local fire department manual attack lines. After first water application, the original diagnostic fire combines with extinguishing actions to form the composite fire. Chapter 17 describes the logic of the interactions.

Time links the diagnostic fire with all fire defenses and risk characterizations.

A diagnostic fire is the "load" against which the fire defense "resistances" are compared. The diagnostic fire may describe conditions for a specific model room interior design, or it may be standardized to represent appropriate fire conditions. When the profession standardizes fire conditions for performance-based analysis and design, those conditions will represent a consistent, rational fire threat for a room. Standardized fire conditions will eventually have the same function as the diagnostic fire described herein evaluating fire defenses and characterizing associated risks.

10.2 Interactive Performance Information (IPI) Chart and the Diagnostic Fire

The cluster of rooms that are used for a diagnostic fire scenario includes a room of origin and any additional rooms through which the flame-heat component can propagate. These same rooms as well as additional rooms through which smoke can propagate are a basis for the smoke-gas component.

The diagnostic fire provides the environment against which fire defenses and risk may be evaluated. The IPI chart provides a structure to describe all parts of a performance-based analysis. Section 1A describes the flame-heat conditions for the room of origin and the cluster of additional rooms.

Top-down evaluations (Chapter 8) are valuable to initially sort out the complex array of barrier–space modules. They provide a rapid means to identify important fire growth hazards that are caused by holes in barrier room separations. This performance-based

Fire Performance Analysis for Buildings, Second Edition. Robert W. Fitzgerald and Brian J. Meacham.
© 2017 John Wiley & Sons Ltd. Published 2017 by John Wiley & Sons Ltd.

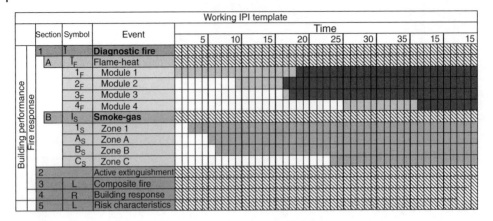

Figure 10.1 Interactive Performance Information (IPI) diagnostic fire representation.

qualitative analysis provides insight into the type and severity of risk characterizations. Integrating these different needs enables the detailed quantitative evaluations to become more effective.

The IPI chart lists each room in the cluster of rooms that form the diagnostic fire. This enables one to show the type of barrier failure to allow the fire to spread to the adjacent room as well as identifying time to flashover for each successive room. Because each room in the cluster is listed, one can use the information to incorporate active suppression. The time-based composite fire conditions enables one to have greater confidence in active extinguishment evaluations.

Section 1B describes the smoke-gas spread from the room of origin. Because smoke moves much faster and through more rooms than the flame-heat, groups of rooms may be combined into zones. A zone may be decomposed into its individual rooms for greater detail, if needed.

Figure 10.1 illustrates an IPI chart that depicts flame-heat and smoke-gas conditions for a diagnostic fire. Spreadsheets allow one to use additional layers to describe supplementary numerical values. These sheets also provide other data and visual records to document useful information.

10.3   Closure

The diagnostic fire that incorporates the room of origin and additional rooms through which the fire can spread is essential for performance-based analysis and design. Because the diagnostic fire includes both flame-heat and smoke-gas components it becomes the focus of fire defense analysis. Although this book focuses on performance analysis, knowledge of fire expectations provides the insight to develop effective risk management programs.

11

Fire Detection

11.1   Introduction

Detection is the event that launches the building's fire response. It is a critical event in alerting occupants for emergency life safety action and initiating manual suppression. Detection can also play a role in other fire defenses by causing actions such as closing doors, recalling elevators, changing operations of air-handling systems, or triggering equipment operation.

There have been huge advances in recent years in the ability to relate detector actuation with fire conditions. Today it is usually possible to tailor a detection system design to its intended function. Thus, one can devise a detection system to actuate for a pre-selected fire size.

This book looks at the other side of the process – analysis. A performance analysis predicts the diagnostic fire size at which a given detection system will actuate. The detection system may be standards complying or not; modern or outmoded; complete or incomplete. Detection may involve detection instruments only, humans only, or both. Equipment evaluations are relatively methodical and orderly, but human interactions involve greater variability and uncertainty.

This chapter discusses detectors, their operating principles, and the transfer of products of combustion (POC) from a fire to detector locations. Chapter 12 discusses the alarm functions after detection. This enables one to separate detection analysis from the actions that are set in motion by the detection event.

Fire Performance Analysis for Buildings, Second Edition. Robert W. Fitzgerald and Brian J. Meacham.
© 2017 John Wiley & Sons Ltd. Published 2017 by John Wiley & Sons Ltd.

PART ONE: AUTOMATIC DETECTION

11.2 Instrument Detection

Fire detectors can provide spot protection, line detection, or volume and space detection. Spot detection occurs when a detector is located in a fixed position and is able to sense selective POCs. Heat detectors and smoke detectors are two common spot detection instruments.

Line-type detection occurs when the detection instrument has a one-dimensional linear sight. For example, in a photo beam smoke detector, a transmitter sends a light beam to a receiver 100 m or more away. An alarm signal is initiated when smoke enters the path of the light beam and obscures enough of the light to correlate with the sensitivity setting of the detector.

Another type of continuous line detection uses open-circuit wires that are held apart by heat-sensitive insulation. When the insulation melts, the wires make contact and initiate an alarm. Line-type heat detection provides continuous coverage along the length of the cable rather than an area protected by a spot-type heat detector. Line-type detection is generally less sensitive (e.g. operates at a higher temperature) than spot-type detection.

Volume or space detection involves active air sampling that uses fixed perforated tubes in the space. Air is continuously drawn through tubes to a detection chamber. When smoke-contaminated air reaches the detection chamber, the small smoke particles are sensed, producing an alarm.

Because aspirating or active air-sampling smoke detectors are more sensitive than spot-type smoke detectors, they are commonly used in clean room environments or areas that house very expensive equipment. Passive air-sampling smoke detectors use standard duct-type smoke detectors that rely on air entering a sampling tube to reach the detector.

11.3 Detection Instruments

The type of detector, its sensitivity, and its response to the POCs in the environment are all parts of a performance analysis.

Detectors are classified according to the combustion products that they are designed to sense. They may be described as:

1) Heat detectors
   a) Fixed temperature
   b) Rate of rise
   c) Rate compensation
   d) Continuous line
   e) Thermo-electronic spot
2) Smoke detectors
   a) Ionization
   b) Photoelectric
   c) Photo beam (light obscuration)
   d) Air sampling
3) Flame detectors

Each of these detectors responds to the presence of a specific fire signature. A *fire signature* is a specific, measurable POC that causes the instrument to be aware of the fire. Detectors will respond to the fire signature reaching a predetermined threshold level of magnitude or rate of change. The fire signature for the specific type of detector being evaluated becomes a part of the diagnostic fire description.

11.3.1 Heat Detectors

Several heat detector types are available. A fixed-temperature heat detector actuates when the sensing element reaches a specified thermal value. On the other hand, a rate-of-rise heat detector responds to a rate of temperature change that exceeds a calibrated value. It is also possible to obtain rate-compensation or rate-anticipation detectors that respond to the temperature of the surrounding air reaching a predetermined level. Combination detectors can respond to both a fixed temperature threshold and a rate of temperature rise.

Generally, a fixed-temperature heat detector is the least costly and most reliable. However, it is also the least sensitive for detecting a fire. Consequently, fixed-temperature heat detectors are placed in locations where heat will not dissipate before reaching the detector, thus delaying actuation. They are also placed in locations where a larger fire at detection is acceptable. Rate-of-rise detectors actuate when the rate of temperature increase exceeds a predetermined level. They usually detect a smaller fire because they are more sensitive to rapid changes in temperature.

11.3.2 Smoke Detectors

Smoke detectors sense airborne POCs carried into the sensing chamber by air movement. Smoke detectors may be identified by two types of operation: ionization and photoelectric. An ionization smoke detector has an effective electrical conductance between two charged electrodes in the sensing chamber. The air between these electrodes has been made conductive by ionizing the chamber with a small amount of radioactive material. When smoke particles enter the sensing chamber, the normal current flow due to the ion migration is reduced. The electronics of the detector sense this decrease in current flow. Ionization detectors are particularly sensitive to invisible airborne POCs, such as those generated by a flaming fire.

Spot-type photoelectric smoke detectors use a light beam that propagates through the air within the detector housing. When smoke particles enter the air between the light source and the receiver, they reflect or scatter the light beam. In either case the amount of light reaching the photosensitive receiver changes, and the electronics of the conductor start the detection response. A photoelectric detector is more sensitive to visible smoke, such as would be generated by a smoldering or a smoky fire.

Generally, ionization detectors are more sensitive to flaming fires near the detector. Ion detectors remote from the fire source do not provide as rapid detection as photoelectric detectors. Photoelectric smoke detectors are more appropriate when visible particles are common. They are more stable in locations near kitchens or furnace rooms.

The reliability of ionization and photoelectric smoke detectors has improved in recent years. The unwanted (nuisance) alarm rate is still high in homes using ionization spot detectors, especially if located in or near kitchens or bathrooms where steam can be expected, but is declining.

It should be noted that multi-criteria devices are also available. These may include combinations of smoke detection, heat detection and CO detection. The objective of the multi-criteria device is that it correlates fire signatures to increase reliability (e.g. a fire can produce smoke, CO and heat, whereas CO coming in from vehicle exhaust lacks the smoke and heat components).

A photo beam (linear beam) smoke detector is commonly used in line detection for large open spaces. A photo beam detector transmits a light source at one end and has a photosensitive receiver at the other end. The distance between the transmitter and the receiver may vary from several meters to a few hundred meters, depending on the type of coverage. When smoke obscures part of the light beam, the receiving device senses less light and initiates the detection response.

An active air-sampling smoke detector uses a piping network of sampling tubes to draw air from the protected space into a detection chamber. Three types of detection chamber are commonly used. One uses a photoelectric light-scattering system that irradiates the sampled air. Another uses a laser source with a particle-counting procedure. Either of these devices can sense smoke particles at very low concentrations. The third system is a cloud chamber type in which air samples are introduced into a high-humidity chamber within the detection device. The humidity of the air is increased and the pressure is lowered slightly. Moisture in the air condenses on any smoke particles, forming a cloud that is sensed by the photo optics in the detection device.

11.3.3 Flame Detectors

Flame detectors provide extremely fast fire detection by "seeing" the radiant energy produced by combustion. Flame detectors may operate in the ultraviolet (UV) or the infrared (IR) wavelength range. The receiver of a flame detector receives radiation and then filters out all wavelengths other than those for which the detector is designed. Detector selection matches the type of hazard with the expected fire type.

Flame detectors that detect radiant energy in a line of sight must be able to see the fire. Consequently, the space between the potential fire source and the detector must be free of obstructions. Flame detectors that are used in high-hazard industrial areas where response must be very rapid require care in design and installation.

11.3.4 Operating Modes

All detectors have either a restorable or a non-restorable operating mode. A restorable detector is one where the detector returns to normal operation when the fire signature is removed. A non-restorable detector is destroyed by the combustion products and must be replaced.

For example, a fusible element heat detector operates by melting the eutectic metal. The detector cannot be restored to its original operating condition after operation. On the other hand, a bimetallic snap-disk detector operates by the differential expansion of two different metals on heating. The difference in the coefficient of expansion of the two metals causes a bending that activates the detector. When the heat is removed, the sensing element returns to its original condition without additional restoration efforts. Smoke detectors can be restored to normal operating condition by clearing the detector of smoke and resetting it from the fire alarm control panel.

11.4   Automatic Detection Analysis

Detection instrument actuation defines the event of fire detection. The fire size at detection is the performance measure. Although the detection event may initiate several fire defenses, detection rather than subsequent actions is the focus of this chapter.

Although building performance analysis starts at established burning (EB), detection can occur before or after EB. The global timeline relates the detection event to EB.

Detector performance evaluations use the following information to quantify the fire size at detection.

- Detection system
  - Detector type
  - Detector sensitivity
  - Detector locations
- Diagnostic fire
  - Room of origin
  - Fire signature (smoke, heat, radiation)
  - Fire growth characteristics
- Building geometry
  - Room size and shape
  - Ceiling height, slope, and smoothness (e.g. irregularities caused by structural members)
  - Path from fire to detector
  - Intermediate barriers to smoke flow (e.g. doors open or closed, transom height, hanging partitions)
  - Natural air movement forces (e.g. windows open or closed, air currents)
  - Mechanical air system forces (e.g. supply or exhaust sources).

An evaluation examines the path from fire to detector using information about the detection system, the diagnostic fire, and building geometry. Then, two sequential time durations are quantified. The first examines the POC movement path from the diagnostic fire to the detector. The second examines detector sensitivity.

The following questions provide a basis to evaluate fire size at detection and show it in Section 2a (MD) of the IPI chart:

- **Given:**
  - *Ignition (IG)*
  - *Time _____ minutes from IG*
  - *Fire size _____*
- **Question:** *Will enough POCs reach the boundaries of the detector to cause actuation with the most sensitive setting (DB)?*

- **Given:**
  - *Ignition (IG)*
  - *Time _____ minutes from IG*
  - *Fire size _____*
- **Question:** *Will the detector actuate (DO)?*

These questions examine both the quantity of POCs that reach the detector and the sensitivity of the detection instrument. Detector sensitivity provides a base from which to evaluate actuation. For example, not all 135° heat detectors actuate at the same fire size. One must be aware that rated detectors can actuate at different fire sizes because of differences in response time index or other mechanical-electrical construction features.

The term "enough POCs" is intended to use as a base sensitivity the lowest concentration of POCs that would cause actuation for the rated (or specific sensitivity) detector. The second question adjusts that value to account for sensitivity differences between generic detectors. This event also uncovers when a mismatch may exist between detector selection and the products of the fire signature.

The two questions require one to be conscious of building-induced time delays. For example, a computer model may identify a fire size at detection. However, detectors may be located in dead air pockets or in locations where windows or heating, ventilation, and air conditioning (HVAC) operations produce air currents that deflect POC movement. These conditions delay actuation and require larger fire sizes for detection.

Detector analysis is particularly suited to a combination of quantitative and qualitative analyses. Each has a different role and gives a different insight into detection performance. Collectively, they can provide a clear picture of the likely fire size at detection.

11.5 Instrument Reliability

A detection analysis of Section 11.4 assumes complete reliability of the detection system. However, one must also examine the system's reliability.

Detection system reliability involves three distinct components:

- *Manufacturing control* – this describes the instrument reliability to actuate when it is removed from the shipping carton. A manufacturer may describe this as a quality percentage for the shipment.
- *Construction control* – this component examines the quality of the system's installation. For example, are all electrical leads connected correctly? Is there a malfunction of circuitry or signals? Will the system operate?
- *Maintenance control* – the long-term maintenance becomes an ongoing part of system reliability.

Instrument reliability and construction quality together define the initial mission readiness. It is possible to have a completely reliable system at installation if *all* detectors are tested and respond at the time of installation. When only a sampling of detectors are tested, the evaluation reflects the confidence that the system has been installed properly and will work.

When the building management has a long-term, effective maintenance plan in place, the reliability of the detectors is likely to remain high. On the other hand, if system maintenance is not a priority, the reliability will deteriorate over time and its operational level will decline.

Detection system reliability can be described by the following three independent assessments:

- **Given:** *Detection system is present.*
- **Question:** *Is the detection instrument reliable (rdi)?*

- **Given:** *Detection system is present.*
- **Question:** *Is the detection instrument installed properly (idi)?*

- **Given:** *Detection system is present.*
- **Question:** *Can the detection system actuate when enough POCs enter the chamber (adi)?*

PART TWO: HUMAN DETECTION

11.6 Concepts in Human Fire Detection

Humans can detect a fire by sensing the POCs associated with feeling heat, smelling smoke, seeing light, or hearing noise. Often a human will sense an abnormal condition, such as an unusual odor or a trace of smoke. However, the cue may not be strong enough to cause the person to believe that a hostile fire actually exists. An individual may often decide to investigate the cause of the abnormal condition. However, this investigation does not define the event of detection.

We define human fire detection as that instant when an individual clearly recognizes or believes that a hostile fire exists. Subsequent actions by the individual are not a part of the detection process.

One may recognize that evaluations of human detection rarely involve a single representation for a building. Variability analyses enable one to examine a variety of feasible alternative scenarios of the fire, human condition and activities, and building design. Analyses for occupied conditions and for the unoccupied status may be augmented by differences in occupant location or condition (e.g. asleep or awake). Enough conditions may be examined to provide a good picture of ranges in performance of human fire detection.

It is common for humans and automatic detection systems to co-exist in the same building. A performance analysis treats each as a separate element. Either or both can detect a fire. Certainly, both together are better than either alone. However, each is examined as an independent part, assuming that the other is inoperative. Unit 4 describes a procedure to integrate the two modes of detection.

11.7 Human Detection Analysis

A qualitative analysis is the only process available to evaluate human detection. Variability analyses can organize the sensitivity studies to examine appropriate "what if" situations.

An analysis for human detection is similar to instrument analysis except that humans have moveable locations and variable conditional states. And, prediction of the human condition is often unreliable. Nevertheless, the effort to obtain the necessary information often produces great value in understanding building performance and risk characterizations.

The following questions are the basis for fire size and associated time at human detection:

- **Given:**
  - *Ignition (IG)*
  - *Time _____ minutes from IG*
- **Question:** *Will enough POCs to cause detection reach the human (HB)?*

- **Given:**
  - *Ignition (IG)*
  - *Time _____ minutes from IG*
  - *Fire size _____*
- **Question:** *Will the human decide that a fire exists (HD)?*

The process is based on an identified fire scenario and known building features. This information may be summarized as:

- Diagnostic fire
  - Room of origin
  - Fire signature (smoke, heat, radiation)
  - Fire growth characteristics
- Building geometry
  - Room size and shape
  - Ceiling height, slope, and smoothness (e.g. irregularities caused by structural members)
  - Path from fire to human
  - Intermediate barriers to smoke flow (e.g. doors open or closed, transom height, hanging partitions)
  - Natural air movement forces (e.g. windows open or closed, air currents)
  - Mechanical air system forces (e.g. supply or exhaust sources.
- Human detection.

The final component of human detection is the individual. This evaluation does not consider a random chance that an individual may be present near the room of origin. Rather, it defines a specific scenario for the human as the detector. For example, the human may be one of the following:

- A receptionist in Room 827 whose condition is awake, alert, and busy with work assignments.
- An occupant in Office 421 who is inattentive because of work immersion.
- A security guard making rounds in an unoccupied building. The guard is awake, alert, and passes through each room on regular 2-hour intervals.
- A nurse on Ward 7E of the Tender Care Hospital.

In addition to the occupant's location, human attributes become part of the analysis. The major attributes are:

- Human condition
  - Physical (i.e. mobile or having restricted movement)
  - Mental (i.e. capable of making rational decisions)
- Human status
  - Awake
  - Asleep
  - Distracted (e.g. working, reading)
  - Impaired (e.g. inebriated, drugged)
  - Fire experience and training.

In many ways, human detection analysis resembles "telling a story" in which the human location and characteristics are clearly defined. Thus, human detection is based on specific expectations rather than chance occurrences. Variability scenarios are clearly integrated with qualitative performance analyses. The package of outcomes can identify conditions that have relevance to building performance over the 24-hour and 365-day continuum of time. Collectively, they provide a good insight into the strengths and weaknesses of human detection for the building, as well as an understanding of the conditions in which the human detection may influence building performance for better or for worse.

11.8 Closure

Fire detection is a key link in the way in which a building performs in a fire. Detection is the event that starts the entire process. Understanding the reliability and ability to predict the fire size and time of detection provides a basis for the confident depiction of a building's performance and risks.

12

Alarm: Actions After Detection

12.1   Introduction

Detection is the event that launches the building's fire responses. After the fire is detected, different individuals, groups, or devices start actions to escape the fire, to put out the fire, or to change the building to alter fire conditions.

Alarm is the common term that describes that part of the communication chain that informs of the existence of a fire and initiates certain subsequent actions. Because each of these actions plays a distinct role in holistic performance, we prefer to avoid the word alarm. In its place we separate the function into discrete components:

- *Alert* occupants.
- *Notify* the local fire department.
- *Actuate* selective building operations.

Although the starting point for these functions is fire detection, each function plays a different role in holistic performance, and each is part of a separate component evaluation. Figure 12.1 identifies functions following detection.

Time is the element that links all parts of a performance analysis, and the time to complete these functions can vary greatly. For example, the time to alert occupants can be almost instantaneous with detection for some installations, yet delayed for others. Notification of the local fire department can be done quickly or involve a long time duration, depending on the building design and operations.

A performance evaluation estimates the time duration between detection and the subsequent events shown above.

Uncertainty exists when human response is required to complete an event. On the other hand, uncertainty is minimal with automated system operations. A variation analysis enables one to understand the range in time durations (and associated performance) for human uncertainty in a building evaluation.

This chapter describes the functional and operational characteristics of alerting occupants, notifying the fire department, and actuating building defenses. The structure for analysis is covered at different locations of *Unit Three: The Analytical Framework*. Chapter 33 discusses "alert occupants", while "notify the fire department" is described in Chapter 27.

Fire Performance Analysis for Buildings, Second Edition. Robert W. Fitzgerald and Brian J. Meacham.
© 2017 John Wiley & Sons Ltd. Published 2017 by John Wiley & Sons Ltd.

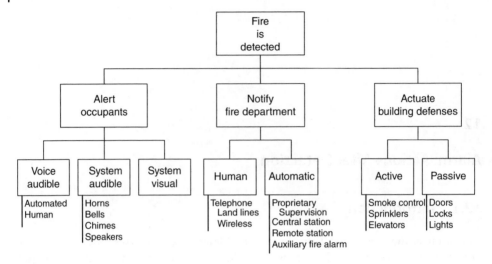

Figure 12.1 Organization chart: alarm actions.

PART ONE: ALERT OCCUPANTS

12.2   Focus on Alert

The *alert occupants* (OA) event informs occupants that a fire has been detected. Occupant actions after receiving a cue that a fire exists can vary greatly. For example, after receiving a cue from a fire, occupants may:

- Investigate to verify if a fire exists or its condition
- Hide
- Attempt to rescue others
- Leave the building.

Other common occupant actions may include:

- Initiate manual fire suppression during the early stages of fire growth.
- Notify the local fire department.
- Alert other occupants to the existence of the fire.

Actions taken by humans can greatly influence building performance. Sometimes the human factor is timely and valuable. In other cases, human actions or inactions are major contributors to fire losses. Any evaluation involving human decision-making under stress requires value judgments for the expectations involving time durations and outcome success.

A disciplined performance analysis must resist the temptation to base decisions on what an individual "should do." This possibility would be included in a variability analysis that methodically examines reasonable alternative actions. Human actions and decisions depend on perceptions of the situation and other factors such as personal experiences with fire, perception of threat, education, training, and the actions of others.

Human responses must be evaluated when they are an identifiable link in a fire performance component. Although human behavior under stress makes evaluation difficult, and the window of uncertainty is broad, the IPI chart can portray the time variations and their effect on component performance. The understanding that evolves provides a more complete insight into fire performance for a building.

12.3   Alerting Occupants

The time duration from initial detection to occupant alert affects risk characterizations and may influence other fire defenses. Thus, a clear understanding of function and operations of the alarm system is a necessary part of a performance analysis.

Occupants are commonly aware of a fire by an audible alarm sound that often occurs at the same time the detector actuates. Sometimes an intentional delay in alerting occupants may be incorporated into the system. Sometimes the alarm signal may use flashing strobe lights as a stand-alone signal or in conjunction with an audible signal. Sometimes another occupant or an outside passer-by alerts other occupants of a fire. Each mode of occupant alert is evaluated for time duration and reliability.

12.3.1 Audible Signals

Audible signaling devices alert occupants of a fire by transmitting a distinctive sound that has little chance of misinterpretation. These devices may be incorporated within the detector (e.g. home fire detector) or they may be separate appliances that transmit an alarm signal from another location. Audible sounds can involve horns, bells, chimes, or other distinctive signals.

The objective of an audible signal is to enable an individual located at a distance from the transmitter to hear the sound and be alerted that a fire has been detected. An evaluation considers the initial loudness of the transmitted sound, its attenuation along the building's path from transmitter to occupant, the sound intensity that reaches the individual, and the intelligibility (if a voice signal). The following question guides an evaluation of the ability of a normal hearing individual to receive an audible signal:

- **Given:**
  - *Fire is detected (Det)*
  - *Time from EB* _____
  - *Audible signaling device location* _____
  - *Target occupant location* _____
- **Question: *Will the occupant hear and recognize the audible signal?***

A critical performance measure is the sound pressure level (SPL) that an occupant hears at the target location. This is a function of decibels (dB) weighted to the frequency of hearing perceptions (dBA). dBA is the unit of measurement in acoustics.

The SPL intensity diminishes as the signal moves through the building. Acoustic surface treatments along the path, as well as the completeness and number of intermediate barriers all affect attenuation. On the other hand, multiple alarm transmitters can overcome this effect and increase the sound level for individuals remote from the fire.

Intelligibility is a function of the audio speaker, volume of a space, and interior finish. It is a measure of how well one can understand the message being transmitted. For example, the signal from a loudspeaker in a subway system bounces off the hard concrete walls and may be difficult to understand. Intelligibility is particularly important in large spaces (e.g. convention centers, high-rise buildings, transit centers) where voice communication is typically used.

12.3.2 Visual Signals

Visual signaling is used to augment audible alarms or to act alone for special conditions. Strobe lights are most common, although other types of alarm lighting may be used.

Visual alerting is most commonly associated with hearing-impaired occupants, as well as industrial and other noisy environments. The main considerations for an analysis include:

- Location of the light transmitters
- Location of the target occupant
- Visual path from transmitter to target occupant
- Light intensity and signaling
- Occupant condition (e.g. awake, asleep, impaired).

The following question guides an evaluation of an individual's ability to receive an audio signal:

- **Given:**
  - *Fire is detected (Det)*
  - *Time from EB* _____
  - *Visible signaling device location* _____
  - *Target occupant location* _____
- **Question:** *Will the occupant recognize the visual signal?*

12.3.3 Human Alerting

A human who has become aware of a fire may alert other occupants in the building. Unfortunately, the lack or inoperability of fire detectors makes this action necessary for some buildings. The person alerting others may be another occupant or an individual who by chance passes the building and observes a fire. Although human alerting of other occupants can be important, we view the action as a chance event that is not a part of a performance analysis.

12.3.4 Nuisance Alarms

Nuisance alarms (often called false alarms) can occur in buildings. When several unwanted alarms occur within the first month of operation, occupants often decide to ignore the signal. This causes delayed response by occupants and may allow dynamic fire and smoke conditions to make escape routes untenable.

Nuisance alarms may be caused by inappropriate detector sensitivity for the building environment. Inadequate maintenance can also be a problem because it commonly prevents or delays actuation rather than causing premature or unnecessary operation.

12.3.5 Operating Modes

Initial actions after detection may be classified as public or private. A public mode alerts all occupants of the detection of a fire. The alerting usually occurs simultaneously with detection or shortly thereafter.

Although a public operating mode normally implies a rapid, complete alerting of building occupants, some modifications may be incorporated into alerting systems. For example, hotels, restaurants, or retail stores may use intentional delays to verify that the alarm is real rather than false.

Delayed alerting practices intend to avoid disturbing occupants by small incidents or unwanted alarms. Techniques include alarm verification (system waits for a second alarm) or positive alarm sequencing, where time is built in for an operator to seek and confirm a fire. Alarm delay can be up to 180 seconds according to some codes. Delayed alarms can allow a fire to grow beyond employee extinguishment capabilities. The delayed alerting combined with exponential fire growth places occupants in greater jeopardy. And the local fire department will encounter a larger fire at first water application.

A private operating mode is common in institutional occupancies such as hospitals or prisons where occupants do not have free movement. Employees are often assigned emergency duties, including notification of the local fire department. Therefore, a private operating mode informs selected staff members of the emergency rather than providing a general occupant alert.

Some occupancies can have a combination of systems. For example, residential, including hotels, may install a local smoke alarm in a room that is not connected to the building's fire alarm system. However, fire detection in common spaces, such as corridors, or sprinkler and manual activations will sound an alarm throughout the building.

12.4 Summary

The alarm function informs occupants of fire detection. However, one should not assume that alarm after fire detection is either instantaneous or certain. A performance analysis examines the process from detection to occupant alert. The method of alerting must be appropriate for the individuals being informed. Also, intentional delays in public operating modes must be examined to identify the fire size when the alert event is completed.

Careful investigation of the complete fire–fire suppression process will provide understanding of the importance of any delays. In a similar manner, careful scrutiny of a private operating mode provides an understanding of holistic performance. Equipment and system reliability is an integral part of the evaluation process.

PART TWO: NOTIFY LOCAL FIRE DEPARTMENT

12.5 Introduction

The fire department notification process starts at fire detection and is completed the instant the local fire service communication center understands the fire's location. The time duration between detection (MD) and notification (MN) is the performance measure.

Notification may be done by human or automated transmission. Time durations, reliability, and uncertainty can vary substantially when the analysis involves humans. Automated methods are faster, more reliable, and relatively straightforward to evaluate. Nevertheless, because many buildings need a human link between the fire and the local fire department, its process becomes an integral part of a performance analysis.

This chapter describes the function and process of fire department notification. Chapter 27 provides an analytical framework. *Unit Four: Managing Uncertainty* discusses ways to handle uncertainties and communicate expectations to others.

12.6 Human Notification (MN)

An individual often must notify the local fire department of a fire's location. Unfortunately, decision-making by individuals in a stressful situation makes time estimates for this assessment somewhat uncertain. Nevertheless, the exercise of estimating time durations between MD and MN provides an insight into building performance. An added benefit is that the analysis often gives clues to potential risk analyses involving life safety.

The process involves the diagnostic fire, the building, and an individual. Given the fire size at detection, the time from MD to MN examines three sequential events:

- How much time will elapse between detection (MD) and an individual deciding to call the fire service communication center (dmn)?
- Given (dmn), what is the added time to actually send the message (smn)?
- Given (smn) what is the time needed for the communication center to understand the location of the fire (rmn)?

Credible scenario analyses are based on understanding how the building works. Most building operations function in a way that a few reasonable scenarios may be identified. This enables one to gain a relatively clear picture of the pessimistic, optimistic, most reasonable, and special situation scenarios. Normally, one can adapt a base scenario with a few carefully chosen "what if" changes to understand the effect of variations.

The estimate of time between MD and MN is based on the following questions:

- **Given:**
  - *Fire is detected (MD)*
  - *Time from EB* _____
  - *Fire size* _____

- – *Location of individual* _____
- – *Location of fire* _____
- **Question:** *Will the individual decide to call the fire department (dmn)?*

- **Given:**
  - – *The individual decides to call the fire department (dmn)*
  - – *Time from EB* _____
  - – *Fire size* _____
  - – *Location of individual* _____
  - – *Location of fire* _____
- **Question:** *Will the individual send the message (smn)?*

- **Given:**
  The individual sends the message (smn)
  Time from EB _____
  Fire size _____
  Location of individual _____
  Location of fire _____
- **Question:** *Will the fire department receive the message accurately (rmn)?*

Because the events are single value and sequential, one may use the time steps of the IPI chart to establish a duration between MD and MN.

12.6.1 Decide to Notify the Fire Department (dmn)

Fire detection occurs when (a) an instrument actuates or (b) an individual senses enough products of combustion to believe that a fire exists. Initially, we will first examine the scenario that an instrument detects a fire and an occupant is alerted by the alarm. The occupant then notifies the fire department.

A detection analysis will have provided the following information:

- The building size, function, and architectural layout.
- The diagnostic fire.
- The fire size at detection by a detection instrument.
- The fire size when a human becomes aware of the fire.

An MN analysis augments that information with the following knowledge:

- Occupant characteristics of the people who live and work in the building:
  - – Status: awake; asleep; work; restrained; specific activity (if important).
  - – Condition: mental; physical; social impairment (e.g. alcohol, drugs).
- Physical impairment (e.g. sight, hearing):
  - – Mobility.
  - – Activity at the time of detector alarm.
- Familiarity with the building and the general way the building works.
- Occupant activities, locations, and movement expectations on a 24/7 basis.
- Any specialized fire training or instructions that occupants may have been given.
- Special factors related to the evaluation about ongoing operations.

Appropriate scenarios are identified to estimate time durations between fire detection and a decision to notify the fire department.

12.6.2 Send the Message (smn)

The terminal events for sending the message are the decision point (dmn) and making the connection with the fire service communication center (smn). The time duration for this action is affected by:

- Ability to locate a telephone
  - Landline telephone
  - Smartphone
- Condition of the caller
  - Mental
  - Physical
  - Emotional
- Fire environment
  - Smoke
  - Heat
- Perceptions of danger
- Ability to send message
- Correct telephone number.

Many smartphones now contain GPS information that can help to identify the location of the device (therefore, the fire). Also, occupants may take photos on their smartphone to send or upload. This can be helpful in documenting location, size of fire along a time line, and similar data.

The analysis examines the process from the individual's location when the initial decision to call the fire department is made to the location for sending the message. The elapsed time duration may involve many associated factors of the scenario. Variability "What if" analyses are usually easy to recognize and time duration alternatives can be rapidly estimated.

12.6.3 Message is Correctly Received (rmn)

The final action in human link notification is the correct receipt of the message (rmn). Although this may seem trivial, the situation is not always as "automatic" as it may appear. Communication between the caller and the dispatcher is not always smooth.

Communication centers often handle a wide range of calls in addition to fire emergencies. Police, medical, and a variety of other emergencies must be prioritized and correctly routed during periods of high volume. Also, the caller may be stressed and it is not uncommon for incorrect addresses to be transmitted. Factors that often affect this communication include:

- Accuracy of the message
- Language coordination
- Intelligibility of the message
- Protocol of the communication center
- Training and instructions
- Condition of the caller
  - Mental
  - Physical
  - Emotional.

Although many of these factors may seem very subjective, attention to them often uncovers difficulties that may not have been anticipated. Time duration estimates can vary because of communication center delays during periods of high activity.

12.7 Discussion

Fire department notification by an individual provides a small but significant link in the holistic performance of any building that does not have an automatic sprinkler system. A variety of scenarios can be examined for the human link between the building and the local fire department. After an initial comprehensive scenario analysis, most variations may be easily recognized and quickly estimated. This enables one to bracket possible time durations to recognize any conditions that deserve special attention.

Fire performance must consider the entire year. In addition to normal operating conditions, several other scenarios are common:

- When the building is unoccupied, no one is available to notify the fire department except a chance passer-by. Because passer-by actions are not considered for performance evaluations, the time from MD to MN would be excessively large. Therefore, a building will normally have an occupied scenario and an unoccupied scenario.
- When the building is partially occupied by overtime workers or a cleaning crew, fewer individuals are available to notify the fire department. The time durations can become large because of greater distances to travel, more unfamiliarity with the building, and potential difficulty in entering some spaces. While cell phones are ubiquitous, they are not universally available.
- Often, occupants believe that a local alarm to alert occupants is simultaneously transmitted to the fire department. This is not always the case, and a performance analysis would determine if the alarm were local or transmitted elsewhere. We examine conditions that can delay notification.
- When an individual rather than a detection instrument detects the fire, the potential for delay is greater. Often, the individual may decide to extinguish a small fire first. Or alert other occupants before notifying the fire department. Or flee. Or... One may examine relevant scenarios to understand alternative outcomes better.
- Unique buildings pose unique problems. A performance analysis examines the building and occupant characteristics for conditions that may delay fire department notification. For example, board and care facilities often have occupants whose behavior does not conform to expectations. Apartment houses have a variety of conditions that lead to longer times for MN. Industrial facilities often experience a delayed notification trait.

The significant building performance measure is time. The question about notification is not *if*, but *when*. Assumptions about occupant behavior are often inaccurate because humans make decisions based on their perceived environment and background. Although enormous uncertainties exist, an analysis of the human link between detection and fire department notification often exposes significant potential weaknesses. The gap in knowledge involves both human decision-making and physical capabilities under abnormal conditions. Nevertheless, experience has shown that systematic analyses of scenarios will counterbalance the uncertainty with acquisition of a clear insight into the building's operational effectiveness. Understanding the nature of potential problems enables one to develop more effective procedures for a risk management program.

12.8   Automated Notification Services

Four types of automatic transmission equipment are associated with notification of the local fire department. One involves a type of alarm transmitted to a supervisory station maintained by the building owner. This type of protection is called a *proprietary supervising station system* and is frequently used in hotels, large offices, larger department stores, and apartment buildings having a security staff. Two other types of third-party notification involve signal transmission through a *central station* or a *remote supervising station*. The fourth type of notification is an *auxiliary system* that transmits a direct signal to the fire service communication center. The principles of operation and basic functions of each are described in the following sections.

12.8.1   Proprietary Supervising Station System

A proprietary fire alarm system serves a single property or a group of non-contiguous properties under the ownership of a single corporate entity. The alarm and signaling devices connect to a supervising station where trained personnel responsible to the owner of the protected property are in constant attendance. This type of system is widely used in large offices and other commercial and industrial organizations that monitor and act on a variety of emergency services.

Besides the fire alarm function, a proprietary supervising station may monitor security and other controls. For example, it may trigger fire suppression systems or control other building functions, such as occupant alert, elevator recall, certain emergency operations, heating, ventilation, and air conditioning (HVAC) and smoke control, and emergency communications. The proprietary supervising station monitors trouble signals, guard services, and actual fire alarms.

A variety of requirements are specified for the construction of the facility and the design, installation, testing, and maintenance of equipment. Personnel training and standard operating procedures are important for prompt and reliable actions in an emergency.

Buildings that use a proprietary supervising system are normally large and complex. Floors and subsections are separately zoned to identify locations more specifically. Modern equipment is sophisticated and can identify specific locations with digitally coded information. This type of information can reduce the time to fire detection and enable the local fire department to locate the fire more rapidly.

The time duration for retransmitting a fire alarm signal normally ranges from 15 to 45 seconds. Some common conditions that create longer delays in retransmission to the local fire service communication center are:

- The attendant may send a guard to investigate before calling the fire service communication center.
- The attendant may notify the owner of the emergency before transmitting the signal to the local fire service communication center.
- The attendant may attempt to verify that a signal means a fire and not a false alarm by awaiting the actuation of other detectors.

12.8.2   Central Station

A central station is a facility operated by a person or company whose business is to furnish a variety of services relating to fire protection systems in protected properties. Normally the central station is remote from the protected building, and a single facility will service many buildings from a geographical region. The company providing central station services is listed by Underwriters Laboratories (UL) or Factory Mutual Research Corporation (FM Global). The local authority having jurisdiction (AHJ) verifies the character of the installation by a UL certificate or an FM Global placard. A minimum of two individuals must staff the station at all times to ensure prompt and continuous attention to all signals. To be listed, a variety of quality and redundancy conditions must be incorporated into the system. The main conceptual difference between a proprietary system and a central station is that the proprietary system is under the ownership and control of a single corporate entity. On the other hand, a central station services many different clients.

A central station takes action when signals are received from the water flow alarm of sprinkler systems, actuation of special hazard suppression systems, manual fire alarm boxes, and fire detectors. If the building has a guard service, the central station also monitors the periodic reporting of the guard. The signals may be of operational supervision, trouble, or actual alarms. When a signal reaches the central station, a prescribed set of procedures take place.

Central station services involve a variety of activities for the subscriber on the premises and remote from the premises. On the premises, the central station company is responsible for installation of the equipment and its testing and maintenance. In addition, the company provides a runner to reset the equipment manually when required. A central station company monitors all signals from the protected property, manages the entire system, keeps records, and retransmits fire alarm signals to the fire service communication center. Highly protected risk (HPR) industrial facilities and commercial facilities having property or operational continuity of high value normally use a central station.

When an alarm actuates, the operators are expected to retransmit the alarm without delay to the fire service communication center. After providing notification to the local fire service communication center, the subscriber is notified by the fastest available means.

One may expect retransmission of an alarm to a fire department to be completed within 15–45 seconds. One can estimate this event more accurately by studying records of past alarm retransmissions.

Although the record of listed central station facilities is good, possible causes of delay include:

- Contacting the owner before retransmitting the alarm to the fire service communication center.
- Poor operator training resulting in "missed" or incorrect responses to an alarm.
- Leaving a system in a test mode after testing is complete.

12.8.3   Remote Station

A remote station fire alarm system transmits the signal from a building fire alarm system to a remote location, usually a fire service communication center. Personnel at the remote location are expected to take appropriate action. The remote station facility normally receives signals from properties of several owners. In this context it is similar to a central station. However, there are several important distinctions between the two types of system.

Often a local fire department is unwilling to receive routine supervisory or trouble signals from buildings because of personnel or operational constraints. However, the local fire department may be willing to allow all alarm signals to be received at a location acceptable to the local AHJ. UL or FM Global is not required to list the other supervising facility. Nor is the facility required to provide the redundancy of a central station. However, there are specific requirements to comply with the standards for remote supervising station fire alarm systems.

Trained personnel are required to be present at all times, recognize the type of signal received, and take prescribed actions. An alarm signal must be retransmitted directly to the fire service communication center, whereas trouble and supervisory signals may be handled differently.

The retransmission of an alarm to the local fire service communication center should take 15–60 seconds. Although the reliability of remote stations can be excellent, it is not always as high as that of central stations. Some factors that can influence the reliability and speed of alarm retransmission include:

- The attendant may notify the owner of the emergency before transmitting the signal to the local fire service communication center.
- The attendant may be asleep when the alarm is transmitted; this delays retransmission.
- Poor operator training may result in "missed" or incorrect responses to an alarm.
- A system may be left in a test mode after testing is complete.

12.8.4   Auxiliary Fire Alarm System

An auxiliary fire alarm system transmits a fire alarm actuation directly to the local fire service communication center. This system is the fastest and most reliable means of fire department notification. However, it can be installed only if the community has a public fire alarm reporting system and the AHJ grants permission for the connection and accepts all the equipment. If the community does not have a public fire reporting system, the owner cannot install an auxiliary fire alarm system.

A public fire alarm reporting system is most commonly associated with the street box for reporting fires by the public. These publicly accessible boxes are frequently located at or near the main entrance of schools, hospitals, and places of public assembly. The AHJ may designate locations for additional boxes. The public system may be used if the distribution cables and wires and the power supply and circuitry are available for signal transmission. It is possible also to use coded wire or radio reporting systems or telephone reporting systems. The major issues for these systems are that the community has an approved public fire alarm reporting system available and uses a direct connection circuit.

Auxiliary fire alarm systems must be installed and maintained by the owner of the building. Normally, only fire alarms are used in auxiliary systems, although other public emergency calls may be incorporated if they do not interfere with the transmission and receipt of fire alarms. When other types of supervisory or trouble signals are included in the system, a remote station system is recommended.

The time from detector actuation to the community fire service communication center is instantaneous. The reliability is extremely high, although the following factors may reduce the reliability if they are present.

- Poor maintenance of cable plant.
- Weather-related conditions that adversely affect the cable plant.
- Damage to auxiliary boxes due to traffic accidents.

12.9   Discussion

The automated systems for notifying the local fire department after detection are fast and reliable. Normally, any method will notify a communications center in less than 1 minute. Nevertheless, both time and reliability among the systems will differ.

Reliability is associated with the operation of a signaling system. Although more complex than detection systems, the same process as described in Section 11.5 may guide the analysis. This examines the following three components:

- Manufacturing control – this describes the system components when they are received from the manufacturer.
- Construction control – this component examines the quality of the system's installation.
- Maintenance control – the long-term maintenance becomes an ongoing part of system reliability.

System and construction quality collectively define the initial mission readiness. It is normal to have a completely reliable system at installation because of acceptance testing. Continued reliability depends on an effective, documented, and inspected maintenance plan.

PART THREE: BUILDING SYSTEM INTERFACES

12.10   Release Services

A third type of action after detection produces changes in normal building features. A detector actuation can interact with other building services to alter their actions. These actions are commonly called *release services* because they trigger changes in building service operations to affect the fire and fire defenses.

The most common release services are as follows.

Elevator recall.   During a fire, elevators can stop at a fire floor because of heat shorting electrical wires or the call button acting as a pseudo-smoke detector. This can occur whether or not a call is made to the fire floor. When this occurs, occupants in the elevator are placed in danger.

Perhaps a more significant application involves fire department control. Fire fighters use elevators to transport equipment and personnel. This mode of transportation becomes unavailable if the elevators are in service to transport occupants.

When elevators are recalled to the ground floor at fire detection, the fire department can control movement either for their work or to direct occupant egress. Fire department keys are used to override automatic controls.

HVAC controls.   During a fire, air-handling equipment may continue operating, shut down, or transfer into an emergency mode to pressurize some floors and exhaust other floors. It is possible to interface HVAC controls with the detection system to change operations to a more desired mode.

Visual alarm signal appliance.   This device, commonly called an annunciator, is intended to be positioned so that fire fighters can locate the fire alarms that actuated. It is normally desirable to lock in the initial detector to identify the fire location more accurately. Because there are a wide variety of annunciators, a performance analysis examines actual operations.

Door controls.   It is possible to lock or unlock selective doors upon detector actuation. This may be useful to control movement or to enable emergency forces to move through barriers more easily.

Lighting.   Emergency lighting is important to enable occupants and emergency forces to move through a building more easily.

Automatic suppression.   A wide variety of special hazard suppression systems and some pre-action sprinkler systems open the water supply or other valves to initiate agent application.

Release actions can be useful to put a building into emergency mode at an early time. A performance analysis is aware of the effect of release actions to evaluate the holistic behavior.

13

Fire Department Extinguishment: Arrival

13.1 Introduction

A building fire continually changes from ignition to extinguishment. Interactions of fire, extinguishment methods, people, and the building produce a unique blend of conditions that change minute by minute. Sometimes the combinations produce a small fire that results in slight risk and minor damage. Other situations can generate a large fire and horrible outcomes.

Figure 13.1 shows the only two ways to put out a hostile fire: automatic sprinkler suppression and fire department extinguishment. When a building has no automatic sprinkler system, the local fire department is the only active way to extinguish a fire. Because most buildings in the world rely only on the local fire department for active extinguishment, this component is an essential part of any performance analysis.

It is anticipated that most readers will have no experience and little knowledge of fire department operations. Nevertheless, an understanding of how a building can help or hinder manual fire extinguishment is important to fire safety engineering. This does not mean that a fire safety engineer must have experience as a fire fighter. It does mean that a fire safety engineer should have a basic understanding of fundamental fire fighting operations so that he or she can recognize how certain building features will affect those operations. Therefore, the principal audience for this description of manual extinguishment is the engineer rather than operational fire service officers.

13.2 Organizing the Topic

A holistic performance analysis provides a means to understand how a building will behave from ignition to extinguishment. The understanding enables one to integrate the roles of building architecture, fire defenses, and occupants with the dynamic changes of the fire. *Time* links all parts of this dynamic process.

Chapters 13–15 describe building analysis for manual fire extinguishment. The analytical framework of Chapter 28 structures the procedure. Collectively, these chapters describe a way to estimate manual fire extinguishment outcomes for any site-specific building. An evaluation identifies the influence of building design decisions on manual fire fighting activities as well as providing an awareness of fire ground procedures for engineers with no fire fighting experience.

Fire Performance Analysis for Buildings, Second Edition. Robert W. Fitzgerald and Brian J. Meacham.
© 2017 John Wiley & Sons Ltd. Published 2017 by John Wiley & Sons Ltd.

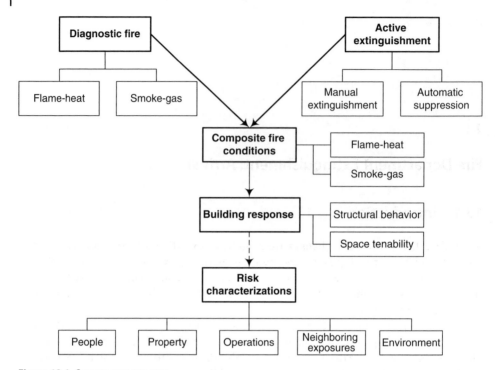

Figure 13.1 System components.

Chapter 13 addresses several background topics. To give perspective, *Part One: Manual Extinguishment Overview* outlines the complete process. Because each community provides public fire protection in the manner that is believed to be best for its unique needs, *Part Two: Community Fire Departments* discusses common fire department organizations and equipment. *Part Three: Community Fire Response* discusses communications center operations that provide the link between the building and the start of fire department actions. Part Three also describes ways to estimate time durations from fire department notification (MN) to first-in and subsequent response at the site (MR).

Chapter 14 describes initial fire ground activities from first-in response (MR) to first water application (MA). Analyzing the time to apply first water to a fire provides an insight into ways the building and fire department interact. Chapter 15 discusses operations from first water application (MA) to extinguishment (ME).

Fire ground operations focus on life safety (search and rescue of occupants and fire fighter safety), fire extinguishment, and asset protection (preventing fire extension and damage to exposures). The incident commander must manage these operations, make decisions, and allocate resources.

A manual extinguishment analysis does not encompass all fire ground decisions and actions. Rather, it considers only fire suppression. An evaluation compares diagnostic fire conditions with critical events from detection to extinguishment. The examination enables one to recognize interactions between the building, the diagnostic fire, and local fire department operations.

PART ONE: MANUAL EXTINGUISHMENT OVERVIEW

13.3   The Role of the Fire Department

The role of the fire service is that of a change agent. When a fire department arrives at a burning building, it takes a dangerous situation and improves the safety of occupants; puts out the fire or controls it to allow a safe burnout; and tries to prevent the fire from extending to other parts of the building and to neighboring exposed structures.

The ability to change a situation depends on the fire conditions at arrival, available community resources, building and site conditions, and experience of the incident commander. A building fire can be a very complicated situation. The fire ground commander must manage the available resources to search for and rescue occupants, locate and extinguish the fire, protect important assets, and protect adjacent properties exposed by the fire. These tasks must be accomplished within a brief time period for a situation in which information is incomplete, often inaccurate, and dynamically changing throughout the incident.

A manual extinguishment analysis evaluates building and site conditions that affect fire suppression. The analysis does not evaluate local fire departments in isolation nor does it examine all of the other activities that take place on the fire ground. Rather, its function is to understand and identify the influence of the building's site, architecture, and fire defense features on manual suppression.

13.4   Building Analysis Overview

The world has more experience in manual fire suppression and less quantitative data with which to evaluate a building's influence on fire fighting than any other part of the entire fire safety system. Nevertheless, one can gain a clear understanding of a building's performance for manual suppression using a systematic framework for analysis.

Here, we outline the general process to provide a context with which to organize the associated information. An analysis for local fire department extinguishment uses the following components:

1) A specific diagnostic fire scenario starts the analysis. This fire provides a measure against which the manual suppression components are compared. Chapter 10 describes diagnostic fire scenarios.
2) Estimate the time duration for Part A: Ignition to notification. These activities include:
   a) Fire size at detection (MD) (Chapters 11, 26)
   b) Notification of the fire department (MN) (Chapters 12, 27).
3) Estimate of the time duration for Part B: Notification to arrival (Chapters 13, 28). The events are:
   a) Dispatch response companies (MS) (Chapter 13)
   b) Emergency response arrivals at the site (MR) (Chapter 13).
4) Estimate the time duration to complete Part C: Arrival to extinguishment. These events include:
   a) First water application to the fire (MA) (Chapter 14)
   b) Control the fire (MC) (Chapter 15)
   c) Extinguish the fire (ME) (Chapter 15).

Figure 13.2 shows a single value event logic diagram of the major events. Each of Parts A, B, and C is the responsibility of different individuals. Collectively, they must be integrated to provide a clear understanding of manual suppression.

Figure 13.3 shows a representative timeline that identifies the sequence of these events. These major events combine influences of the diagnostic fire, the building and site, and the local fire department. Time is the factor that links all of the relevant parts.

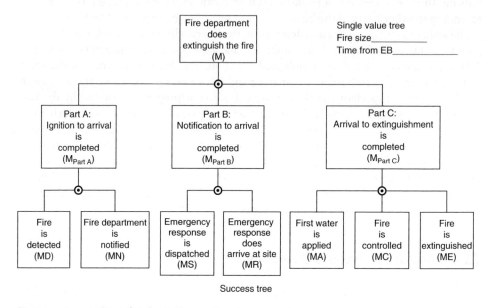

Figure 13.2 Event logic: fire department extinguishment (M).

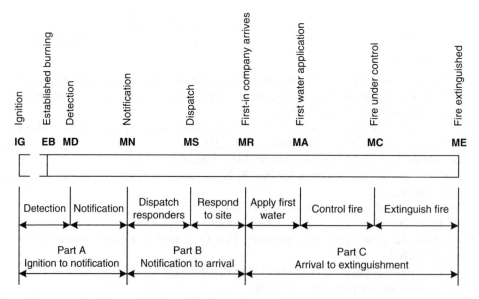

Figure 13.3 Timeline: fire department extinguishment (M).

13.5 Part A: Ignition to Notification

The evaluation for Part A: Ignition to notification estimates the diagnostic fire size at the time the community fire service alarm center (i.e. "fire department") is notified of the building's location. This segment is under the management control of the building owner or an occupant.

The timeline duration for Part A: Ignition to notification is established by evaluating the following events:

- *Detection (MD).* The building's detection system is evaluated to establish a fire size at detection. This information provides the time estimate from established burning (EB) to detection (MD).
- *Fire department notification (MN).* Fire department notification may be done by human actions alone, automated procedures, or a combination of human and automated actions. Each building provides unique conditions that influence the results. Uncertainty of actions can range from substantial to insignificant depending on the situation and conditions.

A methodical analysis of the detection and notification package provides valuable information about the building's role in manual suppression.

13.6 Part B: Notification to Arrival

Local practices establish the community's response at the fire scene. These practices reflect the community's political and economic decisions for fire emergencies.

The time duration between notification (MN) and arrival (MR) is defined as the department response time. Although first-in company arrival is an important event, the analysis calculates the time for company arrivals for first and subsequent alarms.

Figure 13.4 expands the segment from MN to MR into its component parts. An evaluation focuses on three distinct time intervals:

- *Alarm handling time* – the time to process the message and decide which fire companies to dispatch.
- *Turnout time* – the time for the companies to mount and start the apparatus.
- *Travel time* – the time between apparatus wheels crossing the fire station threshold to stopping at the site.

Each community handles fire response in its own unique way. Although each community has different standard operating procedures (SOPs) for alarm handling, turnout, and company responses, local processes are well defined. A systematic analysis enables one to estimate time durations with confidence.

We separate the process into two events:

- *Dispatch the response (MS).* The community alarm center must determine the appropriate company response for the emergency. The message to the responding companies (MS) completes this event.
- *Apparatus arrival (MR).* The time at first-in arrival defines the completion of the response event. However, the process also calculates arrival of subsequent responding companies.

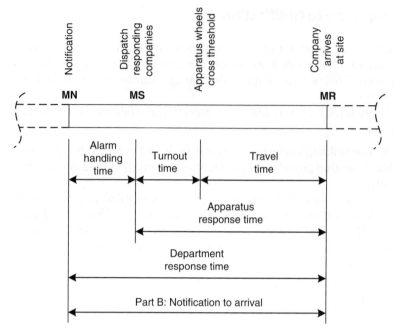

Figure 13.4 Timeline: fire department response.

The Part B analysis estimates fire size and conditions at the time of first-in arrival. The type of fire department, number and type of apparatus, and number of responding fire fighters are known at that time.

13.7 Part C: Arrival to Extinguishment

Manual fire suppression requires fire fighters to locate, attack, control, and extinguish a fire. Although local fire fighting terms may differ, the concepts remain the same. For example, the term *confine* is often used in extinguishing descriptions. It usually denotes the activity of preventing fire extension by confining the fire within a specific building area. We use this term to be synonymous with a fire that is "under control."

The terms *control* or *under control* mean the incident commander understands the fire conditions well enough to believe that building and fire fighting resources can eventually extinguish the fire without further extension. Although the fire could be a raging inferno when the incident commander declares it to be under control, the commander is confident that the fire can be confined and eventually extinguished without further spread to other parts of the building.

Of the many fire ground activities that must be managed after arrival, we evaluate only fire extinguishment. Other fire ground activities, such as search and rescue and protection of exposures, are not a part of extinguishment analysis except in ways they may affect the number of fire fighters available for active suppression.

When the first water application can extinguish the fire, one need evaluate only the initial fire attack. Control and extinguishment take place at nearly the same time. However, when the first attack line is not adequate because the fire is too large or the building creates difficult conditions for extinguishment, additional hose lines and different extinguishment tactics may be necessary.

Water application disturbs the natural fire and causes dramatic changes in fire conditions. We use the term *composite fire* to describe this turbulent mix of conditions. After first water application, the diagnostic fire scenario is no longer a good measure for fire expectations. Nevertheless, the original diagnostic scenario may have additional value to understand extension routes and potential fire conditions.

Eventually, the fire will be extinguished by the local fire department. An analysis will provide a basic understanding of how the building can work with the local fire department to put out a fire.

The Part C analysis initially estimates diagnostic fire conditions when first water could be applied. The knowledge gained during this analysis establishes understanding for fire control and extinguishment expectations. Three major events comprise a Part C analysis:

- *First water application (MA).* The evaluation estimates the time to apply first water to the fire. This trial analysis examines locating the fire, stretching an attack line from an attack launch point (ALP) to the fire, establishing a water supply, and applying water. The outcome gives a good understanding of whether the first water application will be able to extinguish the diagnostic fire.
- *Fire control (MC).* When the first water application is not able to extinguish a fire, additional hose lines, combined with building layout features, are integrated into fire suppression tactics. Normally, these additional measures take place during the initial attack.

  When a fire is "under control," the incident commander believes the fire will not extend beyond defined boundaries. In addition, he or she understands generally what the fire will do during the remaining fire fighting operations. Although the term fire control identifies a "benchmark" of the fire's history, it is often somewhat imprecise. Nevertheless, the concept is useful in evaluations.
- *Fire extinguishment (ME).* The building and fire conditions may enable the fire to be extinguished quickly by an aggressive attack. Or the process may have a long duration and need defensive measures. The integration of diagnostic and composite fire conditions, the building architecture, and local community resources enable one to get a "feel" for general outcomes.

PART TWO: COMMUNITY FIRE DEPARTMENTS

13.8   Fire Department Organizations

Communities provide the type of fire services that seem appropriate for their local fire problem. Staffing may be full-time, paid-on-call, or volunteer fire fighters. Or it may be a combination of these. Suburban and rural communities generally use volunteer or paid-on-call fire fighters. Urban communities normally use paid fire fighters.

Full-time fire fighters continuously staff the fire stations and respond to all fire and many non-fire emergencies. Paid fire departments also have fire prevention, inspection, and code approval responsibilities. Fire stations are normally staffed 24 hours a day with full-time fire fighters, although the number of personnel can vary greatly among communities.

Volunteer and paid-on-call fire fighters are employed outside the fire service but respond to emergency calls. They participate in regular training programs to build and maintain their firefighting skills. Paid-on-call fire fighters earn wages for the time that they respond to emergency calls. Volunteer fire fighters do not receive compensation for response to calls. Paid-on-call and volunteer fire fighters normally experience a narrower variety of incidents than full-time fire fighters.

A variety of mix'n'match staffing combinations can exist in local fire departments. Some communities have only full-time fire fighters. Others have only volunteer or paid-on-call fire fighters. Many have a nucleus of full-time fire fighters that are augmented by volunteer or paid-on-call fire fighters for emergencies beyond the capability of the on-duty staff. Some communities may have only a full-time chief and a few part-time fire fighters. They rely on volunteers or paid-on-call fire fighters to staff fire responses. The number of on-duty fire fighters and organizational types can vary greatly among communities.

In recent years, local fire departments more commonly rely on automatic and mutual aid agreements with neighboring departments. Smaller communities with inadequate financial resources can combine tools, equipment, and personnel to handle larger incidents. Although these agreements provide resources to handle larger emergencies, coordination and management on the fire ground and within the communities requires careful attention.

Over the years, fire departments have acquired increasingly more responsibility. Today, fire departments have become the central emergency management agency for a community. Thus, while fire fighting may be a dominant image, the role of a fire department has changed to reflect a broad range of emergency operations.

The organization of the community fire department influences the type of response, its speed, and the number of personnel who can attend the emergency. SOP will vary enormously from community to community. It is essential to understand the local staffing, equipment responders, the type and amount of aid that is available from neighboring communities, and SOP instructions when analyzing the building's performance for manual extinguishment.

13.9   Fire Companies

A variety of equipment is available for a fire department to do its work. A company is considered to be the apparatus and the fire fighters that staff it. Thus, when we speak of engines, ladder trucks, rescue vehicles, and other specialized pieces of apparatus, the staffing to accomplish their functions is integral to the description. Specific customs, equipment, procedures, and terminology vary between countries, geographic regions, and communities. However, the basic concepts and functions are relatively consistent.

13.9.1   Engine Company

The function of an engine (also called a pumper) company is to put water on a fire. Pumps on the engine take water from a supply source and push it through attack hoses for fire fighters to apply to a fire.

Staffing

An engine company has a driver who usually becomes the pump operator after the apparatus stops at an attack site. The company has an officer who directs operations and fire fighters to do the work. Usually, the officer also participates in the work.

Staffing is usually viewed as an officer, driver (pump operator), and two fire fighters. Some engine companies have a staff of two or three. Others may have six or seven.

Fire fighters think in terms of tasks. Engineers think in terms of time. All tasks associated with putting water on to a fire must be completed before water can be applied. Thus, the number of people available to do the work is important because of its effect on time durations. Knowledge of company staffing is essential information.

Attack Hose Lines

The water flow path from engine pump to fire is provided by hose lines and a building's standpipes. The route may be direct or circuitous. It may be long or short. It may involve delays created by architectural obstacles or it may have relatively few impediments to movement. However, fire fighters must do the work of stretching an attack line to make a continuous path for water flow.

Water Supply

The water that a pump pushes through the attack line can come from one or a combination of sources.

On-board Tanks   Engines carry a tank of on-board water. On-board tanks may carry from 1100 liters (300 gallons) to 3800 liters (1000 gallons). The usual on-board tank size is 1500 or 1900 liters (400 or 500 gallons). An initial attack usually starts with the on-board water. If this water is insufficient, it is supplemented from other sources.

Hydrants   Underground piping and hydrants are a first choice to supplement on-board water when they are available. Hydrants fed by municipal or private water systems can supply water directly to the attack pumper or the water may be moved through one or more relay pumpers to reach the attack pumper.

Drafting   When an available body of surface water, such as a pond, stream, or cistern is available, the water may be used to supplement on-board water. Drafting sources may be supplied to an attack pumper through relays from other apparatus.

Mobile Water Supply (MWS)   Site locations that do not have underground water pipes or sources to draft water must transport water from longer distances. Shuttle operations using water tender trucks that discharge into a portable reservoir supply the water. Documentation of demonstrated shuttle water supply capabilities is available from the insurance rating organization that evaluated the local fire department.

13.9.2 Ladder Company

A ladder company (also called a truck company) does everything on the fire ground except apply water to the fire. Sometimes they do that too. A ladder truck provides equipment to the fire scene to handle a wide variety of specialized needs. For example, a ladder company may provide assistance with forcible entry, search and rescue, ventilation, and property protection by moving and covering the contents with salvage covers.

Ladder companies also carry ladders to use for occupant escape or an engine company to enter the building to stretch attack lines. A ladder company often has the capability to deliver an elevated master stream in a defensive fire fighting operation.

Some ladder trucks may also have pumps and on-board tanks that give them the capability to relay water from supplemental sources to attack pumpers.

Staffing

Ladder company staffing can vary considerably among communities. An officer and three fire fighters usually form an efficient team. However, it is common for a ladder truck to be staffed with as few as two and as many as seven fire fighters. If a ladder company is not available, engine company personnel must do the other fire ground tasks.

13.9.3 Specialized Companies

While engine companies and ladder companies are a principal focus of fire suppression, many fire departments have additional, more specialized equipment. Some of the more common support companies include the following.

Rescue Company

A rescue company or rescue squad provides a combination of engine company and ladder company functions. Rescue companies are often viewed as elite, fast-response units that can conduct an initial search and rescue or fight the fire. The actions depend on the needs at the specific site at the time of arrival. Often the rescue company will have an officer and five fire fighters, although the number may vary. In some departments the staffing could be as low as two firefighters assigned to a truck company officer.

Quint

A quint is a combination vehicle that has been gaining popularity in recent years. A quint combines the usual engine capabilities of pumper, booster (water) tank, and hose with an aerial device (usually 17–23 m; 55–75 ft) and ladders. When several quints are at a fire scene, they have the flexibility for each arriving vehicle to assume a different sequential role in the fire fighting process to adapt on-site needs more easily.

Special Purpose Apparatus

Besides engines (pumpers), ladder trucks (trucks), and rescue vehicles, many fire departments have a variety of special-purpose vehicles for specialized emergencies.

For example, local departments may have special vehicles to deal with hazardous materials, water rescues, additional hose supply, breathing air replacement, or forest fires. Water tender trucks to transport water are common in many rural communities. The equipment reflects local community needs and resources.

13.9.4 Emergency Services

The role and perception of fire departments have been changing significantly. A generation ago, fire departments were perceived to have only fire fighting functions. Today, communities recognize that the fire department is a local emergency management agency. Many fire departments provide emergency medical services of varying types to a community or region. In addition, the local fire department is the first service considered for emergencies such as collapse, rescue, terrorism, and flooding. The range of emergency services is broad and increasing in scope and complexity.

13.9.5 Response Information

Each community staffs and equips its fire department to respond to perceived local needs and budgets. Within any individual community, the number of vehicles, their storage locations, the staffing, and the number of vehicles assigned to a first alarm or to multiple alarms will vary substantially. Some communities initially send a single engine to the scene. A common first response for other communities is two engines, possibly one ladder truck or quint, and a chief. Sometimes a rescue company augments this response. Still other cities may dispatch four engines, two ladder trucks, one rescue, a battalion chief, and 37 fire fighters to a first alarm response.

A building analysis for fire extinguishment uses a scenario to evaluate the situation. Although an evaluation may appear to be complex, the process can be systematized to accommodate a wide variety of conditions and outcomes.

A building analysis is based on available local fire department response and operating procedures. This requires information concerning the local fire department. Fortunately, community fire departments have systematic organization and operating procedures. Also, local fire departments are very willing to describe their operations and procedures, and usually welcome discussions relating to building fire fighting situations.

13.10 Building Fire Brigades

The term *fire brigade* can have many meanings. Outside of the United States and Canada, most local fire fighting departments are called fire brigades. In that context, a fire department and a fire brigade are the same organization for our purposes.

However, in the United States, the term fire brigade can have a different meaning. Often, a local industrial or institutional building may have a group of employees who are cross-trained to fight fires within the facility. Not all organizations have them. However, when a building fire brigade does exist, the fire suppression capabilities can be evaluated using the same process as for local fire department extinguishment.

The amount and type of equipment, the organization and command structure, regularity of training, and fire fighting experiences influence the effectiveness of a building fire brigade. The capability can vary from a "paper" organization with little training, experience, and ability, to a skilled group that can provide rapid, effective action. If it does exist for the site-specific building, it is important to perform an objective evaluation of fire brigade capabilities.

The building fire brigade fulfills two functions. One involves putting out a fire while it is still small. The second function is to change the fire growth and control the size that will be encountered by the local fire department. The local fire department becomes a mutual aid resource in this situation.

If a building fire brigade does exist, a fire scenario is evaluated in two parts. One examines the initial actions of the building fire brigade. The second evaluates the outcome after the local fire department arrives on the scene.

Occupants can have an influence on an incipient fire. However, in a systematic analysis, occupant extinguishment is not a part of an evaluation. We place occupant fire suppression activities into the "fire prevention" part of a building analysis. Thus, occupant extinguishment becomes a part of the "prevent established burning" analysis. This segmentation maintains a methodical analysis of building performance.

PART THREE: COMMUNITY FIRE RESPONSE

13.11 Fire Department Response Time

The time duration between MN and MR is defined as the fire department response time. Although first-in company arrival defines the termination time for MR, the analytical process is the same for all responding companies. Figure 13.5 shows the major events.

Although determining a fire department response time for a site-specific building is not complicated, evaluations require attention. For example, one must recognize that comments like "We respond to all buildings within 3 minutes" require skeptical examination. The statement may be given with complete sincerity. However, examining details, checking audio communication recordings, and calculating travel times for remote locations often disclose differences in interpretation.

Priorities and procedures of the local emergency communication center affect dispatch time. Response time for many fire departments is often perceived as the "travel time" after an engine leaves the station. Turnout time can vary enormously for different types of department organizations and standard operating procedures.

13.12 Communications Centers

The role, scope, and operation of emergency communications centers are evolving rapidly. Historically, the scope of operations could vary from "mom'n'pop" part-time civilian dispatchers for small volunteer fire departments to fire fighter-manned

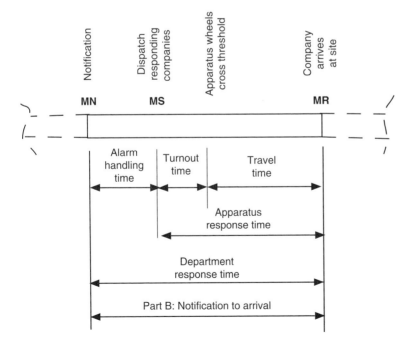

Figure 13.5 Timeline: MN to MR.

departmental rooms to sophisticated communication centers that handle all community or regional emergency services.

Today, this vital function is evolving toward city or regional consolidated communication centers. During the past 20 years the sophistication and speed of equipment, information availability, and information processing have transformed communication operations. And, changes are continuing at an accelerated rate. Understanding how a community communications center works can give an insight into the time and reliability of alarm processing time.

The incorporation of modern equipment seeming to offer never-ending improvements is not without growing pains. The human element remains important. Although jargon relating to each emergency service has rich meaning, civilians unfamiliar with certain terms can misinterpret the words. Continuing personnel training is essential. Also, communications center equipment compatibility with cell phones and other electronic devices requires attention. However, the role of community communication centers is an important link in the manual suppression analysis.

Each communication center operates somewhat differently when providing the important functions of a building fire performance analysis. Nevertheless, there is some commonality of activity even when standard operating procedures are tailored to the center's local needs. The operational program is detailed in a *communication plan* of the center. This document describes all aspects of emergency communication, and many needs for a building analysis may be obtained from this plan.

The communication plan describes such information as the response policy and priority hierarchy. Routine emergencies may involve different response priority during complicated periods when police or medical emergencies occur simultaneously with a fire emergency. Also, a plan will describe regular responses, actions with alarm companies, and mutual aid. Generally, a communications center plan will provide much useful information relating to a building's performance analysis.

13.13   Alarm Handling Time

Notification (MN) is the instant at which the community fire service alarm center is aware of the fire and its location. This is not the instant when an individual transmits the message. Nor is it when a device or third party becomes aware of the fire (e.g. automated systems). For example, a telephone call to 911 or a fire detection alarm transmitted to a central station or to a remote station does not complete notification. It is the instant that the Public Safety Answering Point (PSAP) receives and understands the location and nature of the fire emergency.

Alarm Point
The alarm point occurs at the instant when the *fire service* alarm center (PSAP) becomes aware of the building fire and its location. The alarm point is synonymous with notification (MN).

Company Alert Point
We define the alert point as the instant when the fire emergency is transmitted to the responding fire companies. In full-time fire departments, the alert point is usually the

instant when a bell, horn, or other device is sounded at responding fire stations. In some communities, the alert point is the instant when the responding fire fighters receive the first sound of a radio or telephone message calling them to the fire emergency. In other communities the alert point may be the sounding of a siren at a central location. The respondents and the method of transmitting the message can influence time durations.

Alarm Handling Time

The alarm handling time is the time duration between notification and the dispatch of responding companies. Because each community has its own protocol for identifying and alerting responding companies, the alarm processing time can vary. In some locations the alarm processing time may be as little as 20 or 30 seconds. In other communities, as much as 2 minutes or more could elapse before responding companies are alerted. Time durations of about 1 minute are common.

Emergency calls are frequently recorded. Analyzing audio tapes can identify time durations and potential difficulties that may arise. Also, many community communication centers show these data in computerized reports. The data are generally available, but may require some inquiry to obtain them.

Dispatch the Fire Response

The dispatch event (MS) is often synonymous with the alert point. Some communities may have slightly different procedures. However, the time interval between MN and MS is completed when the communication center dispatches (sends) specific responding companies to the fire scene.

13.14 Turnout Time

The *company turnout point* is the instant when the wheels of a responding apparatus cross the threshold of the fire station. The turnout time is the interval between the alert point and the turnout point. Turnout time can vary quite significantly among communities depending on the type of fire department and the departmental SOP.

When fire fighters live in the station, turnout time is often 20–40 seconds during the day and 30–60 seconds at night. The turnout time for paid-on-call and volunteer companies depends on the department SOP. Two common procedures are often used. In one method, responding fire fighters are alerted of the fire emergency and as few as two fire fighters will drive the engine from the station directly to the fire scene. Fire fighters individually drive their own vehicles to the fire ground, some often arriving before the first apparatus. When the department has a driver on duty or when the driver lives and works near the fire station, the turnout time may be short. When the driver must travel to the station, the turnout time is longer.

A second common procedure has responding fire fighters travel to and assemble at the fire station. After enough fire fighters have responded to staff the company, the apparatus leaves the station. This procedure may involve a greater turnout time. Some volunteer companies have members living in the stations to reduce turnout time.

A complete performance evaluation requires analysis of the departmental SOP for turnout and response to the fire scene. This involves the first engine and all additional responding fire apparatus. Some communities have only one engine. Others have several engines and additional types of emergency vehicles. Automatic aid and mutual aid agreements are common among neighboring communities in many jurisdictions.

Understanding local community resources and procedures is important. Time durations for the first company arrival as well as a complete response can vary greatly. Community and mutual aid time and resources may be influenced by other emergencies such as brush fires, electrical storms, wind storms, heavy snow, and other natural or man-made emergencies that produce a high frequency of emergency calls. The fire grows exponentially while individuals move linearly, and a few minutes can dramatically change suppression tactics and damage for a building.

13.15   Travel Time

The *arrival point* is the instant at which the first-in unit stops at the building site; travel time is the time duration between the turnout point (when the apparatus wheels cross the station threshold) and the arrival point (when the apparatus wheels stop at the building).

Different pieces of apparatus will have different travel times because of station locations and different travel speeds. Because an effective fire attack normally requires several fire companies, one can identify travel times for all responding vehicles.

Although travel time is easy to estimate, several variables may influence results. One knows the distance between the station and the site, as well as the route difficulty, terrain, and natural impediments to travel. For example, limited access highways, bridges, and railroad crossings often limit movement along particular routes. Regional planning departments often have transportation computer models that calculate travel time durations. Limited experimental testing has indicated that these models are useful and accurate. They are an important resource for performance analyses.

Most pieces of fire equipment are large, heavy, and cumbersome to drive. Travel speeds are affected by corners, hills, and unimproved roads. Traffic congestion may slow travel time when gridlock occurs, although time of day can give some predictability for traffic jams. However, ordinary automobile congestion often does not reduce travel time significantly because private vehicles usually clear the way for emergency vehicles.

Weather can influence the travel time. Fog-reduced visibility, ice, and heavy snow can reduce the speed of emergency vehicles. Nevertheless, one may construct reasonable travel time estimates relatively easily.

The only additional travel time element involves site access. The *respond to site* event is completed when the wheels of the engine stop and fire ground operations begin. There are situations when the building may be visible, but site access problems, such as gated entries, insufficient turning radii, speed bumps, difficult terrain, and narrow, incomplete, or blocked streets, delay vehicle placement at the site. Site access is an important part of the response analysis.

13.16   Response Time Analysis

The time estimate between notification (MN) and departmental response (MR) is usually easy to determine. Although procedures may vary among communities, confidence in results is usually high.

We divide this time duration into two parts. The first studies the alarm handling procedures of the fire service communication center and selects an appropriate interval from initial notification (MN) to dispatching the first alarm companies (MS).

The second part examines standard response procedures by the local fire department. The first segment involves actions before the apparatus wheels cross the threshold of the fire station. The time durations are sensitive to fire fighter locations, time of day, and standard operating procedures. The second part calculates travel time from wheels crossing the fire station threshold to the wheels stopping at the fire ground.

The following questions form the basis for establishing the time interval between MN and MR:

- **Given:**
  - *Fire department is notified (MN)*
  - *Time from EB* _____
- **Question:** *Will the responding fire service communication center dispatch the first alarm response (MS)?*

- **Given:**
  - *Fire service communication center has dispatched the first alarm response (MS)*
  - *Time from EB* _____
- **Question:** *Will the responding (first-in or other) apparatus arrive at the site (MR)?*

Although quantification of the time duration from MN to MR is easy to compute for each responding unit, one must examine conditions carefully. Casual examination has shown that the response time between MN and MR can vary greatly, often more than "advertised."

14

Fire Department Extinguishment: First Water (MA)

THE FIRE FIGHTER AND THE ENGINEER

Before we get lost in a sea of new concepts and techniques, let's pause to review building performance. More specifically, we will focus on the building's performance for manual suppression.

Buildings are seldom designed with manual fire suppression as a conscious goal. Perhaps that added design dimension might not have enough value for its cost. For whatever reason, building designers specify traditional code and standards requirements to provide for manual fire fighting needs. The fire department is expected to deal with the existing building during any fire incident.

Most buildings in the world must rely on the fire department for active fire suppression. A performance analysis evaluates how much damage and what types of risk to expect. For example, will a fire cause little damage and small risks? Or, will the fire produce major damage and high risks? Or will the results fall between these extremes? How much? Where? A performance analysis must make predictions – *before the fire*.

When fire fighters arrive at a burning building, they are often confronted by a dangerous, complex situation. Their role is to make the situation better than it would be without their efforts. Fire fighters are skilled professionals. They use routine procedures to put water on the fire. Because each building fire is different, standard operations are adjusted to fit site-specific needs.

The incident commander assesses fire conditions (the "size-up") and issues orders, e.g. "Engine 1 – stretch an attack line to the fire," "Engine 2, supply Engine 1" or "Ladder 3, prepare the attack route." Fire fighters instinctively apply their skills and tools on the building to accomplish the commands. To complete the assignment, fire fighters will use what works and avoid what doesn't work. Details for accomplishing orders are not specified on the fire ground. They just get done because training and experience prepare fire fighters for many situations. However, the sequence of necessary tasks to apply a hose stream for any site-specific building evolution can be easily described, if one wishes to do so.

The fire safety engineer analyzes a building for its expected performance – before the fire. In order to compare fire propagation and water application, the engineer must calculate time durations to complete selected fire department operations. Calculations must incorporate many variables such as staffing, equipment, terrain, weather, and

Fire Performance Analysis for Buildings, Second Edition. Robert W. Fitzgerald and Brian J. Meacham.
© 2017 John Wiley & Sons Ltd. Published 2017 by John Wiley & Sons Ltd.

building design features. Calculations must also examine uncertainties to give windows of time to complete important events. And, calculations must be fast and be easy to use.

An important part of a performance analysis interfaces fire fighting actions and engineering predictions. Fire fighters think in terms of tasks. Engineers analyze in terms of time. We must link these two ways of thinking. Although fire fighters are not engineers, and engineers are not fire fighters, a performance analysis links the two professions. We are trying to establish a connection to understand what to expect in the site-specific building.

Chapters 13–15 discuss building analysis for fire department suppression. A major part of these chapters provides an awareness of fire ground operations. They are written for the fire safety engineer who has no fire ground experience. Collectively, they attempt to build a bridge between the two professions.

However, another part of the information involves quantification. The emphasis for Chapter 14, Parts Three and Four describe ways to calculate the time to apply water to a fire. In these sections we change the focus from informational awareness to analytical procedures.

Two types of calculation are used. The first treats water supply functions as mini-construction projects. Supply water procedures are ideally suited to critical path methods (CPMs). These well-established techniques have been used in project management for over half a century. We adapt these methods to calculate time durations for supplying water to an attack pumper and the attack launch point (ALP). A network that connects defined modular tasks describes the work also serves as a framework for calculations.

Advancing attack lines through a building uses a different analytical technique. Hose movement operations are associated with architectural segments that define the building's functionality. A fire attack route links architectural segments. The cumulative time to complete sequential modular tasks through the architectural segments establishes the time for water application.

Network construction for these two calculation methods is well defined, rigorous, and easily understood. If one were to examine only the network sequencing, a good understanding of building and fire fighting compatibility would emerge. However, their value is greatly magnified when quantification is introduced.

Published documentation to quantify modular tasks used with the networks does not yet exist. However, all is not lost. Modular task time durations can be established in several ways. One involves using videos of fire training exercises. The tasks are so common in fire fighting evolutions that completion times are readily obtainable.

Another technique involves expert opinion of experienced fire fighters. Although time is not a focus of attention on the fire ground, site-specific variables that affect the time to complete tasks are clearly recognized. Experienced fire fighters can adjust video results to provide realistic time estimates for different conditions.

However, a permanent, citable solution can easily be created. Many years ago the construction industry needed a better way to estimate costs and time for contract proposals. The industry systematized unit time and cost data for construction components. Modular units incorporated variations such as the number of workers, equipment, materials, work conditions, and experience. The data became an important tool for preparing construction estimates.

The construction industry created unit cost and time estimates for defined construction components. This did not diminish the role of contractors or affect job security.

Quantity "take off" did make estimating faster, easier, and more accurate. Today, several publishers produce expanded cost estimating documents. Values are adjusted annually to reflect up-to-date information as well as changes in construction that may occur.

Fortunately, documentation to establish reasonable time durations for fire fighting tasks is much easier to create. A research plan to document the task information will be described elsewhere. It will show that individuals, local fire departments, or professional organizations can systematically and inexpensively establish time durations for fire fighting tasks. However, until consensus time durations are published, one must estimate values with the best means available.

Systematic organization and technological applications simplify much uncertainty in manual fire suppression. Modern technology provides means to determine time durations for evolutions. Spreadsheets, computer programs and digital catalogs of modular task data produce calculations quickly and easily. Removing computational drudgery allows the engineer to make dynamic comparisons and become more creative in understanding performance in buildings.

Part Three: Water Supply Analysis and *Part Four: Interior Fire Attack Analysis* both describe engineering techniques for evaluating the time to put water on a fire. The process links tasks and time into a network analysis. Rather than describing procedures in isolation, we attempt to put them into a building context.

The scenario that we selected does not explain how the fire should (or would) be fought. The primary goal is to illustrate a process for organizing events and calculating outcomes of applying water to a fire. This may require a little tolerance. For example, fire fighters will have to resist the need to say, "This isn't the way we would fight this fire." Engineers will have to learn enough about fire ground operations to identify tasks that must be put together to complete an evolution. Both groups will use a new way of thinking.

We are presenting an analytical technique. The process uses terms that may not be common either to the fire safety community or to the fire service. However, techniques are easy to apply and modern technology can simplify quantification. But all of the tasking data do not yet exist – at least not in citable publication form. Nevertheless, a framework for consistent analysis does exist, which is what we present here.

The fire safety world is evolving from prescriptive codes to performance design. Analysis is a necessary companion to design. Fire departments are an integral part of fire safety for buildings, and manual fire suppression is an essential component of holistic fire performance. Chapters 13–15 provide an introduction to this component.

14.1 Introduction

Each building fire is different. The fire and its environment are constantly changing. Fire ground operations are diverse and require different tasks that must be coordinated and adapted to needs for phasing in or phasing out of action. Available staffing and equipment may not fit with specific fire ground needs. Assignments are based on information that can be incomplete, inaccurate, and changing. Tactical adjustments may be needed as better information becomes available. Nevertheless, fire ground operations have procedures that can be organized into an analytical structure. Of the many operations that take place on a fire ground, we will analyze only one – fire extinguishment.

This analysis does not include the incident commander's decision-making functions relating to size-up, allocation of resources, and selecting the best strategy based on available information. Nor does it discuss search and rescue or occupant egress with or without fire fighter assistance. Instead, this chapter describes techniques to calculate supply water quantity and time sequencing. Also, the chapter describes procedures to estimate the time to advance an attack line through any specific building path.

Knowing the time it takes to apply water, combined with the diagnostic fire scenario provides a means of understanding the building's role in fire extinguishment. Quantitative analyses are combined with knowledge obtained from qualitative estimates to establish a sense of proportion for understanding performance.

Time is the common factor in all scenario analyses. The Interactive Performance Information (IPI) chart can compare time-related performance outcomes.

PART ONE: AN OVERVIEW OF MANUAL EXTINGUISHMENT ANALYSIS

14.2 The Process

The location, size, and movement of the fire influence strategy and tactics of fire fighting. When the fire is small, fire fighters can extinguish it quickly with an offensive attack. When the fire is large, the fire department establishes defenses against further fire growth. The terms "small" and "large" are not defined by fire size alone. Rather, operational decisions and outcomes are based on a combination of fire size and growth characteristics, available fire fighting resources, and building architecture.

Figure 14.1 shows a representative diagnostic fire and an associated fire suppression timeline from ignition to extinguishment. The diagnostic fire for manual fire suppression need only show time to flashover and approximate heat releases for multi-room propagation. The sequencing of rooms and time durations to reach flashover are important to fire fighting operations.

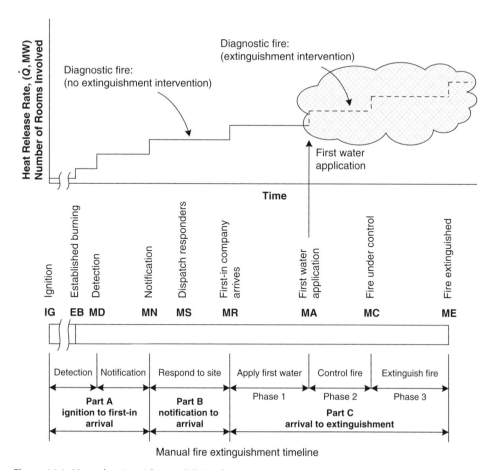

Figure 14.1 Manual extinguishment (M) timeline.

The dashed and shaded part of the diagnostic fire is intended to signify that fire conditions will change after initial water application. At that time, the initial fire environment is upset and composite fire conditions are created by the interaction of the fire, water application, building layout, and ventilation. Composite fire conditions and the resulting building environment continually change until extinguishment occurs. Although diagnostic fire conditions are not accurate after first water application, the descriptor continues to have uses throughout the analysis.

Part A: Ignition to notification describes the building's function in notifying the local fire department. This analysis estimates the diagnostic fire size at the time of notification. *Part B: Notification to arrival* identifies the community response to the fire incident. This part describes available fire fighting resources and estimates the fire size at the time of first-in fire company arrival. *Part C: Arrival to extinguishment* examines actions to complete the extinguishment process.

The diagnostic fire size and conditions at the time of first-in arrival identify a starting point for Part C. This part examines fire suppression activities after the fire department assumes control of the fire ground.

We divide Part C into three phases, even though the boundaries between the phases in an actual fire are often blurred:

- *Phase 1* examines the theoretical fire size when water can first be applied to the fire. Fires that cannot be extinguished with an initial attack require more resources and tactical decisions to control and extinguish the fire. Nevertheless, a Phase 1 evaluation provides substantial understanding about what one may expect during the continuum of fire suppression in the building.
- *Phase 2* examines the fire, building conditions, and fire fighting resources around the time that first water can theoretically be applied. This knowledge provides a basis for assessing whether the fire can be extinguished with the initial attack. If not, the analysis provides valuable information about fire extension, barrier effectiveness, and fire fighting resources.
- *Phase 3* compares conditions identified in Phase 2 with available resources to confine, control, and extinguish the fire. The outcome of Phase 3 is an understanding of the potential extent of the fire before extinguishment is expected.

Often, the Phase 2 information becomes clear when conditions at theoretical first water application are known. Therefore, analysis of Phase 1 gives valuable information about the building's ability to work with the local fire department to reduce damage. The results of the Phase 1 analysis provide a basis for Phase 2 decisions and Phase 3 outcomes.

14.3   Phase 1: Initial Water Application (MA)

After the fire department is aware of the fire and building location, a response is dispatched. The size and type of response depend upon community resources and standard procedures. The first fire company arrives at the site and is followed by the other first alarm responding companies. Fire ground operations begin at first arrival.

Upon arrival, the first-in company begins to acquire information about building access and the general situation. Based on this preliminary knowledge, initial actions are undertaken. As additional units respond and command is transferred to senior officers, the fire ground becomes more organized.

Fire ground operations are a team function because many tasks must be coordinated within a relatively short time period. Officers size up the situation in order to develop a plan for handling the incident. A working fire deals with uncontrolled burning. It usually requires: (1) search and rescue; (2) fire suppression; (3) protecting assets; and (4) protecting exposed property. When possible, these are done simultaneously. Otherwise, the critical fire ground needs are prioritized and completed when possible. After initial fire extinguishment, overhaul operations take place to ensure no hidden fire remains and complete extinguishment has been achieved.

The major activities in a Phase 1 analysis for initial water application (MA) are:

- Find the fire.
- Establish supply water using on-board water and on-site water to supply the attack lines.
- Stretch an attack line and apply first water to the fire.

Although these events are independent, they are carried out simultaneously. The attack line cannot apply water until all tasks have been completed. Time is the measure that links all activities.

14.3.1 Find the Fire

There are two parts to finding the fire. One is locating the floor of origin, and the other is identifying a specific location on the fire floor. Dense smoke often obscures visibility and it becomes difficult to locate the specific fire area (seat of the fire). Accurate information on the fire location is needed to place attack lines and direct hose streams effectively.

The task of locating the fire can range from trivial to very time-consuming, depending upon the fire and the building. Fires in small buildings or fires that are visible from the outside are easy to find. However, it can be time-consuming and difficult to locate a fire in windowless buildings, below-ground areas, upper floors in high-rise structures, large open spaces, and large or complex buildings.

The fire floor and general fire area can be located by:

- Human communication, usually an occupant.
- The building's *textural alarm signaling device*, commonly called an annunciator.
- Observations of a fire officer scouting the building.

An occupant or a security person often provides information about the fire and its location. This is common during periods in which the building is occupied. Alternatively, an officer can learn of the fire's location from a building's fire alarm system annunciator. Technically called a textural alarm signal appliance, this device identifies the location of the fire. Modern devices provide specific text to identify the device in alarm with a time stamp to indicate the order in which the devices actuate. Some may also include a graphic interface panel with the floor plans or a cross-section view. If a building has an annunciator, the unit should be positioned for easy recognition. Often the visible display panel is near the entrance or in the lobby. Large or complex buildings may locate the annunciator in a dedicated room or a fire command center. At other times it may be located so unobtrusively that it cannot be easily found.

When an annunciator is used, the display should give a specific, direct, and unambiguous location of the fire. For example, if the first detector actuated on the south wing of the seventh floor, an indicating light should identify that location. Unfortunately, some annunciators are not that specific. Additionally, the location may be difficult to read or ambiguous. Therefore, part of a performance evaluation examines the location and quality of the annunciator and the likelihood of its identifying a clear, unambiguous fire location.

If the fire officer is not directed to the fire by an occupant or an annunciator, a reconnaissance search must be initiated. Clues, such as the lowest smoky floor level, are a start. However, if an air-handling system spreads smoke throughout the building because it does not automatically shut down, smoke clues can be thwarted. A flaming fire visible to the outside gives a strong indication, if observation is possible. In any event, the fire officer uses available clues to scout out the fire location. Locating the fire can sometimes be difficult and time-consuming because visibility may be poor and locked doors may delay access to building locations.

A performance analysis traces the process for locating the fire. This often exposes potential weaknesses in the building design. For example, the seat of a fire on a floor of origin may be difficult to locate because smoke obscures vision, often to complete blackness. When this occurs, the reconnaissance slows down significantly. Thermal imagers help. Resources may be directed to apply agent to the suspected area. The result may be the start of an effective fire attack or it may produce nothing. Thermal imaging is valuable in locating fires when visibility is obscured or within wall cavities and above ceilings.

A crew of fire fighters stretching a hose line sometimes accompanies an officer trying to locate the fire. This has two advantages. One is the safety of being able to return along the hose line path if conditions become too untenable. The second is that the line can be in place faster if the fire is along an appropriate attack route. However, a disadvantage is that it takes longer to replace a hose line than to place it initially. Wrong explorations can cause delays in attacking the seat of the fire.

14.3.2   Establish a Water Supply

Two sources provide a water supply for fighting fires. One is the on-board (booster) water that can be used to initiate a fire attack before a permanent water supply can be established. The second is the sustained water that is provided by hydrants, static water supplied by ponds, cisterns, and rivers, and mobile water sources (MWS) (i.e. tender operations).

A water supply evaluation analyzes the on-board source and the sustained water source separately. These two sources are combined to produce a time-based picture of supply water quantity and pressure from arrival to extinguishment. Each water supply line is analyzed as a separate sub-project of the complete operation. An IPI chart coordinates all activities and displays time durations and water quantity available for fire fighting operations.

Construction management techniques can be adapted to quantify time durations for all conditions and available staffing. Because these tools are so common and well established, computer programs are available to aid in organization and calculation. Although initial time estimates must now be based on video analysis and experience, these estimates can easily be replaced by task measurements similar to those of construction management.

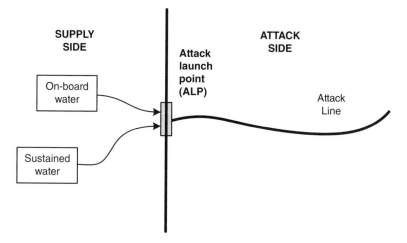

Figure 14.2 Attack launch point.

14.3.3   The Attack Launch Point

A quantitative analysis estimates the time to apply first water to the fire. The analysis separates the process into:

- The time to stretch an attack line and discharge water on the fire.
- The time to supply the attack line with water.

The ALP is a key concept in making the analysis tractable (see Figure 14.2). The ALP is a specific location that separates (i.e. connects) water supply and fire attack operations. Each attack line has its own ALP. The ALP location depends on the building, entry point locations, standpipe location (if any), and fire suppression procedures. Common locations for an ALP are:

- Discharge port of an attack pumper
- Building standpipe outlet
- Wye connection at the end of a flexible supply line from a pumper.

Figure 14.3(a) shows an ALP as the discharge port of an attack pumper. In this illustration, on-board water is augmented by the sustained supply water that moves from a hydrant through a relay pumper to the attack pumper. The attack line starts at the pumper discharge port.

Figure 14.3(b) shows an ALP at an upper floor standpipe outlet. In this illustration, the sustained supply water starts at the hydrant. Supply hose moves the water to the attack pumper and through another supply hose to the standpipe connection. Supply water moves up the standpipe to the level where the attack line is connected. The ALP becomes the upper floor standpipe outlet because that location separates supply water from the start of the attack line.

Figure 14.3(c) illustrates another way to assemble the components. Here, another supply line from the attack pumper to the building has a wye connection at the end. At that location, two attack lines are connected at the wye. This illustration uses two ALPs (ALP 1 and 2) to designate that two attack lines can start from the same location.

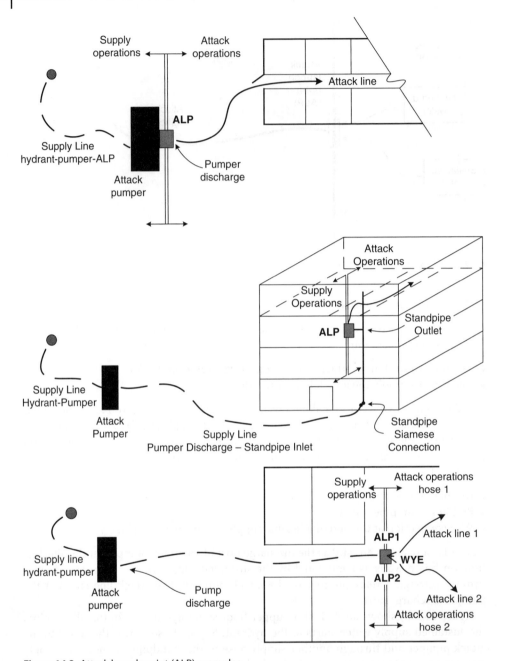

Figure 14.3 Attack launch point (ALP) examples.

The location of the ALP is functional rather than fixed, and the simplicity of the ALP concept belies its importance in quantitative analysis. The ALP is an essential boundary because the analysis of supply water operations is distinctly different from that of attack line evolutions. Water supply can involve many different combinations of water, pumper,

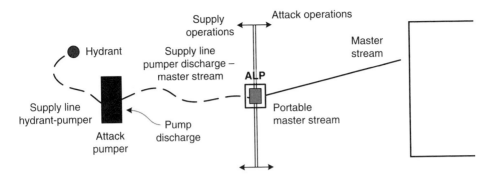

Figure 14.3 (Cont'd)

hose, and other piping. The quantification is dependent on the staffing, equipment, site conditions, building features, and fire ground needs. Nevertheless, each unique setup can be modeled with well-established construction management procedures that are described later in this chapter.

14.3.4 Interior Attack Lines

The goal is to estimate the time first water can be applied to a fire (MA). This time duration combines the time to stretch the attack hose with that for establishing supply water. Although each analysis requires a separate computation, many tasks are done concurrently.

An analysis assumes that an interior fire attack is feasible. In certain situations, such as upper floors of high-rise buildings, an interior attack is the only fire attack available. In other buildings an interior fire attack may not be possible because of fire conditions or access.

Each attack line originates from a specific ALP and moves through the building. Each attack path is evaluated separately. Often, the attack path is evident and can be identified easily. Sometimes alternative paths may be available for an attack route. When this occurs, each can be evaluated during "what if" variability analyses. The IPI chart enables one to compare time durations.

14.3.5 Critical Fire Conditions

A Phase 1 analysis gives substantial knowledge about the building and diagnostic fire. It is usually apparent when an initial attack will be able to extinguish the fire quickly and easily. It can also be clear when conditions exceed initial attack capabilities. Additional information must be obtained when the situation lies between these extremes.

Fire departments regularly use an offensive mode to extinguish the fire quickly. A defensive mode is used when the fire size or building conditions are not favorable for an aggressive attack. Eventually, the fire ground commander will select an offensive mode or a defensive mode as an extinguishment strategy. Critical fire conditions describe the building and fire situation that cause a commander to change from offensive to defensive operations.

Critical fire conditions are not a specific quantity. Rather, this refers to a set of conditions that would indicate that the fire fighting strategy should be defensive. Therefore, the critical fire will be different in one building compared with another. Additionally, critical fire conditions may differ at various locations within a single building.

Critical fire conditions involve the diagnostic fire, the building plan, barrier conditions and locations, ventilation opportunities, the number of available hose lines and their positions, staffing, water supply, equipment, and building stability. Fire fighter risk is another factor in the decision. The Phase 1 information does not determine the strategy. However, availability of the substantial package of information from a Phase I analysis gives more complete understanding of outcomes.

14.3.6   Extinguishing the Fire

Conditions that influence offensive or defensive mode decisions of a critical fire size give a sense of the losses that can be expected at extinguishment. Chapter 15 discusses fire control and extinguishment.

14.4   Summary

The evaluation of a fire scenario from ignition to extinguishment gives a clear picture of the way in which a building's design affects outcomes and the changing nature of risk during the fire.

A Phase 1 evaluation provides a good understanding of the influence of building and site features on the fire size at MA. Although the representative performance analysis may be specific to the selected scenario, a substantial amount of information relating to the building and site design will be obtained. Often, a single carefully selected scenario will be sufficient to identify the strengths and weaknesses of the building and site design. Extrapolation from the initial scenario analysis will augment the information. When more knowledge is needed, additional targeted diagnostic fire scenarios can be evaluated.

PART TWO: A BRIEF LOOK AT FIRE FIGHTING

14.5   Initial Fire Ground Actions

When a fire department learns of a fire incident, the responding forces traveling to the site never know exactly what they will encounter when they arrive. At one extreme the call may be a false alarm or a small waste basket fire. On the other hand, it may involve a major incident that threatens life and has the potential to be a major catastrophe.

When the fire department does arrive and begin to take action, a generalized process unfolds. We will attempt to portray activities that commonly take place at a "working fire" or at multiple alarm fires. A working fire requires action by most of the personnel responding to the first alarm. When the situation requires more resources than can be provided by the first alarm response, additional alarms are called. Communities have standard operating procedures that define responses for these situations and for mutual or automatic aid.

The first arrival is usually an engine company. This first-in company places itself in what appears at the time to be the most appropriate position to carry out initial activities. These involve balancing several potentially conflicting needs that include gathering information about building access points, the location and size of the fire, learning if there are any potential life safety threats, establishing a water supply, and stretching attack lines. Shortly after the first company arrives, a chief officer and the other first alarm companies arrive.

The incident commander, also called the fire ground commander or the chief, is in charge of the fire scene. Initially, the officer of the first arriving company assumes this role. However, the position is transferred when a more senior officer arrives. The incident commander is in responsible charge of the scene and has management control of all operations. Experience, knowledge, and training are important to a fire ground commander's educational background.

Timely, astute decisions can make the difference between a good stop and a fire that got away. Fire ground decisions are based on the understanding of the situation at the time. Initial information is often inaccurate, insufficient, and constantly changing. Consequently, the incident commander's allocation of resources and strategy may be adjusted as better information is acquired or the situation changes. Occupant and fire fighter safety and extinguishing the fire are universal considerations in fire ground decision-making.

14.6   Information

The incident commander begins to acquire information before reaching the site, and updates continue until the fire is extinguished. A broad picture of the situation is obtained by a *size-up* that becomes the basis for the fire ground commander's initial decisions. A size-up is a mental model of the entire system that makes up the fire ground conditions. A size-up begins on route with forming reasonable expectations and then becomes more defined on arrival at the site when better information becomes available.

Fire ground operations use available community resources, manpower, and equipment to rescue endangered individuals and to reduce the destructive damage of a hostile building fire. The size-up provides the information on which to base decisions about the strategy and tactics to achieve that goal.

An ideal size-up would include the type of information described below. Obviously, time constraints of required early action preclude collection of all the information initially. When complete information is not available, the fire ground commander makes decisions based on the available knowledge at the time. The major questions that a size-up attempts to answer are:

- *Life safety.* What is the occupancy and how many people are in the building? Where might they be located? Are they capable of self-evacuation or guided evacuation? How certain is the information?
- *Building construction.* In buildings of this type, can barriers generally contain a fire to one or few rooms, or do they allow the fire to propagate easily? What is the collapse potential? Can the building construction be used to advantage in controlling the fire and limiting its propagation? Does this building present a safety hazard to fire fighters?
- *Fire.* Where is the fire? What is its size? What is burning, and what is the fuel loading? Are hazardous materials present? Are they involved in the existing fire? Can they become involved if the fire extends?
- *Water supply.* How many hydrants are available and where are they located? How much fire flow is available? Are static water supplies available for drafting? Must tenders supply water? How will the terrain and vegetation affect supply water delivery?
- *Resources.* How many fire companies and fire fighters are available? What other types of apparatus are present or available? Where is the equipment now placed? Should it be relocated? Are additional alarms or mutual aid needed? If additional equipment responds, where should it be placed? Are personnel needed for a primary search and rescue or are they available for fire fighting? How many hose lines can we place and where should we position them? Are fire fighters available for relief, particularly in hot or extremely cold weather and difficult fire-fighting conditions? What fire flow can be delivered to the attack lines and what is available when the water supply has been established?
- *Fire attack.* Where are the building access locations? Are there any natural or man-made obstacles to approaching the building from different sides? Where are the stairwells? Are standpipes available and where are they located? Are they wet or dry? Does the building have an operating stationary fire pump and water storage? How do we gain access to the floor of origin? What fire attack routes are available? How clearly are they recognizable? Will doors have to be forced to reach the fire? How many? What size and how many hose lines can be advanced with the staffing available? How long will it take to stretch the hose lines? Must we position hose lines at critical locations to defend occupants, fire fighters or valuable property from fire extension? Will these positions help or hinder fire attack or control? Where and how can we ventilate?
- *Environmental conditions.* What is the time of day? What are the weather conditions (e.g. wind, extremely hot or cold, storms, precipitation)? What site conditions will influence movement (e.g. mud or snow)? How will the wind conditions affect fire fighting? What are the smoke, heat, and visibility conditions within the building? What are the heat and visibility conditions near the fire?
- *Exposures.* Are there any other external buildings, materials, or wild land areas exposed to the fire? Should we protect those exposures before or after attacking the existing fire? Are there internal exposures that should be protected?

- *Building services.* What fuels are present, where are they stored, and where is the fuel shutoff located? What is the electrical system and where is the shutoff? Is emergency generation available? Do we have emergency lighting? Where? Does the HVAC system continue operation, shut off, or shift to an emergency mode of operation at detection? Can its operation be changed to help fire fighting? Where are the controls and how are they used? Do elevators remain in operation or are they recalled at detection? Is there a fireman's key for emergency use? What other building services information is available that would influence life safety or fire fighting?
- *Sprinklers, standpipes, and fire pumps.* Is the standpipe wet or dry? How will the standpipe design affect fire fighting? Does standpipe water need to be augmented from the building's fire department connection? Does the building have a sprinkler system at the fire location? Would a fire department connection to the sprinkler siamese augment water quantity and distribution to enable the system to operate better? Are stationary fire pumps available and working?

We may note that these questions – and a few more – are addressed in a building performance analysis. However, the incident commander manages the emergency with the information that is available at the time. Although better decisions might be made with more information, the fire will not wait. As it continues to burn, the situation changes.

The fire companies take initial actions according to experience, pre-incident building information, preconceived operational planning, and immediate needs. The fire ground commander will modify strategy to reflect conditions that are evident or that change during the fire.

Communications and the chain of command are essential features in fire ground operations. New information about conditions inside the building must be transmitted to the command area. Communication is crucial to update status reports, deploy resources, and to learn the outcomes of any actions.

14.7 Pause for Discussion

The size-up information described above represents an ideal package of knowledge that a fire ground commander would like to have in order to deploy responding forces effectively. Clearly, acquisition of some items would be time-consuming to obtain during the fire. On the other hand, much of the information could be determined during a pre-incident performance analysis and reinforced with in-station training exercises.

In recent years, the number of working fires has been decreasing. This results in fire service officers having less opportunity to build experiences with which to make fire ground decisions. Experience is an important part of a fire officer's education. The shortfall in fire ground experiences may be partially filled with in-station training using performance-based fire scenario simulations.

A partnership between the fire service and the fire safety engineer can pay rich dividends for each. The fire service can give engineers the needed estimating skills as well as insight into a range of "what if" conditions that influence outcomes. Engineering analysis can help to identify realistic diagnostic fire conditions to improve knowledge and reduce uncertainty. A whole range of modern technology innovations can integrate pictures, videos, and three-dimensional (3D) simulations of building conditions that may be encountered.

In-station training for specific building fire simulations is analogous to war games training in military education. While these concepts have practical applications, they are not part of this book.

14.8   Manual Fire Fighting

Fire fighters are knowledgeable craftsmen at their trade. Similar to master carpenters, plumbers, electricians, iron workers, and other trades, the fire fighter has acquired proficiency in fire ground activities that require skill, training, and ingenuity. And, all activities must be done during a very short time period. Although little time is available for contemplative thought, somehow the tasks get done. Competence, judgment, and rapid reaction are all part of the skill set. Time for thoughtful deliberation is a luxury not generally available on the fire ground.

The practice of fire fighting has changed dramatically during the span of a single human generation. A generation ago, the traditional measure of fire department effectiveness was associated with response time. A rapid response enabled the fire to be attacked when it was small and weak. When it worked, this was a reasonable assumption considering larger fire department staffing at that time. A "good stop" could be achieved. When it didn't work, we had a "big one."

Although respiratory protection for toxic and irritant gases in a smoke environment had been available for most of the last century, it was rarely used. Early protection involved masks that used filters, fresh air hose, self-generating air canisters, or stored compressed air. The devices were so heavy and cumbersome that restriction of movement discouraged their use. This condition was dramatically changed in 1976 when a new type of self-contained breathing apparatus (SCBA) was introduced to the fire service. The new devices were small, reliable, lightweight, and theoretically held 30 minutes of clean air. The new SCBA was readily accepted and is now almost universally used by the fire service.

Introduction of the SCBA caused protective clothing to change in form and function. Forty years ago, turn out gear was designed to shed water. From rubber boots to coat to helmet, the primary purpose was to keep the fire fighter dry. Any insulation was intended to keep the wearer warm in cold weather, rather than to provide protection from the heat of a fire. Because the SCBA enabled a fire fighter to move through smoke, clothing needed to be redesigned to provide insulation and protection from the heat of a fire.

The protection from smoke and heat provided by the new SCBA and protective clothing (personal protection equipment, PPE) now enables fire fighters to move relatively easily through hostile environments. Routine fire fighting a generation ago used difficult to move 2½ in (64 mm) hose lines. These larger hoses have been replaced with smaller lines [commonly 1¾ in (44 mm)] having greater mobility.

The combination of personal protection and mobility now allows more aggressive fire attacks than could be used a generation ago. This is a mixed blessing. While fire fighters can become more aggressive with fire attacks, they can also find themselves in unexpected danger. Knowledge of buildings and the fire environment is of greater importance now than it was in earlier generations.

In addition to the personal protection and water delivery improvements, better two-way radio communication improves information-processing and decision-making between fire fighters and the fire ground commander. While the fire suppression activities of locate, attack, control, and extinguish remain valid, capabilities to do these actions have improved dramatically.

14.9   No Two Fires Are Alike

If we were to view fire fighting in military terms, the fire is the enemy and the building and site would be the terrain on which the battle is fought. The fire department is our army.

Often, early in military officer training a "trick question" is, "How would you fight this enemy?" The unaware would answer the question. The more astute would give the correct answer, which is, "It depends on the terrain and the situation."

For a similar question, an appropriate fire officer's response would be, "It depends on the building and the situation." Each building and fire present different challenges and complications. The strategy for dealing with problems and the tactics for solving them will be different for each building. Thus, one often hears that "No two fires are alike."

It may be accurate to recognize that no two fires, just as no two wars and no two battles, are alike. That observation does not mean that the process cannot be analyzed in a systematic, methodical way. Just as with the military, the building fire lends itself to decomposition using modular, interactive parts. Organizing available parts to evaluate any specific fire situation becomes a reasonable task. The results provide a good understanding of what can be expected with fire department operations.

14.10   Summary

The role of a fire department is that of a change agent for conditions that would otherwise occur without its intervention. In addition to putting out the fire, a fire department may rescue occupants, save important assets, protect neighboring structures, and generally improve otherwise deteriorating conditions of a burning building. Fire department intervention changes the state of a building fire to make the outcome better and safer than it would otherwise become.

A building fire can be a very complicated situation. The role of the fire department is to stabilize an out-of-control emergency. Small, slow fires present different situations than do large, fast fires. Occupied buildings involve different priorities than unoccupied buildings. Although the fire department must take the situation and building as it exists, it can't turn back the clock. Damage that has occurred is gone. People who have died cannot recover. Corrective actions are not instantaneous and are constrained by limitations of resources, the building and site, and fire conditions.

The incident commander must balance many urgent needs with available resources. This often requires prioritization. Life safety is a primary concern, and putting out the fire is a close associate. Often, these activities complement each other. At other times, available resources are insufficient to do both simultaneously. Decisions must be made for a dynamically changing situation with available information of variable accuracy, using the building design and community resources. The outcome depends on the "terrain and the situation." The greater the knowledge of these factors and the diagnostic fire "enemy," the better the chance of a good outcome.

Although the fire service has evolved over the years with resourcefulness and ingenuity in fighting fires, the building itself plays a major role in the success or failure of manual fire fighting. The world has more experience in manual fire suppression and less quantitative data with which to evaluate a building's influence on fire fighting than any other part of the entire fire safety system. It is anticipated that this condition will change as knowledge improves computational methods based on a rigorous analytical framework.

PART THREE: SUPPLY WATER ANALYSIS

14.11   Introduction

Time links all parts of a performance analysis. The (local) time duration from arrival (MR) to first water application (MA) is the performance measure for $M_{Part\,C}$, Phase 1. This time is added to the time durations to complete M_{PartA} (EB to MN) and $M_{Part\,B}$ (MN to MR).We compare the global time from EB to MA with the diagnostic fire to estimate the fire size at theoretical first water application.

The incident commander is in responsible management control of the fire scene. Although this includes the entire operation, here we focus only on fire suppression actions.

Two interrelated components dominate fire suppression strategy and tactics. The first may be described as operational management. The second involves moving water onto the fire. Both play important roles on the fire ground, and both affect the building–fire department analysis.

Constraints influence fire ground decisions as well as the general complexity. For example, life safety of occupants and fire fighters is pre-eminent. Often, safety considerations influence otherwise feasible actions. Community resources determine the availability of people and equipment. Occupational safety and health regulations, minimum staffing requirements, and the prudence of providing back-up protection influence suppression decisions.

The water application component involves selecting fire attack routes, estimating the time to place attack lines, putting water on the fire, and determining water flow (gallons/minute or liters/minute). Time can be calculated for any proposed evolution with any specific number of fire fighters. IPI chart organization gives a good picture of the fire and multiple attack and supply line sequencing.

Part Three: Supply Water Analysis describes procedures to calculate the time and quantity of supply water delivered to the ALP. Part Four: Interior Fire Attack Analysis describes procedures to estimate the time to stretch interior hose lines from the ALP along interior attack routes to the fire. Examining outcomes and extrapolating results for alternative conditions give an understanding of the building's ability to work with the local fire department.

14.12   Scenario Analysis

We illustrate the calculation process with a continuing example that is interwoven with concepts, analytical tools, and associated calculations. Each part can be evaluated as a separate unit. Eventually, they are combined to provide a complete picture of the operation.

The objective for this set of examples is to demonstrate analytical procedures rather than to represent good fire ground decisions. The process can be adapted to incorporate any attack or supply evolutions on the fire ground.

Operations have been selected to demonstrate specific techniques. For example, company arrivals have been phased to illustrate how an IPI chart can integrate different situations. The evolutions illustrate ways to structure different tasks.

Selective information for each example is introduced as needed to avoid premature complications. Operations and time sequencing have been selected to demonstrate techniques. Time is emphasized because it is the single factor that links all events.

Example 14.1   Outline a plan to calculate the time from first-in arrival to first water application for a fire in the southwestern (SW) part of the third floor of the building shown in Figure 14.4.

The fire safety engineer defined the diagnostic fire earlier in the analysis. This provides a time-based estimate of fire propagation and smoke movement. In addition, estimated times for $M_{Part\ A}$ (EB to MN) and $M_{Part\ B}$ (MN to MR) have been calculated. We delay disclosing these values until they are needed in later examples.

The engineer determined the first alarm response to be two engines, one ladder truck, and a battalion chief. Engine E1 is the first-in unit. Engine E2 and the battalion chief arrive 3 minutes later. Ladder L3 arrives 4 minutes after E1.

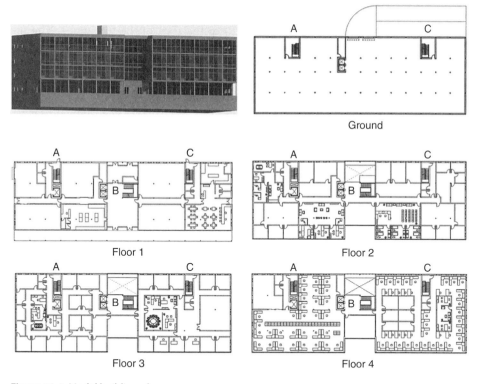

Figure 14.4 Model building plans.

Solution:

At Engine E1 arrival, the first-in officer and fire fighter crew must assess the situation, identify the fire location, determine building access points, assess life safety risks, establish a plan of action, and start initial operations.

Much information of Section 14.6 can be filtered out quickly. Here, let us assume that MN occurred during late evening. Most of the building population will not be present, and an alarm is assumed to have evacuated the building. This makes life safety problems different than during normal working hours. Assume also that an occupant tells the officer that the fire is on the west wing, third floor. No fire is showing from outside observation.

The officer of first-in Engine E1 becomes the initial incident commander. This assignment is transferred when the battalion chief arrives 3 minutes later. The E1 officer gives a situation assessment at the transfer of command. Arrivals of Engine E2 and Ladder Truck L3 complete the first alarm response team.

Although many fire ground activities take place when the fire department arrives, we will examine only fire suppression. A plan to calculate a time for first water application to the fire involves the following activities:

1) Examine potential fire attack routes and select one for analysis. After knowledge has been obtained from the selected scenario, other alternatives may be examined, if necessary.
2) Select the ALP.
3) Identify supply water sources.
4) Estimate any time delay from arrival (MR) to the start of the fire attack. The fire attack may start at arrival or may be delayed by initial information acquisition or other emergency needs. The IPI diagram can incorporate any delay in the start of fire attack into global time.
5) Establish time durations to deliver supply water to the ALP. This analysis will have two parts: delivery of on-board water to the ALP, and delivery of sustained water to Pumper E1 to augment the on-board water.
6) Calculate the time duration for an attack line using the route selected in step 1 to move from the ALP to the fire.
7) Calculate the time for first water to reach the fire.
8) Coordinate time lines with the IPI chart.

Example 14.2 Describe several possible fire attack routes in the building of Figure 14.4, and select one for analysis.

Solution:
Figure 14.5 shows the position of first-in Engine E1 and plans for Floors 1 and 3. We start with knowledge of the fire's location on the SW end of the third floor. Several alternative attack routes are possible. Among the choices are:

1) From E1 through the main lobby, up the central stairs to the third floor. From the third floor lobby along the corridor near the street, to the fire (Alternative 1).
2) From E1 through the main lobby, up the central stairs to the third floor, from the lobby along the more circuitous corridor route to the fire (Alternative 2).
3) From E1 through the first floor to a standpipe located in Stairwell A. Connect a supply line to the fire department connection (FDC) on the first floor and connect an attack line at the second floor landing ALP. Move the attack line from the ALP to the third floor landing and along corridors to the fire (Alternative 3).
4) Through the lobby and corridor to Stairwell A. Then, not using the standpipe, move up Stairwell A to the third floor through corridors on the west side of the building (Alternative 4).

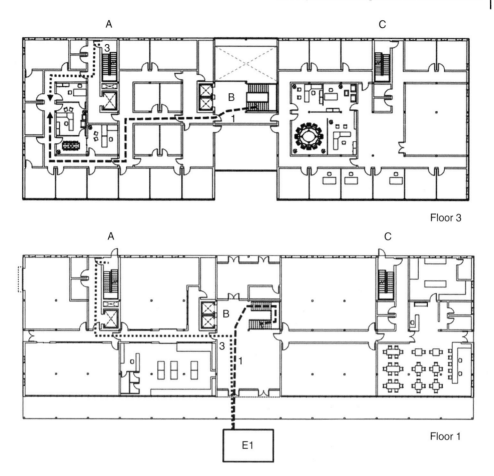

Figure 14.5 Model building: Floors 1 and 3.

Additional variations may be used, such as dropping a rope from a stairwell and lifting the hose rather than carrying it up the stairs. A window may also be used for this movement. Although a seemingly large number of alternatives may appear possible, they include variations that use many of the same tasks. Discussions with the local fire department often help to uncover ways of thinking that affect decisions.

Discussion:
The only difference between Alternatives 1 and 2 is the attack route on the third floor. Supply water is the same. Variation analysis can determine the time difference for alternate attack routes, if useful.

When standpipe connections are used, attack lines are normally connected one floor below the fire floor. If the standpipe in Stairwell A were used, the ALP would be at the second floor landing. Attack lines would stretch to the third floor, through doors, and along one of the corridors to the fire.

Standpipes are important for interior fire fighting. However, the decision is often influenced by whether the standpipe is wet or dry. Wet standpipes have water under pressure, and may be put into service more quickly. Dry standpipes have more operational problems and must be supplied with water. Fire fighters usually prefer to stretch hand lines for the first few floors rather than to use a dry standpipe. Above about four stories, all standpipes are commonly used.

One may note that alternatives are often modifications that have common parts. We will select Alternative 1 as an initial base for timeline calculations.

Example 14.3   Select an ALP for Alternative 1 of Example 14.2.

Solution:
We will select the ALP as a wye at the first (ground) floor landing of Stairwell B. A 2½ in (76 mm) flexible supply hose will connect E1 to the wye. A single 1¾ in (44 mm) attack line will be attached at the ALP and stretched to the fire on the third floors.

14.13   Supply Water Analysis

Supply water to the ALP comes from two sources:

1) *Water stored on board pumpers.* In these examples, we will use only the water carried on the attack pumper as this source. When necessary, other units may transfer some on-board water before a permanent sustained water supply has been achieved.

2) *Water that is delivered to the site from other sources.* Sustained water is the continuous site water that is moved to the attack pumper. The most common source of sustained water in urban areas is underground piping and hydrants. However, static sources such as ponds, lakes, streams, and cisterns also provide sustained water. In addition, water tender trucks that relay water from other locations (mobile water supply, MWS) can provide sustained water.

Water movement through each supply line is analyzed as the separate "mini" construction project. The results are combined in an IPI chart to describe a phased time-related supply process.

The following examples describe the process. Example 14.4 discusses pumping on-board water from E1 to the ALP. Example 14.5 describes supply water delivery from Hydrant H1 to Engine E2. Example 14.6 describes relaying supply water from E2 to E1.

Example 14.4   First-in Engine E1 positions itself at the entrance to the building, as shown in Figure 14.6. Describe the process to deliver on-board water from E1 to the ALP at the first (ground) floor landing of Stairwell B.

Solution:
Major work activities to supply on-board water to the ALP are as follows:

1) Arriving Engine E1 stops in position and driver prepares pump for operation.
2) Fire fighters remove a flexible 2½ in (63.5 mm) supply hose and necessary tools from the apparatus bed and drag the hose from E1 to the ALP at Stairwell B in the building lobby.
3) Fire fighters disconnect the supply hose at pumper and connect the hose to pumper discharge port.
4) Fire fighters connect a 2½ in to two 1¾ in wyes at the ALP.
5) Operator pumps on-board water to ALP.

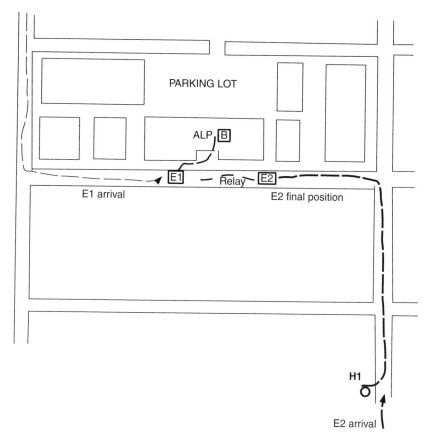

Figure 14.6 Model building: site plan. ALP, attack launch point.

Example 14.5 Supply water can be delivered from H1 to E1 in several ways. In this scenario, incoming Pumper E2 will connect to Hydrant H1 and lay a 3 in (76 mm) supply hose to a position near the building, as shown in Figure 14.6. Then, a relay will be established to move water from E2 to E1.

Describe the work activities to move water from Hydrant H1 to Engine E2.

Solution:

The major work activities are as follows:

1) Arriving Engine E2 stops at H1.
2) Fire fighter alights from E2, removes a 3 in supply hose and tools, and prepares H1 for water discharge.
3) E2 lays hose to deployment position at building.
4) E2 driver prepares pump for operation.
5) Fire fighters alight E2, disconnect 3 in line from hose bed and connect supply line to intake port.
6) Fire fighter opens hydrant, and water flows from H1 to E2.
7) E2 begins pumping supply water from H1.

Example 14.6   Describe the work activities to relay water from E2 to E1.

1) Engine E2 stops in final position.
2) Fire fighters alight E2 and connect a 3 in relay hose to E2 discharge port.
3) Fire fighters drag relay hose from E2 to E1.
4) Fire fighters connect relay hose to E1 intake port.
5) E2 operator pumps water from E2 to E1.

14.14   Supply Water Discussion

Examples 14.4–14.6 describe three common operations for moving water. The objective is to introduce general work activities for the evolutions. Before describing ways to calculate time durations, we note that work activities can be adjusted to meet site-specific needs.

In Example 14.4, fire fighters have a clear path from E1 to the base of Stairwell B. If a closed door were present, an additional activity would be inserted to open the door. The method of opening would involve a time duration that must be incorporated into the analysis. Any work activity that requires time to complete a task is incorporated into the description.

Although it was not part of our description, Engine E1 could have taken Hydrant H1 upon arrival and then lay the supply hose to the building. This eliminates the need for relay from H1 through E2 to E1. However, it also prevents the opportunity to discuss combining multiple operations.

Example 14.6 describes laying a large-diameter supply hose between E2 and E1. Supply hoses are heavy and require much human effort and time to move large distances. It would be possible for E2 to drive to E1, make a relay connection and then return to its final position shown in Figure 14.6. If this were the expected operation, those activities would be identified and tasks would be incorporated into the analysis.

Often, several alternatives are available to achieve a goal. The selection depends on fire ground conditions and initial decisions of the officer in charge. We have identified a base scenario for analysis. Modifications to describe alternative scenarios can be easily evaluated.

The objective is to calculate time durations for each evolution and combine the results to get a composite time. Although the processes in the above examples are simple to describe, they are not sufficient for rigorous analysis. Before returning to these examples to calculate time durations, we will discuss some concepts and tools that structure an analysis.

14.15   Project Analysis

Site and water source conditions can vary greatly at a fire scene. Different communities have their own operating procedures, types of equipment, and numbers of fire fighters available to do the work. Analytical procedures must be able to accommodate broad operational variability.

The analytical process of establishing a water supply for fire fighting is similar to constructing buildings, bridges, ships, airplanes, or roads. Although projects may differ, tools for managing complex, one-of-a kind construction projects are the same. One needs merely to adapt existing project management tools to fire fighting applications. An added benefit is that there are already many computer programs and information technology tools available to reduce computational time and add analytical breadth and flexibility.

The next several sections describe project management concepts for analyzing supply water evolutions.

14.16   Task Modules

An analysis of supply water operations is considerably simplified when one recognizes that all tasks can be grouped into one of two modules.

- *Positional modules* are tasks that involve actions performed at a single location. For example, connecting to and opening a hydrant, preparing a pump to deliver water, and coupling or uncoupling a hose are actions that may be performed within a relatively small space.
- *Distance movement modules* are tasks that involve movement of hose or water from one location to another. For example, laying a supply hose from a hydrant to an attack pumper, laying a hose from an attack pumper discharge port to a fire department connection (FDC), and movement of water from a pumper to an ALP are distance movement tasks.

Each task in an evolution is either a positional module or a distance movement module. Each evolution combines appropriate task modules to describe the specific scenario.

14.17   Time and Tasks

Each supply water evolution requires the completion of a defined set of specific tasks. Regardless of available staffing, equipment, or site conditions, all tasks must be completed before water can move from a source to the ALP. The fire fighter must do the fieldwork. The engineer must:

- Identify the modular tasks needed to complete the work.
- Model a network to represent task coordination and sequencing to conform to staffing and site conditions.
- Estimate the time duration for each task.
- Calculate network time durations to complete the work.
- Coordinate intermediate results into holistic descriptions.

This process is identical to typical project management needs that link disparate tasks requiring sequence and time coordination.

Fortunately, the application for fire fighting is tractable because of two factors:

1) Interactive project management tools are available to organize, quantify, and integrate. The tools that relate to fire fighting are:
   a) Work breakdown structure (WBS)
   b) Network construction and precedence grid

   c) Estimating time durations

   d) Calculating network start and completion times for evolutions and their combination

   e) Using IPI (Gantt) charts to coordinate, understand, analyze, and describe the outcomes.

2) Evaluating mini-projects to move water through discrete paths from Point A to Point B becomes relatively straightforward with network models that connect modular positional and distance movement tasks. The time to complete tasks reflects the conditions that exist.

A systematically organized inventory of task information could reflect widely varying field and staffing conditions. In many ways, developing analytical networks is analogous to construction using a group of "standard Lego blocks" that represent each modular task and the conditions that affect its time to complete. One assembles the modules to represent the component being evaluated.

14.18 Variability

Although positional and distance movement tasks are functionally modular, the time to complete them depends on field conditions and staffing. The major factors that influence time variations are as follows:

- *Staffing.* Fire fighting is a team function. Supply water delivery is task-dependent. Some tasks must be done sequentially while others may be done concurrently. One person can easily do some tasks. Other tasks may require more than one person, or they may be done faster when several individuals are available to do the work.
- *Equipment.* The number of engines, ladder trucks, and other vehicles can influence the strategy for vehicular placement needed to supply water and attack the fire. For example, when fewer pumpers are available, relay operations and tandem supply operations become less efficient or impossible. Because one can identify the time sequencing of arriving units, calculations for supply water delivery may be estimated with confidence.
- *Distance.* The time to complete distance movement tasks is a non-linear function. That is, the time to move an initial x ft is different than that needed to move the same x ft later in the task.
- *Terrain.* Slopes, surface materials, vegetation, landscaping and other ground conditions affect maneuverability and the time to complete distance movement tasks.
- *Weather.* Wind, snow, ice, mud, and extreme heat and cold conditions affect time durations to complete tasks.
- *Hose size.* The hose size affects time to complete tasks. A supply hose is of a larger diameter than attack lines, and the work to move a large-diameter hose requires more time and effort.
- *Fire fighter training and experience.* Firefighter familiarity with the operations and their frequency of doing them influence the time to complete tasks.
- *Fatigue.* Tiredness of fire fighters affects speed of completing work.

The time to complete positional and distance movement tasks must be estimated. At the present time, no accepted values have been published. When a systematic acquisition and publication of time estimates for modular fire fighting tasks becomes available, the data would include the variables noted above and their combinations. For now, we must rely on the opinion of individuals with fire service experience or training exercise videos. Fortunately, most modular tasks are routine and common to fire fighting operations.

14.19   General Analysis

We illustrate supply water evaluations with an integrated set of examples.

The procedure for calculating time durations for a supply water evolution involves the following:

1) Select a path to move supply water from one location (A) to another (B).
2) Use a work breakdown structure (WBS) to identify the positional and distance movement modular tasks needed to complete the evolution.
3) Identify task precedence to complete the project. Staffing affects activities and the order in which they are done. A precedence grid describes task sequencing.
4) Construct a network to describe task events and their sequencing.
5) Estimate time durations for the network events.
6) Calculate the time duration to deliver water from Point A to Point B.
7) Determine the water flow at the ALP.
8) Display the task durations and sequencing on an IPI chart. This provides visual perceptions of interactions as well as useful information and data.
9) Select the time duration for water supply to be established.

14.20   Work Breakdown Structure

The first step in completing a WBS for supply water evolutions is to identify the tasks needed to complete the project. These tasks are grouped into logical components that are decomposed to a level that provides good understanding and relatively simple quantification. Example 14.7 describes the WBS for the evolution of Example 14.4.

Example 14.7   First-in Engine E1 positions itself at the entrance to the building, as shown in Figure 14.6. Example 14.4 described the general work activities to deliver on-board water to the ALP at the base of Stairwell B. Describe the modular tasks to complete this evolution.

Solution:
The tasks to deliver water to the ALP are described below. The actions within each task are provided for information.

Each task is labeled by letter, given an identification descriptor, and designated as a positional (P) or a distance movement task (D). The fire fighters who do the work are identified.

  Operation: Pump on-board water from E1 to ALP.
  Crew: Driver (F1), officer (F2), two fire fighters (FF3, FF4).

ALP is located at the base of Stairwell B.

A) Set E1 to begin pumping operations.
   (Stop at position; set brakes; disengage drive shaft; engage pump; drop tank water to
      pump; exit cab; set chocks.)
   Set E1 – F1 (driver); positional task (P)

B) Don SCBA and any remaining PPE.
   Prep 1 – F2 (officer), FF3; FF4; positional task (P)

C) Remove supply hose and tools from E1 bed.
   (Remove 2½ in supply hose, necessary hand tools, and a 2½ to two 1¾ in wye from
      E1 bed.)
   Prep 2 – F1, F2, F3, F4; positional task (P)

D) Drag 2½ in supply line from E1 to the ALP.
   [Drag supply hose to Stairwell B (ALP).]
   Hose 1 – FF3, FF4; F2; distance movement task (D)

E) Prepare E1 to pump.
   (F1 continues removing 2½ in supply line from E1 bed until enough hose can com-
      fortably reach the ALP; F1 removes an additional length to comfortably reach the
      pump panel; disconnect the supply hose and drag the last section to the pump
      panel; connect the 2½ in hose to the pump output port.)
   Prep 3 – F1 (driver); positional task (P)

F) Connect wye to 2½ in supply line at Stairwell B.
   Conn 1 – F2 officer positional task (P)

G) Signal to E1 to charge 2½ in supply line.
   Signal 1 – F2, F1; positional task (P)

H) Acknowledge signal to charge 2½ in supply line.
   Signal 2 – F1, F2; positional task (P)

I) Pump operator opens pump to charge and pressurize supply line.
   Pump 1 – F1; positional task (P)

J) Water moves from E1 to ALP.
   Flow 1 – distance movement (D)

Figure 14.7 shows the tasks of the WBS in a hierarchical diagram for easy recognition.
Grouping of tasks in this type of diagram makes communication with others easier.

The WBS only identifies the tasks to complete the project. The tasks are discrete and
must completely describe the needed activities. The numbers of fire fighters and their
assignments, the sequencing of tasks, and the time to complete the tasks are not considerations in the WBS.

14.21   Task Precedence

Task sequencing is identified after completing the WBS. Although staffing is not a consideration for describing tasks to complete a project, the number of fire fighters available
to do the work now becomes important.

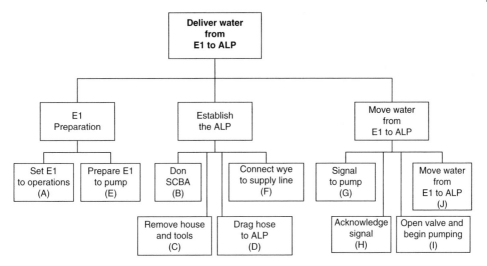

Figure 14.7 Example 14.7: work breakdown structure (WBS).

Available staffing influences the sequence of operations. When staffing is reduced below an optimum number, the order of activities may change because fewer fire fighters must do multiple tasks in a different sequence. When more fire fighters are available, some tasks can be done simultaneously rather than sequentially. Although each supply water project uses the same tasks, their sequencing is adjusted to reflect the number of fire fighters available to do the work.

A *predecessor* is an event that occurs *immediately* before another event. A *successor* is an event that occurs *immediately* after another event.

Table 14.1 recasts the WBS tasks described in Example 14.7. The information is organized as follows:

Column 1: Each task of the WBS is given a letter designation. Although these designations are not essential, the letters are convenient for network logic checks.

Column 2: Each task is given a shorthand descriptor for convenience in showing activities on network nodes. Firefighter designations are shown because staffing is important in establishing event logic precedence as well as to recognize that task assignments are correct.

Column 3: A short description of the task. The symbols P or D signify positional or distance movement modules.

Column 4: The time estimate to complete each task.

Column 5: Predecessors identify the tasks that immediately precede the WBS tasks of Column 3.

Column 6: Successors identify the tasks that immediately succeed the WBS tasks of Column 3.

Although it is not necessary to show both the predecessor and successor tasks, network logic checks are easier when both types are available.

Time durations to complete each task are also shown in Table 14.1. Values are based on video analysis and expert opinion. The distance between Engine E1 and the base of Stairwell B is estimated at 180 ft (54.9 m). Although this sequence of tasks is

Table 14.1 Example 14.7: work breakdown structure (WBS)

Operation: *Pump on-board water from E1 to ALP*
Crew: *Driver (F1), officer (F2), two fire fighters (FF3, FF4)*
ALP is located at base of Stairwell B

	Symbol/FF	WBS activity (task) description	Time duration (min)	Predecessor	Successor
Start	Start	Arriving Engine E1 stops in position at the building	–	–	Set E1 (A) Prep 1 (B)
A	*Set E1* FF1	Set E1 to begin operations (P)	1.3	Start	Prep 3 (E)
B	*Prep 1* F2, FF3, FF4	Don SCBA and PPE (P)	1.8	Start	Prep 2 (C)
C	*Prep 2* F2. FF3, FF4	Remove supply hose and tools (P)	1.4	Prep 1 (B)	Hose 1 (D)
D	*Hose 1* F2, FF3, FF4	Drag supply line from E1 to ALP (D)	2.0	Prep 2 (C)	Conn 1 (F)
E	*Prep 3* F1	Prepare E1 to pump (P)	1.2	Set E1 (A)	Signal 2 (H)
F	*Conn 1* F2	Connect wye to supply line (P)	0.7	Hose 1 (D)	Signal 1 (G)
G	*Signal 1* F2.F1	Signal E1 to charge line (P)	0.2	Conn 1 (F)	Signal 2 (H)
H	*Signal 2* F1, F2	Acknowledge signal to charge line (P)	0.1	Signal 1 (G)	Pump 1 (I)
I	*Pump 1* F1	Open pump (P)	0.3	Prep 3 (E) Signal 2 (H)	Flow 1 (J)
J	*Flow 1*	Water moves from E1 and arrives at ALP (D)	1.0	Pump 1 (I)	End
End				Flow 1 (J)	–

SCBA, self-contained breathing apparatus; PPE, personal protection equipment; ALP, attack launch point.

straightforward, occasionally complex evolutions can arise. A precedence grid helps to identify logical errors.

A precedence grid is a matrix that shows sequential relationships. Figure 14.8 shows the precedence grid for Table 14.1. An X in a box above the diagonal signifies a successor event. Events below the diagonal show predecessor events, although it is not necessary to show both. Consequently, only values above the diagonal are used.

A precedence grid provides a rapid way to recognize conflicts in logic and potential errors. Incomplete logic is indicated when any row or column does not have a mark. When incomplete logic occurs, the precedence grid will only identify that an error has been made, not the location of the error.

	A	B	C	D	E	F	G	H	I	J	End
Start	X	X									
A	—				X						
B		—	X								
C			—	X							
D				—		X					
E					—			X			
F						—	X				
G							—	X			
H								—	X		
I									—	X	
											End
End											—

Figure 14.8 Example 14.7: precedence grid.

14.22   Network Construction

A network organizes the WBS tasks into a logical framework that shows operational sequencing, incorporates time durations, and structures calculations. Networks were introduced in Section 3.6.

Figure 14.9 shows an activity-on-node network for the logic of Table 14.1 and Figure 14.8. This network allows one to trace the paths to ensure that the logic accurately reflects the expected operations.

In this example, the signal must be acknowledged (H) before pumping begins. This creates a gap in logic. A gap in logic is corrected by inserting a dashed line (dummy activity) to allow the descriptive flow to continue, as noted between events (E) and (H) in Figure 14.9.

When an activity has more than one immediate predecessor (say X and Y), and another activity has predecessors of either X or Y (but not both), a logical interruption can arise. When this situation occurs, a dummy activity is inserted into the network to allow the logical flow to continue. We show a dummy activity by a dashed line connecting the nodes. A dummy activity has the following attributes:

- Zero time is needed to complete the dummy activity.
- The dummy activity is inserted to provide a logical flow for analysis.

14.23   Network Calculations

Networks provide both a visual portrayal of an operation and a framework from which to calculate completion times. Figure 14.10 shows the network with time durations for the positional and distance movement tasks from Table 14.1.

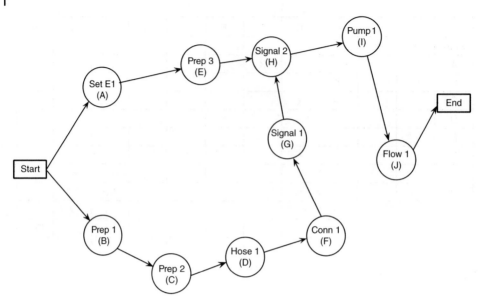

Figure 14.9 Example 14.7: critical path network.

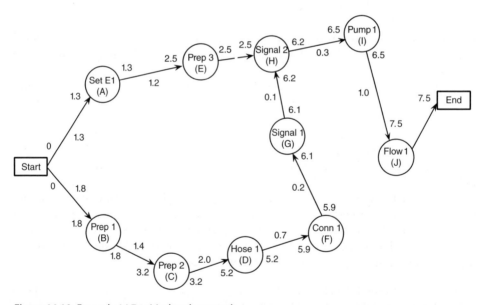

Figure 14.10 Example 14.7: critical path network.

The usual convention shows the time to complete an activity at the middle of the branch. The start time for a branch is shown at one end and the completion time at the other.

The time to move through any path is the sum of time durations along each branch. However, at each node, a subsequent activity cannot start until all dependent activities at the node have been completed. For example, in Figure 14.10, the dependent events leading to complete Signal 2 (Event H) were calculated as 2.5 and 6.2 minutes. Events before 6.2 minutes that would have been completed would remain idle until 6.2 minutes. The start of water flow (Event J) is at 6.5 minutes and water would reach the ALP and complete the evolution in 7.5 minutes.

Calculations keep track of the sequential time durations along each path. Each modular task can be shown on an IPI chart for visual and recorded times. Other information, such as fire flow, can also be incorporated. IPI displays are shown in Section 14.27 for the group of examples.

14.24 Variation Analysis

Normally, during or after an analysis, questions arise such as "what if we did..." There are three common types of variations. The first would retain the same number of fire fighters, but would change conditions such as weather, distance, or hose size. This type can also change building features such as including doors or other conditions that increase or decrease the work tasks. These variations merely change time of existing tasks or the number of WBS tasks to complete the project. In either case, differences are easy to evaluate.

A second type of variation changes the number of people available to do the work. For example, what would be the difference in time for water to reach the ALP if the number of fire fighters were reduced or increased? Although the WBS tasks remain the same, sequencing may be different. Therefore, both time durations and the network construction will change to evaluate these changes.

A third type of modification makes operational changes for the evolution. This is often handled by an add-on to existing tasks, as illustrated by Example 14.8. More complicated changes may require a completely new set of WBS and sequencing selections.

Example 14.8 The evolution described by Example 14.7 established the APL at the base of Stairwell B. The supply water for this evolution is completed when water reaches the wye that is the ALP.

The attack side of the process will connect a 1¾ in attack line at the ALP for deployment up the stairs and onto the third floor. The officer decides to connect the 1¾ in attack line before delivering water to the ALP. How does this change the time duration to supply on-board water to the ALP?

Solution:
All of the tasks of Example 14.7 remain. However, in this situation fire fighters must return to E1, take two 100 ft lengths of 1¾ in hose and appropriate tools, return to the ALP, and attach the attack hose before signaling to charge the supply lines. The following tasks must be added to the WBS and Table 14.1:

K) Return to E1 to get hose and tools.
    (FF3 and FF4 return from the ALP to E1 to get hose and tools.)
    Ret 1 – FF3, FF4; distance movement task (D)
L) Remove hose and tools from E1.
    (FF3, FF4 remove two 100 ft lengths of hose and appropriate tools from E1.)
    Prep 4 – FF3, FF4; positional task (P)

M) Return to ALP
   (FF3 and FF4 return to ALP with attack hose and tools.)
   Hose 2 – FF3, FF4; distance movement task (D)
N) Connect attack hose to wye at the ALP.
   Conn 2 – F2; positional task (P)

The addition of these tasks requires a slight revision of sequencing. Table 14.2 shows all of the tasks, time durations, and sequencing. Factors that have been changed for this operation are shown in bold type. Figure 14.11 shows the network. The time for water to reach the ALP is now 12.3 minutes.

Discussion:
Although the time to provide water to the ALP has increased by 4.8 minutes, the actual time to apply water to the fire will decrease because the operation of collecting the 1¾ in attack hose was simultaneous with some supply activities. This change reduces overall time to apply water to the fire. Essentially, the 4.8 minutes to return to E1 and move the attack hose and tools to the ALP represents pre-movement time discussed in Section 14.34.

14.25   Additional Examples

Examples 14.4–14.6 describe an integrated set of operations to establish on-board and sustained supply water. Examples 14.7 and 14.8 complete the picture for on-board water. Examples 14.9 and 14.10 illustrate hydrant and relay operations. Operational combinations are shown in the IPI display of Section 14.27.

Example 14.9   Example 14.5 described taking Hydrant H1 and delivering supply water to pumper E2. Prepare an analysis to determine the time duration from arrival at H1 to water delivery at E2. The distance from hydrant H1 to the final position of Engine E2 is 900 ft.

Solution:
Table 14.3 shows the information needed to construct a project network and calculate the time duration. The hierarchical WBS diagram that organized the WBS tasks of Column 3 and the precedence grid to ensure correct logic are not shown. Figure 14.12 shows the network and calculations to estimate the time duration. Here, water is delivered to E2 9.6 minutes after initial arrival at H1.

Example 14.10   The water will be relayed from Engine E2 to Engine E1, as shown in Example 14.6. Calculate the time to establish a relay supply. The distance between E2 and E1 is 160 ft.

Solution:
Table 14.4 shows the information needed to construct a network and calculate the time duration. The hierarchical WBS diagram to organize the WBS tasks of column 3 and the precedence grid to ensure correct logic are not shown. Figure 14.13 shows the network and calculations to indicate the local time duration to deliver water from E2 to E1 is 14.2 minutes after the stop of E2.

Table 14.2 Example 14.8: work breakdown structure (WBS)

Operation: *Connect 1½ in attack line to wye and pump on-board water from E1 to ALP*
Crew: *Driver (F1), Officer (F2), two fire fighters (FF3, FF4)*
ALP is located at base of Stairwell B

	Symbol/FF	WBS activity (task) description	Time duration (min)	Predecessor	Successor
Start	Start	Arriving Engine E1 stops at building	–	–	Set E1 (A) Prep 1 (B)
A	*Set E1* F1	Set E1 to begin operations (P)	1.3	Start	Prep 3 (E)
2B	*Prep 1* F2, FF3, FF4	Don SCBA (P)	1.8	Start	Prep 2 (C)
C	*Prep 2* F2, FF3, FF4	Remove supply hose and tools from E1 bed (P)	1.4	Prep 1 (B)	Hose 1 (D)
D	*Hose 1* F2, FF3, FF4	Drag supply line from E1 to ALP (D)	2.0	Prep 2 (C)	Conn 1 (F) Ret 1 (K)
E	*Prep 3* F1	Prepare E1 to pump (P)	1.2	Set E1 (A)	Signal 2 (P)
F	*Conn 1* F2	Connect wye to supply line (P)	0.7	Hose 1 (D)	Conn 2 (N)
G	*Signal 1* F2, F1	Signal E1 to charge line (P)	0.2	Conn 1 (F)	Signal 2 (H)
H	*Signal 2* F1, F2	Acknowledge signal to charge line (P)	0.1	Signal 1 (G)	Pump 1 (I)
I	*Pump 1* F1	Open pump (P)	0.3	Prep 3 (E) Signal 2 (H)	Flow 1 (J)
J	*Flow 1*	Water moves from E1 and arrives at ALP (D)	1.0	Pump 1 (I)	End
K	***Ret 1*** FF3, FF4	Return to E1 to retrieve attack hose (D)	**1.1**	**Hose 1 (D)**	**Prep 4 (L)**
L	***Prep 4*** FF3, FF4	Remove attack hose and tools from E1 (P)	**1.6**	**Ret 1 (K)**	**Hose 2 (M)**
M	***Hose 2*** FF3, FF4	Carry hose and tools to ALP at Stairwell B (D)	**1.9**	**Prep 4 (L)**	**Conn 2 (N)**
N	***Conn 2*** FF3	Connect 1½ in hose to wye at ALP	**0.9**	**Hose 2 (M)**	Signal 1 (G)
End				Flow 1 (J)	–

SCBA, self-contained breathing apparatus; PPE, personal protection equipment; ALP, attack launch point.

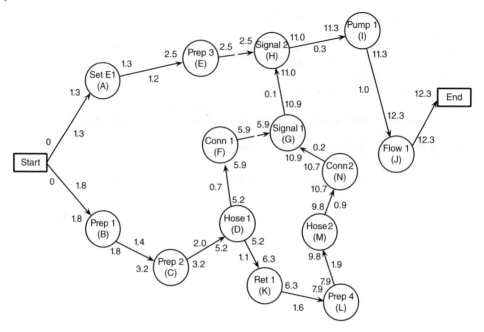

Figure 14.11 Example 14.8: critical path network.

14.26   Levels of Detail

A WBS does not list all details of activity to complete a task. However, tasks that require a recognized functional action must be discrete. For example, the construction industry would define a task as "construct a 20 ft (6.1 m) concrete masonry unit wall 8 ft (2.4 m) high." Or "hang a solid core door." The WBS for those tasks does not identify the numbers of workers to do the job or the construction details that make up the task. However, the functional task and its role in project completion are easily recognized.

Fire fighters are skilled at their craft. They understand the details of their profession, complete routine tasks easily, and adapt to unusual or changing conditions as effectively as conditions allow. Although the fire safety engineer should be generally aware of routine procedures on the fire ground, he or she need not know all task details. For example, in Table 14.4, Task A (prepare H1 for water discharge) is identified. The engineer should be aware that this positional task of preparing a hydrant is a necessary task to move water from a hydrant to the ALP. The engineer need not identify completion details, such as:

- Alight the apparatus and retrieve a hydrant bag and other needed equipment.
- Take equipment to hydrant.
- Return to the engine to take the appropriate hose for supply water.
- Pull (or carry) hose to hydrant.
- Temporarily secure the hose until the supply line is complete.
- Complete preliminary activities of take off hydrant caps to survey for damage or debris, open and flush hydrant, and shut off hydrant.
- Dress hydrant ports with appropriate gates, caps, wyes, siameses, etc.
- Connect supply hose to hydrant.

Table 14.3 Example 14.9: work breakdown structure

Operation: Deliver water from Hydrant H1 to Engine E2 using 3 in supply hose
Crew: Driver (F1), Officer (F2), two fire fighters (FF3, FF4)
ALP is located at base of Stairwell B

	Symbol/FF	Activity (task) description	Time (min)	Predecessor	Successor
Start	Start	Arriving Engine E2 stops at H1	–	–	Prep 1 (A)
A	*Prep 1* FF3	Remove 3 in supply hose from E2 and tools to take H1 (P)	1.9	Start	Prep 2 (B) Lay 1 (C)
B	*Prep 2* FF3	Prepare H1 for water discharge (P)	1.7	Prep 1 (A)	Signal 2 (H)
C	*Lay 1* F1, FF4	E2 lay supply hose to final position (D)	2.4	Prep 1 (A)	Set E2 (D) Conn 1 (F)
D	*Set E2* F1	Set E2 to begin operations (P)	1.3	Lay 1 (C)	Prep 3 (E)
E	*Prep 3* F1	Prepare E2 to pump (P)	1.2	Set E2 (D)	Signal 1 (G)
F	*Conn 1* F2, FF4	Disconnect supply line and connect to E2 intake port (P)	1.4	Lay 1 (C)	Signal 1 (G)
G	*Signal 1* F2, FF3	E2 signals FF3 at H1 to open hydrant (P)	0.2	Conn 1 (F) Prep 3 (E)	Signal 2 (J)
H	*Signal 2* FF3, F2	FF3 at H1 acknowledges signal (P)	0.1	Signal 1 (G)	Open 1 (I)
I	*Open 1* FF 3	Open H1 (P)	1.3	Signal 2 (H)	Flow 1 (J)
J	*Flow 1*	Water flows from H1 to E2 (D)	1.0	Open 1 (I)	Pump 1 (K)
K	*Pump 1* F 1	E2 pumps water from H1 to E2 (P)	0.2	Flow 1 (J)	End
End				Pump 1 (J)	–

ALP, attack launch point.

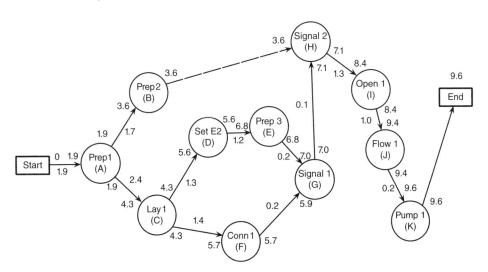

Figure 14.12 Example 14.9: critical path network.

Table 14.4 Example 14.10: work breakdown structure

Operation: *Relay water from Engine E2 to Engine E1 using a 3 in supply hose*
Crew: *Driver (F1), Officer (F2), one fire fighter (FF4)*
Distance between E2 and E1 is 160 ft

	Symbol/FF	Activity (task) description	Time duration (min)	Predecessor	Successor
Start	Start		–	–	Prep 1r (A)
A	*Prep 1r* F2, FF4	Remove hose and tools from E2 bed (P)	1.4	Start	Conn 1r (B)
B	*Conn 1r* F2, FF4	Connect relay hose to pump discharge of E2 (P)	0.8	Prep 1r (A)	Lay 1r (C)
C	*Lay 1r* F2, FF4	Drag relay hose from E2 to E1 (D)	12.0	Conn 1r (B)	Connect 2r (D)
D	*Conn 2r* F2, FF4	Connect relay hose to pump intake of E1 (P)	1.0	Lay 1r (C)	Pump 1r (E)
E	*Pump 1r* F1	Pump water from E2 to E1 (P)	0.2	Conn 2r (D)	Flow 1r (F)
F	*Flow 1r*	Water flows from H1 to E2 (D)	0.8	Pump 1r (E)	End
End				Flow 1r (F)	–

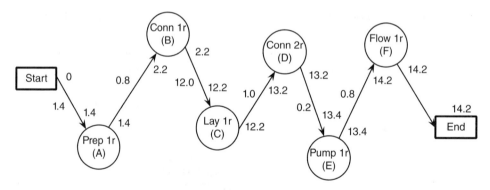

Figure 14.13 Example 14.10: critical path network.

Although the engineer should be aware that a group of activities is part of each task, the details may be combined for analytical purposes. The WBS needs only to describe the tasks as "prepare the hydrant for operation." Thus, task descriptions incorporate the detailed work for hydrant operations. All positional or distance movement modules for water supply operations are described as functional tasks.

14.27 Time Coordination

Examples 14.7–14.10 describe time durations to complete several operations. The "local time" for these operations is easily integrated into a "global time" in the IPI chart.

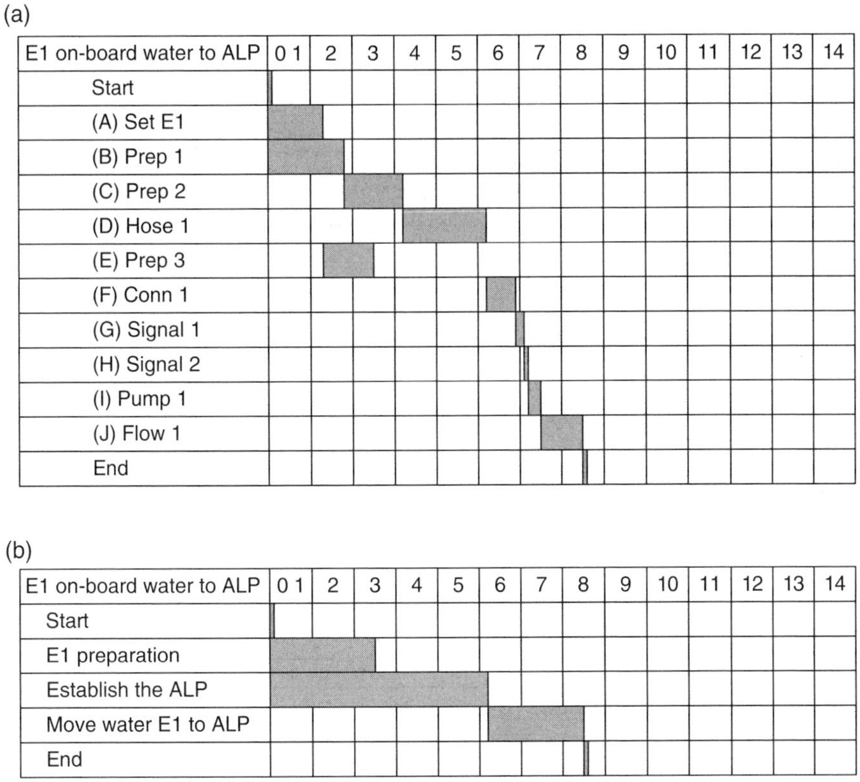

(a)

E1 on-board water to ALP	0 1	2	3	4	5	6	7	8	9	10	11	12	13	14
Start														
(A) Set E1														
(B) Prep 1														
(C) Prep 2														
(D) Hose 1														
(E) Prep 3														
(F) Conn 1														
(G) Signal 1														
(H) Signal 2														
(I) Pump 1														
(J) Flow 1														
End														

(b)

E1 on-board water to ALP	0 1	2	3	4	5	6	7	8	9	10	11	12	13	14
Start														
E1 preparation														
Establish the ALP														
Move water E1 to ALP														
End														

Figure 14.14 Example 14.7. (a) Interactive Performance Information (IPI) chart work activities. (b) IPI chart consolidated activities.

One may show all task durations on individual spreadsheets and later combine the major components into a composite descriptor. The consolidation is usually reserved for the global time IPI chart. Local time charts enable one to examine in greater detail tasking needs, staffing, and the influence of building and site features on time durations. The portrayals also provide a visual portrayal of the operations.

Figure 14.14(a) shows the local time relationships for each of the tasks associated with the network calculations of Figure 14.10. Figure 14.14(b) consolidates activities to describe the major components described in Figure 14.7.

Hydrant and relay operations are shown on Figures 14.15(a) and (b). Figure 14.16 shows a coordinated picture of the three operations. One can recognize that time duration for each major task on a single IPI sheet. Thus, this scenario estimates 28 minutes from arrival of E1 to delivery of sustained supply water at the ALP at the base of Stairwell B.

Time is the focus of these examples. However, water flow calculations can also be included in the IPI information. These operations relate water sources, available community equipment, and staffing.

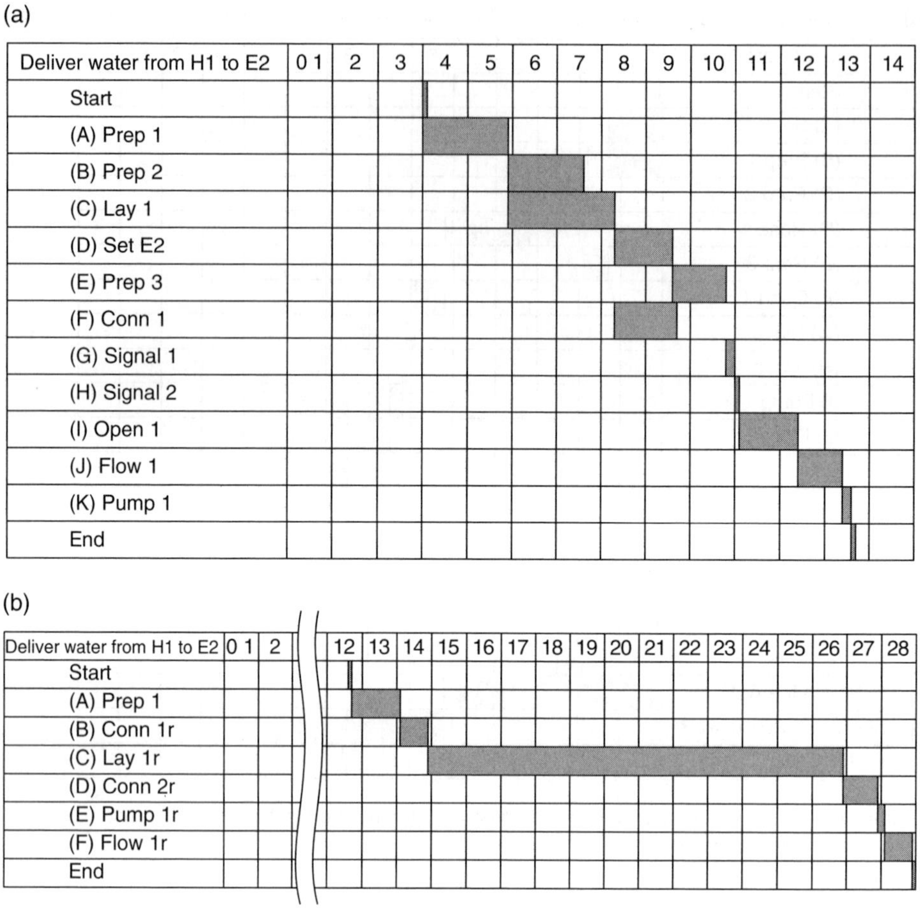

(a)

Deliver water from H1 to E2	0 1	2	3	4	5	6	7	8	9	10	11	12	13	14
Start														
(A) Prep 1														
(B) Prep 2														
(C) Lay 1														
(D) Set E2														
(E) Prep 3														
(F) Conn 1														
(G) Signal 1														
(H) Signal 2														
(I) Open 1														
(J) Flow 1														
(K) Pump 1														
End														

(b)

Deliver water from H1 to E2	0 1	2		12	13	14	15	16	17	18	19	20	21	22	23	24	25	26	27	28
Start																				
(A) Prep 1																				
(B) Conn 1r																				
(C) Lay 1r																				
(D) Conn 2r																				
(E) Pump 1r																				
(F) Flow 1r																				
End																				

Figure 14.15 Example 14.8. (a) Interactive Performance Information (IPI) chart work activities. **(b)** IPI chart consolidated activities.

E1 on-board water to ALP	0 2	4	6	8	10	12	14	16	18	20	22	24	26	28
E1 Reaches building														
E1 on-board water to ALP														
Water from H1 to E2														
Relay water from E2 to E1														
Sustained water at ALP														

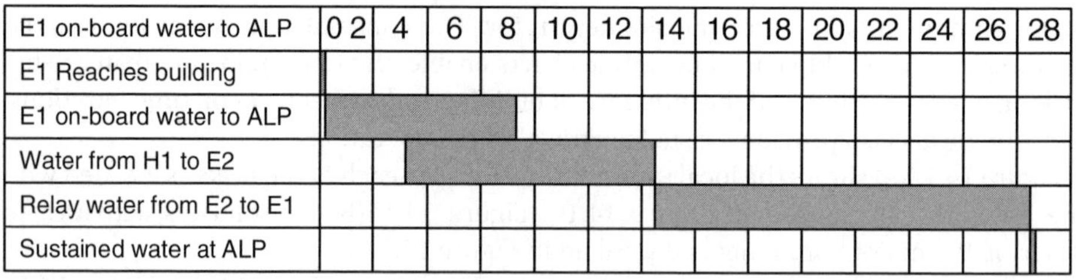

Figure 14.16 Interactive Performance Information (IPI) chart consolidated activities.

14.28 Discussion

Examples 14.4–14.10 were structured for a purpose. That purpose was to illustrate the use of construction project management techniques for fire ground applications. CPMs are ideally suited to calculate time durations for supply operations and to display interactions on the IPI chart. Although we have shown outcomes in local time, the IPI chart converts local conditions to global times.

One may easily recognize that other, and better, evolutions would be more appropriate for this fire scenario. For example, the ladder truck equipped with a pump (i.e. "quint") arriving 4 minutes after Engine E1 could do the relay operation. This would save 14.2 minutes of that operation as well as substantial physical labor for fire fighters. Supply hoses are heavy. Nevertheless, Engine 2 could lay the relay supply line upon arrival, avoiding relay operations. These better ways to do the work can be analyzed using the same CPM techniques.

Critical path method techniques are applicable for all types of water supply operations. For example, an analysis could involve standpipe operations. Wet pipe, dry pipe, and pumping operations are easily modeled. One need only identify the appropriate positional or distance moving tasks to calculate outcomes for any defined evolution.

The key to fire suppression analysis is the ALP. This concept enables one to separate supply functions from attack operations. Each uses a different technique to quantify its evolutions. However, the components may be combined to determine the time duration to match fire ground conditions.

We may also note that fire ground work continually changes. For example, after completion of a water supply operation, only pump operators maintain their positions. Fire fighters are reallocated to other needs, such as stretching additional attack lines. They can provide back-up protection to fire attack personnel or deploy other attack lines to prevent fire extension.

PART FOUR: INTERIOR FIRE ATTACK ANALYSIS

14.29   Introduction

The ALP separates supply functions from attack operations. This separation enables one to consider incident command decisions, work activities, and calculation methods as discrete functions. Although these components are interactive on the fire ground, an opportunity to examine each as an independent activity provides an insight into building features that affect performance.

A performance-based analysis has the luxury of time to examine the fire extinguishment component in leisure. The room of origin and the time-based diagnostic fire and smoke conditions inside the building are known. Floor plans can be examined. Building entry locations and fire fighting aids can be identified. Community resources and response times are known. The challenge is to combine the information to achieve a realistic understanding of what to expect during an actual fire.

Constructing a fire attack timeline forces one to understand how building features influence interior fire fighting. For example, how does the architectural plan affect the time to apply water to the fire? Are suppression aids located effectively? Which barriers are important to an incident commander in controlling a fire?

Part Four describes a method to estimate the time to move an attack line from the ALP to a fire. As with other fire ground actions, the method of analysis is separated from management decisions of what actions to take. Although knowledge about the fire is important to making management decisions, we temporarily restrict this information in order to focus on calculation methods

14.30   Overview of Stretching Interior Attack Lines

The goal is to estimate the time to stretch a hose along a defined attack route and apply water to the fire. Building obstacles, environmental conditions, and available staffing to do the work influence the time to stretch attack hose lines and apply water. We want to understand how the building affects the time to apply water to the fire.

An attack route is a chain of architectural segments from the ALP to the fire. Positional or distance moving modular fire fighting tasks are associated with each architectural segment along the route. Time durations to complete modular tasks are estimated, and the cumulative sum locates the nozzle at any time.

A methodical procedure provides the mechanics to analyze any fire attack route. First, a route is selected and sequential architectural segments along the path are identified. Then, positional or distance movement fire fighting task modules are associated with each architectural segment, and time to complete each module is estimated. A spreadsheet is useful to order and compare alternative routes.

An evaluation of the time to stretch an attack hose line uses the following procedure:

1) Identify the room of origin and diagnostic fire scenario.
2) Describe the time-related diagnostic flame-heat and smoke-gas propagation.
3) Identify building entry points and potential attack routes.
4) Select the ALP and a specific fire attack route.

5) List the architectural segments along the attack route.

6) Identify positional and distance movement modular tasks that are associated with stretching a line through each architectural segment.

7) Identify the number of fire fighters available to do the work.

8) Estimate the visibility and other environmental conditions (from the diagnostic fire) that may be expected at each architectural segment along the attack route during the time fire fighters will be at the location.

9) Note any tasks that may be done simultaneously to facilitate hose movement.

10) Estimate the time duration for each positional or distance movement task module along the path of architectural segments.

11) Estimate the pre-movement time delay for fire fighters and equipment to reach the ALP before starting on the attack route.

12) Construct a timeline for movement from the ALP to the fire.

13) Construct a timeline from first-in arrival (MR) to water application for the selected route.

Descriptions of new concepts in this procedure are discussed below.

14.31 Task Modules

Stretching a hose line from an ALP to the fire involves connecting discrete tasks in a linear chain. All tasks associated with stretching an interior hose line can be grouped into two modules.

Positional task modules are tasks that involve actions performed at a single, relatively small working area. Forcible entry through a barrier or pausing to connect additional hose lengths illustrates actions that may be performed within a relatively small space.

In addition to the delay caused by forcible entry, positional modules may involve tasks such as:

- Don or change SCBA and mask or other PPE.
- Make hose connections.
- Pause to make way-finding decisions.
- Pause to investigate possible safety concerns.
- Pause to await back-up hose line placement when environmental conditions indicate excessive fire fighter danger during the attack.
- Delay for return to engine to obtain additional tools or hose lengths.

Some positional tasks may reduce time to stretch a hose line in comparison with alternatives that do not use the help. While a time delay is never recovered, future movement can be faster when additional fire fighters are available to do preparatory work, such as:

- Ease a hose around corners or obstacles.
- Collect hose slack to facilitate advancing lines.
- Hoist hose in a stairwell.
- Open barriers prior to arrival of an attack line. For example, a truck company may prepare the attack route for water company movement.

Distance movement modules are tasks that involve movement through architectural segments. For example, stretching a hose line along a corridor, up or down stairs, or around corners illustrate distance movement tasks.

14.32 Architectural Segments

Every building plan can be divided into architectural segments. These segments can be linked sequentially to describe a fire attack route. Architectural segments are:

- Corridors
- Turns with linear movement after the directional change
- Stairs up
- Stairs down
- Stair turns
- Large rooms
- Sections in rooms containing obstructions (e.g. large boxes, furniture, stacks, or aisles) that require hose lines to weave around the obstacle.

Architectural segments are the barrier–space modules that have been used throughout this book. Virtual barriers are particularly useful to divide architectural segments along attack routes. We insert a virtual barrier at every location where a modular task or a time function changes. Combined with fire fighting modular tasks, this segmentation simplifies estimating time durations along the chain (i.e. attack route segments).

The length of distance movement segments is important because of the time it takes to traverse the distance. The first few feet of pulling a hose are easier than the last few feet. The time to stretch a hose line along any path is not a linear function.

Corners are important in moving charged and uncharged hose lines. Hose connections can bind at corners. Stretching a charged hose line around a corner can be difficult. When a hose kinks or is bound at a corner, a fire fighter must return and ease the line or dislodge a kink or a hose fitting at a corner. A fire fighter located at corners can help to advance the line. If that level of staffing is not available, delays will occur. A rule of thumb indicates that three corners (turns) are considered the maximum for a single company to maneuver a charged hose without substantial delays.

Stairs offer similar problems. In addition to 180° turns, the physical effort to carry hoses and other equipment up stairs causes fatigue. Standpipes are often installed in stairwells or medium- to high-rise buildings to shorten the time needed to advance an internal attack line to a fire.

Large rooms having several doors leading to potential attack routes can cause delay in traversing the distances and in requiring decisions on the proper door to use. A mistake in selecting the correct door may not be readily apparent, and retracing steps to select another route takes time.

Clutter and contents obstructions, including collapsed goods or materials, can impede the stretching of hoses if it becomes necessary to move the hose around the obstacles. Connections can get hung up while moving through the maze. Charged lines are difficult to move.

14.33   Architectural Obstacles

Every building has architectural obstacles that delay or stop hose movement. Doors and locks are the most common obstacles in a fire attack route. When an attack route encounters another type of delay, such as a pause to select among several possible paths, a positional task is used to represent the time delay.

After a fire attack route has been selected, one identifies architectural obstacles along the path from the entry point to the fire. An architectural obstacle is a building feature that requires time to move a hose line through or beyond it. Architectural obstacles include:

- Doors and locks.
- Hazards that use a pause to make a decision (e.g. floor hole, laboratory, etc.).
- Locations that use a pause to make a way finding decision.
- Clutter and contents impediments that require a change in speed or direction.

The type of door, its hardware, wall construction around the door, direction of swing, and status (locked or unlocked) affect the time to stretch a hose line. The barrier must be crossed. If keys are not available for locked doors, forcible entry will occur. Forcible entry requires appropriate tools, staffing, techniques, and time.

Hazards can cause an individual to pause and consider personal mortality and safety. Electrical hazards are generally recognized. When one encounters a "laboratory" sign, biological, chemical, and nuclear hazards come to mind. The room may be a computer or materials testing laboratory. On the other hand, it may contain chemical or biological toxins. Caution is usually exercised, and this can influence the fire's outcome and fire suppression. Buildings may have holes in floors. When visibility is poor, these conditions and disorientation can pose significant hazards.

14.34   ALP Pre-movement

Preparation to connect to the ALP is a time delay at the start of the process. This activity may be simple, such as alighting from the engine, collecting hose and tools, donning SCBA and PPE, and moving to the ALP. On the other hand, the preparation may involve moving equipment, supplies, and people to a staging area near the ALP.

Elevator movement to transport fire fighters and equipment is not a part of stretching an interior attack line. However, elevators reduce time by moving people and equipment. Elevator recall and the control of their subsequent use by the fire department can greatly influence the time to agent application in tall buildings. Elevator use is incorporated in the pre-movement time duration from first-in arrival to ALP connection.

Pre-movement preparation normally precedes connection to the ALP, although some work may be done simultaneously with early stretching of an attack line. Discussion with the local fire department can identify their SOP for the building being evaluated, so that preparation time durations can be incorporated appropriately.

14.35   Multiple Attack Lines

Stretching multiple attack lines at a fire has many advantages. Sometimes the fire environment is too dangerous for a single hose line, and a back-up line may be used to protect attacking fire fighters. Several attack positions enable a fire ground commander to select the most appropriate fire stream when the situation becomes better understood. Also, multiple hose lines can defend barriers and prevent extension to adjacent rooms while a primary attack line is extinguishing a room fire.

The number of hose lines and available staffing influence the tactics of fire suppression. As a rule of thumb, one engine company stretches one hose line and can extinguish one room. Two engine companies with two hose lines are needed for two fully involved rooms. Although this does not discuss room sizes and extinguishing capabilities of hose lines, it gives a sense of proportion for understanding fire suppression needs.

Each attack line is analyzed as a separate activity. A Phase 1 analysis focuses on the first attack line applying water to the fire. This analysis provides a good insight into the compatibility of a building to complete fire extinguishment. When multiple attack lines are used in an initial attack, the IPI chart can identify sequences and times for fire flow.

14.36   Variables

Variables that influence the time to complete distance moving and positional tasks are as follows:

- *Staffing.* Fire fighting is a team function and completion time is influenced by the number of fire fighters available to do the work.
- *Environmental conditions.* Heat and visibility significantly affect the speed of movement.
- *Distance.* The time to complete distance movement tasks is a non-linear function involving length of travel, staffing, environmental conditions, and hose size and condition.
- *Hose size.* The attack hose size affects maneuverability and the time to complete tasks.
- *Hose condition.* A hose may be uncharged, charged, or charged and flowing. Each condition affects both speed and ability to move a hose line.
- *Equipment.* Forcible entry tools and safety requirements influence the time to complete tasks. Protective fire fighting gear (PPE) affects speed of movement.
- *Fire fighter training and experience.* Task completion speeds are influenced by fire fighter familiarity with the operations.
- *Fatigue.* Fire fighter fatigue delays completion times.

The building looks very different under normal operating conditions and lighting than it does during a fire. Environmental conditions that influence visibility, heat, fatigue, and safety can significantly influence fire ground operations. Visibility is the most common environmental problem inside a burning building because dense smoke can reduce visibility to nearly zero. This significantly affects the time to accomplish tasks needed to attack a fire.

We define a "macro" location as the general fire area while a "micro" location describes the seat of the fire. Smoke conditions are often so dense that a nozzle man cannot see the fire. Fire fighters may need to listen for the fire or to feel the heat to direct hose streams. Thermal imaging cameras are valuable tools to locate open or hidden fires. Ventilation can raise the smoke level and help to locate a fire in a smoky environment.

Heat from the fire and heat retained inside protective clothing increases discomfort and fatigue. Fatigue can occur because the operations are physically and emotionally demanding, motion is restricted, and the work over short time durations is vigorous.

Staffing influences the time to agent application and the type of offensive or defensive strategy. Coordination and specific task completions are necessary to apply water to a fire. Adequate personnel can stretch a hose line in a specific period of time. Fewer fire fighters require more time to do the same job. Because a fire grows independently and exponentially, the added delay may change conditions for an aggressive interior attack. When this occurs, the department may be forced into a defensive mode, causing a longer fire duration and greater damage.

Constraints to interior fire fighting operations can influence operations. For example, the capacity of SCBA air tanks normally provides about 20 minutes of active use. When the fire is not extinguished within the SCBA time limitation, replacement crews may be needed.

Another constraint relates to hose distances. An engine company routinely carries two lengths of hose to the ALP connection. When the actual distance to the fire exceeds this length, more hose must be obtained. This increases the time for agent application. Consequently, path distances from ALP connections involve attack line time constraints.

14.37 Time Estimates

Fire fighters think in terms of tasks. Engineers analyze in terms of time. The tasks to stretch a hose along any specific attack route are constant. The time to complete the tasks varies with the factors noted earlier. Eventually, the fire safety engineer needs a citable source to estimate time durations for the modular tasks. Until this is prepared, video and expert judgment must reflect the evolution and its conditions.

Each of the relatively small number of distance movement and positional modules for interior hose line operations is associated with architectural segments and obstacles. Factors that affect time variations, such as those described earlier, may appear to be large and have conditional uncertainties. However, the positional and distance movement tasks are relatively small in number, and quantification is manageable.

Time estimates for the positional and distance movement modules must currently be provided by the opinion of individuals experienced in fire fighting. Until published data become available, one must rely on expert judgment or video analysis for time estimates.

14.38 Attack Route Analysis

Example 14.11 describes the process and discusses features that affect the time to stretch an attack hose and apply water to the fire. Example 14.12 illustrates the analysis.

Example 14.11   The fire area and an attack route path are shown in Figure 14.17. Discuss conditions to define an attack route evaluation and identify the architectural segments and architectural obstacles that comprise this attack route.

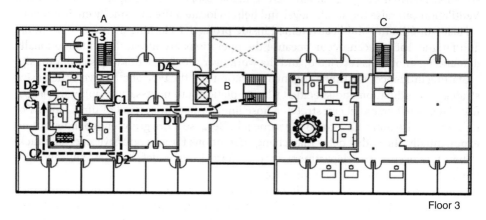

Figure 14.17   Model building: Floor 3.

Solution:
We discuss several features involving water movement analysis.

ALP Connection   Example 14.7 evaluated the time to supply water from Engine E1 to an ALP at the base of Stairwell B to be 7.5 minutes. This estimate was based on water reaching the ALP without having the attack line connected. If we decide that fire fighters from Engine 1 will connect and advance the attack line as in Example 14.8, a pre-movement time must be identified. Example 14.8 estimated 5.5 minutes for E1 fire fighters to return to E1 to collect hose and tools and return to the ALP. However, this evolution is included with the ALP. The actual time saving is 1.8 minutes and an equivalent pre-movement time becomes 3.7 minutes.

   To digress briefly, if a second attack line were to be connected to this ALP, the scenario would define who would do the work and associated time durations. The pre-movement time estimate to the ALP would be relatively simple to determine, and the IPI chart keeps all of the timelines organized.

   Examples 14.7 and 14.8 selected the ALP at the base of Stairwell B. If the location were on the second or third floor landing, the additional distance becomes a simple extension of distance movement modules.

   If a standpipe had been installed in Stairwell B, fire fighters would have a choice to use it or to carry attack lines up the two floors. A dry standpipe would require laying a line to supply water from E1 to the standpipe. In addition, the standpipe must be prepared and charged. This is not likely for a fire within about three floors. On the other hand, if a wet standpipe had been installed, the attack line could be connected and city water pressure with possible stationary pump augmentation would be available at the standpipe ALP. An additional supply line could be connected to boost pressure, if needed. Pre-movement time to attach the attack line to the standpipe ALP is determined by the usual means.

Thus, the ALP enables broad alternatives of supply and attack combinations to be evaluated with the same basic technique.

Attack route   The attack route starts from the ALP, which is located at the base of Stairwell B. The initial architectural segments are defined as each run of stairs. Virtual barriers are used at each change in direction. Thus, four segments would define the movement to the third floor landing.

An alternative evolution could have fire fighters drop a rope from the third floor landing to the ALP. The attack line could be attached and pulled to the third floor landing. The time to travel to the third floor, drop the rope, tie the attack line and hoist it would replace the time to carry an attack line from the ALP to the third floor landing. Any action can be incorporated into an analysis as long as it is defined and activity times are estimated.

Using the original operation of carrying attack lines up each segment of stairs, we identify architectural segments to the fire, which is located in the enclosed suite of offices in the west wing. The architectural segments are defined as follows:

A1: Stairs – first floor to intermediate landing.
A2: Stairs –180° turn from landing to second floor.
A3: Stairs –180° turn second floor to intermediate landing.
A4: Stairs –180° turn intermediate landing to third floor.
A5: Segment third floor landing to Door D1.
A6: Cross threshold of Door D1
A7: < 90° turn at C1 along corridor to Door D2.
A8: Open Door D2.
A9: Cross Door D2 threshold
A10: Continue along corridor to corner C2.
A11: 90° turn at C2 then travel to fire area.
A12: Open door D3 to fire room.
A13: 90° turn at Corner C3 into fire room.

Discussion:
Each architectural segment or obstacle along the route is identified. Virtual barriers (not shown) define corridor segments.

All architectural obstacles along an attack route are identified. In this case, doors D1 and D2 and door D3 to the fire room are (possible) architectural obstacles. When the doors are open, they cause no delay in making the transition. When they are closed, a positional module describes the time needed to cross the obstacle. Although this example describes Door D2 as closed and the other doors open, we identify an architectural segment boundary where doors exist, whether open or closed. The professional time to calculate outcomes does not change. However, the flexibility to incorporate "what if" variations is enhanced.

Similar to the supply water analysis, the first step merely identifies the attack route to be analyzed. Factors that affect time durations, such as staffing, visibility and heat, hose size and condition, and the time to complete distance movement or positional modules do not influence the segments. Attack route segments and obstacles are defined only by the architectural plan.

Movement along or through segments A10–A12 depends on conditions. If doors are closed and the fire remains within the office suite, movement may take place. On the other hand, the attack may occur at Corner C2 if the barrier to the office suite is breached. These conditions may be handled with "what if" variation analysis, so the advantage of incorporating this possible situation helps to provide a better picture of time durations.

Example 14.12 Assume that Door D2 is closed and all other doors along the attack route described in Example 14.11 are open. Prepare a table using the architectural segments and obstacles to estimate the fire attack time. Identify the positional and distance movement modules and estimate time durations to apply water to the fire.

Solution:
Table 14.5 organizes the architectural segments and obstacles of Example 14.11. Information for the analysis and value selections is discussed below.

Discussion:
This first-order analysis uses local time from connection at the ALP. Any pre-movement time to carry hose and tools to the ALP and to make connection must be included either at the supply to the ALP or after connection to the ALP. Here, we start attack time after the connection described in Example 14.8. Local time is converted to global time in the IPI chart.

Judgmental estimates have been used to illustrate the process. A simple research program could provide documentation for quantitative values. The incremental values reflect the conditions of the scenario.

Hose directional changes affect ease of movement and duration times. While an uncharged hose can be moved relatively easily, distances and the number and types of turns significantly affect the work needed to move charged hoses. Therefore, hose size and staffing are important for scenario time estimates.

This example uses an officer (F2) and two fire fighters (FF3, FF4). FF3 is the nozzle operator and F2 is the back-up fire fighter. FF4 speeds advancement by handling a variety of activities such as collecting hose, removing kinks, easing hose around corners, and forcible entry. If no one is available to do this work, time durations may be considerably extended because fire fighters must delay advancing the nozzle to fix the problem. These sequential (step function) time delays must be integrated into the timeline.

Heat and smoke conditions of the diagnostic fire are important to advancing an attack line. For example, when the environment is clear, an uncharged line can be advanced relatively easily. However, when conditions become bad, fire fighters charge the line and movement slows. In addition, fire fighters move in a crouch walk or a crawl for low visibility environments. When the attack line is charged and flowing, movement takes even longer.

An important part of timeline development is mixing and matching conditions and time. The mechanics of combination are not shown here, but rapid, effective estimates can be modeled. The model matches the force to pull the nozzle with unit results for each of the variables noted in Section 14.36. This requires a computer model to match appropriate segments. Although the model is relatively simple, hand calculations may be used to cut and paste appropriate segments for compatible boundary conditions.

The incident commander's management decisions have not been incorporated because this example focuses on the mechanics of analysis. However, fire fighter safety is a concern. During fire conditions that are clearly manageable, one hose line

Table 14.5 Example 14.12: attack line information.

	Architectural segment	Operation/FF	Distance	Environment	Hose condition	Increment time (seconds)	Cumulative time (seconds)
	START at ALP	F2 FF3, FF4	—	Clear	Connection to ALP complete	—	0
A1	Stairs from ALP to second landing (D)	Carry hose FF3, FF4	8 ft	Clear	Uncharged	2	2
A2	180° turn; stairs from landing to second floor(D)	Carry hose FF3, FF4	8 ft	Clear	Uncharged	4	6
A3	180° turn; stairs from second floor to landing (D)	Carry hose FF3, FF4	8 ft	Clear	Uncharged	9	15
A4	180° turn; stairs from landing to third floor (D)	Carry hose FF3, FF4	8 ft	Clear	NA	15	30
	Positional delay (P)	Pause to collect hose at landing FF4	—	Moderate stratified smoke	Uncharged	Work duration, 80 seconds Positional delay, 20 seconds	50
A5	Third floor landing to Door D1 (D)	Advance Hose FF3. F2	30 ft	Moderate stratified smoke	Uncharged	12	62
A6	Cross D1 threshold (P)	Cross threshold FF3, F2	—			0	62
A7	Door D1 to Corner C1 to Door D2 (D)	Advance hose FF3, F2		Moderate stratified smoke	Uncharged	0 s	62

(*Continued*)

Table 14.5 (Continued)

	Architectural segment	Operation/FF	Distance	Environment	Hose condition	Increment time (seconds)	Cumulative time (seconds)
A8	Open Door D2 (P)	Forcible entry, FF4	–	Moderate smoke	Uncharged	45	
	Positional delay (P)	Pause to charge hose FF3, F2	–	Moderate smoke	Charging hose	70	132
A9	Cross threshold Door D2 (P)	Door open FF3, F2	–	Heavy smoke	Charged	0	132
A10	Door D2 to Corner C2 (D)	Advance hose FF3, F2	20 ft	Heavy smoke	Charged	40	172
A11	90° turn at C2 to fire room (D)	Advance hose FF3, F2	25 ft	Heavy smoke	Charged	60	232
A12	Open door D3 to fire room (P)	Forcible entry FF4	–	Heavy smoke	Charged	50	282
A13	Turn C3, apply water (D)	FF3, F2	–	Heavy smoke, heat	Charged and flowing	4	286

FF, fire fighter; ALP, attack launch point.

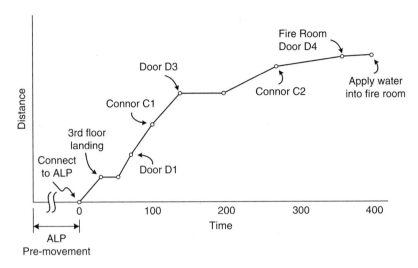

Figure 14.18 Example 14.12: timeline.

may be sufficient. However, when uncertain or hazardous conditions are present, the incident commander may delay attack until back-up hoses are available or additional attack routes have been established. This allows the fire to extend until water can be applied. Calculation procedures for multiple attack lines follow the same procedure as that for the first line. The IPI chart identifies the status of all attack lines at any instant of time.

Environmental conditions caused by the fire affect time to move through architectural segments. This example illustrates ways to incorporate the fire environment by adjusting time durations (Column 5). Environmental conditions also influence the decision of when to charge hoses. It is possible to estimate the evolution for non-fire conditions and then adjust time durations to give a window of time that reflects different environmental conditions that may be present.

Figure 14.18 shows a graph of positional and distance movement tasks from the ALP to the fire. Positional tasks are a step function while distance movement tasks show a slope. Collectively, these tasks describe the movement of Table 14.4. Different graphs can be compared to visually recognize the effects of features such as locked doors or hose collection on times to apply water to a fire.

PART FIVE: PHASE 1 ANALYSIS

14.39   Introduction

$M_{Part\,C}$ (arrival to extinguishment) examines three phases to understand the fire and fire extinguishment. Phase 1 identifies the fire situation when first water can *theoretically* be applied. Phase 2 uses situational knowledge from Phase 1 to estimate boundaries where the fire may be controlled. Phase 3 estimates if the fire will extend beyond these boundaries before it is finally extinguished. Collectively, the combination of these phase evaluations give a good picture of the limit (extent) of the fire at the time of extinguishment.

A Phase 1 analysis provides an enormous amount of information about the building's ability to work with the local fire department. The results of our continuing example with additional knowledge gained from the diagnostic fire and $M_{Part\,A}$ and $M_{Part\,B}$ will illustrate the use of this information to establish a Phase 1 knowledge base. We develop this knowledge by examining relationships among three major parts of the situation. These parts are: (1) the diagnostic fire; (2) the building and site design; and (3) fire department resources involving hose stream locations, time durations, and water flow.

Each of these parts has an interactive effect on performance. Changing one factor normally causes differences in several parts. Thus, performance evaluations involve dynamic and interactive rather than linear "cause and effect" outcomes. Qualitative estimates of changes are often sufficient to develop the understanding. Sometimes selective quantitative estimates can augment qualitative estimates to get a better sense of proportion for performance.

14.40   Phase 1 Comments

The initial Phase 1 assessment considers the diagnostic fire size at theoretical first water application. The time for theoretical first water application is the cumulative duration of times for detection (MD), fire department notification (MN), first-in arrival (MR), and first water application (MA).

In our continuing example, assume the engineer estimated the time for MN as 5 minutes after EB (i.e. $M_{Part\,A}$ = 5 min). Assume further that alarm handling and travel time is estimated at 7 minutes after EB (i.e. $M_{Part\,B}$ = 7 min). Therefore, one could compare the diagnostic fire and manual suppression timelines to establish the fire size at 12 minutes after EB. Although we will eventually show this information, let us delay that knowledge until after we discuss a few aspects of building design and fire fighting.

Examples 14.4–14.8 estimated that first water application could occur 17.5 minutes after first-in arrival (MR). We could identify the fire size at that time (MA = 12 + 17.5 = 29.5 minutes after EB) to get a sense of whether a single 1¾ in line could extinguish that fire condition.

However, those figures are misleading. Let us pause to discuss a more realistic estimate of the time for MA.

14.41   Calculating Time Durations

A network can describe every water movement evolution on the fire ground. Tasks to complete an evolution are constant, regardless of the number of fire fighters available to do the work or the conditions under which they work. However, the time to complete the tasks can vary enormously with field conditions. A performance evaluation must calculate appropriate time durations.

The purpose of Examples 14.1–14.12 is to describe techniques for evaluating time durations to complete fire ground operations. The time to complete any water supply operation is established with traditional project management procedures. One needs only to identify the relevant positional and distance movement modules and develop a CPM network to describe the operation. Although CPM procedures are well documented and widely available, today one must estimate modular times with video analysis or expert opinion.

Individual companies or fire departments can establish modular times in a digital library. Perhaps a professional organization will decide to publish modular time data. Its collection is easy and inexpensive. Nevertheless, even with less than confident tasking estimates, network applications can provide an enormous insight into the building and site features that affect the time to apply water.

14.42   If…

In our continuing journey to learn from a Phase 1 evaluation, we may recognize a few features about our model building. Before examining the diagnostic fire, we can describe some potential situations.

Figure 14.19 shows our building in perspective and size. Although the building is not large, it is common in size and occupancy for most communities throughout the world.

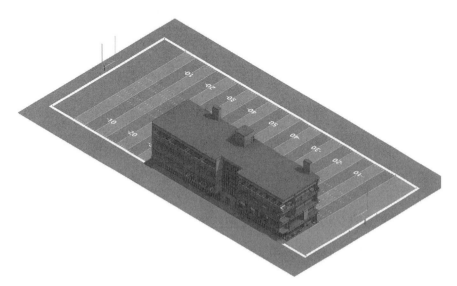

Figure 14.19 Model building perspective.

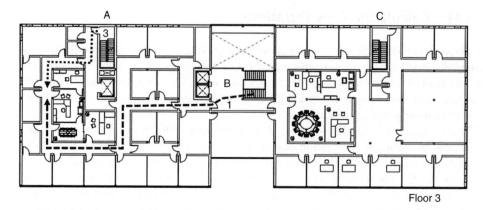

Figure 14.20 Model building: Floor 3.

Figure 14.20 shows the third floor plan. The diagnostic fire scenario is located in the interior suite of offices.

If... the door to the fire room (1) is closed, fire fighters may enter the suite and open the fire room door to apply water. If... the door swings inward, it is easy to force, but difficult to close when safety is needed. If... the door swings out, it is difficult (and time-consuming) to force open, but easy to close.

Also, if the door to the room of origin were closed, only leakage ventilation would be available and the fire would grow slowly. This causes conditions that are conducive to backdraft conditions. If... a backdraft explosion occurs here, the fire fighters may be killed.

If... the doors within the suite were open, fire could propagate to other rooms and the entire suite could reach flashover. Restricted ventilation would remain a problem unless the corridor door were open or the door or wall were of glass and shattered from the fire in the suite. If... fire extends to the corridor, smoke will be dense throughout the entire floor and fire fighting tactics will change. Multiple hose lines would be certain.

The interior suite poses some difficulty for venting the fire. Fire extinguishment often involves "pushing the fire out of the building." This requires an exterior opening and a route for fire movement. Windows in an exterior office could provide that route, although hose lines may be in the path of heat movement.

If... the room of origin were an exterior office, venting and fire attack become easier.

If... the fire were in the suite (1) and flashover had occurred, and venting is difficult, and insufficient staffing and attack lines were available, the incident commander would be likely to attempt to set up a defensive line at the location noted by the heavy dashed line.

If... insufficient hose lines are available to control the fire to the west of this line and allow fire fighters to attack and extinguish the fire, a second line of defense may need to be established.

If... space is adequate to establish an exterior elevated tower at the West side, this water application may be used. However, a defensive interior control will be necessary to prevent the elevated tower from pushing the fire toward the building center. One must understand the effectiveness of important barriers.

It appears that, at a minimum, the central suite of offices will be lost. Perhaps also some outside offices on the west end. Perhaps the fire can be controlled along the defensive line indicated and the owners only lose the third floor west end. Plus there would be extensive smoke damage on upper floors and water damage on lower floors.

The success of controlling the fire will depend on sufficient staffing and effective water application – and appropriate venting. The control evaluation (Chapter 15) depends on all of these features at needed times. Fortunately, all of the tools exist to calculate staffing and equipment arrival times, maneuvering supply and attack hoses into position, and water delivery. The IPI chart can organize the information to give a picture of conditions at any time until extinguishment.

14.43   What If...

Although this book discusses performance analysis rather than design, it is difficult to ignore the ways in which the situation could change. Frequently, the performance knowledge acquired from the Phase 1 analysis enables one to extrapolate outcomes to understand the significance (and costs) of adjusting the fire, the building, or the fire department.

For example, what if...:

...the engine company staffing were reduced from an officer and three fire fighters to an officer and two fire fighters?
...the engine company staffing were increased to an officer and four fire fighters?

These changes would significantly affect the time to deliver water to the ALP and stretch attack lines for this scenario. The analytical techniques provide a way to calculate time differences for these alternatives and their effect on risk characterizations.

Also, what if...:

... Engine E2 did not drag the relay line by hand, but used an engine to place the line?
... the community response were three engines, one ladder, and a battalion chief?
... Engine E1 took Hydrant H1 on its way to position in front of the building, and Engine E2 positioned itself for rapid establishment of a second or third attack line?
... better (or worse) supply water sources were available?
... a wet standpipe were installed in Stairwell B?
... a dry standpipe were installed in Stairwell B?
... the fire in the suite involved two rooms?
... the room of origin were an outside office rather than the inside suite?
... doors D1, D2, D3, and D4 were closed (e.g. common to provide security to the west end of the building?
... the incident commander became aware of a life safety concern for an occupant?

Enough techniques exist to calculate time durations, incorporate the actions on an IPI chart, and understand the effect of these changes on the outcome of fire department suppression. When a digital database becomes established and additional computer models "cut and paste" unitized modular time increments for water movement activities, calculated results become available very quickly. Professional time can be devoted to understanding fire performance and its management.

14.44   The IPI Chart

The IPI chart provides a visual display of the interaction of all components that affect fire department extinguishment. The example of Chapter 28 shows a completed IPI chart for a complete analysis.

14.45   Summary

A pre-fire performance analysis allows an engineer the luxury of pausing to assess the conditions at the earliest time that the local fire department can put water on the diagnostic fire. This provides an opportunity to examine the building and site features that contribute to the situation.

Certainly, many factors are established. For example, community fire department organization, staffing, and dispatching procedures are generally known. Distances and water supply sources are difficult to change. Terrain, weather, and other environmental conditions enter into the evaluation, but these conditions are identified for a particular scenario.

Nevertheless, a Phase 1 manual suppression analysis allows one to gain an insight into those features that help or hinder the local fire department from having a good stop and minimizing threat to life and property. It allows one to examine variational "what if" changes that could make a difference in outcomes. This enables one to recognize realistic variations in outcome expectations. One can separate "Pollyanna" estimates from "Murphy's law" outcomes.

The client can get a realistic understanding of what to expect if a fire does occur. This enables one to create better operational risk management planning. It also allows one to make more cost-effective recommendations for building changes, if they seem appropriate.

Although the analysis is deterministic, uncertainty is clearly a part of an evaluation. The time estimates are more associated with first-order approximations. Nevertheless, the results do provide a clear picture of the influence of the building and the fire department on time durations. *Unit Four: Managing Uncertainty* discuss ways to deal with uncertainty. The analytical structure is independent of the numerical accuracy of quantification.

15

Fire Department Extinguishment: Control and Extinguishment

15.1 First Water Applied... Now What?

What luck! We can predict the outcome of the fire in the clean, dry comfort of our offices. Our complexity is very different from that of the incident commander. We can reduce our uncertainties with the luxury of additional time to gather more or better information. The incident commander is dealing with management under uncertainty in its purest form. Time to make decisions is short, and the on-scene information is a moving target because of a constantly changing situation.

Nevertheless, we must find a way to use the understanding acquired during Phase 1 to predict the following:

- Can the fire be put out quickly, or will it propagate to additional spaces?
- If it continues to grow, where can the fire department set limits of "to here, but no further."
- Is the incident commander potentially facing a good stop, or one that got away? How does the building design affect these outcomes?
- Are occupants or fire fighters being placed in unusual jeopardy until the fire is put out?
- How much of the building will be destroyed before the fire goes out?

Our goal is to use available information to answer these questions. We want to understand how the building and site design can work with the local fire department.

15.2 The Engineer and the Incident Commander

We pause for a moment to look at the roles of the fire safety engineer and the local fire department in fire extinguishment analysis. The engineer is neither skilled at putting out building fires nor at managing a stressful fire scene that has the potential for life loss or extensive damage. The local fire department is not able to quantify diagnostic fires or analyze holistic building performance.

The fire safety engineer can be most effective by providing information to the incident commander. This may be difficult because the two professionals are not at the fire scene at the same time. In addition, they may be separated by geography. Yet at the interface of their professions in terms of discussing the building fire outcomes there is a unique opportunity for interactive communication.

Fire Performance Analysis for Buildings, Second Edition. Robert W. Fitzgerald and Brian J. Meacham.
© 2017 John Wiley & Sons Ltd. Published 2017 by John Wiley & Sons Ltd.

The incident commander can manage the fire ground better when reliable information is available to fit the needs. The engineer can design buildings more effectively with better knowledge of fire ground operations. The interface opens the possibility of a dialogue that can increase understanding for both.

The fire ground is a complex scene. The fire, the building environment, occupants, and the fire department all interact in ways that cause – and are caused by – dynamic changes. We describe the role of a fire department as that of a change agent for conditions that would otherwise occur without their intervention. In addition to putting out the fire, a fire department may rescue occupants, save important assets, protect neighboring structures, and generally improve otherwise deteriorating conditions. Fire department intervention changes the state of a building fire and its outcome. The building design has a role in that process.

Fire departments are paramilitary organizations. They are similar to armies fighting on the battlefield. "Our" army has an incident commander who develops a strategy (i.e. what to do) and tactics (i.e. how to do it). Different types of responding units (i.e. "special forces") select appropriate equipment and to use it to carry out the work. While the incident commander directs operations, the responding units do the work. Fire ground control and communications are essential during the operations.

All local fire departments have limitations. Sometimes, these may be caused by building and site conditions. Problems can also occur because the department may have inadequate resources or insufficient training or experience for the specific emergency. Sometimes, problems occur because incomplete fire ground information makes appropriate selection among different alternatives difficult. At times, decisions are the result of urgent fire ground priorities taking precedence over essential priorities.

The incident commander must prioritize needs based on knowledge of the situation and the availability of resources. For example, if occupants need protection from fire propagation, the commander may direct a fire attack to provide temporary occupant protection – even if a more efficient hose line placement would be elsewhere. Sometimes a potentially successful offensive attack may be abandoned to avoid placing fire fighters in unusual jeopardy. Situations are rarely "black and white," and a mental model (size-up) helps to prioritize decisions involving balancing resources and dangers.

The role of the fire safety engineer is to understand how building design features affect manual extinguishment performance. This skill is acquired by evaluating selective events in the framework. Evaluating important events is difficult and filled with uncertainty. Nevertheless, the intellectual effort to understand conditions and organize them into a framework to analyze performance provides a vehicle for understanding. Practice and discussion of opinions with others enhance the skill.

15.3 Pause to Review Available Information

Figure 15.1 again shows a representative diagnostic fire and timeline for the complete manual extinguishing component. The completion of Parts A and B indicates that two rooms have flashed over and a third room is becoming involved at first-in arrival.

Phase 1 estimates the fire conditions at the time that first water can be applied. One can be relatively confident with this assessment because the diagnostic fire conditions

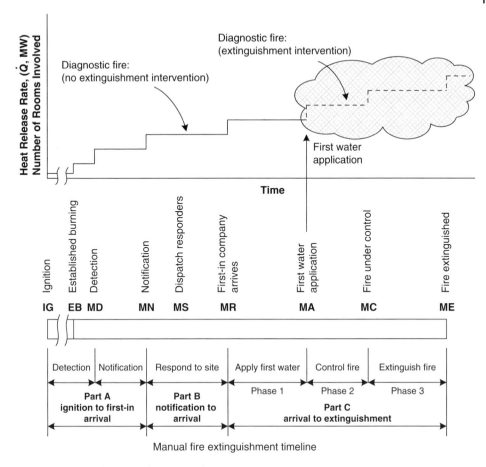

Figure 15.1 Manual extinguishment timeline.

are known and we can estimate reasonably well the time to apply first water. The systematic logic provides a framework for thinking.

Evaluations are more difficult after first water application (Phase 1). We must now use understanding on which to base our predictions. Fortunately, we do have information to guide our evaluations.

Let us briefly review two aspects:

1) What we want to know:
   a) Phase 2 – Can the initial Phase 1 attack put out the fire quickly? Where can the barriers and available fire fighting resources establish a defense to control additional fire extension? How can available resources put out the fire?
   b) What will be the extent of the fire before it is extinguished? What are the expected risks to the exposed?
2) What we know at the end of Phase 1:
   a) The building and site plans.
   b) The diagnostic fire for flame-heat, smoke-gas, and sequential times for flashover in the cluster of relevant rooms.

c) Fire size and conditions at first-in arrival (MR).

d) Fire conditions at first water application (MA) (Phase 1).

e) Number and type of fire apparatus at the scene (including multiple alarm responses); time sequence of arrival; number of fire fighters available for suppression; number of fire fighters available for other fire ground work.

f) Supply water quantity and the time sequence for delivery to the attack launch point (ALP). Each supply water path may be calculated by procedures as illustrated in Chapter 14.

g) The number of attack hose lines that are available and their likely placement. The fire flow and time to place each attack line in service may be calculated using the procedures described in Chapter 14. Placement locations may need some care in selection. Variation "what if" analyses are helpful here to bracket time and water application alternatives.

h) An Interactive Performance Information (IPI) chart showing time relationships for various layers of detail involving fire suppression components and their availability.

i) Variation "what if" analyses using the IPI chart to bracket uncertainty with water application times.

Pre-incident training exercises can use technology capabilities to portray time-lapse changes for fire conditions; attack hose line placement and fire flow; and barrier effectiveness at stopping fire extension. Information technology can simulate many variations to create visual awareness of alternative tactics.

Building information is combined with expected fire department actions. Discussions with local fire officers can give additional insights into potential problems and decisions that are likely to be made at the scene.

15.4   Phase 2 Assessments

Manual fire suppression requires fire fighters to locate, attack, control, and extinguish a fire. Phase 2 examines the status of fire conditions and fire suppression at the theoretical time when first water is applied. Knowledge about the number of available hose lines and their placement, ventilation conditions, and the building's ability to affect fire movement is available. It should be relatively clear if rapid extinguishment with the first attack is possible. Other conditions will not be as evident, and a Phase 2 study will indicate conditions that the incident commander can incorporate into other considerations to make a decision.

The Phase 2 analysis may not be as methodical as depicted earlier. The process is often fuzzy and evolving. Nevertheless, the incident commander must make a decision about how to fight the fire with the available resources and building conditions. The choices are an aggressive offensive attack, a defensive mode to stop extension at some point, or a combination for some situations.

The initial decision may or may not be independent of whether to mount an exterior or interior attack. As more information is acquired and the success or failure of initial operations becomes apparent, the fire ground command strategy may change to fit the needs of the situation.

15.5   Offensive Attack

If the fire ground commander believes the fire can be safely confined and extinguished without further fire extension, an offensive attack would be used. An offensive mode of fire fighting involves laying attack hose lines and aggressively applying agent to the fire. When this can be done, the fire is often extinguished relatively quickly. When the fire is small enough, a single hose line can extinguish it.

There are two mechanisms used to extinguish fires small enough to be put out by direct attack. One is to apply a fine water spray into a confined space. The water converts to steam, prevents air from continuing the combustion process, and smothers the fire. To use this form of extinguishment, the fire fighters must have the modern personal protective equipment clothing, often called bunker gear or turnouts. This clothing protects against thermal injuries, lacerations from debris, and head injuries from falling objects. The self-contained breathing apparatus (SCBA) enables a fire fighter to breathe in this atmosphere.

A faster and more effective means of extinguishment "pushes the fire out of the building." When a room can be ventilated to the outside and water is discharged, the expanding steam and fire gases drive the fire out of the building. Among the many benefits of ventilating a fire at the proper time is that visibility is improved and the hot combustion products are pushed out of the building.

Envisioning the fire being "pushed" is a useful concept in evaluating buildings. To some extent, the fire department can control the direction in which the fire will move. If people or valuable property (e.g. company financial records or data storage) are threatened, the direction in which the hose lines are placed and the building openings through which the fire is pushed become important. When multiple hose lines are utilized, communication between fire crews is important. Fire movement requires a planned, coordinated effort. The building architecture provides fire attack routes and also contributes to the fire movement.

15.6   Defensive Fire Fighting

In defensive fire fighting, fire fighters attempt to control the fire and limit its extension. Then, the fire can be worn down until it is weak enough to be put out.

A defensive stand is usually done from outside the building. Heavy streams pour water into the building until the fire begins to die down. Although the building is frequently a total loss, the process does reduce the likelihood of fire spread to neighboring buildings exposed to the fire.

Sometimes defensive fire fighting is done inside the building. Fire fighters may try to use an offensive attack, but find it unsuccessful and change to a defensive mode. Frequently, the fire fighters must "defend the barriers" to prevent fire from extending into adjacent rooms. When fire fighters can prevent fire spread beyond identifiable barriers, the fire is contained. This condition is often described as the fire being "under control" because its spread has been stopped and the incident commander understands that it will be subsequently extinguished.

Interior defensive fire fighting to control spread is important in multi-occupancy buildings to protect interior exposures from the fire. Not only is the "defend the barriers" useful for exposures on the same floor, but the technique is essential for multi-story buildings to prevent upward fire propagation.

Sometimes both offensive and defensive fire fighting may be used at the same time. The decision depends on the "situation and the terrain." The distinction between offensive mode and defensive mode may seem a bit fuzzy at times, but damage and continued dangers until the fire eventually goes out are clear.

15.7 Barrier Functions in Fire Fighting

Barriers, or the lack thereof, are important in fire fighting. Decisions relating to the mode of fire attack and the ability to control and extinguish a fire are influenced greatly by barrier locations and status.

A barrier is any surface that will prevent or delay flame or smoke movement in a fire. We view all potential movements of fire, water, and people in three dimensions. Normally, we consider floors, ceilings, and walls as barriers. Openings such as doors and windows are considered part of the barrier, regardless of the presence or status of protectives.

Viewed in terms of their functions in fire fighting, barriers are:

- *Directional channels*. Barriers channel the movement of flame-heat and smoke-gas in the building. This may be beneficial or detrimental depending on directions and what fire fighters will experience.
- *Protection*. Barriers can give protection to fire fighters from the fire and its products of combustion. This protection may provide time to set up fire fighting actions.
- *Deflection*. When ceilings are rigid and not frangible, solid stream application can be effective to break up water into droplets. This gives more rapid extinguishment because the water droplets have a larger surface area to mass ratio and can be converted to steam more easily, similar to sprinkler discharge. On the other hand, barriers may prevent effective application of water by blocking hose streams.
- *Ventilation*. Barriers can be opened to release flames, heat, and smoke to make the building environment more compatible with fire suppression. For example, ventilation improves visibility to enable the seat of the fire to be located in a smoky atmosphere. A more common use of ventilation involves releasing the fire out of the building. The combination of water application and ventilation can be very effective to direct fire movement when the actions are well coordinated.
- *Orientation*. Walls enable fire fighters to maintain a sense of location when advancing a hose in low (or no) visibility.
- *Danger*. One must also anticipate the potential for disaster caused by holes in floors and falling overhead debris. In low-visibility environments when fire fighters must crawl to advance, holes in floors may not be seen until too late. Similarly, falling sheetrock, suspended ceiling materials, or room contents (e.g. tall storage frames) may occur with little or no warning.

A barrier may help or hurt fire fighting operations. A Phase 2 assessment enables one to recognize the effect of the tactics being considered. This is often more obvious in the comfort of an office analysis than during fire fighting operations.

The lack of barriers can also influence fire fighting decisions. For example, in a large open space, attack from one side will push the fire toward unobstructed spaces. If it is not possible to push the fire back into the already burned parts of the building, this fire fighting tactic will cause the fire to spread further.

15.8   Exposure Protection

The threat of fire extension to exposures is an important consideration on the fire ground. This is particularly important when a defensive mode of fire fighting is required. Two types of exposure must be considered:

- External exposures
- Internal exposures.

External exposures are the most easily recognized. When a nearby structure is exposed to the radiant heat from a large building fire, the heat transfer can cause spread to additional buildings. The vulnerability of neighboring buildings to fire spread is something the incident commander must consider with defensive fire fighting. Water may be applied to buildings that are not yet involved. Also the direction of heavy stream applications may be influenced by the vulnerability of neighboring buildings.

When there is the possibility of fire extension to exposures, the fire ground commander must decide if the best strategy is to aggressively fight the fire or to protect the exposures from fire extension and allow the fire to burn itself out. A third approach protects exposures and also applies water to the burning fire, but not in an aggressive attack. These latter strategies are part of defensive modes of fire fighting.

Internal exposures are common, particularly in multi-tenant buildings. Fire in one section of a building can spread to other parts. The incident commander may have some influence over the fire's directional movement depending on the direction of the attack.

Tall buildings may allow fire propagation through the floors or through windows above the fire floor. Staffing is needed to defend these barriers from fire extension. The tactic of assigning fire fighters above a fire is not generally recommended because of safety concerns regarding floor collapse and upward fire spread. However, awareness and consideration of fire spread to uninvolved sections of a building are part of the incident commander's mental model of fire control.

15.9   Constraints

Available resources and operational constraints enter into fire ground decision-making. These may be viewed in terms of site, fire fighting resources, and attack constraints.

Site constraints relate to the ability to access a fire from any outside direction. Many buildings are so close to other structures or have restricting terrain features that it is impossible to approach the building from adjacent sides. It is not uncommon for buildings to be unable to be accessed on one, two, or even three sides. Thus, defensive fire fighting becomes difficult or impossible. Also, some buildings, such as hotels have large canopies and landscaping at a front entrance that make it difficult to place ladder trucks or engines at effective locations. A site examination can reveal difficulties in providing certain types of fire fighting evolutions.

Fire fighting resources relate to staffing, water, and equipment. The Phase 1 analysis will reveal any weaknesses in these factors. For example, if water is expected from a stream or pond, summer temperatures or prolonged drought conditions may dry up the source. Similarly, winter temperatures in cold climates may freeze over the source. Even in locations with underground pipes and hydrants, fire flow may be insufficient or unavailable for some high-demand needs. Generally, the water system's pipe size and fire flow information will identify potential difficulties for some needs.

Staffing and equipment must be considered in the extinguishment analysis. The time durations at which fire fighters can respond to the scene, their numbers, and the equipment seems to be constantly changing. For example, volunteer and call companies are often short-handed during the day because of work locations and commitments, but will have sufficient staff in the evening or weekends. Budget constraints influence the number of fire fighters that will respond. Large fires require many people.

Attack constraints relate to the personal protection and equipment in interior fire fighting operations. One is the capacity of SCBA air tanks. Normally, 20 minutes of use is a reasonable estimate. If the fire is not extinguished within the time limitation provided by the tanks, replacement crews and tanks might be needed. Again, staffing available at the site remains an important part of an evaluation.

A second constraint relates to actual hose distances. An engine company will often carry two lengths of hose to the standpipe or supply hose connection. The length of hose segments is important information. For example, if the hose lengths are 15 m (50 ft), additional lengths must be obtained when the actual distance exceeds 30 m (100 ft). This takes additional time, which increases the time for agent application. The stretch from 12 m (40 ft) to 18 m (60 ft) is very different than the stretch from 27 m (90 ft) to 34 m (110 ft). Consequently, fire attack path distances from ALP connections to the fire location are important.

In summary, the building and site features have a major influence on the mode of fire extinguishment and its effectiveness. Many become evident in the Phase 1 analysis, and others are discovered during the Phase 2 assessment. The intention is to ensure that no one will be surprised at the outcome of a hostile building fire because of unexamined assumptions.

15.10 Critical Fire Conditions

The critical fire condition describes the building situation where the incident commander decides to use a defensive mode of fire fighting rather than an offensive mode. The critical fire condition is not a specific number. Rather, it is a set of conditions that would indicate that the fire must be fought in a defensive mode. Therefore, fire sizes will be different in one building compared with another. Additionally, they may be different within a single building.

The critical fire condition is based on the diagnostic fire, the building architecture, barrier conditions, ventilation opportunities, number of available hose lines, building stability, and fire department resources. Usually, a Phase 1 analysis clearly indicates if an initial attack will be able to extinguish the fire quickly and easily. It will also be clear about when conditions exceed the initial attack capabilities. Often, the conditions do not indicate either of these extremes and additional information must be obtained.

Phase 1 information is only part of the relevant conditions that influence the incident commander's decision. At some time during the fire, the incident commander will decide upon an offensive mode or a defensive mode of extinguishment. The critical fire condition describes the building and fire conditions that enable a commander to make this tactical decision. The availability of Phase 1 information before the fire can give an indication of likely expectations.

15.11 Fire Control (MC)

The declaration that the fire is "under control" is an important event in the fire extinguishment process.

When a fire is "under control" it is contained to a condition whereby the incident commander understands what it will do until it is extinguished. Usually, a fire is considered under control when its spread is stopped even though there may be regions of uncontrolled burning within the fire area.

During a rapid, offensive attack, control and extinguishment are very close together. In defensive fire fighting, extinguishment may occur much later than the time when defensive boundaries are identified:

- **Given:**
  - *First water application has occurred (MA)*
  - *Boundaries to contain fire movement have been identified*
  - *Time from EB ____ minutes.*
- **Question:** *Will the fire department control the fire to these boundaries (MC)?*

15.12 Fire Extinguishment (ME)

Eventually the fire will go out and the fire department will have contributed to this outcome. We use the term extinguishment to indicate that the visible fire has been suppressed. It is possible for hot spots to remain or small fires to exist in concealed spaces. Overhaul occurs after extinguishment to ensure that all fire conditions have been eliminated.

The completion of manual extinguishment of visible flaming is often expressed as blackout, an old fire fighting term. Other terms are also common, such as knockdown and tamp-down. Many fire fighters define knockdown as the elimination of the major flame condition, and extinguishment as the end of overhaul. Whatever term is used, eliminating major flaming (except for hot spots) is the event being evaluated.

The building is a major determinant of manual suppression success; the number of hose lines, their positioning, and venting are others. The fire will eventually be put out. The question is one of fire extension. The event is expressed as follows:

- **Given:**
  - *Fire control has occurred (MC)*
  - *Boundaries for fire control have been identified*
  - *Time from EB ____ minutes.*
- **Question:** *Will the fire department extinguish the fire without extension (ME)?*

The success of this event is not that the fire will eventually go out, but whether the fire department will extinguish the specified fire before it extends to a size larger than the boundaries defined in fire control. In other words, given that the agent application (MA) component is successful, will the building/fire department partnership extinguish the fire before it extends to larger sizes?

The format for this question is single value. The response is related to the time and specific fire control area. If the fire cannot be extinguished within the specific control area, a new area is selected. The response is always negative until the anticipated time of extinguishment has been reached.

15.13 Summary

A manual extinguishment analysis estimates the fire size and conditions at extinguishment. An evaluation provides information about conditions and risks within the building during the complete scenario from established burning (EB) to the time of extinguishment (ME).

A single scenario analysis for manual extinguishment often gives sufficient information about building conditions to enable one to characterize risks. When additional information is needed, supplementary scenarios can be evaluated. These may often be done by extrapolation of initial information without formal analysis. Even when additional scenarios must be examined, much information of the first analysis remains valid. Thus, "what if" variation analyses are often relatively easily evaluated.

Although quantification and decision-making involve subjective judgment, the physical conditions of event outcomes provide a good basis for decisions. Because scenario analyses are strongly influenced by time durations, the IPI chart analysis provides a means to order scenarios and conditions. Vertical columns at specific time durations may be examined to understand the range of conditions that exist at that time. This enables one to make judgments with greater confidence. A better insight into the effect of the building on fire fighting is likely to be obtained if the local fire incident commander participates in the scenario discussion.

16

Automatic Sprinkler Suppression

16.1   Introduction

Although concepts of sprinklers and similar early devices were conceived and constructed in the early 19th century, Henry Parmelee developed the first practical and extensively used sprinkler system in 1878. The first automobile was built at about the same time (1886), and "mass production" of 13 cars of common design first occurred in 1896. Since their beginnings, the technological development of sprinkler systems has at least paralleled, and perhaps even exceeded, the evolutionary technology of the automobile. New developments and improvements in sprinklers and sprinkler system technology continue to appear at a rapid pace.

Over the years, the automatic sprinkler system has had an outstanding record of success in controlling or extinguishing unwanted fires. Nevertheless, there have been cases in which buildings protected by an automatic sprinkler system have sustained major losses. Questions arise as to what factors produce success, what flaws induce failure, and how one can compare the quality of different systems.

16.2   Sprinkler System Performance

A performance analysis evaluates a component for its ability to perform the function for which it was intended.

Observation of an automatic sprinkler system in a building often conveys such great confidence for fire safety that a performance examination may become cursory or careless. "Why bother?" is a common attitude, because sprinkler system experiences seem to cover many sins of omission or commission.

The first National Fire Protection Association (NFPA) standard in 1896 was written for automatic sprinkler systems. This standard has evolved over the years from a few pages of guidance into substantial volumes involving design, installation, water supply and distribution, maintenance, pumps, specialized fuels and storage. In addition, books are written about how to understand, interpret, and design using the standards. These important fire safety systems are well represented by an industry that is interested in successful performance.

Fire Performance Analysis for Buildings, Second Edition. Robert W. Fitzgerald and Brian J. Meacham.
© 2017 John Wiley & Sons Ltd. Published 2017 by John Wiley & Sons Ltd.

What can go wrong? Common attitudes are:

- "The insurance company must have checked the design or else it wouldn't write a policy."
- "If the AHJ [authority having jurisdiction] didn't approve the sprinkler design quality, the building would not get a certificate of occupancy."
- "If the fire department didn't regularly certify the system, the building would be closed."
- "If conditions under which the sprinkler system had been originally designed were changed, the owner would ensure that functional quality would be maintained."

Even with all of the statistical success of sprinkler systems, these concepts are not accurate. Some systems are clearly of excellent high quality; some are marginal; and some will not work when needed. Can a performance analysis distinguish the quality of sprinkler system in a building? The answer seems to depend on the examination of details.

There are six types of information for an existing sprinkler system:

1) Observable and clearly known.
2) Determined by testing.
3) Able to be calculated.
4) Obtained from design drawings and calculations, if available.
5) Inferred from common practices and the other acquired information.
6) Information that is not known and cannot be determined.

All information has an acquisition cost and requires time. The value added to the performance evaluation must be balanced against its cost. How much is enough becomes an important professional decision.

A general sense of quality can often be acquired by observing certain important features of an existing system. Depending upon the importance of the performance needs for the building, additional layers of knowledge can be obtained by attention to strategic areas that relate to performance sensitivity. Selectivity of focus can be valuable to gain a better knowledge of expected performance.

Part One: Sprinkler Systems discusses sprinkler systems and their operations. The intent is to develop an awareness of functional operations and have a commonality of language. *Part Two: Sprinkler Performance* describes the factors that relate to system reliability and suppression. The topics are organized around performance thinking. Observational and analytical evaluations each provide a different, yet complementary understanding of a system's performance. Chapter 29 describes the analytical framework to guide a sprinkler system's evaluation.

PART ONE: SPRINKLER SYSTEMS

16.3 Sprinkler Extinguishment

A large number and variety of sprinklers are available for installation in ceilings or on sidewalls. Each type will fuse (i.e. open) at a specified heat condition and deliver a particular water droplet size, density, and distribution pattern onto the fire. A sprinkler can be designed for generalized occupancy conditions or for a specified fire and its room environment. To envision conventional sprinkler operation, let's describe a possible scenario.

Assume that you are able to observe the interior of a relatively large room with a number of sprinklers in the ceiling. Ignition and established burning (EB) occur in a fuel package and a fire plume rises and begins to develop a hot layer at the ceiling. As the fire continues to burn, more heat and higher temperatures accumulate at the ceiling.

Now consider a sprinkler such as that of Figure 16.1. The sprinkler frame (a) is connected to a pipe. The water in the nozzle (b) is held back by a cap (c), which in turn is kept in position by levers (d). A fusible link (e) maintains the levers in position. The link has a heat-actuating element (f) that softens when heated. When the temperature reaches a predetermined level, the link separates, causing the levers to disengage and the cap to be blown off by the water pressure in the nozzle. Water flows from the nozzle and strikes the deflector (g), causing a pattern of water distribution to the floor below.

Now return to the room fire. As the fire grows, the hot-layer temperature at the ceiling continues to rise. When a link absorbs enough heat energy, it fuses and releases the cap from the sprinkler. Only sprinklers near the fire will fuse, allowing the water to impinge on the deflector and discharge a pattern of water droplets onto the fire. Many fires are extinguished with the operation of only one sprinkler, thus reducing the fire and water damage substantially. Although additional sprinklers are sometimes needed, most fires are controlled by the operation of only a few sprinklers.

Occasionally, fire conditions and sprinkler capabilities are mismatched. If the fire can continue to grow, more sprinklers will fuse. The water supply now becomes important, because when too many sprinklers open, water pressure and quantity are reduced. This may allow the fire to grow beyond the capability of the sprinkler system.

Some useful characteristics of typical sprinkler system operations are as follows:

- Each sprinkler is independent. Only sprinklers fused by the heat transfer from the fire will operate.
- Sprinklers normally do not fuse quickly in a fire. Depending on the ceiling height and type of fusible link, one would normally expect the sprinkler to fuse as the fire grows to 350–500 kW [about 1.2–1.5 m (4–5 ft) in height]. We may note that the common fire size at sprinkler actuation is between the enclosure point and the ceiling point of the realms of fire growth (Section 6.6). It is possible to install fast-response sprinklers that fuse when the fire is smaller.
- Most commonly, only one or a few sprinklers will open during a fire because of the speed and effectiveness of suppression.

After the sprinklers fuse, water continues to flow until the sprinkler system is shut down. Water damage in unoccupied buildings can be reduced drastically by the installation of water flow alarms connected to the local fire department or to a supervisory service.

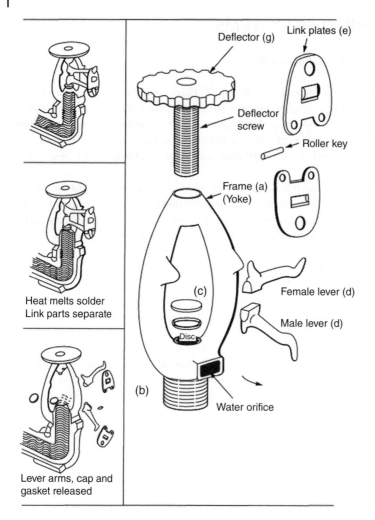

Deflector (g)
Link plates (e)
Deflector screw
Roller key
Frame (a) (Yoke)
Female lever (d)
(c)
Male lever (d)
Disc
(b)
Water orifice

Heat melts solder
Link parts separate

Lever arms, cap and
gasket released

Figure 16.1 Sprinkler. *Source*: Reproduced by permission of Insurance Services Offices, Inc.

16.4 The Sprinkler System

The sprinklers described in the previous section are only one component of a complete sprinkler system. Other parts of the system supply enough water at sufficient pressure and duration to control or extinguish the fire. This is accomplished by the water supply, the water distribution system, pumps and other operational devices, where needed. The automatic sprinkler system may be organized into five major components:

- The water supply and distribution system that brings water to the building.
- The control system of valves, pumps, and other devices that link the water supply system to the internal building sprinkler piping system.
- The building's piping distribution of risers, mains, and branches.
- Sprinklers that discharge water to the fire.
- A monitoring system for trouble and alerting alarms.

Figure 16.2 shows a sprinkler system schematic. The water supply reaches the building through a feed main. A water control valve is needed to shut off water for system maintenance. Water control valves must be open for water to flow to a fused sprinkler. The water flows vertically through a riser to a feed main to a cross main to the branch lines on which the sprinklers are connected. A variety of pumps and other devices may be installed to maintain or adjust pressure.

16.5   Types of Sprinkler Systems

The previous simple description of sprinklers and the sprinkler system provides a basic organization from which variations can be introduced to meet specialized design requirements. Building needs and designs can vary substantially. The sprinkler system must be able to adjust to the functional needs of building operations and still perform the role for which it was intended.

Building sprinkler systems are designed to address specific hazards that will exist in buildings. For example, a sprinkler system for an apartment or office building will have different requirements from those of an unheated warehouse or an industrial operation in which a large, fast fire could occur.

Sprinkler systems may be classified as *wet pipe, dry pipe, deluge,* and *preaction.* The appropriateness of the installation features, as well as evaluation techniques for reliability and design effectiveness of a specific installation are site-specific. Our goal is to provide a basic understanding of sprinkler systems with which to structure evaluations.

Wet-pipe Systems
The piping for a wet-pipe sprinkler system is filled with water under pressure. When a sprinkler fuses, water can immediately flow from the orifice. This is the most common type of sprinkler system in buildings that do not have freezing potential or other special needs.

Dry-pipe Systems
A dry-pipe system is used in spaces that are subject to freezing or in spaces where a greater water flow control is desired. The piping contains air or nitrogen under pressure, rather than water. Compressors maintain the necessary pressure. The water supply is held back at a dry-pipe water control valve located in a heated space. When a sprinkler fuses, the release of air causes a pressure drop that automatically opens the dry pipe valve and releases water into the system. However, if the air in the pipes is not removed quickly, water movement can be delayed. To avoid this condition, quick-opening devices (QODs), called accelerators or exhausters, can be installed to quickly expel air and gases in the pipes, allowing water to reach the open sprinklers more quickly.

Deluge Systems
A deluge system can apply water over a relatively large area by supplying water to open sprinklers simultaneously. This type of system is used for special types of hazard where the immediate application of large quantities of water during a short time is needed to control a potentially rapidly expanding fire. Flammable liquid storage facilities, aircraft hangars, and extra hazard industrial operations might install a deluge system.

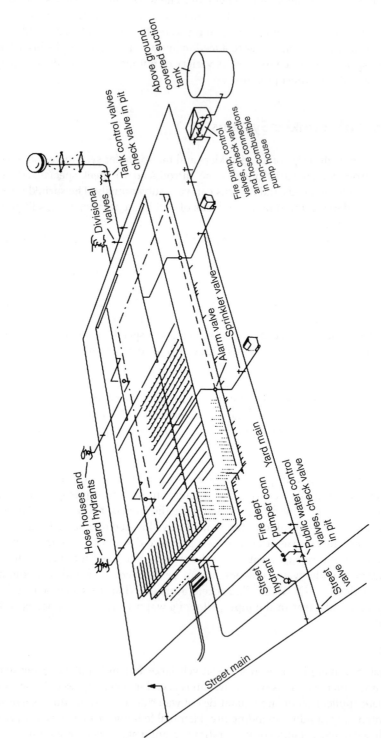

Figure 16.2 Sprinkler system. *Source*: Reproduced by permission of FM Global.

A deluge system is usually actuated by a heat detection system. When a heat detector actuates, water is released by a deluge valve and fills the dry-piping system to operate. A variety of special control devices, such as water control valves, releasing mechanisms, and supervisory equipment, are available for installation.

Preaction Systems

A preaction system is used where room contents are very sensitive to the discharge of water. Library stacks or valuable exposed files illustrate concerns that owners may have for an inadvertent discharge of water. A preaction system is a dry-pipe system with an additional control linked to a sensitive fire detection system. If the sprinkler fuses but the detector does not sense a fire, then no water will flow. On the other hand, if the detector senses a fire, the water control valve is opened to permit initial water flow through the dry pipes. However, if the sprinkler does not fuse, no water will discharge. A preaction system will activate and discharge water only when the sprinkler fuses and also the detection device senses a fire.

Specialized Designs

Other designs may be devised by combining sprinkler system components to meet specialized needs. For example, rather than using a conventional automatic sprinkler system on a ship, an automated system may be used. This design uses a conventional sprinkler system with a manual opening of the water control valve. Creative automatic suppression designs can meet distinctive needs for specialized functions or hazards.

PART TWO: SPRINKLER PERFORMANCE

16.6 Organization for Thinking

Often, sprinklers are observed in a building. It is assumed that the sprinkler system will work during a fire, and we place another building into the "safe" category. Little additional thinking goes into the decision. However, is this an appropriate conclusion?

This conclusion can sometimes lead to disaster. Can we recognize what can distinguish between success and failure? Let's organize our thinking to improve powers of observation and deduction.

Sprinkler systems installed and maintained in accordance with the NFPA or FMGlobal standards perform admirably. However, many sprinkler systems do not meet these standards. Or differences in conditions occur between the initial sprinkler installation and today. Or the building may have changed use and building renovations may have inadvertently compromised the sprinkler system. Or inattention to proper maintenance may have compromised the system. Or... the list goes on.

Qualitative and quantitative analyses complement each other well for the performance evaluations of sprinkler systems. Each provides information that is not covered by the other. While variability studies can give some additional insight, they are not as useful in sprinkler system applications as routine qualitative and quantitative analyses.

A performance analysis for a sprinkler system examines three major components:

1) The diagnostic fire
2) The reliability of the system
3) The design effectiveness of the system.

We define the reliability of a sprinkler system as the likelihood that water will discharge when a sprinkler fuses. We call this event, agent application (AA). AA is evaluated only once for any zone of coverage.

We then examine the installation's design effectiveness for the diagnostic fire. The design effectiveness component answers the question, "Given that agent application (AA) occurs, will enough water be delivered at an appropriate pressure and with a suitable distribution pattern to control the fire?" This addresses the system's operational effectiveness (AC).

Published sprinkler standards do not describe the fire that is used for design. Rather, they base quantification on a hazard classification. This is easier and generally suitable for most installations. However, sometimes a mismatch can occur between fire growth speed and sprinkler response. A performance analysis attempts to recognize conditions where the fire and operational effectiveness may be incompatible.

16.7 Agent Application (AA)

An external water source delivers a quantity of water under pressure through the underground piping system on the property. Water that eventually enters the system's piping network is controlled at the building by a group of valves, pumps, and other devices. These water control system components influence the system's reliability. However, other conditions may also affect the ability to flow water from a fused sprinkler.

Valves act to control the water flow from external supply pipes to the internal sprinkler pipes. Valves are used in all systems and must be open to allow water to move into the sprinkler piping network. Water control valves can be unintentionally closed because of inattention or ineffective supervision. Many installations have several sectional control valves. All of the valves, public and private, that control the feed of water in a zone must be open. During routine maintenance and daily building operations, fire protection needs may be neglected. This neglect is a common cause of closed sprinkler valves.

Sprinkler water supplies are usually separated from the building's potable water because public health authorities have concerns about the possibility of contaminated sprinkler water flowing back into the potable supply. Backflow can occur when sprinkler water pressures are greater than the domestic pressures. One-way check valves often address the problem satisfactorily. Reduced-pressure backflow preventers are sometimes used.

Backflow prevention devices are controversial between fire protection engineers and public health authorities. Improperly specified or designed backflow preventers can reduce pressure or prevent water from reaching the sprinkler, thus reducing the reliability of the system. Fire protection professionals feel that backflow preventers reduce sprinkler reliability and water flow, whereas health officials feel these devices are always necessary to avoid waterborne diseases.

Quick-opening devices, such as accelerators and exhausters for dry-pipe systems, improve the speed with which water can be applied to a fire. Dry pipe systems with QODs are not as reliable as wet-pipe systems that do not need such equipment.

Detection in preaction systems adds another device which needs periodic maintenance to avoid malfunction. All devices that affect the water discharge from a fused sprinkler contribute to agent application reliability.

Another potential problem that arises more frequently with dry-pipe systems than with wet-pipe systems is blockage of the pipes. Pipe scale and debris can prevent water from flowing through the piping system to the open sprinkler.

Pumps perform several functions, such as pushing water through the piping system, increasing water pressure, and maintaining high static pressures for operation or water delivery. They can influence both agent application (AA) and design effectiveness (AC).

Fire pumps must function under conditions more severe than the routine operational conditions for other kinds of pump uses. Fire pumps operate only under fire conditions or test conditions. This infrequent use may not have the maintenance attention of other operational pumps in constant use. Fire conditions may interrupt electricity or fuel delivery for pump operation. If pumps are an important part of the sprinkler system, the reliability of the pumps, controllers, drivers, accessories, and power for emergency conditions is a part of an evaluation.

In summary, if a sprinkler fuses and calls for water, a number of system components must function to deliver the water to the location of the fire. A reliability analysis includes proper system design combined with a regular inspection and maintenance program.

16.8 Agent Application Events

When a sprinkler fuses, we expect water to reach the location and discharge onto a fire. We define this as the reliability (AA) of the sprinkler system.

A performance analysis examines two types of impediment to flowing water reaching the sprinkler. These are:

- **Given:** *Fire has started.*
- **Question:** *Will all supply valves be open when the sprinkler fuses (vaa)?*

- **Given:**
  - *Fire has started*
  - *All supply valves are open*
  - *Sprinkler has fused.*
- **Question:** *Will water reach the sprinkler (waa)?*

Agent application is the only major part of a performance analysis that does not change with changes in diagnostic fire conditions.

16.9 Operational Effectiveness Observations

Given that water discharges from an open sprinkler (AA), an evaluation next looks at the system's operational effectiveness (AC). Operational effectiveness examines the interaction of the diagnostic fire and sprinkler water discharge. The diagnostic fire describes the fire's rate of heat release, plume momentum, and speed of fire growth.

These diagnostic fire values may be shown in the Interactive Performance Information (IPI) chart, Section 1A. It is often useful to include the associated floor area with the elapsed times and changes in diagnostic fire conditions. To give a perspective for the room or general area of sprinkler effectiveness, each specific floor area may be associated with diagnostic fire conditions and time durations from the IPI chart. The performance analysis examines the success of sprinkler control (or extinguishment) for each of these floor areas (or times).

We generally describe successful sprinkler operations as "control," rather than "extinguishment." The concept is that sprinkler success will prevent the fire from extending until it is completely extinguished by manual extinguishment. Because most sprinkler systems have water flow alarms connected to a fire department or a monitoring service, responding units can easily complete the extinguishment process. Although we normally use the word "control" for sprinkler operations and "extinguishment" for fire department operations, the outcome is functionally the same.

The AC analysis considers four potential failure modes addressed by the following questions:

- **Given:**
  - *System is reliable (AA)*
  - *Fire size* _____
  - *Floor area* _____
  - *Time from EB* _____
- **Question:** *Will the nozzle(s) open before the fire extends beyond the fire size being examined (fac)?*

- **Given:**
  - *System is reliable (AA)*
  - *Nozzle(s) open*
  - *Fire size* _____
  - *Floor area* _____
  - *Time from EB* _____
- **Question:** *Will the discharge density be sufficient to control the fire (dac)?*

- **Given:**
  - *System is reliable (AA)*
  - *Nozzle(s) open*
  - *Fire size* _____
  - *Floor area* _____
  - *Time from EB* _____
- **Question:** *Will enough water continue to flow and control the fire (cac)?*

- **Given:**
  - *System is reliable (AA)*
  - *Nozzle(s) open*
  - *Fire size* _____
  - *Floor area* _____
  - *Time from EB* _____
- **Question:** *Do obstructions affect water spray application to the fire (wac)?*

Each of these potential failure modes examines a different aspect of sprinkler system performance. Successive single value evaluations examine the continuum of sprinkler performance. Evaluations start at the first sprinkler operation and continue with additional sprinkler actuations until the fire is controlled or until it is clear that the performance weakens and the fire cannot be controlled. The IPI chart shows the status of all events.

Qualitative observations identify features that reduce sprinkler system effectiveness. A trained eye can recognize many of the features that lead to less effective fire control. Calculations can augment this information with quantitative measures. Together, these two aspects of the fire and the system can give a better picture of sprinkler control expectations. We briefly describe these concepts here and expand the discussion in subsequent sections.

Sprinklers Fuse (fac)

The evaluation of this event starts with examination of a specific fire size or condition. For example, what fire size will cause initial sprinkler fusing? Does the room have pockets of combustibles that can cause a fast fire to extend to larger sizes before the sprinklers fuse? This can cause a situation where the fire "outruns" the water discharge.

Some sprinkler systems have impaired sprinklers that are corroded, painted, taped, or have bagged links. Some sprinkler links are protected from the heat of a fire by obstructions that deflect heat. In some buildings, the ceiling height is too large for heat to cause sprinklers to fuse at an appropriate time. All of these features are recognizable to visual inspection and are more common than we wish.

Density Sufficient (dac)

This event relates water application and droplet sizes with the diagnostic fire. In an effective design, water droplets are sufficiently large to penetrate the fire plume and extinguish or control the fire. This evaluation requires that sprinkler discharge characteristics be examined.

Continuous Flow (cac)

Sprinkler design areas identify the water flow that is needed for the hazard and the number of sprinklers within that area. This event examines the number of sprinklers that can be supplied by the available water flow. It is not uncommon for combustion conditions to be different from those that were assumed when the sprinkler system was first designed. For example, cellulosic materials may have been the basis of an initial design. However, if the manufacturer changed to petroleum-based materials with different combustion characteristics, the fire may grow rapidly and cause more sprinklers to fuse. This reduces the quantity and pressure from those originally anticipated.

The water supply could have been marginal when the system was initially constructed. For example, existing community water mains may initially have been inadequate for the hazard. It is possible that the hazard classifications may have been adjusted to reflect available water supply, thus avoiding the necessity to increase water main sizes. The cac event examines the number of sprinklers that can be supplied by the available water.

Water Spray Obstructions (wac)

Obstructions to effective water spray from the sprinkler to the fire can reduce suppression effectiveness. Obstructions may be caused by building construction or by building contents. For example, building renovations often install new walls or ceilings with little regard for the location or spray pattern of sprinkler discharge. Also, horizontal surfaces such as tables, and vertical objects such as space dividers or tall fuel packages can protect fuels from water spray until the fire grows to less manageable sizes. Visual observations are the best way to recognize conditions that compromise sprinkler operations.

16.10   Sprinkler Fusing (fac)

There are literally hundreds of different sprinklers in use. All of them use essentially the same concepts. The differences are associated with the specialized features for the components. Some of the features that influence performance evaluations are described below.

The link-and-lever mechanism of Figure 16.1 is a common type of releasing device. The links are held in place by a solder that has a lower melting point than the parts that are joined. For sprinklers the most common temperatures at which the solder will fuse are 57.2–76.7°C (135–170°F). However, to accommodate different environmental conditions, sprinklers can have a range of temperature classifications up to 340°C (650°F). Temperature ratings are stamped on the sprinklers with associated color codes for visual recognition. Different temperature ratings for the solders reflect needs for the ambient temperature conditions in which the sprinklers will be placed. Most ordinary hazards, such as offices or residences, use 57–74°C (135 or 165°F) sprinklers. Industrial sprinklers may require higher temperatures when operations take place in heated environments.

Frangible bulbs are another common heat-actuating element. These devices use a liquid that partially fills a small glass bulb. When the temperature rises, the liquid expands, the bubble disappears, and the pressure rises. This causes the glass bulb to break, releasing the cap on the nozzle. When the cap is released, the sprinkler operates in the normal manner. The operating temperature is determined by the volume and type of liquid in the tube. Color codes indicate temperature classifications.

Although a temperature rating can give a sense of the fire size at fusing, the temperature rating alone can be misleading. Sprinkler sensitivity also varies with the mass, size, and shape of the solder; the temperature difference between the sprinkler and the surrounding environment; and the velocity of the fire gases that flow by the sprinkler. Frangible glass bulb sprinklers are affected in a similar manner.

In other words, all 74°C (165°F) sprinklers do not fuse at the same fire size, other conditions being equal. The response time index (RTI) gives a measure of the sprinkler link sensitivity to its actuation response. The RTI is a constant for each sprinkler. It is directly proportional to the mass and specific heat of the heat-actuating element and inversely proportional to the convective heat transfer coefficient and surface area of the element. The larger the mass of the solder, the less sensitive it is to fusing, regardless of the temperature rating. Fast-response sprinklers have an RTI of less than 50. The RTI for standard-response sprinklers may vary from 80 to 350, with 100 being a more common value.

Interest in sprinkler actuation technology relates to its relationship with fire spread. Sprinkler systems can have a relatively broad range of performance characteristics. A performance evaluation attempts to answer the question, "Will the sprinkler system control a fire of x kW involving y sq ft of floor area?" The implication for this question is, "Will the links fuse before a fire grows beyond this size?" Fire size and growth speed are compared with the response sensitivity of the sprinklers to estimate the answer.

Sprinkler actuation is also affected by factors associated with fire size and room geometry. The hot layer takes longer to develop with high ceilings. This gives the fire more opportunity to grow before sprinklers actuate. High-RTI sprinklers will not fuse as rapidly as low-RTI sprinklers. Sprinklers with inappropriate link temperatures may not fuse fast enough to apply agent before the fire grows to a larger size.

Sprinklers that are corroded, painted, taped, or bagged take longer to actuate. Sprinklers that are protected from the heat because of location or obstructions that prevent the fire gases from reaching the heat-sensitive element have a delayed actuation. Evaluation methods described in Chapter 29 identify the major factors that are associated with this assessment. Fortunately, most of these can be recognized by observation. If a new design is anticipated, one may include appropriate requirements in the specifications.

The sprinkler fusing event (fac) examines sprinkler response sensitivity and the fire's behavior. If the first sprinkler does not control the fire, more sprinklers must fuse to control the fire. The event examines fac for incremental fire sizes until the system controls the fire or fails.

16.11   Water Discharge (dac)

When a sprinkler fuses, water is discharged through the nozzle and impacts on the deflector. The water pressure and the deflector design determine the size of the water droplets and the distribution pattern of the water spray. Factors that influence the operational effectiveness of the water discharge component include the sprinkler type, water droplet sizes, the water density, fire characteristics, and the room size.

Sprinklers may be *upright, pendant,* or *sidewall.* When upright sprinklers are used, the piping system is visible. Water is directed upward to the deflector, which redirects a percentage of this water downward in an umbrella spray pattern to the fire. A sprinkler is often used in the pendant position. This is usually installed to hide the piping system in the ceiling, or sometimes because of design needs.

When a sprinkler is in the pendant position, the water is directed downward to the deflector. A percentage of the water is redirected upward to strike the ceiling and again fall to the fire in a spray pattern. Upright and pendant sprinklers are not interchangeable because their designs are distinct and the spray patterns will differ.

Sprinkler water spray extinguishes a fire by several actions. One is the absorption of heat from the fire. Fire is an exothermic chemical reaction. When the heat generated by the fire is less than the heat loss due to heat transfer and water evaporation, the fire will recede. The opposite occurs when the heat generation is greater than the losses. Heat absorption by the phase transformation of converting water to steam is a major heat loss from the fire and a significant contributor to fire control. In small, confined rooms, large amounts of steam contribute to the displacement of air (oxygen) from the fire. Small water droplets are more efficient for steam generation than large water droplets, because water evaporation is greater due to the more exposed surface area per volume of droplet.

Water from a sprinkler can also pre-wet unburned fuel adjacent to the burning materials. Before a fire can pyrolyze that fuel, the water must be removed by evaporation. The time needed to dry the fuel contributes to the delay in fire spread. In addition, the water spray cools the burning fuel to cause slower fire growth.

Any sprinkler water application to a fire will produce each of these mechanisms to some extent. It is possible to design the sprinkler to produce water droplets appropriate for the expected fire conditions. For example, a water mist may be appropriate for small rooms in which the escape of gases is limited. The small droplets evaporate more easily, creating steam and choking the fire.

On the other hand, a water mist would be ineffective in a large room with fuels that produce a rapid heat release because the fire pressures would drive the steam away before oxygen depletion could occur. In this type of situation, it is more effective to use a fast-response sprinkler and discharge of large droplets that can penetrate the fire plume to the seat of the fire. For most situations, an appropriate choice is a standard sprinkler that bridges between these extremes.

This discussion introduces the concept that all sprinkler systems do not have the same quality and performance effectiveness. The system design should be appropriate to the hazard. Room size, the rate of heat release, and the fire plume momentum are important in selecting the appropriate sprinkler response for the hazard.

Sprinkler standards provide guidance for automatic sprinkler design. Briefly, in the usual design process, one identifies the type of hazard in the space. Then a preliminary piping layout for the building is selected and the hydraulically most demanding area is identified. Each sprinkler in this design area must have a discharge flow rate at least equal to a selected, defined water rate application. This rate is defined as the water application density expressed in liters/minute per square meter (lpm/m^2) or gallons/minute per sq ft (gpm/ft^2). The water supply must be sufficient to provide the required density and duration of flow. Water distribution is a principal component of the sprinkler standard.

The water density describes the volume of water that discharges from the sprinkler to the fire. Discharge density is usually expressed in terms of liters/m^2 (gallons/ft^2) of floor area. This failure mode cannot be obtained from observations. However, reviewing any available design records may give the application density. Also, investigation of design area and hazard classification at the time of installation may give some indication of likely design density.

The water density event estimates whether water quantity and distribution will control the design fire. Often, the density is expressed in terms of a required design density (RDD) and an actual design density (ADD). These densities provide an indication of the ability of a sprinkler system to control a fire within the design zone.

16.12   Water Flow Continuity (cac)

A water flow analysis determines the quantity, pressure, and duration of water for the number of open sprinklers selected for evaluation. This component considers the source of the water and the quantity available at the site. In the schematic of Figure 16.2, the water quantity and pressure available at the first water control valve are the base from which to start. Then pressure changes due to control valves, backflow preventers and check valves, pumps, manual fire streams, fire department connections, risers, feed mains, cross mains, and branches can be incorporated to determine the pressure and quantity of water delivered to the sprinklers being evaluated. The collection of this information gives an insight into the water supply and the sprinkler system for the building.

Water Supply

The water supply and external distribution pipes to a building influence the number of sprinklers that can be supplied in addition to other demands for manual extinguishment and the duration of water flow. A water supply may be broadly classified as public or private. A public water supply uses a system of underground pipes to transport the water from the source, such as a reservoir or well to the point of use, which is defined here as the boundary to the site. A private water supply processes and stores water for normal or emergency use on the site and moves it from the site boundary to the first building water control valve. In some cases, an independent public water supply may be supplemented by private water supplies.

Public water sources and their underground piping determine the quantity and pressure that can be delivered at the time of use. Community water quantity and pressure will vary with the quality, size, and layout of the pumping and underground distribution system, pipe elevations, and other local demands at the time of use.

Private water sources provide quantity and pressure for normal and emergency use at the property. There are several types of water systems, all of which have provisions for producing, treating, and storing a sufficient quantity of water for normal and emergency use. The water source may be a well, pond, lake, or stream. Elevated tanks are frequently used for the dual purpose of storage and maintaining adequate pressures.

The piping distribution on the site is a part of the complete sprinkler system. This piping system transports the water from the public or private supply to the building. The piping system can have a single feed to the building or a looped feed that provides multiple entry points and supplies. These alternatives can influence the dependability of the flow continuity. For now, we merely recognize that the site's piping system is part of an evaluation.

Control Equipment

Water control valves, backflow preventers, quick opening devices (QODs), and pumps were discussed in Section **16.7**. The focus at that time was their reliability in delivering water to an open sprinkler. Now the focus changes to determine the quantity and pressure of water for the number of open sprinklers that are being examined.

Many devices produce friction losses that reduce system pressure. Pumps do the opposite by increasing pressure and volume into the piping system. The capacity of the pumps, controllers, the zones for which they pump, and the fire's duration are parts of a water flow evaluation.

When the water volume and pressure for combined sprinklers and fire department hose streams are inadequate, fire department connections (FDC) are provided. Fire department connections enable the fire department to pump water directly into a sprinkler system. This water can augment low-pressure conditions and override closed valves or failed QODs. The FDC availability, locations, and fire department water needs are part of water flow evaluations.

A continuous flow of water is necessary until the fire is extinguished. For those conditions where the fire is extinguished quickly by the sprinkler system, water supply needs are modest. When a large fire is controlled but continues to burn, the water supply must be maintained.

16.13   Obstructions (wac)

The final part of operational effectiveness looks at the disruption of sprinkler discharge patterns by obstructions. The influence of obstructions can vary greatly. At one extreme, obstructions may provide a relatively minor interference to water distribution. At the other extreme, obstructions can have a major effect on the sprinkler system performance, including contributing to a sprinkler system failure. This event investigates the significance of obstructions to controlling a fire.

A sprinkler system may not have had any major water spray obstructions when it was installed. However, as buildings go through periodic, routine renovations, features such as new walls or air-conditioning ducts may be installed. Lack of care in construction and inspection can influence sprinkler spray patterns. In an existing building, conditions that shield discharge are relatively easy to recognize. In proposed new installations, the sprinkler design standard incorporates requirements relating to installation practices when obstructions occur. Diligence is needed to ensure that water spray obstructions are not introduced during the construction or renovation process.

In addition to the more obvious obstructions of new partitions and ducts, other construction features, such as lighting fixtures, cable trays, beamed ceiling construction, columns, and soffits, can shield a potential fire from the sprinkler discharge pattern. Interior design involving items such as workspace dividers, privacy curtains, cabinets, and tables also affect performance by shielding a fire from the water spray.

Shadow areas caused by high-piled storage introduced after the sprinkler system was installed can also shield a fire. Take, for example, a storage facility sprinkler system originally designed for low-piled cartons in non-combustible containers; subsequently, technology using highly combustible container materials and different inventory practices introduced high-piled storage in the building. These changes significantly affected the sprinkler system in two ways. First, the diagnostic fire characteristics were changed dramatically. This affected the ability of the sprinklers to fuse (fac) in a timely manner as well as the RDD (dac). Secondly, by shielding a fire, sprinkler water spray patterns were different from those initially anticipated. This influenced the water spray application (wac).

The obstruction influence event (wac) provides an opportunity to assess the effect of shielding on sprinkler suppression. Fortunately, this event can be easily assessed by qualitative observations.

16.14   Operational Effectiveness Guidelines

The IPI chart enables one to "stop the world" and examine conditions at any specific instant in time. This is analogous to studying a series of photographs displaying the related conditions. Thus, questions appropriate for each specific event can be examined for the same conditions.

Table 16.1 provides a summary of the more common factors that affect performance of the critical events.

Table 16.1 Common factors affecting performance of the critical events.

Operational function	Factors that alter effectiveness
Nozzle opens before the fire extends beyond the area under study (fac)	Fire (type, rate of growth); temperature of the link inappropriate for the design fire; response time index (RTI) inappropriate for the room size, ceiling height, or design fire; sprinkler link protected from the heat of the fire; sprinkler link corroded, painted, bagged, or taped; sprinkler skipping
Enough water continues to flow to control the fire (cac)	Demand at point of use (number of sprinklers open, expected density, expected pressure); water supply adequacy (quantity, pressure, peak demand period variations); changes in water supply since original design; influence of pipe corrosion; external distribution system (pipe size), type (e.g. loop or dead ends); building connections (single service or multiple service); water availability at the building; storage tanks and towers; pumps (adequacy, reliability); power supply; disruption to external or internal water distribution system (earthquake, explosion, flood)
Water discharge density is sufficient (dac)	Design fire characteristics (HRR, fire plume momentum, speed of growth, fire size); water density needed to control or extinguish the fire (actual delivered density is ADD, required delivered density is RDD); water density available (quantity, pressure); water droplet size and characteristics; room size and container characteristics (size, ceiling height, ventilation, shafts)
Water spray is not obstructed (wac)	Fuel protection; high-piled storage; obstructions below sprinklers (vertical, horizontal); privacy curtains; space dividers; cabinets; tables; beams, girders, trusses; sloped ceilings; columns; new construction (walls, ductwork, ceilings)

HRR, heat release rate; ADD, actual design density; RDD, required design density.

16.15   Analysis and the IPI Chart

The IPI chart is an interactive tool for performance analysis. The IPI chart becomes a repository for information that formed the basis for evaluations, quantification, and decisions. The chart enhances understanding and communication with others about the phasing in and phasing out of different components.

However, the IPI chart is also valuable to maintain a perspective of the many different parts that interact during the examination of different conditions in an analysis. Because a performance analysis examines functionality and interactions, rather than compliance with code and standard provisions, the quantitative and qualitative interfaces at specific times provide a valuable insight into actions.

16.16   Auxiliary Equipment and Other Conditions

Although the topic of this chapter is automatic sprinkler suppression, one should be aware that there are other items of equipment and other considerations that will lead to a more complete understanding of a sprinkler system. One of the more valuable auxiliary devices is a water flow alarm with connections to a supervising service or to the local fire department. This device serves as an automatic fire alarm (notification), as well as a way to reduce potential water damage by enabling the fire department to shut down the sprinkler system after extinguishment. This is particularly useful in installations that are unoccupied for periods of time.

Test connections to do routine maintenance should be located to reflect the system's performance demands. Valves and equipment necessary for operation must be protected against freezing, explosions, flooding, earthquakes, and windstorm. Also, when water is applied to a fire by sprinklers, by the fire department, or by both, it is advisable to design a drainage system to allow the excess or unwanted water to be removed with a minimum of ancillary damage.

16.17   Partially Sprinklered Buildings

Many buildings have sprinkler systems that protect selected spaces rather than providing complete building coverage. For example, corridor-only and exit-only sprinkler systems are often found in buildings. Sometimes much of a building may be sprinklered except for small areas that have contents deemed sensitive to water damage. Whether the logic and arguments are good or bad, a performance evaluation assesses what is, not what should be.

The goal of evaluating a partially sprinklered building is the same as that for any building, i.e. to understand its fire performance and tell the story of what will happen for diagnostic fires located in different rooms. A partially sprinklered building requires at least two and sometimes a few more rooms of origin for analysis. If a fire starts in a non-sprinklered area adjacent to a sprinklered area, the analysis looks at the barrier effectiveness and capacity of the sprinkler system for a fire that would move (often massively) into the sprinklered space. The sprinkler system must contend with a very different fire than would occur if the fire originated in the sprinklered room. Its suppression effectiveness may be significantly eroded. The questions in the evaluation process remain the same, but the answers and expected performance may be substantially different.

16.18   Fire Department Mutual Aid

The fire department can be a valuable asset for sprinklered buildings. Sprinkler systems are designed to control, rather than extinguish, a fire. Nevertheless many sprinkler actuations do extinguish the fire. For those fires that are controlled but not extinguished, manual methods complete the task.

The local fire department usually completes the extinguishment process and turns off the water to prevent excess water damage. A water flow alarm connection to a supervisory service or the fire department is an important part of this sprinkler control. The fire department also can assess the situation and provide assistance to occupants or to other needs of the building.

In those cases where the sprinkler system may not control the fire, the fire department attacks the fire using manual fire fighting. Major fire fighting efforts may be required if the sprinkler system has failed to function or if it was substantially inadequate. Fire department connections to the sprinkler system can augment sprinkler water pressure and quantity. The number and locations of fire department connections become part of a building evaluation.

16.19   Automatic Suppression

The term "automatic suppression" is often used to mean types of automatic systems other than the automatic sprinkler system. They may be described as follows.

- *General area extinguishing systems.* These systems protect a large area, usually an entire building. The automatic sprinkler system is the most common type of general area fire protection. Total flooding foam systems may also be considered a general area extinguishing system.
- *Special hazards extinguishing systems.* These systems are sometimes called "spot" protection. Their function is to protect identified special hazard areas within a building. For example, special hazards equipment can include:
  – Flammable liquids in dip tanks, oil quenching tanks, or metal cleaning equipment.
  – Spray booths using flammable paints and finishing materials.
  – Test facilities where flammable liquids or gases are used or stored.
  – Commercial cooking involving deep fat fryers or grease accumulation in ducts and equipment.
- *Barrier defense equipment.* Automatic equipment, usually involving sprinkler hardware, may be used to prevent fire extension through a barrier or opening. Illustrations of this type of equipment include:
  – Water curtains
  – Exposure cooling systems.

Almost all automatic suppression systems can operate without the need for human intervention. However, situations may exist where it is desirable to have manual operation of fixed extinguishing systems. This is relatively rare in buildings, but more common in ships. A fixed extinguishing system that requires human activation may be described as an automated extinguishing system, and its analysis combines automatic and manual extinguishing activities.

The term "automatic suppression" in this book is used to mean general area automatic sprinkler system fire protection. The operation and framework of special hazard protection are described in Chapter 23.

16.20 Closure

Understanding how site-specific details affect an existing sprinkler system performance is often difficult for those with relatively little experience with sprinkler technology. Nevertheless, acquiring observational skills to accompany calculation methods becomes a valuable asset in recognizing sprinkler system performance. Surprisingly, many potential problems can be recognized with strategic observations. Confidence increases as one gains experience in visual examinations of a variety of sprinkler systems.

Performance evaluations can be useful for new buildings as well as for existing buildings. Sprinkler design for new buildings creates relatively little uncertainty when the system is designed in accordance with established standards. Performance is specified and, when the system is installed in accordance with recognized standards, the success has been outstanding. Trust in design and installation becomes the norm.

The major areas of performance attention for new designs involve recognizing potential atypical conditions. These frequently relate to the diagnostic fire. For example, light hazard conditions appropriate to some uses allow larger sprinkler spacing. Some fuel packages can produce a strong, fast fire growth. This type of fire in combination with high-RTI sprinklers can allow the fire to grow faster than the sprinkler can fuse. This may cause too many sprinklers to open and challenge the water supply. A clear description of diagnostic fire conditions can call attention to this type of situation. Thus, performance thinking is useful to supplement conventional standards design procedures.

When an existing sprinkler system must be evaluated, the assessment changes focus. Sprinkler systems in existing buildings can deteriorate over time as uses change and the building undergoes renovations. If clear attention is not given to maintaining a sprinkler system, its quality may diminish. The organization of reliability analysis and operational effectiveness guides systematic attention to important performance features.

17

The Composite Fire

17.1 Introduction

A *diagnostic fire* is the natural fire that starts in a room of origin and propagates through barriers in a progression of sequential rooms. The diagnostic fire assumes no extinguishment intervention. This natural fire scenario is based on room contents, barrier openings, and the architectural design.

The diagnostic fire continues until active extinguishment methods apply first water and disturb the combustion process. At the time of first water application, the diagnostic fire changes characteristics and subsequent numerical estimates are no longer valid.

A *composite fire* describes the continuum of fire conditions from ignition to extinguishment. A composite fire has two parts. One uses the regular diagnostic fire from ignition to first water application. The second describes fire characteristics from first water application to extinguishment.

Figure 17.1 shows the major components of fire performance and risk analysis. The composite fire is the base from which to estimate building performance and risk characterizations.

17.2 The Fire Limit (L)

A performance evaluation examines outcomes for selected fire scenarios. This gives an understanding of the building's behavior and associated risks.

The fire damage that is sustained up to extinguishment is an important measure of building performance. The term *limit* (L) describes the extent of fire (i.e. flame-heat) damage at extinguishment.

Only two types of active extinguishment are possible: fire department extinguishment (M) and automatic sprinkler control (A). We evaluate each of these as an independent event. After the initial water application, complete extinguishment may be rapid or lengthy depending on the situation. The composite fire combines the fire and extinguishment at each time step. Evaluation of the limit (L) gives an understanding of sequencing and extinguishing effectiveness.

Figure 17.2 shows an event logic diagram for the major active extinguishment components. The single value ("photograph") of these events enables one to show time sequencing performance on the Interactive Performance Information (IPI) chart.

Fire Performance Analysis for Buildings, Second Edition. Robert W. Fitzgerald and Brian J. Meacham.
© 2017 John Wiley & Sons Ltd. Published 2017 by John Wiley & Sons Ltd.

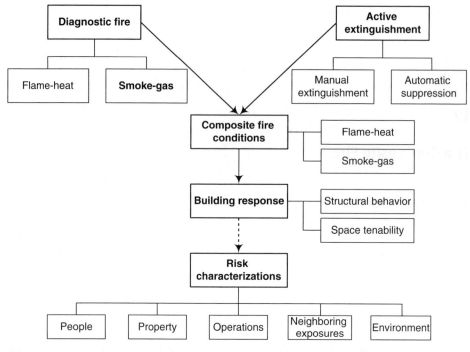

Figure 17.1 System Components

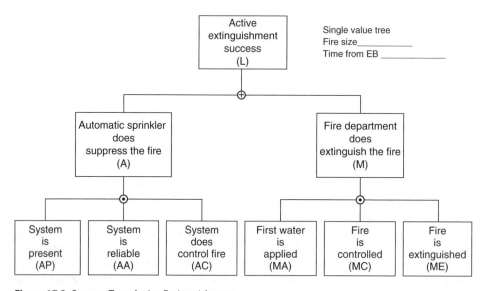

Figure 17.2 Success Tree: Active Extinguishment.

We note that both types of extinguishment are conceptually the same process. Details of nozzle placement and water supply may differ, but both require that water be applied, that the fire is controlled, and eventually extinguished. Analytical details establish time durations and the relationship between the fire and the extinguishing agent.

At each time step, each of these events may be evaluated by the following questions:

- **Given:**
  - *Fire size* _____
  - *Floor area* _____
  - *Time from EB* _____
- **Question:** *Will the automatic sprinkler system control the fire before it grows to a larger size?*

- **Given:**
  - *Fire size* _____
  - *Floor area* _____
  - *Time from EB* _____
- **Question:** *Will the local fire department extinguish the fire before it grows to a larger size?*

The limit (L_F) combines the results of these two components.

17.3 Composite Fire

The composite fire combines the diagnostic fire with fire extinguishment actions at each time step. The IPI chart, Section 3 provides a structure to order and understand fire and extinguishment interactions from ignition to extinguishment.

The composite fire describes the flame-heat and smoke-gas conditions with which the building performance may be evaluated and risks characterized. The flame-heat component examines fire extinguishment performance. The smoke-gas event examines transport of products of combustion through the building. Each moves through a building by different routes and at different speeds. When the flame-heat component is extinguished, smoke-gas generation ceases.

The IPI Section 3A describes the status of the flame-heat component (L_F). Section 3B describes the status of the smoke-gas movement (L_S) from the fire through the building. L_F and L_S terminate when the fire is extinguished. Until extinguishment, we describe relevant information about the fire and smoke in the cells of the IPI chart. The information from these cells provides the basis for evaluating structural behavior and target room tenability.

17.4 Theoretical Completeness

The analytical framework described in this book has a rigorous theoretical base. All components of the framework and the associated IPI chart have been carefully organized to enable one to examine each component in isolation. Influences of other components that affect performance may be identified with the vertical columns of the IPI chart. The single value quantification ("photographs") is sequenced with the horizontal rows ("video") to incorporate dynamic interactions.

A building analysis is deterministic and site-specific. However, the origin of this performance framework is Nelson's General Services Administration (GSA) "Appendix D"

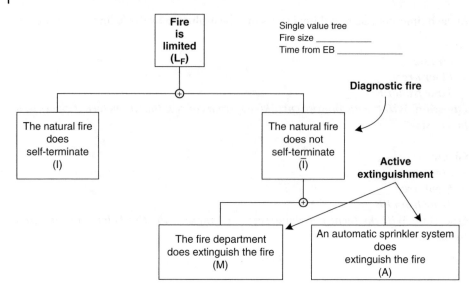

Figure 17.3 Success Tree: Limit of Flame Movement.

Building Firesafety Criteria. Although Nelson's original document evaluated components in terms of "probability of success," the underlying base was actually the degree of belief philosophy, rather than the classical or Bayesian philosophies. The degree of belief probability was commonly used at that time to describe expected behavior rather than a statistical distribution. However, none of the users were aware of this distinction at that time.

During these early days of performance-based analysis and design, the procedure was called the systems approach. The use of probability in systems analysis enabled one to organize the complex system of building fire safety more clearly. Performance evaluations used "probabilistic" curves to describe dynamic behavior over time. This enabled one to use the newly emerging disciplines of systems analysis and failure analysis into structuring the building fire safety analytical framework.

In the interest of theoretical completeness, the logic tree of Figure 17.3 portrays the limit of fire movement (L_F). Thus, the fire may be limited (L_F) by fire self-termination or by active extinguishment by the local fire department (M) or an automatic sprinkler system (A). The OR gates are logic gates describing "either/or."

This theoretical completeness is inherent in the analytical framework of this book, although we simplify descriptions for component evaluations. Therefore, the natural, non-self-terminating fire is the diagnostic fire, and fire department extinguishment and automatic sprinkler control comprise the active extinguishment. Although the composite fire is the focus of evaluations, a more complete theoretical structure is shown in Figure 17.3.

17.5 Summary

When the fire (i.e. load) and water application (i.e. resistance) interact, the fire either continues to burn (usually at a reduced intensity) or it is extinguished. Sometimes extinguishment is fast and complete with little damage. Other fires require long, arduous activities in which substantial damage occurs. Usually, the process lies between these extremes. A performance analysis determines what to expect. The IPI Section 3 documents the time-related outcomes.

The cessation of flames defines success in manual fire extinguishment (ME). After the main fire is out, overhaul operations ensure the complete extinguishment of all hot spots.

If the building transmits an alarm to the fire service communication center during sprinkler actuation (A), the local fire department responds. After arriving, the fire department shuts down the sprinkler system and extinguishes any residual fire. When sprinklers are not shut down after the fire is put out, continued sprinkler operation can cause additional water damage.

The limit (L_F) of flame-heat movement is a prime indicator of building performance. Its primary value is the systematic analysis of the interaction of the fire, the automatic sprinkler system (if present), and the fire department. Chapter 30 identifies the framework for analysis.

18

Materials, Codes, Standards, Practices, and Performance

18.1 Introduction

Structural analysis and design for fire conditions is the most advanced component of holistic fire safety performance. And yet structural applications are rarely used in building design. The problem may be more cultural than technical.

In fire protection tradition, the term "structural" is associated with prescriptive fire endurance ratings that are associated with the building occupancy and height and area dimensions. These parameters are used to prescribe fire endurance ratings for columns, beams, girders, floor assemblies, wall assemblies, roof assemblies, and opening protectives. An awareness of these topics will provide a context for the discussion of structural fire performance in Chapter 19.

This chapter provides awareness for traditional practices by organizing the discussion into five parts:

- *Part One: Building Anatomy* describes elements of building construction and gives an overview of fire behavior for common construction materials.
- *Part Two: Historical Perspective* discusses aspects of the built environment and fire endurance in building codes.
- *Part Three: Fire Endurance Test* describes the standard fire endurance test and interpretations of results.
- *Part Four: Fire Severity Theories* discusses relationships of fire endurance testing and room fires.
- *Part Five: Transitions* summarizes the traditional building code treatment of structural fire requirements in a context of modern structural calculation methods.

Building construction and professional structural design methods intersect with traditional regulatory fire requirements when considering fire effects on structural elements. Calculation-based engineering methods and traditional building code approaches to structural fire requirements view these elements in very different ways. This chapter discusses traditional regulatory practices to provide a context for calculation-based structural design for fire conditions in Chapter 19. We hope the discussion will help to provide a bridge between traditional practices and performance-based calculations.

Fire Performance Analysis for Buildings, Second Edition. Robert W. Fitzgerald and Brian J. Meacham.
© 2017 John Wiley & Sons Ltd. Published 2017 by John Wiley & Sons Ltd.

The standard fire endurance test is so closely associated with structural fire practices that one should understand this test to interpret its performance. Fire severity is a concept that links the standard fire test and room fire conditions. This link provides a useful historical context for the functions of modern calculation methods.

Fire safety engineering is a worldwide profession. Unfortunately, these historical descriptions are based primarily on North American culture. It is hoped that others who are more globally aware will fill in the gaps and deficiencies that are certainly present.

PART ONE: BUILDING CONSTRUCTION

18.2   The Structural Frame

A building structure is an assembly of framing elements that provide the skeletal support, spatial separations, and the building enclosure. These elements establish the form and spatial organization for a building:

1) **Horizontal elements**
   a) Slab or deck
   b) Joists or purlins
   c) Beams
   d) Girders
   e) Trusses
   f) Non-bearing ceilings
2) **Vertical elements**
   a) Columns
   b) Bearing walls
   c) Non-bearing walls

These terms are routine in the building industry, and their use in describing the structural system and building envelope can convey information about its potential behavior in a fire. Figures 18.1 illustrates some of these elements.

The term "non-bearing" indicates that the assembly does not support any loads other than its own weight. Interior partitions that divide a larger space into smaller rooms are usually non-bearing. Because most interior partitions in modern buildings are non-bearing, they can be disassembled and reassembled at other locations to provide maximum interior design flexibility.

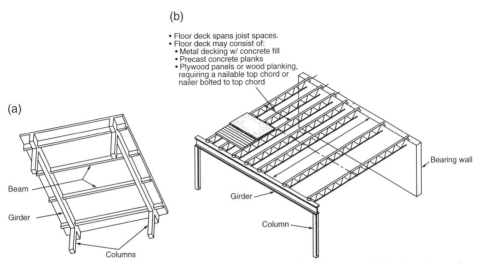

Figure 18.1 Structural systems (Ching & Adams; © 2001 John Wiley & Sons, Inc. This material is used by permission of John Wiley & Sons, Inc.).

Exterior walls that do not support floors or roofs are non-bearing. These non-bearing enclosures that keep out the weather and maintain the interior environment are often called "curtain walls." Curtain walls may be constructed from a variety of materials and forms. Large expanses of glass are a common, although not exclusive characteristic. This type of construction became popular around the time of World War I and today it is almost universal in high-rise buildings.

Suspended (non-bearing) ceilings, often called membrane ceilings, provide several attributes in construction. They are attractive and hide unsightly floor construction, electrical services, and duct work. A membrane ceiling can delay or prevent heat from a fire from becoming excessive in the void (interstitial) space between the ceiling and the structural floor. Membrane ceilings can be important to structural fire design.

The floor slab (floor deck) is designed to span between supports called joists or beams. Beams and joists perform the same function. When spacing is approximately 1.2 m (4 ft) or larger, the supporting members are usually called beams. The term "joist" refers to uniformly and closely spaced members.

Beams or joists support the floor deck (slab). Beam or joist spacing is the basis for determining deck thickness and strength. Joists and beams are, in turn, supported by girders or a bearing wall. Roof support members are usually called purlins or rafters. Columns or bearing walls support the horizontal framing system.

The truss is another important structural element. A truss is an articulated structure composed of slender, lightweight members connected in the form of triangles. The members act in either tension or compression and are arranged to carry loads efficiently. Truss members are most commonly made from structural steel or wood. Light, evenly spaced steel trusses are called open-web joists or bar joists.

In the hierarchy of structural framing, the slab or deck is supported by beams or joists; the beams or joists are supported by girders; girders are supported by columns; when girders are omitted, the members are supported by a bearing wall.

The most common structural materials are wood, structural steel, reinforced concrete, prestressed concrete, and masonry. Materials and framing systems can be mixed and matched in a variety of ways. The selection of materials and the manner in which they are put together fall within the domain of the structural engineer.

The structural system design is often based on dimensions, geometry, and architectural style and use. Local preferences or materials availability may also influence the choice. Although a large number of combinations may seem to exist, a framing system is logically arranged and easy to trace.

18.3 Material Behavior in Fires

The heat of a fire adversely affects all building materials. Some materials are affected at lower temperatures or at shorter time exposures than others, and one should be aware of a material's sensitivity to elevated temperatures. The fire behavior for a few of the more common construction materials is described in the following sections.

18.3.1 Structural Steel

Structural steel is a dominant material for providing strength in modern building construction. High strength and stiffness, ductility, the ability to form a variety of shapes and dimensions, and ease and speed of erection are among the qualities that steel possesses. Structural steel may be used as the skeleton framework for buildings or as the tensile reinforcement in reinforced concrete.

Structural steel is non-combustible and does not contribute fuel to a fire. Nevertheless, one should not develop a false sense of security of its durability in a fire, because structural steel also loses strength and stiffness when subjected to elevated temperatures.

Yield strength is the most significant parameter for load-bearing capacity. As the temperature in the steel rises, the yield strength declines and its ability to support loads is reduced substantially. The stiffness is reflected by the modulus of elasticity, E, also decreases with an increase in temperature.

Structural steel also expands when heated. When expansion moves a column or bearing wall out of alignment, premature collapse may occur.

18.3.2 Concrete

Concrete is a mixture of cement, sand, aggregate, and water. When exposed to heat, concrete loses water and also strength. Because different aggregates affect fire performance, the strength at elevated temperatures is compared on a percentage basis of the normal temperature strength.

Although concrete structures usually give a sense of fire security, elevated temperatures can adversely affect performance. As with other construction materials, heat causes concrete members to expand. The load-carrying capacity of reinforced concrete will increase when construction features restrain the ends against thermal expansion. However, when thermal restraint is low, as at building exteriors, the expansion may cause cracks to open and supporting members to deform excessively.

Concrete has a high compressive strength but very low tensile strength. Because flexural loads cause compressive and tensile forces simultaneously in a member, the tensile loads must be carried by structural steel reinforcing bars, often called *"rebars."*

Reinforced concrete is a composite that takes advantage of each material's strengths by locating reinforcing steel in positions of maximum efficiency. In beam design, the reinforcing bars are placed in positions of tensile stresses.

Concrete has a tendency to spall, which is the disintegration or falling away of part of the concrete surface. This exposes the steel reinforcement and interior regions of the concrete. In a fire, an explosive build-up of steam causes spalling because the expanding water cannot be moved rapidly enough to the unexposed surface.

Spalling of the protective concrete cover can reduce the member strength. In normal weight concrete, siliceous aggregates spall much more readily than calcareous aggregates. Lightweight aggregates perform much better for fire resistance. Sharp edges spall more readily than rounded or beveled edges.

From a barrier viewpoint, concrete can crack relatively easily to allow a hot-spot ignition into the adjacent space. This cracking does not necessarily lead to an imminent massive failure or collapse. Reinforced concrete in continuous structures usually has collapse resistance beyond the appearance of first cracking.

18.3.3 Concrete Masonry Units

The manufacture of concrete blocks, often called concrete masonry units (CMUs) is a very common use of plain concrete in buildings. These blocks are normally made as solid bricks or of larger hollow blocks. CMUs can be manufactured in a variety of sizes and shapes. The void spaces of the hollow blocks may be left unfilled or they may be filled with sand or another inert material to increase the fire endurance.

18.3.4 Prestressed Concrete

Sometimes, steel reinforcing wires are stressed in tension before a reinforced concrete member is fully cured. After the tension of the wires is released, the concrete is placed under compressive stress before any additional loads are applied. Prestressing increases the economy of the structural member and may be used with either precast structural members or cast-in-place members.

Prestressing normally uses concrete of a higher compressive strength than other reinforced concrete structures. Also, the prestressing wires are of a much higher strength than normal rebars. Although prestressing wires have a greatly increased tensile strength, they rapidly and permanently lose strength at elevated temperatures and must be protected from the heat of a fire.

The thermal and structural properties of the concrete are essentially the same for both types of construction. However, the fire behavior may be quite different. For example, the natural camber caused by the prestressing operation may cause the member to bow and explode upward at collapse. Because prestressed concrete structures are often precast and erected without monolithic continuity, stress redistribution does not provide additional resistance to collapse.

18.3.5 Wood

Wood is commonly used for joists, beams, girders, and columns. Wood also may be used as a sheathing on floors, roofs, or walls to serve as a surface for loading or other materials. The function of the material determines its role in a building evaluation.

Although wood is able to maintain its integrity for a period of time before failure occurs, exposed wood construction augments the fuel of the fire. When wood burns, it forms a char. The char acts as insulation that delays pyrolysis and slows, but does not stop, the continued deterioration of the member. The rate of combustion varies with the depth of char and depends on the species and orientation of the wood. The rate of char varies from about 1.0 mm/min (1/25 in/min) for soft species of wood to about 0.5 mm/min (1/50 in/min) for harder species. An average value of 0.6 mm/min (1/40 in/min) is often used as a general approximation. Thick members provide substantially more time before collapse than do thin members.

18.3.6 Gypsum

Gypsum products, such as plaster or gypsum board (often called sheetrock, plasterboard, or dry wall) perform well in a fire. The gypsum has a substantial amount of chemically combined water. During a fire, the water slowly evaporates, causing the gypsum to calcine and disintegrate into a soft powder. This heating action moves slowly through the material and provides endurance time before the fire resistance is compromised.

Gypsum board consists of a plaster core sandwiched between sheets of paper. There are many types of wallboard, ranging in thicknesses from 6 to 25 mm (¼–1 in). A regular gypsum board is the most common type. For better fire resistance, type X gypsum board is often specified because glass fibers in the plaster mixture hold the calcined gypsum in place longer during a fire.

18.3.7 Glass

Several forms of glass are used in building construction. Ordinary window glass is used as glazing for doors and windows. This type of glass has relatively little resistance to breakage in a fire. To provide greater integrity in a fire after cracking, wired glass is often specified. Thick plate glass or tempered glass used in curtain walls can withstand high temperatures for longer time durations before breaking.

When evaluating glass for fire resistance, the entire structural framing and installation becomes important. For example, if the glazing does not have adequate gasketing, the fire temperatures will cause a metal frame to deform, and any type of glass insert can break readily and very quickly.

Fiberglass-reinforced building products use another form of glass. Products such as prefabricated fiberglass bathroom units and translucent window panels and siding have advantages of economy and aesthetic appeal. The fiberglass does not burn or support combustion. However, the fiberglass acts as reinforcement for a thermosetting resin. Fiberglass products are usually about 50% resin, which is very flammable even when fire retardants are incorporated into the composition. Although the fiberglass itself is noncombustible, the entire unit is very combustible. In addition, it produces substantial amounts of smoke during combustion.

Fiberglass insulation is a third application of glass in building construction. Again, the fiberglass does not burn. However, the glass fibers are coated with a resin binder that is often flammable and can propagate flames when they are in an open, unconfined space.

PART TWO: HISTORICAL PERSPECTIVE

18.4   The Built Environment Around World War I

The built environment in the early years of the 20th century was clearly in a condition that was ready for change. Most buildings were constructed by following a prescriptive building code. Design loads, materials specifications, allowable stresses, serviceability requirements, construction details, and fire protection were integrated into a prescriptive code document. This enabled routine construction to be performed without an engineer of record because compliance with the building code was perceived as providing a sufficient standard of care.

In the few decades on each side of 1900, fire was a major concern for the economy and throughout society. The Second Industrial Revolution was well established, and an increasingly efficient transportation infrastructure encouraged concentration of urban centers for commerce and industry. Worker housing was often clustered close to employment, resulting in large numbers of mixed-occupancy buildings within relatively small geographical areas.

Building regulations focused primarily on structural safety with nominal attention to fire safety and architecture. The codes were poorly written and often loosely enforced. Fire safety regulations and their enforcement were inferior. Concentrations of industrial operations combined with large numbers of worker families in sub-standard housing provided a prescription for disaster – and disaster often occurred.

During this era a number of interrelated developments in building design and construction were taking place. The professional engineering community influenced major changes in both structural engineering and fire safety. Interactions involving architecture, engineering, technology, and commerce had significant effects in the building industry. Although the history of building buildings during this period is fascinating, we will summarize only a few movements in structural engineering and prescriptive fire safety regulations that relate to our topics.

18.5   Structural Practice Around World War I

Although the art of structural engineering had been practiced for many centuries, the birth of its modern era may be identified with the quarter centuries before and after 1900. Knowledge of structural engineering relating to building technology was evolving rapidly by World War I. A framework for thinking was consistently used to understand local as well as holistic behavior. A base of engineering methods was available to calculate forces, stresses, and deformations. A theory of analysis for reinforced concrete structural members and a standard of practice had just become available (1909). Theory and practices for structural steel design were available, although the first standard of practice was not published until 1923.

Building codes established applied loadings for occupancy types. However, the variation among codes was so enormous that cost savings achieved from better design knowledge could be negated by excessively prescribed loading requirements. A Department of Commerce study (1924) and the Uniform Building Code (1927) encouraged more consistent values for local building codes.

Structural engineering practitioners largely consisted of university-educated engineers who understood the general framework for analysis as well as the capabilities and limitations of the available engineering methods. Although technical capabilities were primitive compared with the engineering methods available today, they were adequate to provide suitable engineering services.

18.6   A Century of Evolution

Structural engineering practices that were generally in place by 1930 were nearing closure by 1950 as the profession reached early maturity. A framework existed in which both micro and macro performance could be understood and communicated. Essentially, the same performance-oriented thought process existed throughout the world. Practitioners could explain needs in a way that researchers could understand and develop solutions. The profession could recognize research concepts that enabled the discipline to reach greater levels of understanding. It enabled new buildings to be designed and built with greater confidence in their structural performance. The "bigger and better" building did not pose an unusual obstacle.

During the last half of the 20th century, design standards made dramatic changes as new theories and knowledge provided modern ways of achieving safe design at lower cost and with less professional time. Lower factors of safety became possible as better understanding of behavior grew. As new technology was integrated into design standards, outcomes were calibrated against the performance provided by previous methods. This calibration allowed the profession to clearly recognize how changes would affect their experiences, economic, and design decisions.

Computer, communication, and information technology has been evolving over the past half-century. More powerful calculation methods allow an engineer to understand aspects of structural performance more clearly. This enables the structural engineer to transfer attention from tedious calculations to the better planning and creating of solutions.

18.7   Fire Safety Around World War I

Around World War I, fire safety was a major economic and societal concern. Conflagrations were common in cities throughout the world. For example, Baltimore (1904) lost 80 city blocks; San Francisco as a result of the earthquake and fire of 1906 lost 28 000 buildings; and Chelsea, Massachusetts (1908) lost 3500 buildings. Within the single day of March 21, 1916 three separate city conflagrations (Georgia, Texas, and Tennessee) resulted in the destruction of 2770 buildings.

Examples such as these were common throughout the world, and made fire safety a major societal concern. Fire losses were also a major economic concern for industry and building owners as well as for the insurance industry. Many insurance companies had experienced or were close to bankruptcy.

Human safety from fire was becoming an even greater concern. Large life loss fires, such as the Iroquois theater fire (1903) with 602 deaths and the Collinwood School in Ohio (1908) that cost 175 lives, were but two of many notable tragedies. However, the reality was that most citizens had a great anxiety about fire and safety in the buildings in which they worked and lived, and which they used on a daily basis.

18.8   The Fire Safety Solution

Although political pressures often inhibit enactment of legislation that is restrictive or makes construction more costly, the counter pressures of perceived societal and business risk from fire were clearly evident. Frequent disasters, combined with public outrage and fear, created an environment for political change.

Experience and knowledge about fire safety had been accumulating during the years around the turn of the 20th century. The Factory Mutual (FM) Insurance Corporation (1835) incorporated the concept of rewarding "good citizens" among the factory owners that used effective fire protection measures. FM approvals and standards are widely accepted in the industrial world. In addition, FM established a basic and applied research program in fire safety.

The National Board of Fire Underwriters (NBFU) was formed in 1866 as a trade organization of the insurance industry. Although the early years focused primarily on setting insurance rates, it became evident that a "win-win-win" situation could exist for business owners, society, and the insurance industry by encouraging legislation to ensure more and better fire safety construction practices. By 1892 the activities of the NBFU had become predominantly associated with building codes and education of architects, engineers, and the public to better ways of fire protection.

Underwriters Laboratories (UL) evolved from activities of its founder, William H. Merrill, in conjunction with the Chicago World Exposition in 1893. Merrill used the concept of examining and testing new equipment that was being invented at a rapid pace during the Second Industrial Revolution. Although the initial focus of UL involved the fledgling electrical industry, by 1901 its sphere of influence had expanded to testing a broad range of products related to fire safety. In addition to testing, UL was active in the development of test standards, factory inspections, and information on equipment safety.

The National Fire Protection Association (NFPA) was organized in 1896. Its first standard addressed sprinkler design and installation. From this initial effort, the organization has become the premier standards-making organization in the world. The NFPA is a world leader in a wide variety of fire safety efforts in all aspects of education and application.

Although much has evolved during the past century from these and other organizations, the state of the art of fire technology was relatively primitive at the time of World War I. Nevertheless, fire safety professionals were knowledgeable about conditions that produce unsafe buildings. Good practices could be distinguished from bad practices. The professionals could study disasters and explain why "this should never happen again."

Considering the status of fire technology and the limited number of practitioners and researchers at the time of World War I, the societal fire problem seemed to be best solved by the legislative route. Fire safety professionals, led by the insurance industry, crafted building code regulations that could be enacted and enforced by local jurisdictions.

The organization of codes and the prescriptive language of its specific regulations were brilliantly conceived. This process created the structure of the prescriptive code and enforcement environment that exists today. Large fires and citizen concerns about fire safety have declined significantly during the past century. Building codes and their enforcement have been important to that improvement.

The International Council of Building Officials (ICBO) published its first model building code in 1927. Although there had been a long history of building codes to regulate construction, this code established an organizational structure that continues to be the basis of modern codes.

18.9   Building Code Organization for Fire Safety

Fire safety in building codes is organized around fire endurance requirements. The building's occupancy is identified and its height and area are determined. The code prescribes the fire endurance for floors, partitions, walls, and structural elements from this information. The fire endurance requirements for the main structural elements, specified as standard test hours, are identified early in the building design program.

After the occupancy, use, and fire endurance are known, other fire-related building requirements are identified. Structural requirements are often used as "trade-offs" (i.e. equivalency) when certain good practices are selected. For example, structural requirements may be reduced for many fully sprinklered buildings.

The fire endurance is determined from the results of a standard fire endurance test. The ASTM E-119 test in the United States and the ISO 834 test throughout much of the world are the basis for establishing the fire endurance. Calculation methods are available to substitute for physical test results. These calculations are based on results of previous standard fire test results and should not be construed as equivalent to the calculation methods discussed in Chapter 19.

Structural fire endurance requirements are important to prescriptive fire designs. Because of their importance for code enforcement, one should develop an awareness of the background to fire endurance. This background is also useful to appreciate the performance concepts described in Chapter 19.

18.10   Structural Fire Topics Around World War I

Height, area, sprinkler use, standard fire curve, fire endurance test, and building construction became interconnected in the years around World War I. Ira Woolson, Simon Ingberg, and others in the insurance and fire protection world were active participants in the process.

In 1914, Ira H. Woolson published the results of a survey of fire chiefs in cities in the United States with populations over 20 000. The survey recorded the opinions of the responding 117 fire chiefs (1 in 3 response) with regard to building size, construction, sprinklers, and the ability to control and extinguish a fire. The main conclusions were:

- Manual fire extinguishment is difficult in large buildings.
- "Fireproof" building construction was important in fighting fires.
- Recommended allowable heights and areas in factory buildings (based on this fire service survey) were as shown in Table 18.1.

Table 18.1 Recommended allowable heights and areas in factory buildings (based on Ira Woolson's fire service survey).

Building type	Stories	Area between fire walls (sq ft)
Brick and joist construction, not sprinklered	3	6 000
Brick and joist construction, sprinklered	5	13 000
Fireproof construction, not sprinklered	5	10 000
Fireproof construction, sprinklered	8	20 000

The 1920 NBFU model code cited this source for relating height and area to structural fire endurance. The concept became established for the 1927 ICBO building code. Although attempts have been made over the years to "rationalize" the height and area selections, the values remain based on opinion with little technical justification.

Building construction at the time of World War I was generally either combustible or built of masonry and steel or reinforced concrete. The construction was either "light" or "substantial." The building code attempted to require substantial construction for larger and more important structures.

At that time, the standard time–temperature relationship for the ASTM E-119 fire endurance test was criticized because it did not represent actual building fires. The criticism was so great that both the standard curve and the fire test faced abandonment. However, Ingberg saved the standard fire curve and fire endurance test with an "equal area" theory to relate actual fire test results to the standard fire. The theory and tests are described in Section 18.13. Ingberg's efforts saved the standard fire curve and fire endurance test.

Enough fire endurance testing had been done by that time to recognize that using results of the standard E-119 fire endurance test could provide desirable methods of construction for that era. Thus, specifying the fire endurance generally assured that construction practices of that time would provide a "substantial" structural frame that would help fire fighting. The prescribed values had relatively little to do with fire endurance for actual fires.

18.11   Building Code Observations

The building code, fire standards, and the approval system provided a clearly recognized process for establishing acceptable building fire safety during this era around World War I. Code requirements were prescriptive, and the code expanded to address emerging fire problems that became recognized as building construction evolved.

Codes essentially identified what to do to achieve acceptable fire safety designs. Fire standards described how to do what the codes had prescribed. For example, when a building code specified that sprinklers must be installed, NFPA Standard 13 described how to design and install the sprinkler system. The standards were also written in a prescriptive manner to specify a mixture of installation and design requirements.

Fire compliance requirements were normally specified by the architect and not by an engineer because the perception was that a designer only needed to comply with the code to handle the fire problem. Specialized technical fire knowledge was perceived to be either not necessary or a skill that could be acquired easily. Meeting the code was defined as having a "safe" building. As contrasted with other disciplines, the code, and not a design professional, was responsible for public and economic safety from fire.

An increased awareness and concern about fire safety began to emerge during the 1960s. The World War I focus of protecting against conflagrations and large building losses that had been a central feature of building codes was now directed more conspicuously toward incorporating life safety. Buildings became larger and more complex, and often involved multiple occupancies. Open plan architecture and four-season climate control became attributes for new construction. Recognition was growing that plastics in buildings and furnishings were creating a different type of fire threat than had experienced with cellulosic materials.

The response to these new and different threats and associated risks to life safety was handled with an increasingly regulated building environment. Building codes and standards became more inclusive. Regulations were written by looking at the past to ensure that "those problems" would not happen again. During this time, fire protection equipment and hardware also improved significantly, and the advances were incorporated into improved standards.

Often, technology spawned unforeseen new problems. Unfortunately, many of the solutions incorporated into the organizational framework that had been established a half-century earlier made compatibility with modern buildings difficult.

One may look back with admiration at the prescriptive code and standard system that evolved during the 20th century. The code and standards system, combined with the increasing professionalism of their enforcement, transformed fire safety from the economic and societal anxieties of World War I into a contemporary perception where fear from fire is not high in the public consciousness.

Even with the improvement in public fire safety, the prescriptive code and standard system is not without its shortcomings. Prescriptive building codes may be described as a compilation of good practices that have a weak technical basis. Although they are viewed to be "easy" to administer because of their prescriptive form, control is "difficult" because most individuals in the building industry do not perceive fire as their responsibility. The designer and code official often find themselves in adversarial positions over interpretation of a requirement that interferes with what is perceived as a better, more functional design. Cost, effectiveness, and regulations have different levels of importance between the building designer and the code official. Differences of opinion are rarely resolved by rational, analytical procedures that consider both the "micro" equivalence of an issue along with its "macro" effect on complete building fire safety.

Regardless of opinions or criticisms, the building code is the law. As such, the code official is a "police officer" obliged to administer the law. Consequently, in a prescriptive code environment, code compliance becomes the central concern. Comparative building fire safety and its costs are not an issue. Because a prescriptive building code specifies a wide range of safety features, code compliance is often presumed to be synonymous with a safe building. Unfortunately, this is not correct. In fact, the level of safety provided in a building code is indeterminate because the organization and prescriptive language precludes performance analysis.

The building code system that has evolved since World War I has been a success. It has transformed perceptions and the reality of fire safety from apprehension and economic concern around World War I into an atmosphere of confidence today. The confidence of the public and business may be misguided in many cases because potential fire risks based on performance analysis can be clearly recognized in many buildings. Nevertheless, fear from fire does not have the public concern today as it had in the past.

The building code system that created this success can also be justly criticized for creating excessive and unnecessary added costs. It can also be faulted for imposing questionable and inflexible constraints with regard to architectural design freedom. Nevertheless, the code is the law. However, a part of that law allows use of alternate methods and materials that can demonstrate an equivalent level of protection.

PART THREE: FIRE ENDURANCE TESTING

18.12 Fire Test Interpretations

The generally accepted basis for specifying fire endurance of barriers is the standard fire resistance test, ASTM E-119 in the United States and ISO 834 in most of the rest of the world. These tests are similar, and their differences are insignificant when evaluating field performance.

The ASTM E-119 test protocol notes that its intent is to compare test behavior, and the results should not be construed as suitable for other conditions or fire exposures. The test measures and describes assembly responses to controlled fire conditions. It does not incorporate all of the factors important for actual fire conditions.

Although the E-119 standard notes that test results are not intended to be used for acceptance criteria, there is no other practical way of identifying construction assemblies for prescriptive code enforcement. Therefore, one should keep in mind that fire endurance test results are a way to satisfy prescriptive code requirements and not an indication of the actual time the assembly will resist a building fire.

Perhaps the biggest misconception in the fire and building code community is that the fire endurance rating describes the time that the assembly should last in a building fire. This incorrect perception is one of the reasons that code compliance and building fire performance can be so different. Nevertheless, one may logically assume that a higher-rated assembly may last longer in a building fire than a lower-rated assembly – sometimes. We should realize the actual fire endurance time is often quite different from the rated values.

18.13 The Standard Fire Endurance Test

The ASTM E-119 test establishes fire endurance classifications of floor–ceiling assemblies, roof–ceiling assemblies, walls, partitions, beams, and columns. ASTM E-152 and ASTM E-163 are similar tests for doors and windows. Floor and roof assemblies, beams, columns, walls and partitions, doors, and windows all have their own furnaces and test criteria. Figure 18.2(a) illustrates a floor furnace. The opening measures about 4 m × 5 m (13 ft by 17 ft). A full-size assembly is installed in the furnace, and loads to produce the design stress are applied to supporting members.

A fire is started in the furnace, and the air supply and fuel are adjusted to provide a specified, standard time–temperature relationship (Figure 18.2b) on the exposed side of the assembly. The test continues until the sponsor wishes to stop or until failure occurs. Failure is defined as any of the following:

- Temperature on the unexposed side rises to an average of 250°F (139°C), or a hot spot of 325°F (181°C) above ambient.
- The unexposed barrier surface cracks to allow fire gases to penetrate and ignite combustibles on the unexposed side.
- The assembly cannot support the applied loads.

The first two failure modes are associated with hot-spot ($\overline{T}$) performance and the third mode relates to a massive ($\overline{D}$) performance.

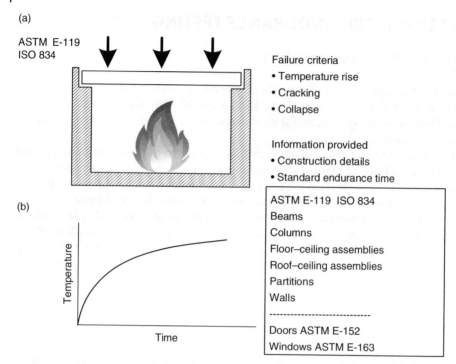

(a)

ASTM E-119
ISO 834

Failure criteria
• Temperature rise
• Cracking
• Collapse

Information provided
• Construction details
• Standard endurance time

ASTM E-119 ISO 834
Beams
Columns
Floor–ceiling assemblies
Roof–ceiling assemblies
Partitions
Walls

Doors ASTM E-152
Windows ASTM E-163

(b)

Temperature

Time

Figure 18.2 Fire test information.

The time is noted when failure occurs or when the test was stopped. This test exposure time is then rounded down to standard fire endurance classifications of 20 min, and 0.5, 0.75, 1, 1.5, 2, 3, and 4 hours. For example, if an assembly failed at 1 hour 59 minutes, it would be given a fire endurance of 1.5 hour. If it lasted to 2 hours 1 minute, it would receive a rating of 2 hours. A test exposure of 2 hours 59 min would also produce a 2 hour listing. Therefore, a fire endurance rating indicates that the assembly avoided all failure modes in the standard test for at least the rated time duration.

The fire endurance classification (in hours or minutes) for restrained and unrestrained test assemblies and the construction details of the assembly are published. Although other information is recorded, neither the failure mode nor information about the assembly behavior during the fire is published. Test results are considered privileged information and additional data are available only from the test sponsor.

The ISO 834 standard is essentially the same as the ASTM E-119 test. Although there are differences in details, the philosophy and general practices are similar.

18.14   Fire Endurance Test Discussion

One function of any standard test is to enable results to be repeatable and reproducible among testing facilities. Although the time-temperature relationship is specified and consistent, its comparative repeatability is questionable because of differences in thermocouple shielding and the manner in which temperatures are controlled. Some furnaces control temperature by adjusting the airflow and others by adjusting the fuel. Some facilities record the fuel amount, whereas others do not because it is not required by the test standard.

The net heat flux determines how hot the specimen actually becomes. Paulsen[1] found significant differences in heat flux among six European furnaces.

Variation in furnace construction, furnace controls, and flame emissivity can cause substantial differences in the heat flux to the specimen. The ability of different furnaces to reproduce results for any specific assembly is suspect. This could provide an opportunity for manufacturers, if they were so inclined, to select the test facility that gives the best results for their materials and construction. Prescriptive code approval is a strong motivation to obtain important fire endurance classifications.

In addition to furnace differences, the thermal restraint of testing contributes to the fire endurance. The thermal coefficient of expansion causes construction materials to expand when subjected to increase in temperature. When assembly elongation is axially unrestrained, structural elements can expand freely as the temperature increases. If the test furnace prevents the ends from expanding, axial forces are induced. The difference in fire endurance can be substantial because these axial forces provide a pre-stressing action. The change in fire endurance can vary from a reduction in endurance time for a few assemblies, to a negligible change for others, to a substantial increase for many. Most assemblies will exhibit a better load-carrying capacity and longer fire endurance time when ends are axially restrained against movement. This increase can be as much as 100% or more. A listing of both restrained and unrestrained fire endurance ratings provides some guidance on behavior.

The standard test is supposed to be representative of the details and dimensions used in construction. Physical-analytical functions are very difficult to scale, and the specimen is prepared as well as possible to conform to construction and the test furnace. However, axial, flexural, and shear stresses and their combinations; size scaling; flexural and lateral restraint; and lateral and local buckling cannot be incorporated simultaneously for a specimen with a 5.2 m (17 ft) fixed length. Construction details and workmanship also influence the correspondence between laboratory and field performance in fire situations.

Heat absorption or transmission through assemblies can differ substantially. During a test, the fuel/air mixture is adjusted to maintain the standard time–temperature criteria near the exposed assembly surface. This means that significantly more fuel is needed for some assemblies to maintain the same time–temperature relationship. In other constructions, such as with combustible materials, the fuel mixture is reduced substantially. A natural fire is not able to adjust its fuel/air mixture as easily as the test furnace.

A building fire produces positive pressures whereas a standard fire test uses a slightly negative pressure to prevent most smoke and hot gases from infiltrating into the laboratory. This allows cooler laboratory air to leak into the furnace, affecting fire resistance. The plenum temperature of floor–ceiling and roof–ceiling assemblies having suspended ceilings influences fire endurance. When cooler air from the laboratory is drawn into the furnace, one can expect better fire resistance.

The purpose of test standardization is to have comparative results among different laboratories. Although the test protocol remains constant, a fire endurance time is not

---

1 Paulsen, O., *On Heat Transfer in Test Furnaces.* Technical University of Denmark, Lyngby, 1979.

the same for all laboratories. Although fire endurance test information contributes to our understanding, one must interpret results carefully when relating test results to actual performance in a building fire.

Test time exposures are intended to compare the fire endurance of assemblies. They should not be interpreted literally for an actual building fire. Other relevant construction conditions may contribute to an early collapse or to an extended endurance under natural fire exposures. When the goal is to understand and describe performance, one must combine fire test information with structural behavior, knowledge of materials at elevated temperatures, compartment fire dynamics, and judgment.

PART FOUR: FIRE SEVERITY

18.15 Introduction

An ongoing recognition has existed during the past century that the ASTM E-119 (ISO 834) laboratory test does not reflect expected field performance. Nevertheless, test results have encouraged desirable types of construction in building regulations. Although the goal of fostering "substantial" building construction was usually realized with the construction assemblies of World War I, this desired objective has not always been achieved with building construction elements that emerged after World War II. One must examine assemblies and their installations carefully when structural behavior is important to performance.

In the early 1920s, Simon Ingberg described an initial linking of fuel in a space and the standard time–temperature curve with E-119 test results. This theory became an early justification for building code specifications. During the 1960s and 1970s, the Ingberg theory was supplemented by additional theories to estimate the real time to barrier failure.

A comparison of these theories is useful to appreciate the role of modern calculation methods for structural performance. Part Four discusses associations with the following topics:

- *Fuel loads.* The combustibles in the room provide a measure of heat energy against barriers and the structural frame.
- *Ingberg theory.* The Ingberg theory defines fire severity and its measurement. It also provides a base against which other theories can be compared.
- *Room factors.* The influence on fire severity of room construction, ventilation, and fuels gives an appreciation for variability in heat energy application for fuel loads.
- *Severity theories.* Several fire severity theories attempt to link the ASTM E-119 (ISO 834) test and associated real times in building fires.
- *Real time estimates.* Different fire severity theories and burning rates provide a means of interpreting test results for estimating the real time to failure.

18.16 Fuel Loads

The fuel load is a measure of the calorific content of the room at complete burnout. It is calculated in terms of equivalent kg (lb) of cellulosic-based fuel per square meter (foot) of area. The floor area is the most common way of normalizing the fuel load, although the bounding surface area may also be used. The following examples illustrate traditional fuel load calculations in a space.

Example 18.1 Calculate the fuel load for the office shown in Figure 18.3. The weights of cellulosic and plastic combustible contents (not the total weight of materials) are shown in Table 18.2. Express the fuel load as pounds per square foot (lb/ft^2) (kg/m^2) of floor area.

Solution:
Fuel loadings are expressed in terms of equivalent cellulosic materials. The heat energy content for wood, paper, and other cellulose materials is about 8000 Btu/lb (4450 kcal/kg). Values for plastics range from 9000 Btu/lb (5000 kcal/kg) to 20 000 Btu/lb (11,100 kcal/kg).

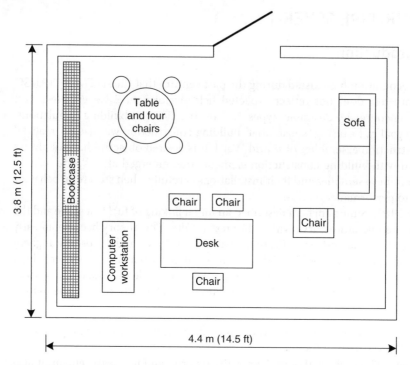

Figure 18.3 Example 18.1 room fire load.

Table 18.2 Example 18.1: weights of cellulosic and plastic combustible contents.

	CELLULOSIC	PETROLEUM/CHEMICAL (kg)
Bookcase & Contents	800 lb (360 kg)	10 lb (5 kg)
Desk and 3 Chairs	50 lb (23 kg)	20 lb (9 kg)
Computer & Workstation	40 lb (18 kg)	50 lb (23 kg)
Table and 4 Chairs	80 lb (36 kg)	10 lb (5 kg)
Sofa	20 lb (9 kg)	30 (14 kg)
Upholstered Chair	10 lb (5 kg)	10 lb (5kg)
Carpet and Underlay		1 lb/ft^2 (5 (kg/m^2)
Occupant-Related Goods	60 lb (27kg)	20 lb (9 kg)
Total	1060 lb (478 kg)	150 lb (70 kg) + Carpet

A common simplification assumes that the heat content of all petroleum-based materials is 16 000 Btu/lb (8900 kcal/kg). Therefore, 1 lb (0.45 kg) of petroleum-based material is assumed to have the equivalent heat energy content of 2 lb (0.9 kg) of cellulosic material.

Fuels may be categorized into building fuels, contents fuels, and occupant-related goods (ORGS), as described in Section 6.9. The fuel load is the total weight of fuels in the room divided by the floor area.

For the office shown above, the fuel weight is:

Equivalent cellulosic fuel weight $1060 + 2 \times 150 + \text{carpet} = 1360 + \text{carpet}$

The fuel load is:

Fuel load $= 1360 / (14.5)(12.5) + 2 = 7.5 + 2 = 9.5 \, \text{lb/ft}^2$ of floor area

Fuel load $= [478 + 2(70)] / (4.4)(3.8) + 2 \times 5 = 47 \, \text{kg/m}^2$ of floor area

Example 18.2   Calculate the fuel load if an interior finish of 3/8 in (9.5 mm) wood panels were applied to the walls. Floor to ceiling is 8 ft (2.44 m). Windows are 4 ft (1.22 m) high.

Solution:
The approximate area of wood paneling is:

Wood panel area $= 2(12.5 \times 8) + 2(14.5 \times 8) - 14.5 \times 4 = 374 \, \text{ft}^2$

If we assume that wood weighs 36 lb/ft³, a slice of 1 ft², 1 in thick, weighs about 3 lb. The total weight of wood in the room is:

Wood weight $= 374 \, (3 \times 3/8) = 420 \, \text{lb}$

The fuel load is:

Fuel load $= 420 / (12.5)(14.5) = 2.3 \, \text{lb/ft}^2$ of floor area.

The total fuel load for this room is $9.5 + 2.3 = 11.8 \, \text{lb/ft}^2$ ($57.8 \, \text{kg/m}^2$) of floor area.

Example 18.3   Express the fuel loading for the room of Figure 18.3 in terms of lb/ft² (kg/m²) of bounding surface area.

Solution:
The equivalent fuel loading of 9.5 lb/ft² (47 kg/m²) may be converted to equivalent pounds of fuel by multiplying by the floor area:

Equivalent (cellulosic) weight of combustibles $= (9.5)(12.5)(14.5) = 1720 \, \text{lb}$

The room surface area is:

Surface area $= 2(14.5)(12.5) + 2(14.5)(8) + 2(12.5)(8) = 795 \, \text{ft}^2$

The fuel load becomes:

Fuel load $= 1720 / 795 = 2.2 \, \text{lb} / \text{ft}^2$ ($10.8 \, \text{kg/m}^2$) of surface area

When we add the fuel of the wood interior finish, the equivalent fuel load becomes:

$$\text{Fuel load} = 2.2 + 420/795 = 2.7 \, \text{lb}/\text{ft}^2 \, (13.3 \, \text{kg/m}^2) \text{ of surface area}$$

18.17 The Ingberg Correlation

Around the time of World War I, knowledgeable individuals in the fire community recognized that some construction assemblies behaved better in fires than others. The distinctions could be roughly ordered with ASTM E-119 test results. However, a link between natural building fires and the standard fire test did not exist. Simon Ingberg was the first to describe a link with the standard fire test, natural fires, and a failure time.

In the 1920s, Ingberg defined fire severity as an area under the time–temperature curve above a base temperature. He proposed an "equal area" theory which states that the fire severity in different fires is equal when the areas under their time–temperature curves are equal. Figure 18.4 illustrates conditions in which fires are defined to have equal severity. Therefore, the equal area theory can relate any natural fire to the standard time–temperature curve.

This theory was based on the results of 10 burnout tests in two different rooms. The test facility was built in 1922. By 1929 the reported tests were increased to 16. Some burnouts required as much as 13 and 20 hours of actual time.

The fire loads were plotted with the associated standard test time using the equal area theory. Two coordinates were plotted for each test fire, one using a base temperature of 150°C and one using a base temperature of 300°C. Figure 18.5 shows the best fit for these calculations.

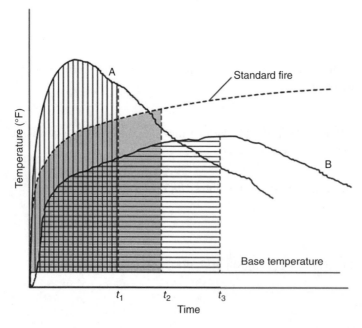

Figure 18.4 Equivalent fire severity.

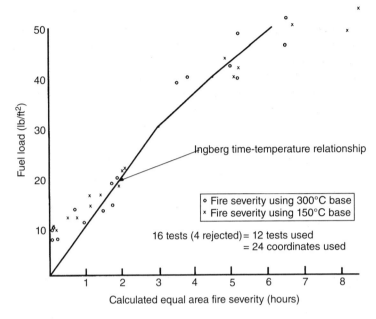

Figure 18.5 Ingberg correlation.

Table 18.3 Tabulated results for Figure 18.5.

Fuel load (lb/ft^2)	1	5	10	20	30	40
Test time	6 min	30 min	1h	2h	3 h	4.5h
10^3Btu/ft^2	8	40	80	160	240	320
kg/m^2	4.9	24	49	98	146	195
MJ/m^2	90.8	454	908	1815	2723	3630

Table 18.3 shows the tabulated results. Although a few additional fuel load studies were done during the intervening years, the first comprehensive fuel load study was published in 1942.

18.18 Room Fire Discussion

The Ingberg definition of fire severity and its relationship between fuel load and standard test times has been used since the 1920s. Although the concepts are simple, many inaccuracies are associated with the theory and its application. Even Ingberg's original paper noted that, although radiation energy was very important, it was not used. However, the theory was the best that could be proposed at the time.

During the 1960s and 1970s, an international cooperative effort was undertaken to understand the post-flashover fire and its effect on structural elements. The program integrated experiment, heat transfer theory, and calculations to increase the understanding of room fire behavior. Some of the more important influences are discussed in the following.

Ventilation

Some flow of air is needed to cause flashover and continued burning. Therefore, at least one barrier in the room must have openings that allow the inflow of air and the outflow of fire gases. The ventilation of the room is expressed as $A/\sqrt{h}$, where A is the total area of all openings that allow flow of air or fire gases, and h is the weighted average opening height, measured from the floor. The ventilation ratio of a room is calculated as $A\sqrt{h}/A_T$, where A_T is the total bounding surface area of the room.

Fuel

Fuel quantity, material type, surface area, and arrangement influence burning characteristics for a fire. Because the heat energy of a fully developed room fire is transferred to all the bounding surfaces, expressing the fuel load in terms of room bounding surfaces, rather than floor area, represents the physics better. Nevertheless, conventional practice continued to use floor area.

Fuel–Ventilation Relationship

Ventilation has an important effect on natural fires. The burning characteristics are often viewed in terms of fuel controlled burning and ventilation controlled burning:

- *Fuel-controlled burning* – when a large quantity of ventilation is available, fuel-controlled burning takes place. The rate of heat release of the fire is limited by the fuel surface available for combustion. Fuel-controlled burning occurs in spaces having large ventilation openings that supply large amounts of fresh air. Temperatures are higher than for ventilation-controlled burning, and much heat energy is lost through the openings.
- *Ventilation-controlled burning* – when the ventilation area is restricted, air supply is reduced and the fire's heat release rate is limited. This ventilation-controlled burning causes fuels to burn more slowly, more inefficiently, at lower temperature, and for longer durations.

Opening Locations

Ventilation opening locations influence the burning duration. The fire "breathes" through an opening by drawing in fresh air and expelling heat and fire gases. When an opening is in a short wall of a long, narrow room, the fuels do not burn uniformly. They seem to burn in phases that move the burning sequentially to the end of the room. This produces longer-duration fires than may be anticipated by a theory that assumes a well-mixed reaction that consumes fuels at a relatively uniform rate. Thus, long, narrow rooms with openings in one end may exhibit a greater fire severity than configurations that have an aspect ratio closer to 1.0.

The demarcation between fuel-controlled burning and ventilation-controlled burning depends on the fuel load and the ventilation ratio. Ventilation-controlled burning is more common in building fires and usually, but not always, produces the most severe conditions for a structure. Therefore, it is usually assumed that ventilation-controlled burning exists in a fully developed fire.

Although higher fire gas temperatures will occur in fuel-controlled burning, the duration of burning is shorter because much of the heat energy is transferred out of the room by the air–fire gas exchange. This reduces the heat energy application against the barriers. Figure 18.6 illustrates temperature relationships for different ventilation conditions.

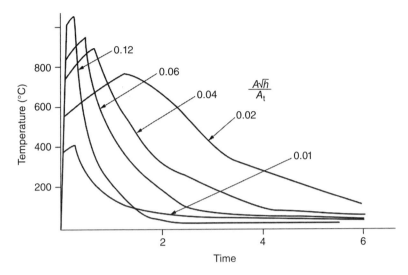

Figure 18.6 Temperature variation and ventilation. *Source*: Reproduced with permission of Lund University.

Bounding Surface Thermal Properties

The insulative qualities of the room have an influence on the time–temperature relationship of the natural fire:

- Barriers in *well-insulated (thermally thick) rooms* do not lose much heat to the outside and the room acts more like an oven. Fires in well-insulated rooms burn hotter and more heat energy is applied to the barriers.
- Barriers in *poorly insulated (thermally thin) rooms* transfer more heat to the outside. This heat transfer reduces the heat energy in the room and fires burn somewhat cooler.

Temperature

Radiation emitted from a hot surface is proportional to the fourth power of the absolute temperature (Stefan–Boltzmann law). Because materials are particularly sensitive to the radiated heat from a fire, higher temperatures cause a more rapid deterioration in construction assemblies than lower temperatures.

18.19 Fire Severity Theories

Research and computational techniques expanded knowledge about fire severity, and several additional theories were developed to relate fuel loads and room construction to the standard fire test. The most common theories are described below.

18.19.1 Ingberg Theory

The Ingberg theory was described in Section 18.17. The equivalent time in the standard test, τ_e, is shown in Table 18.4 or it can be calculated from:

$$\tau_e = 0.0205L \tag{18.1}$$

Table 18.4 Typical Comparisons for Room Conditions and Standard Test Fires.

Fuel load (kg/m^2)	$\dfrac{A_w \sqrt{h}}{A_t}$	Real time (min)	τ NHL (min)	τ Law (min)	τ Pettersson (min)	τ Ingberg (min)
36	0.01	152	81	136	106	44
36	0.07	22	13	54	40	44
36	0.15	10	27	38	28	44
57	0.01	242	133	216	168	70
57	0.07	35	62	85	64	70
57	0.15	16	36	61	44	70
79	0.01	335	193	299	233	97
79	0.07	48	80	118	89	97
79	0.15	23	44	84	60	97

Room construction: size 5 m × 5 m; floors reinforced concrete; walls wood stud and gypsum.

where L is the fuel load in kg/m^2 of floor area ($L < 146$ kg/m^2 or 30 lb/ft^2) and τ_e is in hours.

The Ingberg theory is the simplest way to calculate fire endurance time needed by room fuel contents. The equivalent test time is linearly proportional only to the fuel load. The theory neglects the influence of ventilation, fire temperatures, thermal properties of the bounding surfaces, and room size. Because the increased radiant energy of high temperatures causes greater material degradation than the standard test fire, high-intensity fire conditions make the correlation unconservative.

18.19.2 Law Correlation

Law (1970)[2] analyzed a large number of fire tests involving a variety of room sizes, ventilation ratios, and fuel loads. This correlation is the basis for calculating an equivalent time for a fire test as:

$$\tau_e = 0.022 \frac{A_F L}{\sqrt{A_V \left(A_t - A_F - A_V \right)}} \tag{18.2}$$

where:

A_F = floor area (m^2);
A_V = ventilation area (m^2);
A_t = compartment bounding surface area (m^2).

This equation is an improvement over the Ingberg correlation because it includes ventilation and is based on a relatively large number of tests. However, the compartment thermal properties are not included.

---

2 Law, M, *A Relationship Between Fire Grading and Building Design and Contents*, JFRO International Note No. 374, 1970.

18.19.3 Pettersson Equation

Pettersson[3] expanded on Law's concept to include ventilation height and area as well as a factor relating thermal properties of the bounding surfaces. The equivalent time that an assembly should withstand a fire endurance test is:

$$\tau = 0.31 CL \frac{A_F}{\sqrt{A_t A_V h_V^{1/2}}} \qquad (18.3)$$

where:

h_V = average height of ventilation openings (m);
C = bounding surface thermal properties.

The term C incorporates thermal properties of the bounding surfaces as follows:

$$C = 0.09 \text{ when } 0 < \sqrt{k\rho c} \leq 720 \text{ J m}^{-2}\text{s}^{-1/2} \text{ K}^{-1}$$
$$C = 0.07 \text{ when } 720 < \sqrt{k\rho c} \leq 2550 \text{ J m}^{-2}\text{s}^{-1/2} \text{ K}^{-1}$$
$$C = 0.05 \text{ when } 2550 > \sqrt{k\rho c} \leq 720 \text{ J m}^{-2}\text{s}^{-1/2} \text{ K}^{-1}$$

Equation (18.3) incorporates thermal properties of the barriers as well as the height of the ventilation openings to calculate an equivalent standard fire test time.

18.19.4 Normalized Heat Load

Harmathy[4] developed an equivalent barrier fire exposure that relates standard test results with the destructive potential of a fire. Example (18.2) describes a room fuel load (equivalent heat content) in terms of the bounding surface area. This normalized (i.e. distributed) the heat content equally over the entire surface area. Rather than assuming the heat content to be uniform over all surfaces, the *normalized heat load* (NHL) normalizes the total incident heat flux absorbed by the bounding surfaces. The thermal property that "weights" the surfaces is the thermal absorptivity, $\sqrt{k\rho c}$. The NHL then relates natural compartment fires with test furnace fires.

The NHL is defined as:

$$H' = \frac{1}{\sqrt{k\rho c}} \int_0^\tau q\,dt \qquad (18.4)$$

For a room with a natural fire, H' is the total heat energy released by the fire divided by the bounding surface of the room thermal inertia of the boundary:

---

3 Pettersson, O., *The Connection Between a Real Fire Exposure and the Heating Conditions According To Standard Fir Resistance Tests – With Special Application to Steel Structures.* ECCS, CECM-III-74-2E, Chapter 2, 1970.
4 Harmathy, T.A. The fire resistance test and its relation in real world fires. *Fire and Materials*, Volume 5, Number 3, 1981.

$$H' = \frac{M_f \Delta H_c}{A_t \sqrt{k\rho c}} \tag{18.5}$$

The rate of heat release is influenced by the amount of fuel and its characteristics, the ventilation, and the combustion efficiency. Some of the heat is released through openings and some is unburned. When these factors are included, the value for H' becomes:

$$H' = \frac{(11.0\delta) M_f \times 10^6}{A_t \sqrt{k\rho c} + \sqrt{\phi M_f}} \tag{18.6}$$

where:

M_f = mass of fuel (kg);
δ = combustion efficiency;
φ = ventilation factor.

Equation (18.5) is a more conservative representation for the NHL, whereas equation (18.6) provides values closer to expected fire behavior.

When a compartment fire is compared with a test furnace, the equivalent test time can be calculated. This correlation will vary with the test furnace. For the National Research Council of Canada test furnace, the equivalent time in the test furnace may be expressed as

$$\tau_e = 0.11 + 0.16 \times 10^{-6} H + 0.13 \times 10^{-9} H \tag{18.7}$$

The normalized heat load is a carefully formulated and theoretically sound way to relate compartment fires with standard test results. The correlation can vary with different test furnaces, although the discrepancy may not be significant. The theory relates to thermally thick barriers that absorb all of the heat energy with only a small temperature rise on the unexposed face. It does not apply to thermally thin barriers in which the heat is transmitted to the adjacent room and the temperature rises substantially on the unexposed barrier surface.

18.20   Fire Severity Comparisons

The methods cited above attempt to approximate a real time that is equivalent to the standard test time that would occur at fuel load burnout. All of the equivalency methods relate fuel loads to an equivalent time in the standard fire test. The terminal point is thermal failure for a floor–ceiling assembly. Surface cracking and collapse failure modes are assumed not to occur.

It is important to recognize that these theories relate only to equivalencies with the standard fire test. They do not attempt to relate the fire test to building performance. Although none give sufficient information for a building performance analysis, they do provide useful information for making behavioral judgments.

Newton[5] compared these four theories and the COMPF2 computer program with statistical data for office loadings. The fuel load statistical mean with +1 and +2

---

5 Newton, BA, Analysis of common severity methods and barrier performance. *MSc thesis*, 1994.

standard deviations, and the ventilation statistical mean with −1, and +1 standard deviations were selected from the Culver statistical data. The analyses involved square rooms with 3 m ceilings that varied in area from 6.25 to 400 m^2. The floor and ceiling materials were reinforced concrete. Six combinations of CMU block or gypsum wall constructions were used. The goal was to understand time relationships and comparative sensitivity of these methods with regard to statistical and practical building conditions.

Spreadsheet analyses compare calculations for statistically related conditions. The "real-time" fire duration was calculated using step functions with stoichiometric wood burning with the associated $A_W \sqrt{h}$ ventilation. Table 18.4 illustrates typical results.

The general observations from this limited study of 90 cases using statistical data were as follows:

- The ventilation area is the most influential factor in fire severity. Small openings cause long-duration fires that produce the greatest severity. Large openings cause much faster combustion and heat losses through the openings. This creates a high-intensity, short-duration fire of lower equivalent test duration.
- The fuel load has the expected direct relationship on fire severity in that more fuel creates a longer fire.
- The material properties for thermally thick barriers (here, gypsum and concrete) in the Pettersson and NHL analyses have much less influence on fire severity than does ventilation. The differences between these materials were relatively insignificant.
- Calculation methods are not independent of room size, although limits are not identified. The Ingberg theory is only related to fuel load and is independent of all other factors. The NHL pattern seems to give the most consistent and logical results for all variables. Fire severity increases directly with room size in the Petterson and Law procedures. Larger rooms produce larger equivalent test times.
- The NHL method gives the best indication of equivalent test time. Results are consistent and logical. The NHL method has the strongest theoretical base and the heat energy application includes not only the fuel load but also other factors relevant to compartment fires.
- The NHL method is relatively simple, but cumbersome without a computer program to order the calculations. NHL results involve only heat transmission for thermally thick barriers. Cracking or falling away of materials and the barrier's structural integrity are not a part of the analysis.
- None of the methods provides enough information to determine if a massive ($\bar{D}$) failure or hot-spot ($\bar{T}$) surface cracking occurs.

These analyses did not investigate applicability limits of the methods. The range of applicability for fire calculations is always important and rarely identified.

18.21 Awareness Pause

The fire endurance of barriers is complicated. It may be useful to pause to reflect on information that has been presented up to now. The major concepts are:

- The vast majority of barrier failures are due to openings and weaknesses that are not addressed by the standard fire test and the associated prescriptive fire endurance classification. The most important understanding of barrier performance is gained by qualitative evaluations described in Chapter 8.

- The fire endurance times prescribed in building codes have practical value because they make it difficult to have flimsy construction in large buildings. The fire resistance time durations should not be taken literally. They enable one to make general comparisons about expected behavior between different construction assemblies. Although prescribed fire resistance requirements do result in more stable construction during a fire, assembly performance does not necessarily correspond to the test rating. Fortunately, barriers having structural stability and reasonable construction substance seem to be effective in preventing fire propagation.
- No published investigations of the relationship between actual building performance and standard fire test results have been discovered. Any relationship between test time, fire severity calculations, and field performance is unknown.
- Equivalency theories that relate fuel loads, ventilation ratios, and thermal properties to the fire test are based on heat transmission through the barrier. They do not address material cracking or structural collapse.
- From a practical viewpoint, the fuel load and other conditions that affect burning rarely cause failure in an unpenetrated barrier. Nevertheless, these factors can influence barrier performance.
- The purpose of a barrier analysis is to understand multi-room fire performance of a building. Unpenetrated barriers can be effective in containing a fire during the important (early) time durations of a building fire. Penetrated barriers allow fire to spread through the holes and are the usual cause of fire propagation.

18.22　Estimating Burnout Time

If one could believe that real time replicates the standard test time, it would be a simple manner to use the correlation described in Table 18.3. That is, for every 1 kg/m² (0.205 lb/ft²) of fuel consumed, 6 min would elapse. However, when heat energy is released faster or slower than the standard fire test, real time to failure is different than the test time.

A first approximation of real time can use the stoichiometric (i.e. perfect) burning of wood. This can be calculated as:

$$\dot{m} = 5.5 A \sqrt{h} \tag{18.8}$$

where:

$\dot{m}$ = rate of fuel consumption (kg/min);
A = area of openings (m²);
h = average height of openings (m).

Alternatively, in US units:

$$\dot{m} = 8.86 A \sqrt{h} \tag{18.8a}$$

where:

$\dot{m}$ = rate of fuel consumption (lb/min);
A = area of openings (ft²);
h = average height of openings (ft).

18.23   Influences on Barrier Performance

The objective of a barrier evaluation is to understand the way in which the barriers will perform in a building fire. Often, this can be as simple as deciding that the construction and field conditions will provide more resistance than needed by the fuel load, and no formal analysis is needed. Detailed analysis is important only when conditions indicate that more information is needed.

Some construction practices or conditions may make the barrier stronger or weaker than would be indicated by published fire test information. The influence of the factors is discussed below. The synergism of factor combinations has not been investigated.

Applied Loading
Superimposed loads certainly reduce the $\bar{D}$ performance and may reduce the $\bar{T}$ behavior.

Thermal Restraint
When construction methods restrain structural members against thermal expansion, a prestressing condition occurs. This condition may result in substantial increases in $\bar{D}$ capabilities and some increase in $\bar{T}$ resistance. On the other hand, some assemblies may have no change in performance and others may experience reduction in resistance. The direction and magnitude depend upon the materials and construction details, and may be anticipated with a structural analysis.

Construction
A variety of conditions exist in the standard laboratory test that are not the same as at the construction site. For example, the laboratory can only test statically determinate systems. Modern construction regularly uses continuous construction that provides increased strength against collapse. Continuous frame construction increases the $\bar{D}$ capability and perhaps also the $\bar{T}$ performance because of decreased deflections.

Another aspect of construction involves scale (i.e. size). The laboratory test involves surface areas of 100 ft^2 (walls, partitions) (9.3 m^2) or 220 ft^2 (floors) (20.4 m^2). Limited information seems to suggest that larger surface areas crack more readily than laboratory sizes. It would appear that $\bar{D}$ values should also be lower for assemblies larger than test sizes.

Structural deflections can cause cracking of supported elements, such as slabs, ceilings, and panels. Suspended membrane ceilings often provide thermal protection for floor–ceiling assemblies. When they remain in place, they can be effective. Ceiling systems with designed expansion capabilities seem to behave well. However, when expansion is not controlled, deflections in the ceiling support structure cause tiles to fall prematurely.

Still another concern involves ceiling utility fixtures that allow heat penetration. UL-tested ceiling support structures and fixtures seem to perform well. However, untested fixtures may be suspect when they appear in older buildings or structures in which architectural approved "or equal" substitutes may not have been used.

18.24   Automatic Protection and Barriers

A small number of barriers are protected by automatic suppression equipment to defend barrier openings by discharging agents that delay or prevent movement of flame-heat into the adjacent space. These barriers fall into two groups. The first involves discharge of water onto physical barriers, such as glass. Exterior sprinklers or water distribution on glass panels are of this type. A second situation occurs when a virtual barrier (e.g. escalator opening or open space division) uses a water curtain for barrier protection.

The application of water to barrier glass usually increases the fire endurance. An analysis includes both reliability of the sprinkler system and the effects of expansion of the supporting frame and structure.

The effectiveness of a water curtain to prevent fire extension is controversial. Often, the method is used to provide an equivalency for enclosure requirements of building codes. This avoids viewing restrictions for open architectural spaces. While this alternative to more conventional barriers is convenient from an architectural function and it does provide code approval, performance effectiveness is questionable and should be examined.

PART FIVE: TRANSITIONS

18.25   The Issue

The building industry is approaching a situation where modern performance-based methods are conflicting with "comfortable" traditional procedures. Calculation procedures in structural engineering are clearly superior to the standard fire test practices of the past century. However, the direct replacement of structural requirements of traditional building codes with performance-based calculations has many implications. Before returning to technical subjects, we philosophize briefly about changes.

The transition from a traditional code-dominated culture to a holistic, performance-based culture will be difficult and prolonged, but inevitable. The movement will require decades to complete. Costs, legal issues, tradition, standard practices, vested interests, and the necessity of learning new and different skills will inhibit the natural evolution of the fire safety engineering profession. However, the movement has progressed to a point of no return and the transition will persevere.

In a performance-based engineering discipline, one should recognize the distinction between design standards, equipment standards, installation standards, material standards, and building codes. Equipment, material, and installation standards are necessary to understand what to expect from commercially available products. These topics are discussed more completely in Chapter 19.

Design standards associated with performance methods are specially formulated and based on engineering calculations. Allowable values for the control variables establish margins for safety and serviceability of performance. The role of building codes is to establish the performance expectations for loadings. These design standards establish a level of safety and serviceability. Both are based on, and integrated with, an understanding of engineering methods.

Building design integrates component behavior with holistic performance in a way that enables one to recognize the effect of one on the other. Traditional fire regulations provide a holistic solution. Unfortunately, the effects are not transparent and often not economical. Often, they are found to be ineffective. Nevertheless, the experience of traditional regulatory design provides a valuable foundation for understanding holistic performance. The systematic organization enables one to compare building performance for traditional and performance-based conditions with the same measurements.

19

Concepts in Structural Analysis for Fire Conditions

19.1 Introduction

Performance-based structural analysis for fire conditions requires cooperative interaction between two professions: structural engineering and fire safety engineering. Separately, each profession possesses a set of technical skills that addresses complementary parts of the topic. Together, these parts combine to provide a more complete understanding of structural performance during a building fire.

The target audience for this discussion is the fire safety engineer rather than the structural engineer. The goal is to describe concepts to help the fire safety engineer to become aware of structural engineering.

The information will be organized into four parts:

- *Part One: Building Design* briefly describes the development process, routine building design, and information technology.
- *Part Two: Structural Engineering and Building Design* gives a brief history of the evolution of structural engineering methods, codes, and standards to current practices.
- *Part Three: Structural Engineering* describes concepts of analysis and design in structural engineering. A simple example illustrates these concepts for normal temperatures.
- *Part Four: Structural Performance in the Fire Environment* describes concepts of structural analysis during fire conditions.

19.2 Structural Fire Performance

The structural engineer is responsible for the building's performance under normal design and temperature conditions. In addition to strength and stability requirements, serviceability considerations such as deflections, vibrations, or surface cracking due to excessive movement are within the scope of routine practice.

Structural engineering methods can estimate building movements and collapse potential from a fire. Fire safety engineering techniques can define fire environments with which to evaluate performance.

Fire Performance Analysis for Buildings, Second Edition. Robert W. Fitzgerald and Brian J. Meacham.
© 2017 John Wiley & Sons Ltd. Published 2017 by John Wiley & Sons Ltd.

A structural fire performance analysis estimates the potential for:

a) Localized collapse of structural members or assemblies.
b) Large deformations or deterioration that will not cause collapse, but will require replacement during reconstruction.
c) Conditions that can cause instability and large scale collapse of building sections.
d) Structural deflections that may cause inoperability of doors or cracks in barriers.
e) Conditions that may cause structural collapse during the cooling phase of a fire. Although this type of failure is not common, it can occur in certain situations.

Structural engineers analyze movement and potential failure of structural members. This includes estimating changes in the load-carrying capacity of members caused by heat conditions and time.

Fire safety engineers define appropriate diagnostic fires for structural analysis. In addition to single-room fire environments, the fire safety engineer is responsible for recognizing multi-room fire spread, as discussed in Chapter 8. Depending on the needs of a risk management program, single- and multi-room fires may extend to room burnout or to truncated fire conditions that reflect extinguishment.

The fire safety engineer must understand enough about basic structural systems and practices to communicate effectively with the structural engineer. Although it is not within the professional domain to calculate structural behavior, the fire safety engineer should recognize qualitative response of structural systems and insulation techniques on performance.

PART ONE: BUILDING DESIGN

19.3   The Development Process

Building buildings is a complicated business. Thousands of individuals move through their entire professional careers by phasing into and out of creating buildings.

A developer, who is the individual in responsible charge of a project, manages the entire process. Sometimes the developer is the owner; at other times the developer represents an owner. Although the corporate staff and hired consultants provide information and detail work, the developer makes final decisions on substantive matters. The developer is the prime mover of a building venture.

A hierarchy of contracts and subcontracts coordinates professional services. Among the consultants, the architectural team of architect, structural, mechanical, and electrical engineers is the usual group that creates the plans and specifications. This team may be augmented with an array of specialists such as foundation engineers, site planners, fire safety engineers, interior designers, and others with specialized skills to make the building stronger, safer, more functional, or more attractive. Each design professional must coordinate specialized skills with other professionals. Although each specialist has a certain amount of decision-making autonomy, the collective efforts must be integrated to produce a quality building.

It is useful to have a general understanding of the development process to understand the roles of the architectural team and the fire safety engineer in the process of building buildings.

19.4   Building Design

The developer initially conceives the scope and focus of the building. Coordinated studies examine the physical, legal, and financial feasibility of the project. These preliminary studies define the function and needs of the building and establish a construction budget. The architectural design process normally begins after this pre-development phase is completed and the site is controlled.

The architect assembles a team that normally consists of a structural engineer, a mechanical engineer, and an electrical engineer. The team may be augmented with specialists to deal with needs outside the scope of the team's professional skills.

A building design is evolutionary, involving three principal phases of activity, including preparation of *schematic plans*, *detailed design*, and *construction documents*.

The schematic design phase addresses the critical decisions that affect the quality and functional value of the building. The architect discusses the needs and desires with the developer to gain an insight into how the building should work. Preliminary (schematic) plans and elevations for the building and site are created to address the perceived ideas and needs. These initial sketches are used to judge whether the envisioned ideas have been captured in a practical design. Changes continue until the functional layout satisfies the developer's ideas and needs.

As the schematic drawings undergo evolutionary changes, the mechanical engineer considers heating, ventilation, and air conditioning (HVAC) requirements and the electrical engineer studies power needs. The structural engineer works closely with the architect during the schematic phase to identify structural framing schemes and materials to blend with architectural concepts.

The building evolves during the schematic phase. Creative ideas and "what if" alternatives are easy to examine, and changes are relatively inexpensive to incorporate. The schematic phase is the most creative part of design and requires professional experience, knowledge, and judgment. Effectively, the completion of schematics establishes the building. Although minor changes can be made during subsequent phases of design, any substantial changes are difficult to make and expensive because they affect all members of the design team.

The detailed design phase produces specifics for the building. The architect refines circulation patterns and functional requirements, provides dimensional consistency, and integrates modular assemblies. The mechanical engineer designs the plumbing and HVAC systems. The electrical engineer designs illumination and power circuits. The structural engineer sizes all structural members and examines serviceability and construction compatibility.

Although general concepts for fire safety are considered during the schematic stage, most code compliance requirements are addressed during the detailed design phase. The mechanical engineer is often designated by the architect to deal with fire safety matters. Structural fire requirements would have been identified early in the process and easily incorporated. The electrical engineer would ensure that design and installation comply with the applicable electrical codes. The architect ensures that egress requirements are incorporated into the circulation pattern.

Fire safety is often addressed as a regulatory constraint to be incorporated into routine design. Code compliance is commonly assumed to be the same as a fire-safe building. Although fire safety is usually perceived to be a code issue, the building industry is beginning to recognize that code compliance and actual fire performance can be very different. Needs for risk management in a site-specific building are not addressed by building code regulations.

The detailed design phase completes selections of systems, arrangements, and sizing. The construction documentation phase coordinates the plans and specifications to provide a package of documents from which contractors can estimate the costs and prepare a bid to construct the building. Plans show dimensions and ways to construct the assemblies. Specifications provide information that defines the quality of the building. Specifications describe the materials or systems to use, and often the details and quality of construction.

After evaluating the bids from responsible contractors, the developer selects the contractor to build the structure.

19.5   Information Technology

Computer applications are ubiquitous in the entire design and construction process. Software performs routine engineering calculations. Computers can size framing members and store technical data. The scope of information technology is evolving so rapidly that its influence is making remarkable changes in the way buildings are designed and built.

Time and certainty are the most significant concerns within the entire building development process. Many aspects of the building process are sensitive to completion and decision time durations. Although some uncertainty is inherent to the process, clear and timely information enables the project to proceed more quickly and reduces risks.

Negotiations and decisions are inherent to the process. Rapid, reliable information enables the human brain to understand its holistic impact and make decisions with greater confidence. And human decisions drive the process.

In recent years, building information modeling (BIM) has become an important tool in the design and construction of buildings. The influence of BIM and its digital successors can only grow.

Computer modeling has a significant impact in the schematic process of building design. Floor plans, elevations, three-dimensional models, sections, and visual perceptions may be done quickly and accurately. Changes and "what if" alternatives may be examined rapidly. The computer is changing the way we design our buildings by providing tools to replace procedures that formerly require greater time to complete or modify. Computer applications become an indispensible aid to human understanding and communication. They do not replace human decision-making.

Building information modeling provides many additional services to the building process. Certainly it augments engineering design and construction. Also, the store of data and information enables interior design to be incorporated more easily with visual and dimensional coordination.

This book is about performance-based fire design, and BIM can have an enormous influence on the practice of fire safety engineering. On the one hand, BIM can incorporate the requirements of a building code to enable one to relate code compliance with the architectural design. If the perceived role of fire safety is only a code compliance issue, the task is simplified greatly. However, if aspects of code compliance are not compatible with architectural or functional needs, performance equivalency must be introduced.

The "one size fits all" structure of building codes may be unsuited to the developer's needs or the architect's desires. Fortunately, performance-based fire engineering is compatible with building code equivalency evaluations. Figure 19.1 shows schematically the relationship between traditional code compliance and performance analysis.

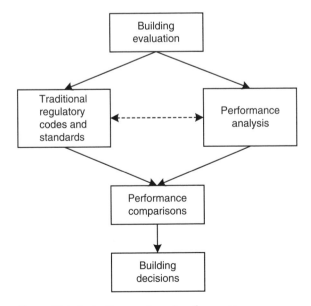

Figure 19.1 Evaluations: code and performance.

Both traditional code compliance and performance-based analysis have similar goals of providing safety from fire. However, the building code identifies compliance, not performance. Performance methods can analyze any building to determine its risk from fire. Both holistic performance and component performance may be examined in detail. Therefore, any proposed building design or existing building, whether code-complying or not, may be compared with suggested changes. The common element for comparison is *time*.

A certificate of occupancy is essential to allow a building to be occupied and used. The authority having jurisdiction and the building owner or the owner's agents must have a meeting of minds about compliance and equivalency. Computer-aided methods such as BIM and ancillary performance-based models provide an ability to rapidly communicate risk-associated conditions with building design features throughout the design process.

PART TWO: STRUCTURAL ENGINEERING AND BUILDING DESIGN

19.6 The Master Builder

The Roman master builder and Architect Vitruvius (first century BC) asserted that a building must exhibit the three qualities of *firmitas*, *utilitas*, and *venustas*. This may be translated to be "a building must be strong and durable, useful, and beautiful." In modern terms, we might express these qualities as *safe*, *functional*, and *aesthetically pleasing*. We might also observe that in ancient times, and also in the modern era, the creation of a building was intended to produce an asset that contributes to helping people live and work better.

From the time of Vitruvius until the latter part of the 19th century, a single individual was normally in responsible charge of building design and construction. This master builder served as architect, structural engineer, and construction manager from start to completion of the project. However, the Industrial Revolution changed the way buildings were designed and built.

19.7 The Rise of Engineering

The Industrial Revolution produced massive synergistic interactions of societal needs, new materials, and analytical techniques which created an explosion of new demands for the building industry. The Bessemer process for making structural steel created a strong, practical material to augment timber and masonry. Railway and highway bridges were being constructed with improved confidence in their stability. New materials and a new way of thinking spawned new building designs. Evolving building techniques and the invention of the elevator enabled skyscrapers greater than six stories in height to be built.

However, the greatest change produced by the Industrial Revolution was the emergence of engineering as a profession. Transportation needs, manufacturing applications, and urbanization created a great demand for technically trained engineers. The Morrill Land-Grant Act (1862) produced a large increase in engineering schools. A practical engineering education replaced the more mathematical and philosophical science studies that had been the earlier basis of technical education.

For conceptual purposes, the 1880s and 1890s may be described as the golden age of engineering. The explosion of new engineering schools changed technical education from natural philosophy to a focus on problem-solving. This produced individuals with different skill sets. Analytical frameworks based on sound theory and practical applications were developed to enable one to understand the ways in which details affected holistic performance. This new engineering way of thinking created innovative solutions for societal needs.

This practical engineering education found a ready market in new technology. Whether it was "chicken or egg," widespread applications of central heating, electricity and lighting, and building techniques became common. All of these resulted in products that enhanced building functionality and construction techniques. Concepts in building design experienced a revolution during the late 19th century.

19.8   The Building

The few decades on each side of 1900 established the organization for creating the built environment of today. Earlier reliance on experience and judgment was augmented by analytical skills to create a better understanding of load and resistance or cause and effect. Engineering concepts enabled experience to be organized, and functionality and safety to be addressed in a different way.

During this era, the master builder was replaced by a team of professional specialists that could focus on details of the process. Although Vitruvius' three qualities of usefulness, safety, and beauty remain central to building design, the process changed the way in which these qualities were implemented:

- *Usefulness is functionality.* Today, we describe this quality as "the building works." The architect creates an arrangement of rooms and circulation patterns to enable communication and operations to be coordinated efficiently. In addition to providing for heat and ventilation needs, the mechanical engineer designs water supply and waste disposal for the building. The electrical engineer creates a lighting environment and provides power circuits to make the building serviceable. Mechanical and electrical engineers bring energy sources into the building and transform them into useful applications.
- *Safety* has traditionally focused on structural stability and fire. A structural engineer addresses prevention of collapse or undesirable deformations. In the 1900s, fire was in the domain of the architect and building code enforcement. Fire insurance organizations became active participants in the process. We will discuss the evolution of these areas later.
- *Beauty* in a building's form has traditionally been in the realm of the architect. Today, architect, owner, or interior designers, either separately or collectively, handle the aesthetics.

Specialization of professional engineers established conditions that encourage responsibility, quality, and innovation. This was not without a few stumbles, although the professions always find a way to overcome their problems. Design practices combined performance concepts with engineering methods for quantification. This professional evolution produced a climate of confidence and creativity.

Traditionally, fire safety has been established by prescriptive regulations. The code and its enforcement, rather than a professional dedicated to the discipline, assumed responsibility for the level of risk provided by a building. Fire safety is now emerging as a performance-based engineering discipline.

19.9   The Emergence of Structural Engineering

The early part of the 20th century was a transition period for the built environment. Before World War I, buildings were normally constructed by following a prescriptive building code that was similar to the prescriptive codes for fire safety that are used today. Design loads, materials specifications, allowable stresses, serviceability requirements, and construction details were incorporated into the same prescriptive code document. Thus, routine construction could be performed without an engineer of record, because compliance with the building code was perceived as providing a sufficient standard of care. This created a situation where the quality of buildings was uneven and innovations were difficult to introduce.

During this period around World War I, the professional engineering community influenced a number of changes that affected building design and construction. Most structural engineering practitioners were university-educated. These engineers understood the general framework for analysis as well as the capabilities and limitations of available engineering methods. Although technical methods were primitive compared with modern engineering capabilities, they were adequate to provide quality engineering services.

Four interrelated aspects of engineering practice were emerging over these years. The growth of engineering standards, modernization of building codes, and professional licensure were necessary, but not sufficient, components. Engineering methods provided the other component to establish the necessary technical understanding as well as a "glue" to integrate the other components.

The integration of these four parts enabled the profession to come of age. The idea of professional licensure rather than self-regulation for engineering services directly related to public use was gaining strength in many parts of the country. The role of building codes began to evolve from holistic regulation to establishing performance expectations by specifying minimum design loads. New construction methods and expanding knowledge of analysis and performance were growing rapidly. Although each part was moving along its own path, their complete integration was accomplished within about 30 years.

The creation of design standards was emerging, with the publication of standards for reinforced concrete (1908), structural steel (1923), and timber (1943). The primary role of a design standard is to identify safety and serviceability. Allowable values for control variables establish margins for safety and serviceability performance. An array of additional standards relating to materials, testing, manufacture, and installation augment the design standards. These ensure a level of quality from commercially available products that is consistent with design assumptions.

The Uniform Building Code (1927) established a modern approach to regulations. The contemporary role of a building code for structural engineering establishes minimum standards of performance for different building occupancies. In addition to performance requirements for fixed and moveable live loads, codes specify environmental loads for snow and ice, wind, and earthquake. They also identify load combinations that the structure must provide. Although a building must be able to support these minimum loads and load combinations, when situations exist that exceed minimum conditions, the engineer must recognize them and use appropriate values.

Engineering methods based on free body concepts (1887) easily moved between analysis and design. These techniques, combined with knowledge of materials and their properties and construction methods were evolving at an accelerated rate. Engineering analysis, combined with performance observations, established the framework for thinking for structural engineering. These engineering methods became the foundation on which codes, standards, and licensure were integrated.

By 1930, most of the ingredients for performance-based structural engineering were more or less in place. To be sure, many gaps in knowledge existed. However, those gaps were filled with judgment values that were based on experience, a sense of proportion, and understanding of the available technology.

The functional organization that was conceptually in place by 1930 was nearing fulfillment by 1950. The integrated parts that comprised discipline-oriented, performance-based design gave confidence in application. These parts were: (1) building codes to specify performance loadings; (2) design standards to establish safety factors, serviceability, and related construction practices; (3) calculation methods to enable internal forces and deflections to be related to structural geometry, materials, and external forces; and (4) licensure to ensure educational and experiential competence of design professionals. The regulatory process recognized that a professional engineer's seal and signature represented an acceptable level of competence, reliability, and integrity.

19.10   A Brief Pause about 1950

The decade after World War II is a convenient place to pause and look at the status of structural engineering. World War II had ended and an enormous pent-up demand was burgeoning throughout the world for new buildings and facilities. New construction equipment (e.g. movable tower cranes), prefabricated assemblies (e.g. lightweight curtain walls), and four-season climate control HVAC installations revolutionized design and construction technology. The demand for taller and larger buildings with greater design flexibility and economy was greater than at any time in history.

Structural engineering had reached early maturity by World War II. The framework for thinking and analytical techniques were the same worldwide. Practitioners, researchers, and the construction industry could communicate with understanding. New buildings could be designed and built with greater confidence in their structural performance. From a structural viewpoint, the "bigger and better" building did not pose an unusual obstacle.

Design standards for structural steel, reinforced concrete, and wood were based on the allowable stress design (ASD). The design process selects a framing system. Code specified loading conditions are applied to the structure. Structural mechanics are used to determine the internal forces that each member and connection must support. Member sizes are selected to support codified loads while ensuring that internal stresses are always less than allowable values specified in the design standard.

Design standards identify the factor of safety. The ASD philosophy specifies an allowable stress that is a fraction of a maximum stress that would produce failure or excessive deformations. For example, the factor of safety in 1950 for structural steel flexural members was 0.66 of the yield point of steel. This factor of safety of 1.5 ensured that flexural members always remained within the elastic range of behavior. Other members and loadings had factors of safety that were appropriate for their conditions.

Although the ASD philosophy is relatively simple, the design specifications stipulated many conditions relating to the range of allowable stress values. For example, the location and type of lateral support affect allowable flexure values. Dimensions of flanges and web affect design considerations. Although ASD bases safety on stress limitations, all of the potential influences that affect structural performance ensured that the member would act in the manner anticipated by the simplifying assumptions used in calculations.

19.11   The Great Leap Forward

The period from the 1950s to the early 21st century saw a major expansion of capabilities in structural engineering. Research flourished and greatly increased knowledge and understanding. More powerful calculation methods and information technology evolved to allow an engineer to understand structural performance more clearly. The computer enabled the engineer to transfer attention from tedious calculations to the better planning and conceiving of solutions. Also, the better understanding of three-dimensional analyses made possible a more economical design for wind forces and enabled space structures to be considered more frequently as design options.

Wind and earthquake loadings became greater concerns. These topics become more formalized, disciplined, and organized during the 1950s. Today, wind engineering and earthquake engineering are recognized sub-disciplines of structural engineering. These topics affect design loads and demonstrate interaction between engineering methods, design standards, and building code performance requirements.

Major research advancements during the 1950s examined structural performance for members stressed in the inelastic range. This research examined the full range of structural behavior from elastic conditions through the inelastic stress ranges to collapse. This was particularly important because elastic design (ASD) methods did not adequately account for the increase in structural capabilities of indeterminate structural framing methods. New methods of construction made continuous (i.e. statically indeterminate) connections more common.

Several descriptive names [i.e. ultimate state design (USD), plastic design, inelastic design] were associated with the theory. Knowledge about structural performance expanded enormously. Design specifications and safety factors were changed to reflect this better knowledge. Of greater importance is the fact that subsequent design specifications have been based on this theory of ultimate (or limit) design of structural members and frames.

Toward the end of the 20th century, design standards underwent more dramatic changes as new theories and knowledge provided other ways of achieving safe design at lower cost and less professional time. The computer influenced research and design enormously. Lower factors of safety became possible as better understanding of behavior grew.

As new technology was integrated into design standards, the new design methods were calibrated against the performance of earlier methods. This calibration allowed the profession to clearly recognize how new methods would affect experience, economy, and design decisions. For example, load and resistance factor design (LRFD) applies different factors of safety to dead and live loads to reflect different levels of uncertainty. These values are used with limit state conditions to avoid failure. ASD applies the factor of safety to member strength and treats dead and live loads with the same uncertainty. Although member sizes using LRFD and ASD may be slightly different in some situations, the two standards have been calibrated to provide comparable safety to the public. Structural engineers are expected to be proficient in all applicable procedures.

Changes to standards normally take place over an extended period of time. Fifteen years to allow new calculation methods to be incorporated completely into practice is common. Some involve longer time frames. An extended period of time to accommodate transitions into the design standards as well as to accumulate experience of performance is important for the profession to gain confidence.

Although this discussion has been based on structural steel design, reinforced concrete and wood have undergone similar evolutions. All of these materials use specifications involving LRFD and ASD. The new ways of proportioning members that have emerged during the past half-century retain the fundamental concepts and confidence in performance of early maturity.

19.12   Structural Design for Fire Conditions

After structural engineering reached early maturity after World War II, research, computational advancements, new technologies, and experience produced rapid and major changes in engineering practice for normal conditions. During that time, knowledge of exterior loads and internal resistances for structural applications also increased. However, during that time period the relative level of knowledge and capabilities in structural behavior for fire conditions increased even more dramatically. While much remains to be done, the growth in knowledge has been impressive.

In 1960, knowledge of compartment fires and structural fire behavior was practically non-existent. Almost all of the research and knowledge was based on standard fire testing and the relationship of test results with compartment fire loads.

During the 1960s the Swedish Institute of Steel Construction sponsored a project under the direction of Professor Ove Pettersson to develop a performance-based design method for structural steel. Professor Pettersson, in cooperation with international research efforts of the *Conseil International du Bâtiment* (CIB) (International Council for Building, now the International Council for Research and Innovation in Building and Construction) developed a rational method for design of structural steel for fire conditions (1972). This research led the way for similar design methods involving reinforced concrete and wood.

Over the next few decades significant progress was made in understanding compartment fires and structural response to fire. The increase in computational capabilities, combined with international research cooperation, made knowledge in both topics grow. Normal theory and concepts of structural mechanics became the foundation for a natural extension to design for fire conditions.

To describe structural behavior, fundamental heat transfer properties for materials subjected to large ranges of temperatures were identified. Properties affecting structural behavior at elevated temperatures were also obtained. The substantial knowledge of structural mechanics at normal temperatures enabled those principles to become the basis for understanding structural behavior at elevated temperatures. Many books describing structural design for fire conditions have been written. Design standards have been established. Although much remains to be learned, progress has been so substantial that one may reasonably estimate structural outcomes for any specified fire environment.

PART THREE: STRUCTURAL ENGINEERING

19.13   Introduction

Structural engineering relates external loads, internal resistance forces, and associated deformations. Geometry, arrangement of supporting members, the way the forces are transmitted through the structure, and the strength of materials form physical relationships. The process guards against failure involving collapse or unwanted and unsightly deformations. Cost, effectiveness, compatibility with other project needs, and efficiency are a part of the process.

Structural engineers use a professional "toolkit" of techniques to design new buildings or to evaluate the structural performance of existing buildings. Both tasks combine the same basic analytical framework for thinking with state-of-the-art professional knowledge. Thinking and associated calculations move back and forth between analysis and design, as each complements the other.

19.14   Beam Analysis

We will use two simple examples to convey a few concepts of structural engineering. These illustrations establish a base for awareness of a way of thinking.

Example 19.1 examines a statically determinate structural steel beam to analyze its internal behavior as loads continually increase up to collapse. Example 19.2 discusses the same problem for a continuous (statically indeterminate) beam.

19.14.1   Simple Beams

Engineering methods relate external loads with the internal strength of a member. Example 19.1 illustrates these relationships for a simple beam.

Example 19.1   Determine the maximum load that can be supported by the simple beam of Figure 19.2.

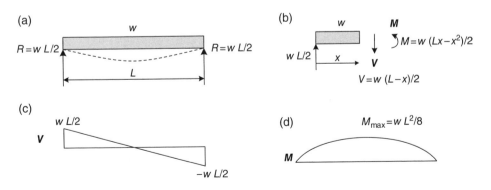

Figure 19.2 Simple beam.

Solution:

Figure 19.2(a) shows the beam and its deflected shape. The free body (b) is used to calculate internal forces and construct the shear and moment diagrams (c) and (d). Simple statics can be used to calculate external reactions, internal shears, and internal moments in the beam. The moment diagram is parabolic for this loading condition with a maximum moment of $M_{max} = wL^2/8$.

These relationships are independent of dimensions, materials, or loading magnitude. Selection of structural materials, geometry, and construction details establishes the load-carrying capacity.

To continue the analysis, assume the beam is structural steel. A stress–strain (σ–ε) relationship for axial-loaded steel is shown in Figure 19.3(a). Stress (σ) is directly proportional to strain (ε) up to the proportional limit. The yield point is near to the proportional limit and indicates the transition from fully elastic to inelastic (plastic) conditions.

The practical limit for functioning performance is the yield point. Beyond the yield point, unlimited deformation takes place with little (assumed no) increase in applied load. The modulus of elasticity (E) is the slope of the σ–ε diagram.

The two segments in Figure 19.3(b) show an idealized structural steel behavior. Structural calculations use this idealized representation rather than the true relationship.

When loads are applied to a beam, each longitudinal fiber elongates or shortens, depending on its location from the neutral axis. When we assume that plane cross-sections before bending remain plane after bending, the stress in each fiber may be determined by relating it to the strain from Figure 19.3. Figure 19.4 shows the progression of strain, stress, and depth of penetration of the elastic and plastic segments for a rectangular cross section as load continually increases.

The bending moment at any section is determined by integrating the force in each fiber times its distance from the neutral axis. Thus, the internal resisting moment at different conditions of maximum stress, σ_y, may be determined as

Elastic range (b):
$$M = \int (\sigma dA)y$$
$$\left(M_{max} = \sigma bd^2/6\right)$$

Extreme fibers reach first yield (c):
$$M_y = \int (\sigma dA)y$$
$$\left(M_y = \sigma_y bd^2/6\right)$$

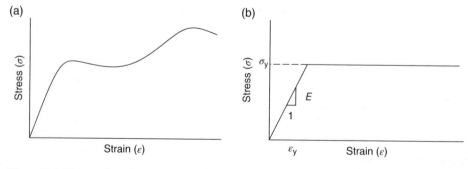

(a)

Stress (σ)

Strain (ε)

(b)

Stress (σ)

σ_y

E

1

ε_y

Strain (ε)

Figure 19.3 Structural steel: stress–strain relationships.

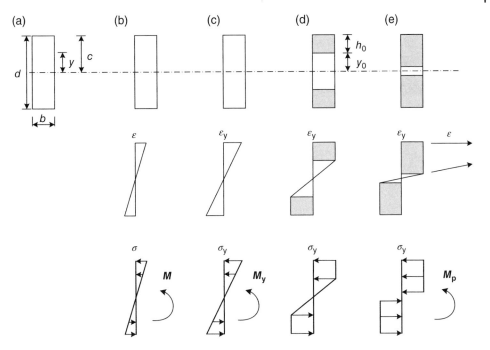

Figure 19.4 Structural steel: elasto-plastic behavior.

Elastic–plastic (d): $M = \sigma_y bh_0 (c - h_0/2) + \int (\sigma dA) y$

Fully plastic (e):
$$M_p = 2[\sigma_y (bd/2)d/4]$$
$$(M_p = \sigma_y bd^2/4)$$

The rectangular cross section of Figure 19.4 shows a 50% increase in moment capacity from the condition when extreme fibers initially reach yield stress, σ_y, to the fully plastic condition when the entire cross-section has reached σ_y. Deflections, rotations, and strains in the fibers can be geometrically related for each level of loading.

Figure 19.5 shows a moment–curvature relationship for the cross-section. If we note the value of M_y as the bending moment when the extreme fibers first reach yield, and M_p as the bending moment when yielding occurs over the entire cross-section, the strength of a beam and the applied load may be calculated at each condition. The term elastic–plastic is used to indicate when some parts of the beam are stressed elastically while other parts are stressed into the inelastic (plastic) range.

The shape factor, $Z = M_p/M_y$, defines the added strength associated with the cross-sectional shape between the condition where first yielding occurs and the fully plastic state. The rectangular cross-section in this example is not an efficient use of material, and a wide flange shape is more practical. While the shape factor for a rectangular shape is 1.5, the shape factor for most symmetrical wide flange beams ranges between 1.1 and 1.2. The shape factor for unsymmetrical sections or other shapes must be calculated.

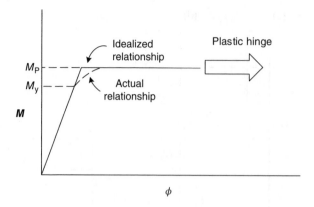

Figure 19.5 Structural steel: moment–curvature relationship.

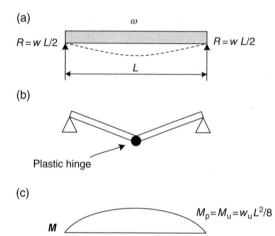

Figure 19.6 Structural steel: collapse mechanism.

Applied loads may be related to the stress conditions. For example, different stages of uniformly distributed loading of Figure 19.2 are designated as:

w = uniformly distributed load;
w_y = uniformly distributed load that first causes stress to reach yield;
w_p = uniformly distributed load that causes the first plastic hinge;
w_u = uniformly distributed load that causes collapse.

A *mechanism* occurs when too many hinges are present and the beam becomes unstable. This instability defines the limit (collapse) load for a beam. The limit for statically determinate beam becomes the load that will cause the first plastic hinge to form, as in Figure 19.6.

The flexural relationship when yielding first occurs may be calculated from:

$$\sigma_y = M_y\left(c/I\right) = M_y/S \tag{19.1a}$$

The maximum moment when a plastic hinge occurs may be calculated as:

$$\sigma_y = M_p/Z \qquad (19.1b)$$

The term S is the elastic section modulus, and Z is the plastic section modulus. Both relate to geometric dimensions (shape factor) of the cross-section.

Discussion If the problem were one of analysis, the beam size (*S* and *Z*) is known, and one can calculate the load, w_y, that will cause first yield or the load, w_p, that will cause the plastic hinge to form. In this statically determinate example, the first plastic hinge causes collapse and $w_p = w_u$. These loads may be compared with the applied dead load and expected live load to determine if failure will actually occur.

If the problem were one of design, the load, *w*, and the allowable stress, σ_a, would be known. A required elastic or plastic section modulus would be calculated and a beam having the necessary cross-section would be selected.

All calculations are based on additional conditions that the engineer must ensure are present. For example, the compression flange must be restrained from lateral buckling by construction details; loads must be applied to avoid torsion; beam dimensions must prevent flange and web failures.

Design specifications and manufactured sizes normally incorporate restrictions to ensure that these failure modes will be avoided. The engineer must recognize when construction conditions do not ensure that these conditions are present. When different conditions exist, the engineer must reduce allowable stresses, change member sizes, or change construction details.

19.14.2   Continuous Flexural Members

Many methods of framing can provide connections with full or partial moment resistance at support locations. We describe this condition in continuous beams and frames as statically indeterminate because reactions cannot be calculated by statics alone.

Statically indeterminate structures are inherently stronger than statically determinate structures, and one can calculate the differences. The useful concept from a fire safety viewpoint is that construction interactions can sometimes make a structure stronger than it may otherwise seem. Conditions that affect added strength are often observable.

Example 19.2 illustrates some concepts that are inherent in collapse loading conditions for indeterminate structures.

Example 19.2   Determine the ratio of w_y to w_u for the center span of the continuous beam shown in Figure 19.7(a).

Solution:
Figure 19.7 illustrates the concept of limit (collapse) loading for a structure. Although one can calculate all conditions, we will describe the range of behavior qualitatively.

Figure 19.7(a) shows a uniformly loaded three-span continuous beam and its deflected shape. Figure 19.7(b) shows the elastic moment diagram. The points of inflection mark locations where the curvature changes and flexural stresses change from compression to tension and vice versa. The bending moment at a point of inflection is zero. While

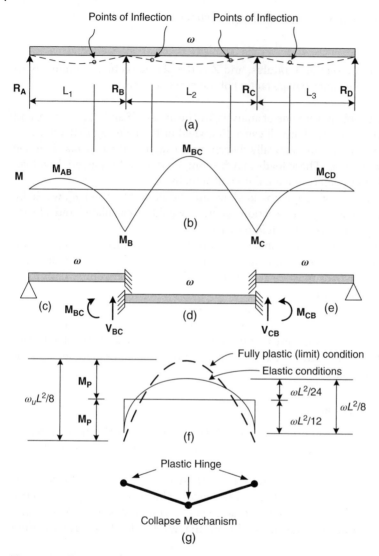

Figure 19.7 Structural steel: progressive load behavior.

these locations can be calculated, structural engineers normally envision deflections of any loaded structure to recognize how the structure resists forces.

The concept of plastic hinges and limit analysis allows one to examine ultimate loading of each span separately. This concept enables every location where continuity exists to be envisioned as a fixed connection. The end moments can be calculated easily.

For example, consider M_{BC} and M_{CB} in Figure 19.7(d). As the applied load continues to increase, these moments first reach M_y and then M_P. Bending moments can be calculated for any beam.

An internal span is represented as having two fixed supports, as illustrated by Figure 19.7(d). This segment must develop three plastic hinges to form a mechanism for collapse, as shown in Figure 19.7(g). A segment having one end fixed and the other end pinned, as shown in Figures 19.7(c) or (e), requires only two internal hinges to form a mechanism. These two types of beam support may be used to evaluate collapse load mechanisms for each individual span of any continuous beam or frame.

Figure 19.7(f) shows moment diagrams for elastic conditions up to M_y. However, a mechanism has not yet developed and additional loads can be applied to the point where a collapse mechanism appears, Figure 19.7(g). The differences in load-carrying capacity may be determined from Figure 19.7(f) as:

$$\text{First yield at ends B and C:} \quad M_y = w_y L^2 / 8$$
$$w_y = 8 M_y / L^2$$
$$\text{A collapse mechanism gives a value:} \quad 2 M_p = w_u L^2 / 8$$
$$w_u = 16 M_p / L^2$$

The development of plastic hinges is sequential up to collapse. As plastic hinges form, the beam is partially elastic and partially plastic. This condition causes internal moments (i.e. resistances) to redistribute. The redistribution of internal forces provides additional strength than would be indicated by purely elastic conditions. For the collapse conditions of Figure 19.7(g), the ratio $w_u/w_y = 2M_p/M_y = 2Z$.

Discussion This simple example illustrates that one can analyze the load-carrying capacity for both statically indeterminate and statically determinate structures. Each loading and geometrical condition requires unique calculations. The process is well developed for normal temperatures.

The advantage of plastic behavior in continuous beams may be expressed as:

Inelastic (plastic) limit load for a fixed end beam (Figure 19.7d): $w_u = 16 M_p / L^2$
Elastic limit for the beam: $w_y = 12 M_y / L^2$

$$w_u/w_y = 16 M_p / 12 M_y$$
$$\text{Ratio} \quad = 4\sigma_y Z / 3\sigma_y S$$
$$= (4/3) f$$

The ratio 4/3 indicates that moment redistribution for a fixed ended beam with uniformly distributed loading can support 33% more load at collapse (limit) than would be calculated when the elastic limit is first reached. Other types of loading and support conditions would produce different numbers. However, values for indeterminate conditions can be calculated, and the values are almost always greater than would be available for the same size beam having simple (determinate) support conditions. This phenomenon is called *moment redistribution*.

The material must remain ductile and have the capacity to allow significant inelastic deformation to form a plastic hinge. In addition, the dimensions of the cross-section must be large enough to avoid local buckling before the collapse mechanism is formed.

A region of localized plastic deformation forms near the location of maximum bending moment. The curvature is much larger at that location than elsewhere. Plastic hinges do not permit free rotation, but resist rotation with a constant moment, M_p.

An awareness concept for fire safety is that continuous (i.e. statically indeterminate) construction is inherently stronger than simple (i.e. statically determinate) construction. If one can develop observational skills to recognize the different ways that buildings are constructed, one can gain a better sense of structural performance, even without the numerical measures that structural engineers can provide.

19.15   Structures and Materials

These descriptions attempt to describe a few concepts of simple and continuous structures, inelastic actions, and moment redistribution. The goal is to form a context with which to discuss a thought process.

Structural engineers make simplifying assumptions to enable calculations to be tractable. An important part of structural engineering practice ensures that details of construction are such that the simplifying assumptions remain valid. Theory, calculations, and construction methods form an integrated package. The way a building is put together is as important, and perhaps more so, than calculations for analysis and design.

This brief description of beam analysis attempts to provide awareness that loadings, structural framing methods, resistances, and deformations are interrelated. How they are combined affects economy and performance.

Structural methods integrate all types of framing components, such as tension members, compression members, flexural members, trusses, and rigid frames. Each member and assembly can be constructed of structural steel, reinforced concrete, wood, or prestressed concrete or combinations. Each member and material has unique behavioral characteristics, engineering properties, and construction needs for effective use. The structural engineer must understand each material, form, failure mode, load and resistance characteristics, and ways to put them together in a building.

Selections of appropriate materials and configurations are based on economy, the architecture, and performance needs. The process attempts to match needs and resources efficiently.

19.16   Structural Engineering

Structural engineering revolves around engineering methods that relate external loads to the internal resistance influenced by geometry and the strength of the supporting structure. Essentially, calculations are based on analysis. Conceptually, even the design process involves a trial-and-error procedure that selects a potential solution and then analyzes its performance. Analysis and design are so intertwined that the structural engineer moves effortlessly between the two.

The analytical framework is universal. Engineers throughout the world use the same methods of analysis to relate failure modes, loads, and resistances. Mathematics, geometry, and strength of materials relate applied forces of any type or direction to

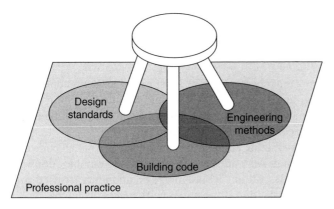

Figure 19.8 Representation: codes – standards – engineering methods – engineering practice.

internal forces that are either axial or shear in effect. The same outcomes are calculated for the same conditions. The term *structural mechanics* is often used to describe the analytical procedures.

These engineering methods are a necessary, but not sufficient, part of practice. Design standards apply factors of safety to reduce failure loads to acceptable working loads. Failure loads are the loads that cause structural collapse or conditions that produce unwanted service conditions such as building sway or deflections that cause cracks in walls and ceilings.

Separate design standards are developed for each major building construction material as well as for common products (e.g. open web joists). Each standard integrates a complete spectrum of possible failure conditions and their interactions into the requirements. Sometimes, design standards are based on different theories (e.g. ASD, USD, LRFD). The structural engineer selects the most appropriate standard for the application.

A third leg is identification of loading conditions to which the structure may be expected to support. Loading conditions address both individual members and the holistic structure. Each member is designed for dead load, live load, and load application conditions caused by the method of framing. Dead load is the weight of the structure itself. The live load is normally provided by local building codes with examination to ensure that larger loads are not needed for the specific building use. Building codes identify not only gravity live loads, but also wind, earthquake, snow and ice loads, and load combinations.

Figure 19.8 illustrates conceptual relationships between engineering methods, design standards, and building codes. Each has an independent role, and all are interrelated with careful coordination by the profession. Engineering methods are universal. Design standards are consistent for construction materials and geographical location. There are several recognized design standards-making organizations throughout the world. Building codes reflect local needs and cultures.

19.17   Structural Engineering and Building Design

Structural engineering has a different role for each phase of building design. The schematic phase is the most creative part of the process. The structural engineer must satisfy architectural needs economically. This involves experience and judgment to examine alternative ways to frame the structure. Architectural compatibility, materials choices and combinations, geometry, construction methods, and economy all interact. Often, calculations involve approximations and overall structural response, rather than detailed analysis.

A building design evolves during the schematic phase. Most major decisions concerning architecture, structural framing, and construction materials selection become established. Normally, only limited quantification is needed to establish critical dimensions such as floor thickness, column locations, and building stability.

The detailed design phase involves sizing all members and ensuring that all code and standard requirements have been met. All load types, directions, and combinations are examined for the various failure modes. Often, preassembled components are used, and the engineer must ensure that their selection causes no unanticipated problems.

The construction documentation phase finalizes dimensional checks and ensures that construction details and specifications agree with the design assumptions. This phase also specifies any instructions with which a contractor should be aware during construction.

Calculations are necessary, but not sufficient, for structural engineering practice. Often, the calculations validate many of the preliminary estimates involving strength and economy. Both individual and holistic performance is integrated in the overall construction. Structural engineers often develop a "sixth sense" for recognizing potential problem areas in a complicated structure.

PART FOUR: STRUCTURAL ANALYSIS FOR FIRE CONDITIONS

19.18 Introduction

The structural engineer is in responsible charge of structural safety. The principal "tools of the trade" are a blend of engineering methods, design standards, and building code compliance. Professional practice applies state-of-the-art engineering procedures to produce economical, functional designs.

Usual practice involves structural performance at normal temperatures. However, during the past half-century, enormous progress has been made in understanding structural behavior in building fires. Although engineering calculations for fire conditions are gaining greater acceptance, one cannot expect all structural engineers to extend structural analysis from normal temperatures to fire conditions. Nevertheless, knowledge and design standards are progressing rapidly, and one may anticipate future structural engineers to become increasingly more skilled in structural fire analysis and design applications.

Today, structural fire design is provided in two ways. The first is the traditional approach using standard fire testing and time-related acceptance classifications. The second extends routine structural engineering to design for fire conditions.

Traditional code compliance is the most common way to incorporate structural fire resistance. Here, the building code rather than an engineering professional assumes the responsibility for performance. The building designer need only specify the appropriate fire endurance classifications relating to occupancy, height, and area. The structural engineer includes the dead load of the fire protection method into structural member design. Otherwise, neither the design team nor the authority having jurisdiction has any responsibility for the fire performance of structural members. ASCE Standard 29 (Structural Design for Fire Conditions) provides ways to calculate fire resistance rating times that are equivalent to results of the standard fire test.

The second approach uses engineering procedures to predict structural performance. The process combines structural mechanics knowledge, heat transfer analysis, engineering properties for elevated temperatures, and room fire analysis into a cohesive framework for analysis and design. Additional research will eventually expand the range of knowledge to one similar to current practices for structural engineering at normal temperatures.

One may describe contemporary structural knowledge for fire conditions as analogous to that of structural engineering for normal temperatures about World War II. Thus, we may infer that the state of knowledge for structural fire design has reached a level of early maturity. Current practices can provide reasonable knowledge of engineering performance, although additional detailed information will emerge.

Here, we identify concepts of structural performance analysis for fire conditions to provide an awareness of the process and general uses with fire safety engineering.

19.19 Outcomes

As a preliminary concept, let us illustrate a characteristic outcome for analytical capabilities. Figure 19.9 shows a representative relationship for the performance of a flexural member. Knowledge of the dimensions, construction materials and support conditions,

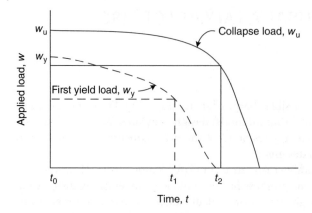

Figure 19.9 Structural fire performance.

fire protection insulation details, and the diagnostic fire enables one to calculate a time-related load capacity for any time during the fire.

The member's load capacity for initial yield, w_y, and ultimate capacity, w_u, can be calculated for normal temperatures. These values are shown at (local) time t_0 for the start of the compartment's environmental changes from normal to fire conditions. The Interactive Performance Information (IPI) chart relates the local time, t_L, for the specific room fire to the global time, t_G from established burning in the room of origin.

The structural strength begins to decrease as the heat of the fire penetrates the insulation and raises temperatures in the member. Decreases in strength are based on changes of mechanical properties with temperature increases. Thus, one can calculate the load-carrying capacity at any time during the diagnostic fire duration. The capacity can be related to the ultimate (also called limit) load that will cause collapse. It can also be associated with conditions that will cause initial yielding that would produce permanent deformation after the fire is extinguished.

After calculating a performance graph, such as that represented in Figure 19.9, one can estimate likely outcomes for different conditions. For example, the time vs. ultimate (limit) capability enables one to recognize collapse conditions at any time, t_2, during the fire. Similarly, examination of the yield capacity at any time, t_1, would identify expectations regarding the need to replace a beam after the fire is extinguished.

A performance graph enables one to understand the effect of different loading conditions. For example, one may predict performance considering the design live load, or the expected live load for movable or immovable live loads, or for the dead load alone. One may even examine other loading conditions of interest, such as transitory live loads or water accumulation from above. If interested, one could also calculate deflections in relation to the time and different loading conditions.

19.20   Pause for Discussion

Structural capabilities, such as those illustrated by Figure 19.9, are based on calculations using structural mechanics, heat transfer through any fire protection insulation, and any diagnostic fire. One may examine flexural members, compression members, tension members, composite systems, and trusses. Additionally, the structural capacity for stress combinations such as in beam-columns in rigid frames may be analyzed. The process can describe performance for structural steel, reinforced concrete, wood, and prestressed concrete.

A performance graph can incorporate a variety of recognized situations that affect structural performance. Even though all required information is not yet available, enough is known to produce reasonable estimates. For example, if column expansion were constrained because of the weight of upper floors or other construction features, the additional thermal stresses would reduce column capacity. Additional capacity reduction occurs for eccentrically loaded columns such as when thermally induced forces cause lateral movement. These strength reductions may be calculated to provide credible estimates.

Structural mechanics can calculate the load-bearing capacity for conditions at normal temperatures. Theory and numerical methods can provide a relatively good understanding of what to expect from recognized building situations during fire conditions.

The loading values may be related to the environment caused by any diagnostic fire. Because the building community has a long tradition with the standard test fire, this time–temperature relationship could be used as the diagnostic fire. Load characteristics at yield and ultimate conditions can be calculated for any interval of time. This enables physical conditions that cannot be replicated by the standard fire test to be used. Other diagnostic fires may be selected to examine more realistic fire conditions. These outcomes can be compared with those that would be expected from a standard test fire. The level of understanding of fire performance for buildings has expanded significantly with modern analytical capabilities.

Although structural analysis for fire conditions does not have the knowledge base of that for normal temperatures, there is enough information to enable one to provide reasonable fire performance estimates. Because any performance curve may be directly transferred to corresponding cells of an IPI chart, one can relate time and structural expectations with other parts of holistic building performance.

19.21 The Process

A structural fire analysis can produce a performance graph analogous to that of Figure 19.9 for any structural member or assembly. An overview of the process involves the following activities:

1) Complete the structural design for normal temperatures and conditions. This activity identifies the structural systems, materials, assemblies, and details of construction for the building. Determine the appropriate member or members to represent structural performance.
2) Select the type of fire protection to insulate the structural member or members and to be compatible with the building's architecture. No fire protection is also an option that may be considered, if useful.
3) Select the diagnostic (or design) fire.
4) Calculate the transient load capability of the structural member for the diagnostic fire. Calculations involve heat transfer from the fire to and through the protective insulation system to the parts of the member that affect its strength. Heat transfer values and mechanical properties of materials over the range of temperatures are the basis of calculation.

Structural mechanics is that branch of engineering that examines structural behavior due to the application of external loads. External forces cause internal forces (stresses) and deformations. Failure modes and instability are an integral part of analysis. Techniques are sophisticated and extensive. Essentially, one may describe structural mechanics as the mathematical and physical analysis of structural behavior.

Structural analysis for fire conditions extends mechanical theory at normal temperatures to predict performance at elevated temperatures. The temperature that the structural element "feels" at any time is central to the analysis. Mechanical properties at elevated temperatures become the basis for calculations.

Heat transfer theory and practices provide the means to calculate temperature changes in structural members. The heat energy moves from a fire through protective insulation systems into and through the structural elements.

Fire environment identifies the fire conditions that affect structural performance. We call this the diagnostic fire. It is also called the design fire for structural design.

Transient load capacity and deflection relationships can be calculated for a specific diagnostic (or design) fire. Theory and calculation procedures for structural mechanics and heat transfer are well developed. Computational capabilities provide ways to determine structural relationships for the transient conditions.

The material properties at elevated temperatures are not yet completely defined for the complete range of temperature variation. However, properties for insulating materials and structural materials are known well enough to provide confidence in calculation estimates. Extensive research in structural mechanics over many years has provided a depth of understanding for performance at normal temperatures. This level of knowledge has not yet been acquired for the range of materials and performance at elevated temperatures.

19.22   Structural Mechanics

Structural mechanics relates external forces to internal resistances and potential failure modes for all types of structures and machines. Although basic theory uses homogeneous, isotropic materials (e.g. steel) to develop relationships, the subject includes all materials and composites.

The principles are based on calculating forces and deformations with regard to structural outcomes as loads increase from small values to failure. The forces and resulting deformations relate to tension, compression, and shear.

Stress is the normalization of force over an area of application. Normal (tension or compression) stress is $\sigma = F/dA$ and shear stress is $\tau = F/dA$. The symbols σ and τ indicate the relationship between the force (normal or tangential) and the area over which it acts.

Figure 19.10 shows the effect of temperature on the σ–ε relationship for structural steel. At elevated temperatures, the σ–ε diagram becomes softly rounded with no distinct yield point. Common practice defines the yield point as the intersection of the actual σ–ε diagram with a line parallel to the initial tangent, but offset by 0.2% ($\varepsilon = 0.002$).

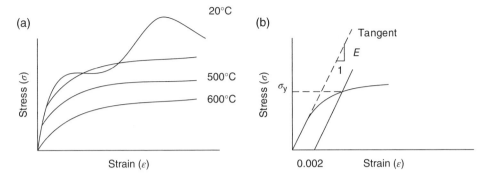

Figure 19.10 Structural steel at elevated temperatures.

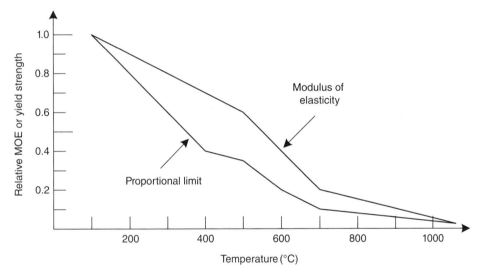

Figure 19.11 Structural steel: mechanical properties at elevated temperatures. MOE, modulus of elasticity. *Source*: Reproduced with permission of CEN.

Graphs of yield stress, σ_y, and modulus of elasticity, E, vs. temperature are shown in Figure 19.11.

Creep describes a condition that produces additional deformation with no increase in load (or stress). Creep is accelerated at elevated temperatures and becomes an important factor in calculations for fire conditions.

The free body diagram enables one to combine types and directions of force and the associated stress. Experimental methods relate stress, σ, and strain, ε, in materials as loads increase from small values to rupture. Fundamentally, these simple concepts become the basis for an enormous field of study.

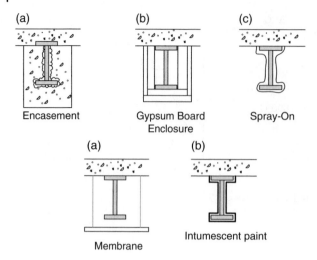

Figure 19.12 Structural steel insulation methods.

19.23   Protection Methods

Structural steel is the backbone of building construction. Although its strength at normal temperatures is excellent, load-carrying capacity decreases rapidly at elevated temperatures. A variety of methods insulate steel from the heat of a fire. The principal methods are: (1) encasement; (2) enclosure by lath and plaster or mechanically fastened gypsum boards; (3) direct applied spray-on materials; (4) membrane protection; and (5) intumescent mastics or paints. Figure 19.12 illustrates these methods.

19.24   Diagnostic Fire

The environment created by the diagnostic fire provides the time–temperature relationship that affects structural performance. Heat from the diagnostic fire is transferred through the protective insulation to the structural member.

Until building regulations establish appropriate time–temperature relationships, the fire safety engineer (or the profession) must identify an appropriate diagnostic fire on which to base structural performance. Figure 19.13 illustrates a few possible considerations.

One could select a relationship that incorporates fire suppression. However, more valuable informative results can be obtained for a fire that is not extinguished before burnout. This enables one to use the results in a variety of different ways, one of which could include early extinguishment. However, understanding what to expect from a fire that is not extinguished provides better understanding of structural performance.

The diagnostic (or design) fire is an important part of a performance analysis. Consequently, identification of regulatory-based diagnostic (design) fires is important for the transition from prescriptive to performance-based design. In the absence of this regulatory guidance, one must rely on professional judgment. Comparisons with the standard time–temperature relationship can provide a base for understanding.

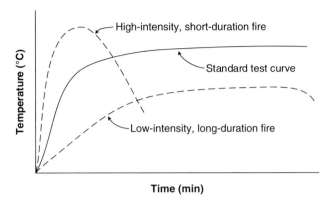

Figure 19.13 Diagnostic fires and structural analysis.

19.25 Heat Transfer

Heat from the diagnostic fire is transferred to the structural member by convection, radiation, and conduction. Each path from fire to structural member is traced through the specific combination of parts that make up the insulation system.

Figure 19.14 illustrates the heat transfer from a fire to a structural member through enclosure protection. Heat is transferred by convection and radiation from the fire, A, to the surface of the gypsum board, B. Heat transfer coefficients lower the temperature at the outside surface of the gypsum board, B. Conduction transfers heat through the board to inside surface, C. Some heat energy raises the temperature of the gypsum board. Radiation and convection transfer heat from C to the structural member, D.

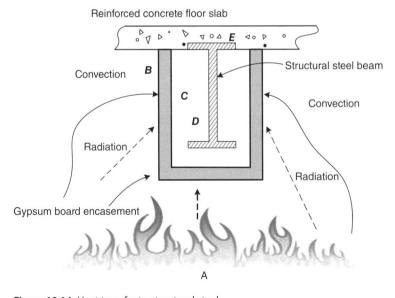

Figure 19.14 Heat transfer to structural steel.

The section factor (A_s/V_s) is the ratio of surface area exposed to heat transfer and the volume of the steel section (per unit length). The value for A_s is associated with the dimensions of the member and methods of construction. Members having less surface area available for heat transfer or large mass (i.e. smaller ratios of A_s/V_s) will experience a lower temperature rise for a given amount of energy than a member having a larger A_s/V_s.

Conduction moves the heat energy through the structural member, causing its temperature to rise. Some of this heat is transferred to other connecting members as well as to the concrete slab at E. This heat movement produces uneven temperatures throughout the beam cross-section.

19.26   Structural Performance

The time–temperature relationship of the fire becomes the base for calculations. The heat transfer from the fire to the member through the protection system produces transient structural temperatures. Thus, the temperature within the structural element may be determined for every increment of time.

Procedures for calculating the strength and stiffness of structural steel components at elevated temperatures are essentially the same as those for normal temperatures. However, the mechanical properties of the steel reflect the values at the elevated temperatures felt by the member. Structural mechanics theory enables one to calculate the deformation and associated stress at any location within the member. Actions such as creep and axial restraint become part of the analysis. Detailed calculations may be done with digital computers. Approximate spreadsheet solutions using lumped mass analysis may also be used to develop understanding of component changes as the fire progresses.

A structural analysis enables one to calculate performance relationships such as that represented by Figure 19.9. Even though a full range of experimental values is not yet complete, enough information is available to provide reasonable estimates of performance.

19.27   Reinforced Concrete

Concrete is a mixture of cement, sand, aggregate, and water. Because the material is strong in compression and weak in tension, steel bars carry tensile forces in structural applications. Combining structural mechanics with experimental physical properties enables one to calculate structural strength and deformation for a wide range of applications. The ability to predict performance in normal temperatures is clearly defined. Calculations of strength and deformation expectations at elevated temperatures are adequate for many applications, and continued research is rapidly expanding this knowledge.

Reinforced concrete structural analysis for fire conditions can develop load and deformation relations analogous to that of Figure 19.9. Here we describe a few of the major factors that affect performance. These factors provide an awareness of structural concepts for calculating load and deformation performance for a diagnostic fire.

19.28 Mechanical Properties

The mechanical properties are determined by compression testing of concrete specimens (usually standard size cylinders). The load vs. deformation describes the σ–ε relationship. Figure 19.15 illustrates the relative diagrams for the structural steel and plain concrete that make up the reinforced concrete composite.

Figure 19.16(a) shows the σ–ε behavior of concrete at elevated temperatures. The symbol f_c is normally used for concrete design. The decrease in ultimate strength, f_c', and modulus of elasticity at elevated temperatures are shown in Figure 19.16(b).

Concrete can be tailored to provide the strength specified for the structural design. Normal strength concrete (NSC) refers to strengths between 2500 and 6000 psi (17.2 and 41.4 N/mm²). High-strength concrete (HSC) provides an ultimate strength greater than 6000 psi (41.4 N/mm²).

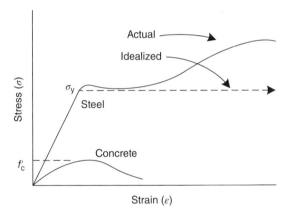

Figure 19.15 Mechanical properties for structural steel and concrete.

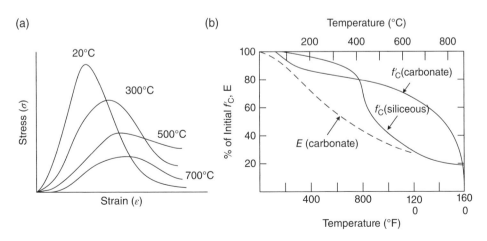

Figure 19.16 Concrete: mechanical properties. *Source:* Reproduced with the permission of Portland Cement Association.

Concrete does not harden or cure by drying. Rather, the cement undergoes a chemical reaction using water to hydrate and harden. Although concrete may appear to be "dry" after it cures, it contains hydrated water to maintain its desired strength. During a fire, this hydrated water moves from warmer to cooler sections. The drying effect causes the cement to lose strength.

A large number of variables affect the behavior of concrete at both normal and elevated temperatures. The properties in Figure 19.16 are shown in general terms or as a percentage of values at normal temperatures to illustrate general behavior. Specific $f_c'-\varepsilon$ diagrams may be determined for the specific concrete conditions used in the structural member.

The types of variables that may affect mechanical and thermal properties of concrete include:

- Cement types – five types of Portland cement are available to tailor the needs of the structure to the chemical environment or structural needs.
- Design strength of the concrete.
- Types of aggregate (siliceous, carbonate, or lightweight).
- Water–cement ratio.
- Curing time and conditions.

Although many variables affect the strength of concrete at normal temperatures, additional factors influence values at elevated temperatures. Among the more important features are the following:

- Concrete has a lower thermal conductivity and higher specific heat than metals. Hydrated water moves through the concrete pores because of the temperature gradient.
- Elevated temperatures lower the compressive strength and reduce the modulus of elasticity. Loss of compressive strength begins at low temperatures and becomes significant at 400°C for siliceous aggregates and slightly higher for carbonate and lightweight aggregates. Most compressive strength is lost at 800°C. A value of 500°C is considered as a limit for practical purposes. Loss in strength is never completely regained after the fire is extinguished.
- Spalling and sloughing refer to a disintegration or falling away of part of the concrete surface. These events reduce strength by exposing the reinforcing steel or the interior regions of the concrete to higher temperatures. In a fire, a build-up of steam causes spalling from expanding water that cannot move fast enough. Explosive spalling is a critical concern for high-strength concrete.
- Creep affects performance at elevated temperatures. The influence of creep is greater after 400°C. Moisture movement, dehydration, and a loss of bonding accelerate the process. Higher stress levels cause increases in creep.
- The rate of coefficient of thermal expansion increases as temperature rises. This produces lateral movement in flexural members and vertical movement in columns. Either of these can affect load-carrying capacity.

19.29 Flexural Members in Reinforced Concrete

The floor slab and structural frame for reinforced concrete are normally constructed monolithically. That is, the forms and reinforcing bars are placed and the concrete is poured as a single unit. Although exceptions may occur, this form of construction provides continuity (i.e. statical indeterminacy) to enable loads to be redistributed when deformations increase.

Figure 19.17 shows a three-span continuous member, similar to that of the steel member in Figure 19.8. The heavy dashed lines in Figure 19.17(a) signify reinforcing steel that is placed in regions of flexural tension. The bending moment, M, can be calculated at any location using the concrete stress block and tensile reinforcing bar stress, as represented by Figures 19.17(c) and (d).

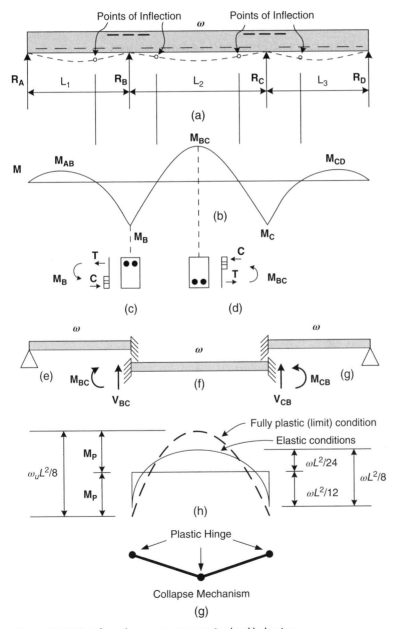

Figure 19.17 Reinforced concrete: progressive load behavior.

Concrete structures are designed to be ductile for normal temperature conditions. That is, failure is caused by inelastic tensile elongation of the reinforcing bars. Increasing rotation causes increased concrete stress, leading to failure.

When a part of the continuous beam deforms too much, loads are transferred in a manner similar to that of structural steel. This moment redistribution continues until beam failure occurs.

Structural mechanics, construction methods, and structural engineering decisions with regard to strength and other factors that affect performance provide the basic information to calculate collapse loads as well as loads that produce intermediate conditions.

19.30 Concrete Members at Elevated Temperatures

The analysis of a reinforced concrete member is similar to that discussed earlier for structural steel. A diagnostic fire is selected. Heat transfer methods determine the temperature at the surface of the member. Then, heat transfer methods determine the transient temperature changes through the cross-section. This produces isotherm contours as shown in Figure 19.18.

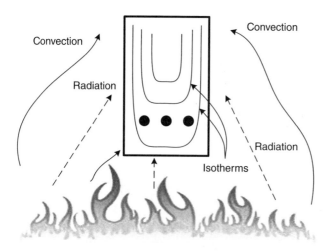

Figure 19.18 Heat transfer in reinforced concrete.

The temperature of the structural steel enables one to determine its stress and deformation. Similarly, the temperature at any isotherm enables one to determine the theoretical stress in the concrete at that location. Only compressive stresses may be used for moment resistance, and these values depend on the temperature. However, approximations can provide a reasonable estimate of load-carrying capacity to produce a representation analogous to that of Figure 19.9.

19.31 Pause for Discussion

The structural analysis for fire conditions discussed above is intended to make the fire safety engineer aware of certain structural concepts. There are tools for analysis and one can calculate load capabilities for any diagnostic fire. Nevertheless, the complexity requires computer analyses to calculate performance. Approximate methods are available to give reasonable estimates. Approximations are often adequate to initially assess whether a problem exists. A more detailed analysis can then be done, if necessary.

In addition to flexural members, column and rigid frame performance can also be calculated. Similar to the discussion for structural steel, one should recognize that the state of the art has reached early maturity. Although there are many analytical capabilities, much more will be done to improve techniques, data, information, and understanding. Nevertheless, a structural engineer who has developed capabilities in structural analysis for fire conditions is able to give reasonable estimates of performance.

19.32   Other Materials

The discussion to this point has described only structural steel and reinforced concrete. However, analytical tools exist for other materials, such as prestressed concrete and wood. In addition, structural assemblies, such as trusses and composite construction, have been studied and performance can be estimated. The capability for structural analysis to analyze a range of construction practices is expanding rapidly.

19.33   Summary

Structural engineering for normal temperatures is performance-based. Broad and sophisticated techniques exist to enable the structural engineer to understand how a structure will perform under most types of loading. Engineering methods are largely defined by structural mechanics, which is a mature engineering field. Design standards integrate engineering methods with serviceability and safety requirements. Building codes identify applied loads and load combinations. As a package, structural engineering practice has reached maturity.

Structural analysis and design for fire conditions are built on engineering mechanics and heat transfer methods. This enables one to examine structural elements and develop a good understanding of what to expect during a fire. Although a good understanding of what one may expect for a holistic response to element failure exists, mathematical techniques are still in the elementary phase. Nevertheless, one can develop an insight into both macro- and micro-structural performance during a building fire. As additional data and calculation techniques build on structural mechanics, our knowledge will expand.

20

Target Spaces and Smoke

20.1 Introduction

The characterization of risk from smoke and toxic gases to people, property, and operational continuity is an important part of a performance analysis. The identification of windows of time during which critical building spaces remain tenable is fundamental to understanding and managing risk.

Time links all parts of a building fire analysis and it is a central focus of risk characterizations. The performance measure for risk is the window of time that a target room will remain tenable.

Humans and critical operational equipment can survive untenable conditions for limited periods of time. The time duration depends on the human condition and the severity of the environmental contamination.

Performance-based risk characterizations do not quantify human death or injury due to exposure to the products of combustion of a hostile fire. Rather, risk characterizations are associated with the window of time that a target space provides tenable environmental conditions.

Chapter 9 discussed diagnostic fires and natural smoke movement in buildings. This chapter focuses on target room evaluations and Chapter 32 describes the framework for analysis. Chapter 22 discusses risk characterizations for humans or data and equipment exposed to untenable conditions.

20.2 Orientation

Figure 20.1 defines the major parts of building fire performance. An Interactive Performance Information (IPI) chart tracks component changes over time due to dynamic fire conditions and the phasing in and phasing out of components that affect behavior. Target space tenability analysis (Component 4B) combines the composite fire with building architecture and other features that affect smoke transport beyond the area of fire origin.

Figure 20.2 represents the physical separation of the fire origin and the target room. A target space analysis is based on:

- The target room location relative to the room of fire origin.
- The definition of tenability in the target room.

Fire Performance Analysis for Buildings, Second Edition. Robert W. Fitzgerald and Brian J. Meacham.
© 2017 John Wiley & Sons Ltd. Published 2017 by John Wiley & Sons Ltd.

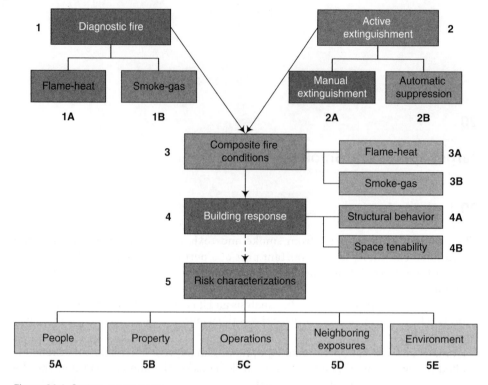

Figure 20.1 System components.

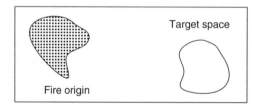

Figure 20.2 Fire and target space separation.

- The three-dimensional array of barrier–space modules that define the paths of smoke movement from the fire to the target room. The modules include the status of barrier openings that affect movement forces.
- The diagnostic fire, smoke characteristics, and natural smoke movement.
- The type and status of openings in the target room.
- Interactions of mechanical equipment, ventilation, and barrier effectiveness that affect smoke movement and concentrations in the target room.
- Fire extinguishment and its effect on the composite fire and its smoke generation and movement.

20.3   Tenability Measures for Humans

A fire produces heat, soot particles of different sizes, water, and a variety of gases, none of which are good for human health. In addition, these products can damage the operation of some equipment and data storage. The definition of human tenability measures and the basis for selection are necessary for performance evaluations.

A wide variety of toxic gases accompany the combustion process and are transported by the same air that carries soot particles that reduce visibility in smoke. Carbon monoxide and carbon dioxide are produced in all fires. Depending on the fuel, many other gases that damage health are present in fires.

Fire tests can measure toxic gases from building and content materials. Toxicity measures and performance criteria for buildings have not been identified. While lethal concentrations for some gases can be established for laboratory animals, levels of exposure doses (including the cocktail effect of gas mixture synergism) that produce human impairment are not known. Although one can recognize that toxic gases are bad for humans, we currently have no measure to quantify performance for gaseous toxicity.

Smoke visibility reduction usually precedes the effects of toxic gases on humans, often by a substantial time. Occupants and building managers can readily understand visibility reduction due to smoke from a fire. In addition, it is possible to roughly approximate visibility in building fires. Consequently, if we establish tenability criteria on visibility distance in smoke, the results will be conservative and one can quantify a building's performance.

Visibility can be affected by the density of soot particles that constitute smoke, or by irritants that sting and cause the eyes to tear. It is possible to have low visibility due to dense smoke and not have eye irritation. Conversely, it is possible to have light smoke with long ocular visibility and be unable to see because the eyes close due to the stinging irritants. Ocular visibility is a more practical measure because we do not know how to estimate eye irritants.

Both normal ambient light and background lighting affect visibility. When all lights are out and the building is dark, one cannot see into the distance whether or not smoke is present. Therefore, we base performance visibility distance through smoke on conditions of normal interior lighting.

This visibility criterion does not identify whether an individual will be hospitalized or killed by the products of combustion. However, the criterion does identify a smokiness level that can quantify performance. Thus, the human tenability criteria may be stated as the ability of an individual to see x m (y ft) in a building fire under normal interior lighting conditions.

Humans can tolerate a mild increase in temperature and some smokiness and toxicity for a short time during building egress or while remaining in a room. Tenability selection will reflect differing life safety needs and may vary with factors that are unique to the building.

Values for an occupant in a defend-in-place room will differ from values for an individual in transit during a building evacuation. A building with uncomplicated circulation patterns may differ from one with a complicated architectural plan where direction-finding indicators are more difficult to recognize. The tenability criteria may change for occupants who are unfamiliar with the building or who may have physical or psychological impairments.

A tenability selection will consider factors such as:

- Building height, floor size, and architectural layout.
- Number of occupants and initial location of a representative occupant.
- The number and complexity of potential egress routes.
- Familiarity of occupants with the building architectural layout.
- Occupant age, mobility, physical and mental condition, and expected activity for the fire scenario.
- Time of day and day of week.
- Pre-fire instructions, education, and training.

A responsible individual must select appropriate values for tenability. Some codes have identified visibility distances ranging from 2 to 10 m (6–33 ft) and different researchers recommend values between 3 and 25 m (10–80 ft) for various conditions.

Some smoke is common and expected in building fires. Light smoke conditions do not generally cause problems. However, too much smoke in the target spaces is unacceptable. Smoke tenability defines the question of how much is too much.

20.4   Visibility in Smoke

The performance measure for smoke tenability is the onset of a defined visibility level.

Visibility is related to soot yield and optical density which can be predicted by computer modeling. A few useful relationships are identified briefly in this section.

The extinction coefficient, which describes the smoke's ability to absorb or scatter light, depends on the optical properties and concentration of soot. Beer's Law predicts the attenuation of light by an absorbing or scattering media. This is expressed as,

$$I_L = I_O e^{-kL} \tag{20.1}$$

where:

I_L = intensity of a light beam after traveling a path of length L;
I_O = original intensity of a light beam;
k = spectral extinction coefficient;
L = path length.

Equation (20.1) can be converted from base e to base 10 using:

$$I_L/I_O = 10^{-DL} \tag{20.2}$$

where D is the optical density. Optical density (OD) can be related to visibility using the extinction coefficient:

$$k = D/2.3 \tag{20.3}$$

Visibility through smoke is the maximum distance at which an individual can see an object. The visibility distance times the extinction coefficient is a constant:

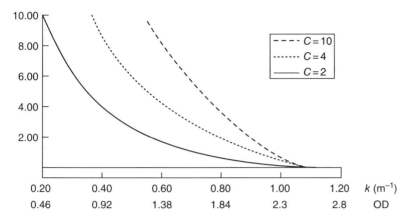

Figure 20.3 Optical density and visibility.

$$Lk = C$$
$$(k < 0.25)$$
(20.4)

or:

$$Lk = C(0.133 - 1.47\ln k)$$
$$(0.25 < k < 1.1)$$
(20.5)

The constant C varies depending on the brightness of the object being viewed. Here are values of C for three types of object:

- Illuminated sign: C varies from 5 to 10
- Reflective sign: C varies from 2 to 4
- Walls, floors, and doors: $C \approx 2$.

Figure 20.3 shows the relationship of visibility distances and optical density using these equations. The value of optical density from a computer program can be converted into an extinction coefficient and then into visibility distances.

20.5   Equipment and Data Storage

Although life safety is the usual focus of attention, fire gases can affect equipment, instrument functionality, and data storage. It is difficult to identify the particular corrosive products and exposures that affect performance. In addition, it is difficult to measure the corrosive products and estimate their residual quantity in rooms distant from the fire origin.

Equipment manufacturers are the best source to identify the effect of corrosive gases on their products. Fire tests can indicate the release of these gases during combustion. If corrosive gases are released, they will be fellow travelers with smoke. If one models the corrosive gases of interest as a mixture with smoke, the onset of a specified visibility level might define equipment tenability. Although correlations may be weak, they do provide a way to describe tenability for equipment or data storage.

20.6   Overview of Target Space Analysis

Target space analysis for smoke contamination is dependent on scenario conditions. For example, the status of door openings and the use of space pressurization can affect time and target space tenability. Quantitative, qualitative, and variability analyses all play different roles in smoke analyses. Thoughtful coordination of their uses can narrow analytical choices and enhance understanding.

The procedure for evaluating the window of time for target room tenability is as follows:

1) Select the diagnostic fire to describe flame-heat and smoke-gas generation.
2) Identify the target room.
3) Construct barrier–space modular networks to define smoke transport paths from the fire to the target room.
4) Identify the scenario conditions and status of barrier openings along the paths of smoke transport.
5) Define tenability.
6) Evaluate the following single-value questions at important time intervals:

- **Given:**
  - *Diagnostic fire is defined*
  - *Target room is identified*
  - *Building geometry and natural forces that affect smoke movement are known*
  - *Mechanical and intentional ventilation forces that affect smoke transport to the target room are known*
  - *Time from EB _____ minutes.*
- **Question:** *Will enough smoke reach the target room boundary to make the target room untenable (Bs)?*

- **Given:**
  - *Diagnostic fire is defined*
  - *Target room is identified*
  - *Enough smoke reaches the target room boundary to make the room untenable (Bs)*
  - *Mechanical and intentional ventilation forces that affect smoke transport in the target room are known*
  - *Time from EB _____ minutes.*
- **Question:** *Will enough smoke enter and remain in the target room to make the target room untenable (Es)?*

An IPI chart (Section 4B) enables one to recognize the windows of time for a target room or a combination of target rooms to remain tenable. The IPI chart also enables one to compare results for different scenarios or conditions to gain a better understanding of performance. Scenario examinations normally include:

1) The diagnostic fire and smoke movement as a separate analysis to provide a basis for comparison.
2) The diagnostic fire with mechanical pressurization and/or exhaust.
3) The diagnostic fire with manual ventilation.
4) The composite fire that combines the diagnostic fire and automatic sprinkler suppression.
5) The composite fire that combines the diagnostic fire and manual suppression from first water application to extinguishment.

An initial scenario examination involving only the diagnostic fire and natural smoke movement provides a foundation for understanding the building's response for smoke tenability. This understanding of natural smoke movement can be combined with qualitative estimates and variability scenarios to identify suitable scenarios for quantitative analyses.

Fire extinguishment can be important for smoke tenability. When first water is applied by the automatic sprinkler system (AA) or the fire department (MA), the natural fire is upset and smoke and burning conditions change. Although smoke generation may initially increase with water application, eventually it diminishes and stops when the fire is extinguished.

20.7 Target Rooms

A target room is remote from the fire origin. Humans may use a target room for transient movement during egress. A series of target rooms may be linked as a means of egress or grouped for operational continuity importance.

Occupancies such as hospitals, nursing homes, and prisons often expect occupants to be defended in place rather than removed from the building. In these cases, each occupant room becomes a target room.

Some buildings have a designed area of refuge when conditions for leaving the building are impracticable. The rooms and the path to reach the area of refuge form a series of target rooms.

Many commercial and industrial buildings have specific rooms that are critical for operational continuity or mission. For example, data storage, accounts receivable, or communication centers often have high importance to a firm's operations. Rooms that contain data or items of equipment that are sensitive to products of combustion are target rooms for risk characterizations.

Although a building may have many rooms, each target room may be isolated for specific attention. Non-critical rooms may be grouped into a "lumped" segment to simplify analysis.

20.8 Barrier Effectiveness

Barrier–space modules were described in Section 8.9. We construct modular chains that describe the paths for smoke movement from the fire origin to a target room. An adjacency matrix may be used to describe room-to-room connectivity and help to identify smoke movement paths for complex architectural plans. The matrix can be simplified to use only the barrier openings that significantly affect smoke movement.

Each module has one or more openings. The size and location of penetrations affect smoke and air movement through barriers. Although barrier leakage and cracks are important, large openings such as doors, vertical shafts, and large ducts dominate initial smoke movement. These openings are used to provide an initial estimate of time durations for smoke to move through modular paths to reach rooms of interest. Because smoke and air move through unpenetrated barriers, path identification can be simplified.

A potentially different scenario is possible for each barrier opening and status. Fortunately, the modular network is valid regardless of the status of openings. The potentially large number of scenarios becomes manageable when preliminary qualitative estimates identify opening conditions that can establish "good" and "bad" outcomes. These qualitative "what if" variability analyses provide guidance for selection of an appropriate scenario to use with quantitative methods.

20.9   Mechanical Pressurization

Mechanical air-handling equipment can pressurize or exhaust selected building spaces. The influence and reliability of mechanical air-handling equipment on smoke movement are part of smoke tenability evaluations.

During a fire, an air-handling system may:

- continue normal operation
- shut down
- change to a defined emergency mode of operation.

If air-handling controls of a heating, ventilation, and air-conditioning (HVAC) system continue in normal operation after a fire starts in a space, the smoke will spread throughout the zone. The rapid smoke spread may cause anxiety among occupants and reduce visibility in rooms far removed from the fire. This smoke can also cause toxic reactions and increase the likelihood that the rooms will become untenable more quickly.

The major role for air duct smoke detectors is to close dampers so that smoky air cannot be forced into rooms. Air systems can also feed fresh air to the fire, causing it to burn faster with greater intensity.

Forced air movement is avoided when HVAC systems shut down during a fire. However, shutting down a system does not prevent ducts from transporting smoke to these rooms if dampers are not present or inoperative. Smoke may be forced through the supply and return air ducts or plenum spaces by fire gas buoyancy, the stack effect, and wind pressures.

The third condition for the HVAC system is that of emergency operation. It is possible to install fans to influence air flow. These fans can be programmed to place some rooms in an exhaust mode and others in a pressurization mode. Operations of this type are described as smoke management systems.

Heating, ventilation, and air-conditioning emergency operations can affect target space tenability. Pressure differentials created by mechanical fans can prevent smoke from entering into a space. When a pressure differential is combined with a physical barrier, a target space may remain tenable for a long time. The time during which doors may be open influences target space tenability.

20.10   Fire Department Ventilation

Fire department ventilation operations can affect smoke tenability in selected spaces. The usual function of ventilation is to raise the smoke level in fire areas so that suppression forces can find the fire and extinguish it more easily. Ventilation can also help in manual fire extinguishment and smoke removal.

Intentional fire department actions or unintentional fire venting can cause vent openings. Heat and smoke release, and pressure differential changes within the building can affect smoke and heat conditions in building spaces. While these influences are rarely incorporated into a smoke tenability evaluation, a series of variation analyses could describe important "what if" scenarios.

20.11   Summary

Forces that affect air movement cause smoke to move through paths from the fire origin to other parts of a building. Smoke movement requires knowledge of factors such as:

- Dynamic smoke characteristics such as temperature, pressure, smokiness, air entrainment, deposition. These characteristics can be identified in IPI chart, Section 1B.
- The type, number and location of barrier–space modules that identify the paths of smoke movement from the origin to the target room.
- Type, location, and status of barrier openings along the paths of smoke movement.
- Natural air pressures that influence smoke movement (e.g. stack effect).
- Status (normal operation, shutdown, or emergency operation) of mechanical air handling equipment.
- Pressurization and exhaust of mechanical air-handling equipment.
- Wind direction and pressures.

A target room tenability evaluation uses scenario analyses to develop an understanding of what to expect. One normally uses all three types of analysis (qualitative, quantitative, and variation) to understand smoke movement and establish a window of time during which target rooms remain tenable.

The process starts with an initial "most reasonable" analysis using only the diagnostic fire to establish a "baseline" of understanding. Changes may be introduced to recognize differences caused by the status of barrier openings and the effect of mechanical air handling operations. A methodical analysis of the scenarios can provide understanding for variations of target room performance.

21

Life Safety

21.1 Introduction

A central part of building fire safety is preservation of life from injury or death.
Life safety integrates three major functions.

1) The building response to the fire scenario   examines the composite fire conditions and describes when target rooms are untenable for humans. Tenability relates to unacceptable conditions of heat, smoke, and structural collapse. Target rooms are designated spaces along an egress path or building spaces that may house individuals that depend on a strategy of defend in place.

The Interactive Performance Information (IPI) chart, Sections 4A and 4B, describe the building response of target rooms with regard to structural behavior and environmental tenability. Because time integrates all components, the IPI chart identifies the windows of time during which specific spaces remain tenable. Essentially, this describes the available safe egress time (ASET) for the spaces.

Chapter 20 discusses concepts for target space smoke analysis and Chapter 32 describes quantification techniques.

2) The time duration during which an individual may occupy discrete spaces of the egress path.   The goal is to identify conditions in which humans and untenable conditions occupy the same space at the same time. Section 5A of the IPI chart describes the pre-movement time required for detection, alerting occupants, and delay in leaving the room, as well as the time to travel along the chain of spaces that define the egress path under study.

Section 5A(c) describes windows of time during which target rooms that house defend-in-space occupants remain tenable. Section 5A(d) provides a dedicated section to note locations for conditions that may contribute to unusual fire fighting hazards.

3) Human response to products of combustion (POCs)   is important in defining tenability. Although this information is not recorded in the IPI chart, it becomes the basis for understanding tenability limits. This chapter describes POCs and their effect on humans.

Fire Performance Analysis for Buildings, Second Edition. Robert W. Fitzgerald and Brian J. Meacham.
© 2017 John Wiley & Sons Ltd. Published 2017 by John Wiley & Sons Ltd.

21.2 Human Reaction to Products of Combustion

The physiological and psychological effects of building fires on humans can be dreadful. Humans are fragile beings when subjected to POCs in a confined space. The major hazards for an occupant of a burning building include the effects of a diminished supply of oxygen, inhalation of smoke and toxic gases, the physical effects of heat and flame, and dangers relating to structural collapse. Although panic is rare, when individuals do not believe there is a way out, anxiety and emotional shock create added problems.

A continuous, ample supply of oxygen is necessary for human survival. Air contains 21% oxygen. When the oxygen is reduced to about 16%, clear thinking becomes difficult and muscular coordination for skilled movements is lost. At levels below 14%, judgment becomes faulty and behavior is irrational. At less than 8–10% oxygen, collapse occurs and death may result within 10–15 minutes unless oxygenation is rapid.

Oxygen deprivation is a common hazard in fire. The fire itself competes for the available oxygen, although this is not a significant cause of oxygen deprivation. Instead, it is the effect of the POCs on the human respiratory system. One of the chief culprits is carbon monoxide, which is the result of incomplete combustion and is present in all building fires. It is a major cause of fire deaths because the hemoglobin in the red blood cells has a much greater affinity for carbon monoxide than for oxygen. Consequently, carbon monoxide replaces the oxygen that the red blood cells normally carry from the lungs to the other parts of the body. The resulting hypoxia (low blood oxygen) can cause death unless prompt treatment is available.

Toxic gases that permeate the building can affect occupants who are a long distance from the fire. These gases can travel through ventilation systems or other openings and channels that exist in all building construction. Doors, stairwells, pipe chases, elevators, and service ducts are but a few of the avenues through which smoke and fire gases can travel. An exposure to concentrations of carbon monoxide alone at 0.4% will be fatal in less than 1 hour. However, much smaller amounts may be deadly when combined with other toxic products. This cocktail effect of gas mixtures is clearly evident, but exposure time and species have not been established.

In addition to the theft of oxygen by carbon monoxide, oxygen can be blocked from the body tissues in other ways. The inhalation of irritant gases in the smoke can cause an outpouring of fluid into the air passages. This causes swelling of their walls, restricting or obstructing the inflow of air to the cells from which it normally passes into the bloodstream. Victims may suffer a wide range of symptoms, such as headaches, nausea, fatigue, and difficult breathing. In addition, individuals may experience confusion, impaired mental functioning, and diminished coordination. This can cause what may be perceived as irrational behavior, such as clawing at a door rather than turning the knob.

Heat in a burning building can cause serious physiological problems. Humans can tolerate temperatures of 65–90°C (150–200°F) for only short periods of time. The tolerance time drops rapidly above 90°C (200°F), and in temperatures above 150°C (300°F) survival time is only a few minutes. These temperatures are easily attained in building fires where temperatures within 3 m (10 ft) of the flames can exceed 150°C (300°F).

The secondary effects of temperatures well below 90°C (200°F) cause other physiological problems. People cannot work in an atmosphere of 50°C (120°F) without protective clothing. Even then, the heart rate increases, causing increased stress. Fatigue, dehydration, heat exhaustion, and heat shock commonly accompany these temperatures. Both climatic conditions and a fire environment can cause temperatures of this level. These conditions contribute to fire fighter fatigue.

Burns due to heat and flames are often incorrectly assumed to be the primary cause of death and injury. Smoke and toxic gases remain the major threat. Besides smoke casualties on the fire scene, nearly half of those who reach the hospital die of respiratory tract damage due to irritant gases. Although fewer fatalities are a result of burns, the pain and disfigurement cause serious long-term complications. Bacteria are *everywhere*, and bacterial infections from burn wounds and damaged lungs are major killers. Shock results from fluid losses out of the burned tissues. This formerly major killer has declined with vigorous intravenous replacement of plasma and other fluids.

Structural collapse of building elements can cause physical injury and restrict movement through the building. Collapse is a concern for building occupants and a particular danger to fire fighters. Deaths and serious injuries occur each year because of unanticipated structural failure. Some of these failures result from inherent building weaknesses, but many are the result of renovations to existing buildings that materially, though not obviously, affect the structural integrity of the support elements. A building should not contain surprises of this type.

Besides being accompanied by toxic and irritant gases, smoke contributes indirectly to death. Dense smoke obscures visibility and irritates the eyes. Consequently, an occupant may become disoriented and be unable to identify escape routes and use them. Reductions in visibility cause occupants to turn back from an intended escape route. At a visibility of about 3 m (10 ft), 90% of occupants will have turned back in a building with which they are very familiar.

Knowledge of fire in a building causes anxiety among its occupants. The combination of emotions, conditioned responses, training, rational action, and physiological and mental impairment due to POCs influences human behavior. To these factors should be added the size, shape, and function of the building, the physical and mental capacity of its occupants, and interior design features that influence decision-making.

Behavior patterns of individuals subjected to sudden, unexplained danger cover a wide range of action. Some people will correctly assess the danger and take prompt action, whereas others may withdraw from the situation. Some may deny the situation altogether and take no action or totally unrelated actions. Still others will take no initiative, but will follow a leader and obey orders if and when they are given.

It is not possible to predict the actions of an individual, but group behavior patterns can be statistically quantified. However, this assumes that the occupants are capable of reacting physically and mentally. Hospitals, nursing homes, prisons, and mental hospitals all house occupants who cannot react with freedom of movement. Part of risk characterization is an analysis of combustion products that can enter target spaces occupied by these individuals.

Special building fire situations often arise. In high-rise buildings, for example, occupants cannot evacuate within a reasonable time. Unless effective areas of refuge can be provided or the fire is extinguished quickly, occupants may be exposed to the toxic combustion products for significant portions of the fire duration. A symptom of oxygen deprivation is that a normally rational individual behaves irrationally. Low levels of carbon monoxide, for example, reduce levels of oxygen available to the brain and other tissues, increasing the potential for irrational behavior.

21.3   Tenability

The definition of tenability is a central issue in a life safety analysis. The tenability criterion is a major decision in a performance analysis. Its selection requires a clear rationale tailored to the building use and architecture and occupant characteristics. Here, we define tenability as a visibility through smoke under normal ambient lighting conditions. Individuals attempting to evacuate a building will re-evaluate their intent and turn back to seek another way out when visibility falls below a certain level. Toxic gases are fellow travelers in the same air and are more hazardous to people than smoke. However, because obscured visibility normally occurs before toxic gases affect humans, its selection is conservative. Perhaps more importantly, it is more readily perceived as an imminent danger by occupants.

A performance evaluation identifies conditions where people and untenable smoke or other fire products can occupy the same space at the same time. *Note that this does not determine that an individual will be injured or killed in a fire.* It merely identifies time durations when people may be subjected to untenable smoke (or other fire) conditions in the building.

21.4   Fire Fighter Safety

Risks for fire fighters are often delegated to the fire service with little thought by others associated with the design, management, or operation of the building. However, the safety of community fire fighters is a factor in a holistic building analysis.

Fire ground operations involve several types of hazard that occur during fire suppression operations, search and rescue of occupants, and overhaul operations. Section 5A(d) provides an IPI location to identify potential life safety conditions that can affect fire fighters.

One type of fire ground operational risk relates to finding the fire and advancing attack lines. The fire environment is hot and smoky. Visibility can vary from passable to nonexistent. Sometimes the time from good visibility to zero visibility can be seconds. It is easy to become disoriented in a visionless situation.

One of the factors of a detailed fire extinguishment analysis involves estimating the time to move through a smoke-logged space. The smoke conditions for spaces along fire attack routes may be described in IPI Sections 1B and 4B. Recognition of conditions that provide poor visibility and high heat conditions provides a preliminary caution for operations in these spaces.

Another fire attack difficulty involves the entrapment of fire fighters by flame movement through overhead voids or through routes around fire hose positioning. The IPI chart, Sections 1A, 4B, and 5A(d) provide an opportunity to document potential sources of entrapment.

Collapse is always a concern for fire fighters. Chapter 19 discusses structural collapse. Particular concerns arise when fire fighters conduct operations above the fire floor. Collapse may occur because thermal expansion can push frangible walls to the point of instability. IPI Sections 4A and 5A(d) can document concerns.

Other potential for fire fighter death or injury relates to hazards that are a part of routine building operations. For example, the presence of unprotected holes and shafts is particularly dangerous during operations where visibility can be almost nonexistent. Also, rooms associated with biological or chemical activities can have hazards that may create health problems in the near or long term after the fire is extinguished. These conditions may be recorded in the IPI, Section 5A(d).

Overhaul operations can provide unexpected hazards. Smoldering fires produce increased quantities of carbon monoxide and other toxic gases. Fire fighters who do not wear SCBA gear are in greater danger from the effects of carbon monoxide and the cocktail effect of other gases. These effects may be delayed until the fire scene is secured and the staff returns to the fire station.

22

Risk Characterizations

22.1 Introduction

Buildings perform a function. That function may be to house its occupants, conduct business management operations, sell merchandise, educate students, store materials, feed and entertain people, or provide space and facilities for a variety of other activities.

Some level of risk is present in all activities. For example, there can be a risk of too much exposure to the sun or too little; eating too much food or too little; driving too fast or too slow. One can extend this concept to recognize that implementation of too many safety requirements may make a project uneconomical to build or operate, whereas too few, or the wrong type of, safety measures provide unsafe conditions. Almost all endeavors have a window of workable conditions and judgment.

A risk characterization describes the expected threat to important functions or objects that are vulnerable to fire. Communication with others not in the fire business involves telling a story about what can happen and why. The goal is to provide a clear understanding of the type and nature of threat to the occupants and operations.

22.2 The Exposed

We characterize risk from the knowledge gained from the performance analysis. Integrating the building's fire behavior with its functional operations provides an insight into the way the building will work during a fire.

A risk characterization does not identify corrective actions to change the threat, nor does it identify appropriate or acceptable levels of risk. Those are functions of risk management. However, a clear understanding of risks associated with the building's functional operations and the building's fire performance is central to developing an effective risk management program.

The main components of risk characterizations are located in Section 5 of the Interactive Performance Information (IPI) chart as:

- Human safety
  - Occupants who are expected to leave the building in a fire emergency
  - Occupants who are restrained, incapacitated, or unable to leave the building
  - Occupants who may prefer to remain in the building
  - Fire fighters conducting emergency operations

Fire Performance Analysis for Buildings, Second Edition. Robert W. Fitzgerald and Brian J. Meacham.
© 2017 John Wiley & Sons Ltd. Published 2017 by John Wiley & Sons Ltd.

- Property protection
  - Ordinary contents
  - Valuable contents
  - Heritage
- Operational continuity
  - Equipment
  - Information and data storage
  - Material supplies
  - Functional spaces
- Neighboring property
  - Exposure losses to neighboring structures
  - Exposure losses to other enterprises within the building
  - Community losses (e.g. jobs, taxes)
- Environment
  - Groundwater contamination
  - Surface runoff contamination
  - Air pollution.

PART ONE: HUMAN SAFETY

22.3 Life Safety

Some buildings may provide superior protection for occupants and fire fighters, whereas others may exhibit very poor qualities. A risk characterization for humans identifies the nature, relative significance, and consequences of what to expect during a fire.

People–building interactions can involve four different roles. One is conventional occupant egress by leaving the building in a fire emergency. A second involves a defend-in-place strategy. This is appropriate for individuals in hospitals, nursing homes, prisons, or other situations where movement is inappropriate or difficult. A third strategy involves moving to a designed area of refuge. Rescue or assistance by emergency personnel can be a part of each option. Awareness of fire fighting operations and their safety is a fourth role of risk characterization.

A life safety analysis determines conditions where people and too many harmful combustion products or conditions might occupy the same space at the same time. Tenability is the threshold for "too many" combustion products.

The most useful and common definition of tenability specifies a maximum distance through which an occupant can see in smoke conditions during normal building lighting conditions. Distances are based on conditions in which most occupants will stop moving along a travel path and attempt to turn back. This definition does not include any reference to death or injury.

Risk characterizations use the window of time that target spaces remain tenable. This identifies an available safe egress time (ASET) for the target space. An analysis relates life safety to ASET. In addition to tenability from smoke-gas, risk includes potential collapse conditions as well as special hazards associated with the building and its operation.

22.4 Overview of Life Safety Alternatives

Figure 22.1 describes alternative actions that may or may not lead to avoidance of death and injury (life safety) to occupants during a fire emergency. Events 0, 1, 3, 17 identify the traditional (successful) evacuation from a building. This may involve either a code-complying means of egress or simply a way out by normal circulation routes.

Events 0, 2, 5 or 6 identify defend-in-place (remain) situations. Here the occupant remains in the room and is either safe from the combustion products (Event 6) or is not safe (Event 5). If the occupant is not safe, rescue may or may not take place, as noted by events 10 and 9. Here "rescue" means by the fire department or by trained nursing or security staff. Rescue by another occupant or passerby is not considered a part of life safety evaluations.

Although designed and effective areas of refuge are rare, in theory they may be appropriate for protecting people in a fire. It is possible in building fires to encounter an accidental area of refuge. That is, an occupant may accidentally reach an area that

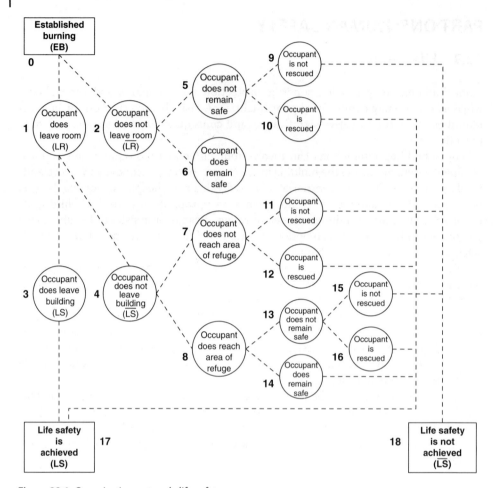

Figure 22.1 Organization network: life safety.

provides enough protection for survival during the fire. An accidental area of refuge that may provide coincidental survival conditions is not considered in a building evaluation. This process does not recognize an area of refuge unless it is a part of a conscious design and becomes the focus of a tenability evaluation.

Paths 0–1–4–7 and 0–1–4–8–14 or 0–1–4–8–13 show the area-of-refuge situation. If the occupant is not safe, rescue may or may not take place, as noted by events 11, 12 and 15, 16. If rescue is a part of a conscious building performance design, only the fire department or a trained building staff are considered. Other occupants may provide rescue aid, but they are not included in an analysis. If a family member is expected to rescue very young, elderly, or disabled individuals, this situation must be identified and those actions become an explicit part of the evaluation. For example, if building occupants are expected to provide rescue aid and this proves to be ineffective, one should not fault the building performance. However, one may question the criterion that occupants must provide the rescue assistance.

22.5   Prescriptive Code Egress

Buildings do not provide unlimited time for egress, and prescriptive codes do not iden-tify any available times. Codes specify good practices to help individuals leave a building safely in a fire emergency. The level of risk is indeterminate, safety is not assured, and alternatives do not have measures for comparison.

A major part of building codes relates to good practices for leaving the building. A *means of egress* is defined as a continuous, unobstructed path of travel through which an individual can move to a public way. A public way is a street, alley, or other land outside the building that leads to a street.

A means of egress has three major components: the *exit access*, the *exit*, and the *exit discharge*. An exit access is the part of a means of egress that enables an individual to move from a location within the building to the exit. In most buildings, the exit access is a corridor leading to the exit stairwell. However, in buildings such as assembly rooms, stores, and galleries, the exit access may lead through aisles or other paths. The exit access must provide free and unobstructed travel. Therefore, intermediate rooms that can be locked are not a part of the exit access.

Depending on the occupancy type and certain building features, such as the presence of sprinklers, codes prescribe maximum travel distances. Codes prescribe minimum dimensions to accommodate a specified number of persons and the character of the activity. Most codes allow a maximum dead-end distance. This dead end is restricted in length because two avenues of possible escape are not available. If the dead end were excessively long, an individual could be trapped and prevented from returning to find another route to an exit.

The *exit* is the second component of a means of egress. In the perception of the layman, an exit is synonymous with the way out. However, from a code perspective, the exit is a special part of the building that connects the exit access with the exit discharge. Although an individual may not have protection from combustion products when traversing the exit access, the exit is viewed as a temporary refuge area that protects an individual for the time needed to move through it.

As a temporary area of refuge, the exit has special requirements to maintain its integrity and tenability. This includes fire-resistant rated enclosures, rated opening protectives, and smoke-proof enclosures for some buildings. The code includes additional safety requirements, such as dimensional limits, guards and handrails, and walking surface specifications. Exit requirements are so distinctly defined that many "exits" to the layman should be considered as "ways out" because they do not meet prescribed regulatory conditions.

Codes permit a *horizontal exit* in some buildings. A horizontal exit permits passage from one part of a building to another part on the same level. There are many specialized requirements, such as fire resistance, refuge area requirements, and egress requirements from the building section that serves as the refuge area. The reliability of the door closer that separates the fire area from the refuge area is normally a requirement.

The *exit discharge* is the third component of an egress system. An exit discharge enables individuals to reach a position of safety outside the building when leaving the exit. Some buildings discharge into a lobby or other building area. When this occurs, additional requirements, including automatic sprinkler protection, are often required for those spaces.

A concern with respect to an exit discharge is the assurance that an individual will not inadvertently travel past a ground floor when a building has floors below ground level. To avoid this, codes normally require stairwells to be discontinuous at the ground floor to force individuals to leave one part of the exit before entering the other part. In addition, stairs are required to distinctly identify the ground-floor discharge.

In addition to dimensional and other special requirements for a means of egress, codes identify the *capacity* of each egress component. The capacity of a component is determined by dividing its width by a factor. For example, in the International Building Code (IBC; USA) the factor for a level component such as a door, corridor, or horizontal exit is usually 0.2, whereas the factor for stairs is usually 0.3. Therefore, the capacity of an 864 mm (34 in) door is 34 in/0.2 = 170 persons although the capacity of a 1118 mm (44 in) stairway is 44 in/0.3 = 146 persons. Similar factors can be found in approved/deemed to comply documents in other countries (e.g. Approved Document B in England).

The final code factor is the *occupant load*, which is the minimum population for which to design the means of egress. The minimum number of people is calculated by dividing the gross area (or sometimes the net area) by a factor based on the building's use and occupancy. For example, the floor area per occupant for a business area (IBC) is 9.3 m^2 (100 ft^2) gross. Therefore the minimum occupant load for a gross office floor area of 2787 m^2 (30 000 ft^2) is 30 000/100 = 300 persons.

Codes in other countries may provide occupant load factors for regulatory compliance or deemed to comply with regulations. If the actual population is larger than that specified, the larger number is used. If the actual number is smaller, the minimum calculated number is used. The sum of each egress component must have a capacity large enough to accommodate the occupant load.

Sometimes one may observe a sign that states "The maximum number of occupants shall not exceed *x* people" or words to that effect. The capacity of the egress components determines this number. For example, if a room has a capacity of 300 people, but the exit doors have a capacity of 180 individuals and the stairways a capacity of 120 persons, the allowable room capacity is only 120 individuals.

22.6   Plans Approval for Prescriptive Code Egress

Building egress is a major design feature that requires approval by the authority having jurisdiction (AHJ). The process normally involves the following steps:

1) The building size and occupancy are established and the occupant load is determined for the different areas on each floor.
2) Exit locations are noted, and the AHJ accepts their remoteness from each other.
3) Exit access distances are checked regarding occupancy and other building features, such as sprinklers.
4) The widths of doors, corridors, stairs, and other parts of the egress components are determined, and the occupant load must be smaller than any component limitation.
5) Other specialized requirements for dimensions, fire resistance, and special characteristics are checked.

When all code requirements are met, the egress system is in compliance and legal.

The building egress system is part of a total architectural circulation plan. A variety of considerations besides fire requirements are a part of a circulation pattern. The earlier discussion referred to code compliance rather than to safety. It is often assumed that a code-compliant design is a safe design. However, the quality of egress systems can vary, and a performance-based risk analysis requires sensitivity regarding the occupants of the building and what they will be doing at different times of the day.

22.7   Overview of Egress Risk Characterizations

A prescriptive code egress system is an inventory of good practices. However, the quality of code complying egress systems can vary greatly, and risk analysis encompasses more than good practices and routes out of a building.

Performance based risk characterizations match target space tenability with people movement. One part examines actions of a representative occupant in a specific room from established burning (EB) to deciding to leave the room and move through an egress path to the exit discharge. This analysis first estimates the pre-evacuation time delay due to building and occupant actions. The second part compares egress movement time durations with windows of time that target spaces along the egress path remain tenable.

A pre-evacuation time delay includes:

- The time duration from EB to detection.
- The time duration from detection to alerting the occupant of the fire and having the occupant understand that a fire exists.
- The pre-movement time delay between occupant recognition that a fire exists and the instant that he or she opens the door to leave the room.

After opening the door to leave the room, the occupant must travel through a sequence of spaces to leave the building. The sequential time durations for the occupant to enter and leave each space are estimated. The conditions whereby an occupant and untenable conditions coincide are identified.

The IPI chart provides a framework for understanding potential egress difficulties. Each cell in Section 4B identifies the tenability conditions in target rooms. Variation analyses enable one to bound tenability for differing status conditions of openings. These conditions are independent of people movement.

Occupant movement is then examined to compare the time at which the occupant will be in each space with the tenability condition of that space. Complications arise when a human and untenable conditions occupy the same space at the same time.

22.8   Discussion

The situation whereby humans and untenable conditions occupy the same space at the same time is described above for egress scenarios. The same concept is valid for all conditions described in Figure 22.1. Thus, defend-in-place or rescue scenarios may be examined using a comparison of the time during which a human will be in a space and the time during which that same target space will be untenable. The smoke generation stops at extinguishment.

The time duration to traverse the sequence of barrier–space modules along an egress path incorporates characteristics of occupant mobility and impairments. For example, an athletic, unimpaired individual exhibits one type of movement. An impaired individual unable to move easily along corridors and stairs exhibits another. Blind, deaf, young, elderly, or frail individuals are other categories. Variability analyses provide an insight into conditions that affect each characteristic group.

The IPI chart helps to envision potential conditions for entrapment of people. The "what if" variability analyses offer an insight into time durations, status of opening protectives, and route selection that affect life safety.

22.9   Pre-evacuation Activities

Three events collectively comprise the pre-evacuation activities of fire detection, occupant alert, and occupant decision. Detection (OD) is usually the same as event MD. We incorporate detection in both places to ensure that the fire size at detection is included. In a rare situation in which MD and OD may be different, the appropriate values may be shown.

The occupant alert event (OA) requires attention to the characteristics, location, and activities of the representative individual. For example, if the alerting process involves sounding a noise, the actions of deaf occupants would be evaluated as a separate scenario from those of unimpaired hearing individuals. Also, the alarm decibel level is affected by the location of the sounding device and intervening architectural obstacles, such as closed doors.

Occupant alert is not the same as fire department notification (MN). The two events may occur simultaneously or in a different order. Some buildings have intentional or unintentional delays in alerting occupants. In addition, the occupant alert may not be the same as occupant recognition that a fire exists. Ambiguous alarms or occupant uncertainty can delay recognition of a fire emergency. Note that the awareness created by a signal or cue does not include the decisions that an individual may make after that event.

After an occupant has been alerted to a fire emergency, different actions can occur. Studies have shown that time delays after a fire alarm are common and often surprisingly long. The common assumption that building evacuation starts at the instant of detection and alarm is a myth.

Attitudes, experiences, cues, and pre-fire emergency training influence the decision to leave the room (OL). This decision is highly personal and is influenced by factors that involve interactions of people, the fire, and the building. Often the time duration is much longer than imagined by individuals who may have superficial assumptions about expected behavior.

The manner in which an individual becomes aware of a fire emergency can influence his or her actions. The ringing of an audible alarm, particularly after a series of false alarms, is not as likely to produce the same response as seeing dark, acrid smoke or hearing an individual shouting, "Fire!" The intensity, volume, pitch, and earnestness conveyed by a voice alarm has a major influence in occupant actions.

Figure 22.2 shows the three major elements of pre-movement activities.

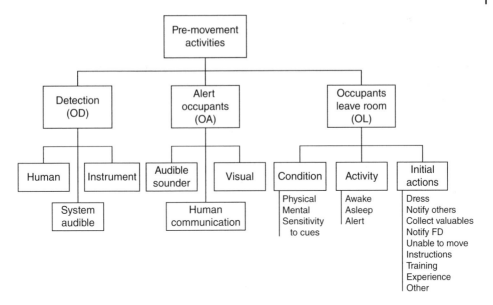

Figure 22.2 Organization chart: pre-movement activities.

The number and intensity of fire cues influences the perception of needs for action. A fire cue may be a level of seeing, hearing, smelling, or feeling the products of combustion, or it may involve the actions of others or of alarm systems. The perception of real danger in uncertain situations seems to be the most important factor. *Recognition* of a threat can be ambiguous and unclear during early stages of a fire. As the volume and intensity of combustion products increase, the uncertainty declines. The time that remains to take effective action is also reduced. Experiences, a predisposed optimistic wish, concepts of personal abilities, and pre-fire training or instructions influence initial actions.

When an individual becomes aware of mild or ambiguous cues, most individuals attempt to obtain additional information. This process of *validation* may take the form of a physical investigation and surveillance, or it may involve questioning other individuals. Eventually the threat becomes defined in the mind of the individual. This *definition* stage involves the recognition of the meaning and structure of the threat, although not necessarily its seriousness.

After an individual defines the threat and knows the nature of the problem, the *evaluation* of personal actions begins. The size and perceived growth of the fire, the building size and function, availability of fire extinguishers, the activity and potential danger to others, and pre-fire instructions are among the factors that influence an evaluation of appropriate actions. The concern and actions of a group of people often influence decisions to a greater degree than do those of an individual in isolation. Possibly the evaluation will lead to a series of actions, usually involving variations of fight or flight. The initial decision of an individual to take specific action is completed with evaluation. Often the evaluation may include a series of if–then scenarios that are useful in evaluating pre-fire instructions and fire program management plans.

After an individual decides an appropriate course of action, the *commitment* stage is entered. At commitment, the individual carries out the decision made during evaluation.

That action may be successful, such as leaving the building safely or fighting and extinguishing the fire. On the other hand, the initial commitment may be unsuccessful and the situation must be reassessed. This *reassessment* may result in continued efforts or in redirecting actions. For example, an individual may initially have attempted to fight the fire, but, being unsuccessful, then attempts to leave the building.

The time frame to complete the decision stages described above could be very brief or it may be relatively long. However, the process is dynamic. The fire grows at an exponential rate. Conditions change more rapidly than is imagined by individuals who have never experienced a fire situation. Anxiety, stress, and levels of activity reflect the intensity and importance of the cues available. The time to define and evaluate decisions can be reduced to seconds when cues are unambiguous and intense. On the other hand, an individual may act in a relaxed manner until the movement of combustion products changes the environment so rapidly that plans must be reassessed and changed very quickly.

22.10   Pre-evacuation Evaluations

Interactive Performance Information chart cells organize time and pre-movement activities. The following questions provide a means of selecting appropriate cells to incorporate the time durations from EB:

- **Given:**
  - *Ignition (IG) has occurred*
  - *Time from EB _____ minutes.*
- **Question:** *Will the fire be detected (OD)?*

- **Given:**
  - *Fire has been detected (OD)*
  - *Time from EB _____ minutes.*
- **Question:** *Will the occupant in Space _____ be alerted to the fire (OA)?*

- **Given:**
  - *Occupant in Space _____ has been alerted (OA)*
  - *Time from EB _____ minutes.*
- **Question:** *Will the occupant in Space _____ decide to leave the room (OL)?*

These questions provide a methodical basis with which to understand pre-movement decisions. The IPI chart enables one to compare human decisions in one location with fire growth in a different location.

22.11   Travel Times

After the occupant opens the door and starts to leave the room (OL), the local travel clock starts. We construct a timeline for the occupant to enter and leave each segment of the egress path. A variety of factors influence the time to traverse the egress route. Among these are people concentrations in the egress route, the distance, path recognition, amount of smoke, and the physical and attitudinal characteristics of the individual.

When the population density is low, individuals essentially move separately and speeds relate to the physical capability of the individual. However, disorientation, anxiety, visibility, building familiarity, and recognition of an appropriate route can affect time durations. The architectural layout of some buildings may be so complex that it is possible to become disoriented under normal conditions. Under conditions of stress, reduced visibility, and unclear directional orientation, selecting an appropriate path often involves trial and error.

When a building has a moderate concentration of people, leaders often provide instructions. Moderate crowds may reduce speeds somewhat. Major problems can arise from congestion at locations where different paths merge. In buildings having zones with large concentrations of people or with constraints to movement, such as narrow aisles and fixed seating, travel speed may be reduced drastically.

Mobility characteristics must be defined. For example, is the occupant able to move 30 m (98 ft) in 60, 90, or 300 seconds? Are there any constraints to movement, such as the need for wheelchairs or crutches? Are there any impediments to movement and decision-making, such as blindness or deafness? Table 22.1 gives a sense of proportion to various speeds of ambulatory movement.

Life safety scenarios may use other characteristics. For example, risk characterizations are different for occupants who are asleep, intoxicated, or under the influence of drugs. They differ for occupants who are mentally incapable of making appropriate decisions in a fire emergency, as well as the very young or the infirm elderly. Should a building design be faulted when a parent leaves infants or small children alone for a short time while socializing elsewhere?

Table 22.1 Travel speed comparisons

Activity	Speed	
	(m/s)	(ft/s)
Comparative movement		
100 meter dash in 10s	10	32.8
Mile run in 4 min	6.7	22
Mile walk in 20 min	1.3	4.4
150 feet in 90s	0.52	1.7
Normal adult walking speeds		
Level passageway	1.3	4.3
Stairways (down)	0.76	2.5
Stairways (up)	0.58	1.9

22.12   Defend in Place

Event 2 in the organizational diagram of Figure 22.1 shows that the individual may not leave the room. This could be intentional because the individual may have decided that their survival chances would be better by remaining in the room than by attempting to leave the building. Or perhaps the individual is asleep or under the influence of drugs or alcohol and hence unaware of the occupant alert. In other situations, occupants may be expected to remain in their rooms for the duration of a fire. Nursing homes, prisons, and hospitals are examples of a defend-in-place strategy for the occupants.

A defend-in-place life safety analysis evaluates the likelihood that an individual will be protected from untenable conditions. The analysis assumes that the room under study is not a room of origin.

A defend-in-place evaluation has the following steps:

1) Identify the room of fire origin and diagnostic fire scenario.
2) Select the target space to be evaluated.
3) Identify the tenability measure for the occupants of the target space. The criteria will reflect the characteristics of the room occupants. The tolerable level of products of combustion for the room might be expressed as follows:
    X ft of visibility in smoke;
    Y_1 parts per million (ppm) of CO dose rate and a limiting acquired dose;
    Y_2 ppm of other specified toxic gases dose rate and a limiting acquired dose;
    Z degrees of temperature;
    Avoid structural collapse.
4) Determine the time duration during which the target room will remain tenable.

22.13   Areas of Refuge

It may be impractical to provide for egress or defend-in-place for certain types of buildings. The concept of an area of refuge is appealing for tall buildings, large buildings, hospitals, and ships.

An area of refuge is a designed and designated area in which occupants may find refuge for the duration of a fire. Few buildings have a designed area of refuge. However, if one were incorporated into the building design, its performance evaluation becomes merely a combination of egress analysis and defend-in-place analysis.

An individual must move from an occupied room to the designated area of refuge. This involves identifying path connectivity to the area of refuge and determining the tenability along the route. The area of refuge becomes a defend-in-place room.

22.14   Fire Department Rescue I

Saving lives is a primary responsibility of the local fire department. This activity in a fire emergency may involve several types of operation. Sometimes the fire department may assign fire fighters to guide or direct occupants to safety. If occupants remain in the building, a search and rescue operation may take place. In this operation, fire fighters methodically search the building to find trapped individuals. If victims are discovered, the rescue may require substantial physical effort. A third way to save lives is to extinguish the fire quickly to stop the generation of combustion products.

Guiding occupants to safety is the least physically demanding and requires the smallest number of fire fighters. Sometimes fire fighters take charge of occupant movement and act as an authority, eliminating the need for occupants to evaluate cues. This operation can efficiently handle large numbers of people. Also included in this category is the traffic supervision and operation of elevators. A fire fighter can control an elevator and use it to manage the transportation of occupants in a building fire. Fire fighters may transport individuals from above a fire to a few floors below the fire, rather than to the ground floor, thus saving substantial time.

Search and rescue is a time-consuming, staffing-intensive activity. Two types of search operations are used: primary and secondary. The primary search is done on arrival, particularly if there is reason to believe that people may be in the building or if a bystander says that occupants remain in the building. A primary search is usually a rapid exploration of rooms to discover if any victims are still in the building. On the fire floor, fire fighters usually move directly toward the fire then systematically search the rooms for victims. On floors above the fire floor, the search usually starts at the point of floor entrance and moves toward the fire.

The primary search may have conditions of poor to no visibility and may involve high heat conditions. Physical strength requirements are demanding because of the low visibility, heat, heavy protective equipment and self-contained breathing apparatus (SCBA), and the forcible entry tools needed to gain entry into locked or blocked areas. Mental stress is also significant. Room searches when there is no visibility are time-consuming and imperfect.

If a victim is discovered in a primary search, their removal can be difficult and time-consuming. The fire fighter must have the physical strength to carry the victim to safety. A general rule of thumb is that the maximum rescue capability of one fire fighter at any single fire emergency is two occupants. This staffing expenditure is for the entire incident. When search and rescue is expected by the fire department, enough personnel must be dispatched to the incident in sufficient time to conduct an effective search and to perform all the other activities that are demanded.

A secondary search is normally conducted after the fire is under control. A secondary search is more careful and is often combined with the overhaul operations. This detailed examination of the building for victims can take a good deal of time if smoke, clutter, and debris are present.

There are multiple needs for available staffing. The fire ground commander must decide the allocation of fire fighters to the needed tasks. When staffing is insufficient, decisions can be difficult.

22.15   Risk Characterizations for Life Safety

Risk characterizations for life safety can take several forms. The most useful describes the maximum pre-movement time that individuals in each room have before they would not be able to leave the building by normal egress routes without encountering untenable conditions along the escape path. This information gives risk managers information into ways to reduce potential risks. It also gives fire departments an indication of possible search and rescue expectations.

The defend-in-place strategy provides information regarding the expected tenability duration for specific occupant rooms. Risk managers may try to improve the situation by changing the performance of one or more of the events in Figure 22.1. This diagram, in combination with the IPI chart, enables one to recognize the sensitive events that may be changed to improve the occupant risks.

The concept of designed areas of refuge has been discussed for many years. This theory combines the internal movement procedures with defend-in-place evaluations. Although the idea is simple and attractive, the application is difficult. An application requires a rigorous analysis, and the acceptance by occupants may be questionable. Nevertheless, the concept is interesting for large buildings in which normal egress is difficult.

PART TWO: OTHER RISKS

22.16   Property Protection

Property, including the building structure and its contents, may be exposed to fire damage or destruction. Many buildings have individual rooms with contents of high monetary value (e.g., information) or irreplaceable historical or sentimental value. The value of this property can be large enough to draw attention when describing a building's fire performance. Some owners do recognize a problem. Others are completely unaware of what may be at risk.

Electronic communications, computers, and controls plus their transmission equipment require a new awareness of property value. For example, fiber-optic interconnections between rooms can have substantial value. A minor incident may damage the cables, incurring a major repair expense. Equipment distant from a fire may be very sensitive to smoke or certain airborne products of combustion. Information and records are valuable commodities.

One may also express property sensitivity (tenability) in terms such as:

- X ppm of HCl (or other corrosive gas)
- Y degrees of heat
- Z amount of smoke
- Water application.

We evaluate high-value rooms in a similar manner to a defend-in-place life safety analysis. That is, the sensitivity of the contents to the type and concentration of combustion products is determined. Then we evaluate the time during which the room will remain tenable.

Risk characterizations include an assessment of the value of information, equipment, and other property in the building. Often only a few rooms of very high value may require attention. A performance analysis gives an understanding of the risk to these pockets of high value.

22.17   Continuity of Operations

Historically, disruption of operational continuity (business interruption) has been important. Operational continuity is becoming even more important in this age of information technology and just-in-time operations. The building code does not address this issue. It is possible for a relatively small fire that does not impact life safety to greatly affect a business enterprise. Functional downtime after a fire can have an important influence on a corporation's economic health. Loss of market share through an inability to provide services or materials can be significant. A risk characterization includes an assessment of the survivability of operations to combustion products.

An operational continuity analysis is similar to that for property protection. That is, one identifies spaces that are mission-sensitive and determines the vulnerability of the equipment or contents to combustion products. Fires far removed from the room of interest can also be a risk. For example, electrical wiring circuits often pass through intermediate rooms. Loss of a room distant from the mission-sensitive room may be insignificant but can interrupt electronic communication severely.

The influence of data storage is another concern for relatively small fires. Often, owners and designers think of redundancy in terms of electronic redundancy. Location can affect equipment redundancy. For example, when backup copies are made, their storage location is important. The development of risk characterizations involves questioning owners and technical personnel to uncover the sensitivity of operations to products of combustion.

22.18 Threat to Neighboring Exposures

The exposure threat from neighboring structures to the building being studied has a role in risk characterizations. Exposure concerns can involve adjacent buildings or they can be within the same building. Risk is not limited by corporate boundaries.

Fire may propagate to neighboring structures by radiant heat energy transfer or flying brands. However, neighboring exposures may also be within the building being evaluated. For example, malls and offices house different business enterprises under the same roof. A fire in one firm can influence business operations in the same building or the same complex of buildings. Risk characterizations consider appropriate scenarios between corporate boundaries.

22.19 Threat to Environment

A building evaluation provides information about the fire duration and damage. When the environment is a concern, information must be gathered about materials that can be involved in the fire. The fire suppression analysis provides a sense of the likely time duration and extent of the fire. Water or other agents used in the extinguishment process will be known and their impact may be examined.

The combination of extinguishing agent and building contents can identify potential environmentally toxic products. Some of the products will be airborne. An environmental engineer with experience in air pollution can evaluate environmental sensitivity to airborne products of combustion. Chemical products created by combustion and extinguishing agent reactions can be evaluated. An environmental engineer with knowledge of toxic products, surface runoff and subsurface transport can evaluate potential environmental risks.

A few fires have gained attention in recent years because of a series of disastrous and expensive catastrophes relating to environmental contamination. An environmental consequence analyses can be included in a building evaluation, if needed. A risk characterization may identify the effect of airborne, runoff, or subsurface movement of toxic products produced by a large fire.

22.20 Closure

The key to understanding risk characterizations for exposed property is to know how the building works. Buildings are constructed to perform functions, and disruption by fire or other natural and man-made threats are not in routine operations. Less attention may be given to disruptive threats because of their infrequency and the immediacy of day-to-day crises. Consequently, the role of risk characterization is to communicate what is at risk and the potential severity. A building fire analysis focuses on what can be lost, not what can be the loss.

Fire performance is integrated with the building's functional operations. One selects potential rooms of origin to relate fire performance to building operations. Combining normal building functions with performance evaluations gives a picture of associated risks. That picture becomes the basis for selecting factors to examine more thoroughly.

23

Fire Prevention

23.1   Introduction

The public and some fire professionals often consider the term fire prevention to signify "prevent ignition." However, performance evaluations extend that concept to include both ignition and established burning (EB). The analysis is separated into two parts: the first examines traditional ignition prevention that relates to the ease of producing the first fragile flame that defines fire ignition; and the second looks at ways in which the initial flames can grow to the size defined as EB.

Traditional hazard analysis examines the conditions that contribute to ignition and initial fire growth. Often, the results are blended with fire propagation and fire defense operations to produce a comprehensive description of the peril. Consequently, one might expect fire prevention to be introduced early in a performance analysis because its failure is the triggering event that sets the entire process into motion. However, we intentionally reverse these topics for two reasons:

1) The primary focus of a performance evaluation is to understand how a building will behave after a fire starts. The frequency or ease of first ignition and the ability to reach EB are irrelevant to the performance of the building after EB has occurred. Often, it is difficult to grasp the true role of fire prevention when the two parts are commingled. Many buildings survive only on the success of fire prevention. It is important to recognize when this condition exists.
2) A risk management program can be developed more easily when one independently evaluates the building's performance after EB. The modular organization of the framework allows one to decouple any one element and examine it in relative isolation. In this way, the costs and effectiveness of any proposed change can be distinguished.

A fire prevention evaluation can start from a fire-free status or it can start from a defined ignition in any designated location. When the parts are uncoupled, one gains a better sense of the threat. The process also gives an understanding of the effectiveness of occupant fire suppression. The objective is to provide a logical analysis that uses available information to evaluate a building's fire prevention program in a way that is useful for risk management.

Chapter 34 describes networks and factors that guide evaluations.

Fire Performance Analysis for Buildings, Second Edition. Robert W. Fitzgerald and Brian J. Meacham.
© 2017 John Wiley & Sons Ltd. Published 2017 by John Wiley & Sons Ltd.

PART ONE: PREVENT ESTABLISHED BURNING

23.2 Prevent EB

The two distinct components of fire prevention are shown as:

- **Given:** *Fire free status (FFS)*
  - **Question:** *Will ignition (IG) occur?*
- **Given:** *IG has occurred*
  - **Question:** *Will the critical fire size (CS) that defines EB be reached?*

Ignition is defined as the appearance of the first small flame. Before this event, the heat energy applied over time may discolor the fuel, or it may cause smoldering. Smoldering can appear briefly and then a flame appears; or smoldering can last for a long period of time with smoke being generated in the process. During the early stages of a fire, flames can appear and then give way to smoldering, and then flames can reappear. The fire can go out by itself at any time. For our purposes, IG is defined as the first appearance of any flame. Status changes after that time, including the fire going out before reaching the size defined as EB, are merely stages in the evolution of the fire.

During this often-unstable initial period of fire growth, either the fire continues to develop and grow or it goes out. When the fire goes out without any external intervention, that event is defined as self-termination (ST). If the fire grows to the critical size (CS) that defines EB, the fire will have moved into the next realm of development.

The factors that influence prevention of IG and EB are discussed in Chapters 6 and 7. Those sections describe realms of burning and factors that influence transition from one realm to the next.

Within Realm 2, from IG to EB, the fire may also go out by the actions of occupant extinguishment (M_p). Occupants can be very effective in extinguishing small fires when they are in the vicinity and have means to put out the fire. The means may vary from a shod foot to covering the flame with an object or using a fire extinguisher. On the other hand, occupants may be ineffective in putting out the fire. One objective of an occupant extinguishment analysis is to be able to recognize conditions that influence each outcome.

23.2.1 Ignition Potential

Ignition is often described in statistical measures. Although statistical frequency is appropriate for some specialized purposes, it does not provide clear performance understanding for a specific building. In this case, an organized, analytical process is more useful to identify problems and correct them, if necessary.

In assessing ignition potential, the building is studied for the location, type, and quantity of potential sources of energy and kindling fuels.

When candidate locations are identified, one assesses whether the heat sources and kindling fuels will come together closely enough for a long enough time to cause an ignition. This examination provides a good insight into operations and potential trouble spots that can cause ignition.

Here, ignition potential is viewed in terms of a process that is based on observation and judgment. If a risk management alternative involves ignition prevention, one can focus on specific modifications that relate to the specific building operations.

The assessment is based on observing building operations to identify the ease and likelihood of combining enough of the factors to cause an ignition. Calibrations may use verbal descriptors, such as red, yellow, or green, or a numerical scale. More sophisticated assessments may use calculations involving distance, heat flux, or other numerical measures. Often, information may be recorded on space utilization plans for initial sorting, documentation, or communication.

23.2.2 Initial Fire Growth

Given ignition, one examines the initial fire growth potential to estimate the ease with which the fragile flames can grow to EB. The critical flame size (CS) that defines EB is either of the following:

- A flame of about 20 kW, or 250 mm (10 in), or knee high, or a wastebasket size.
- A size more appropriate to specific building operations.

A knee-high size (sometimes described as a bucket of fire) is appropriate for EB in most compartmented buildings. This fire size becomes "established" in the sense that fire growth predictions are more easily based on physical principles.

Many ignitions go out shortly after starting. Others continue to grow easily. The EB estimate is based on judgment that enough factors will work together to allow continued fire growth. This stage of fire growth is described in Section 7.3 as Realm 2.

Continued heat flux that impacts on the target fuel after ignition contributes to fire growth. Fire-retardant treatments that reduce combustibility can be particularly effective when the fire is small, such as in Realm 2. However, fire-retardant impregnations, coatings, and paints are not effective forever. Many have a limited life or durability and must be maintained regularly. Nevertheless, if used, they substantially reduce the propensity of initial fire growth.

23.3 Occupant Extinguishment

Occupant extinguishment (M_P) is very dependent on human decisions. These decisions are strongly influenced by the presence or absence of experiences, training, and instructions. Although one may be skeptical about an untrained occupant's suppression skills, an individual who is in close proximity to the fire when it starts can often extinguish it quickly. These ignitions are rarely reported, so statistical evidence is difficult to obtain and suspect in use.

Evaluating occupant extinguishment is analogous to fire department suppression for larger fires. It compares the time for fire growth with the time to get an agent on the fire and put it out. The events that must be completed for successful occupant extinguishment are as follows:

- **Given: *IG has occurred.***
  - **Question: *Will an occupant recognize the fire ignition (MRP)?***

  ↓

- **Given: *An occupant recognizes that a fire exists (MRP).***
  - **Question: *Will the occupant decide to extinguish the fire (MDP)?***

  ↓

- **Given:** *An occupant decides to extinguish the fire (MDP).*
  - **Question:** *Will the occupant be able to apply agent to the fire before it reaches EB (MAP)?*

  ↓

- **Given:** *An occupant applies agent to the fire (MAP).*
  - **Question:** *Will the occupant be able to extinguish the fire before it reaches EB (MEP)?*

Occupant extinguishment normally occurs within such a short time frame that an individual who is not in close proximity to the fire has little chance of extinguishing it. Although a glass of water or a shod foot may be used, a more appropriate analysis focuses on an individual using a portable fire extinguisher.

The maximum fire size that an occupant can extinguish is defined for analytical purposes to be EB. Some skilled occupants may be able to extinguish fires slightly larger than the waste basket-sized fire. However, an untrained occupant using a fire extinguisher would normally have difficulty extinguishing a fire the size of a lounge chair.

The fire size for occupant extinguishment is set at EB for two reasons. The first recognizes that an occupant near the ignition often can take rapid, effective extinguishment action. This is important to a fire prevention program. The second reason avoids commingling manual suppression actions of occupants with those of fire fighters. EB is a convenient, conservative demarcation that can incorporate both forms of manual suppression in a realistic, yet disciplined manner.

23.4 Portable Fire Extinguishers

Portable fire extinguishers are important devices in "prevent EB" evaluations. They can be a valuable first aid measure, or they may be irrelevant. The effectiveness of these devices depends greatly on their location, availability, size, agent type, and the inclination and training of an occupant to use them.

The classification of fires is important to understanding types and characteristics of fire extinguishers. Fires are classified according to the type of fuel involved:

- *Class A* fires involve ordinary combustibles of wood, paper, fabric, and plastics. These are the fuel types in most non-industrial buildings, except for kitchens with cooking greases.
- *Class B* fires involve fires in flammable liquids, such as petroleum products, greases, and gases. These fuels are further categorized with regard to certain flammability characteristics. When the depth of a flammable liquid is greater than 6 mm (¼ in), it is usually contained in tanks or containers. When flammable liquids have no real depth [defined as less than 6 mm (¼ in)] they are not contained and become spill fires or running fires. A third type of class B fire involves flammable liquids or gases that are released from damaged pressurized containers or distribution lines.
- *Class C* fires involve charged electrical equipment. The actual fuels may be Class A, Class B, or Class D. For example, an oil-filled electrical transformer fire involves class B fuels. When the electricity is energized, the fire is classified as Class C.

Similarly, a Class A fire in ordinary combustibles becomes a class C fire when electrical current is present.

- *Class D* fires involve combustible metals such as magnesium, sodium, and potassium.

The extinguisher capacity is important for the size and class of fire. Portable fire extinguishers are rated for their relative extinguishment capacity at a testing laboratory. For example, a Class A fire extinguisher is used by an experienced operator on wood panels above excelsior and on wood cribs. A 1-A rated extinguisher has only enough extinguishing agent for this experienced operator to extinguish a fully involved panel and wood crib once. If enough capacity exists for this experienced operator to extinguish the fire six times, the rating is 6-A.

Class B rated extinguishers are tested by an experienced operator on burning flammable liquids in a flat pan that contains n-heptane floating on water. The flammable liquid surface area that this operator can extinguish with one tank of agent is the rating. For example, a 10-B rating means that an experienced operator is able to extinguish the area of a 1-B rated extinguisher 10 times before the agent in the extinguisher is expended.

The C rating indicates that the agent is suitable for fires in which energized electrical power is present. No rating numerals are used for these fires because they indicate only that the agent can be used with electrical fires. The C rating is used in conjunction with either Class A or Class B fires, or both. An extinguisher rated 6-A, 20-B:C is suitable for class A and class B fires and also can be used when energized electrical equipment is present. The numerical rating indicates its relative extinguishment capability.

Notice that the rating indicates the capability of an experienced operator who extinguishes these fires on a routine basis. An untrained individual will not apply the agent as efficiently as a trained operator. Nevertheless, this fire extinguisher rating system does provide a numerical comparison for the capacity of different fire extinguishers.

Fire extinguishers may be grouped into five different categories based on their extinguishing agent:

- Water
- Carbon dioxide
- Clean agents
- Dry chemical
- Foam.

Dry powder extinguishers are also used for class D fires. All portable fire extinguishers have the following common elements:

- A storage container to hold the extinguishing agent and allow the unit to be mounted or moved.
- The extinguishing agent.
- A means to develop internal pressure sufficient to propel the agent toward the fire.
- Operating features such as hoses, nozzles, and discharge levers.

23.5 Evaluating Extinguisher Effectiveness

Portable fire extinguishers are often viewed as an important fire safety appliance, although the layman may think them more effective than may be justified in practice. Nevertheless, occupant extinguishment (M_p) is an important part of a fire prevention program, and a realistic understanding of performance is important to risk management. However, the evaluation must go far beyond selection of the type and placement of extinguishers.

An occupant manual extinguishment analysis estimates the success of an occupant in extinguishing a fire before it reaches (or slightly exceeds) EB. The success of occupant suppression with portable fire extinguishers can vary from nonexistent to relatively good. An assessment incorporates not only portable extinguishers, but also their availability and the familiarity of occupants with their locations and operation. In addition, factors such as the extinguisher weight and the ability of occupants to reach, secure, and move extinguishers into position before the fire grows too large become important.

To illustrate variations, a library did not want patrons to touch or walk into the fire extinguishers. To prevent unwanted accidents, extinguishers were placed in rarely used exits with the bottom at 2.4 m (8 ft) from the floor. The extinguishers were very safe. The ability of occupants to apply agent before a fire grew too large was questionable. In contrast, portable extinguishers in industrial manufacturing plants are often located where occupants have good access.

The class and size of fire have a great deal to do with the effectiveness of extinguishment. The extinguisher's horizontal stream range relates to the ability of an individual to move close enough to the fire to be able to apply the agent effectively. The quantity and the discharge duration are important considerations in an analysis. Some larger extinguishers have discharge durations of about 1 minute. Other extinguishers will expend the agent within about 10 seconds. There is little time for an untrained individual to develop extinguishing skills when an emergency is being confronted. The most important components of portable fire extinguisher effectiveness are the experience and training of the operator.

23.6 Discussion

The building performance will be the same whatever the cause of ignition. A clear separation avoids obscuring building performance with fire prevention effectiveness. This creates a better understanding of the modes of failure and their outcomes. This separation allows a clearer communication with others regarding the roles and outcomes of failure.

Fire prevention is often an integral part of a comprehensive risk management program. The ability to uncouple fire prevention (prevent EB) from building performance after EB creates a better understanding of the roles of all fire defenses.

One or a combination of ways to use fire prevention information may help to understand its role and communicate better with others.

Statistical Frequency

The frequency of ignition can be identified for a class of buildings similar to the one being evaluated. This information may be useful for comparative purposes to inform others about the regularity of fires. One must interpret the statistics to determine if the term "ignition" really means "established burning." Documentation procedures often unintentionally encourage reporting only ignitions that cannot be ignored. These larger fires may be of such a size that EB is the actual incident being reported.

Initial Growth Scenarios

Selected scenarios may be used for hypothetical ignition locations in order to tell a story to the building management. This procedure starts with "given an ignition" in a specified location. It then examines initial fire growth and occupant manual suppression activities. This technique is useful to describe the effectiveness of different fire prevention programs under consideration.

Ignition Scenarios

One may compare the relative potential for ignition at various locations having differing conditions of heat energy, kindling fuels, and heat/fuel separations. The estimates give an insight into ignition prevention effectiveness and help to order possible alternatives to reduce ignition potential.

Each of these techniques has a role in developing risk management programs that include a fire prevention component.

PART TWO: AUTOMATIC SPECIAL HAZARD SUPPRESSION

23.7   Introduction

Automatic suppression is not usually perceived as a "prevention" component. Some explanation is needed to describe why and how we incorporate automatic special hazard suppression into a performance evaluation.

Special hazard suppression equipment is normally installed to protect a spot area within a larger space. Usually these hazards involve the potential for rapid flame spread such as Class B fires in industrial buildings or food preparation areas in restaurants. Applications such as flammable liquid storage, dip tanks, oil-filled transformers, and deep-fat fryers for cooking illustrate hazards that require special attention.

This type of automatic special hazard protection is not part of an automatic sprinkler system analysis. It is common to have a general area sprinkler system for the entire building, as well as special hazard suppression systems to protect local problem areas within a more extensive operation. This type of automatic protection equipment is installed in industrial or commercial occupancies where each component is analyzed independently. Often, virtual barriers are useful to segment the areas.

Special hazards equipment includes the following:

- Carbon dioxide system
- Dry chemical system
- Clean agent total flooding system
- Water-spray system
- Fine water mist systems
- Foam extinguishing systems
- Explosion protection system.

A special hazards reliability and design effectiveness analysis first considers characteristics of the flammable materials and combustion products as well as the speed and extent of fire propagation. After these potential fire characteristics are understood, we evaluate the reliability and design effectiveness of the suppression system by examining:

- Agent selection
- Detection system
- Release system
- Storage and delivery system.

An important initial determination examines whether the agent is appropriate to the hazard. It is not uncommon to discover a special hazard being protected by an agent that is inappropriate for the fire.

Detection and release systems must account for the expected fire propagation speed and the released combustion products. Special hazard equipment is frequently located in corrosive environments, and long-term maintenance becomes an important consideration in agent delivery evaluations. An additional investigation determines the cost of clean-up and salvage from fire suppression as well as damage to neighboring materials by the extinguishing agent.

23.8   Carbon Dioxide Systems

Carbon dioxide (CO_2) is an inert, non-corrosive, and electrically non-conductive gas. Carbon dioxide extinguishes fires by smothering. The CO_2 gas displaces the normal atmosphere and reduces the oxygen available to the fire to a level where it will not support combustion. While CO_2 may be non-toxic, it can have serious physiological effects on humans at concentrations of more than about 6%. Consequently, humans should not be in confined spaces in which CO_2 is discharged.

Carbon dioxide systems are particularly suitable for fires involving flammable liquids and machinery spaces. They are also useful for food preparation equipment, such as deep-fat fryers, hood and duct systems, and ranges. Because CO_2 systems do not leave a residue and are inert, they often are used in spaces that contain electronic equipment, legal documents and records, valuable clothing, and other materials that could be damaged by other types of extinguishing agent.

Carbon dioxide systems are used for total flooding extinguishment in closed rooms and vaults where openings are small and the gas has an opportunity to remain for a time. Special design features must be incorporated to accommodate spaces where ventilation is greater or where equipment requires a period of time to shut down.

Agent volume and storage requirements depend on the room volume and the design concentration of the CO_2. The storage of CO_2 may be in high- or low-pressure storage tanks or in banks of cylinders, depending on the amount needed.

Smoke, heat, or flame detectors usually actuate carbon dioxide systems automatically. In some spaces, flammable vapor detection or detectors that sense other process abnormalities may be used. All automatic CO_2 systems must also have an independent means for manual actuation. In spaces where humans may be present for extended periods of time, such as electronic computer rooms, manual actuation may be the primary discharge actuation.

In addition to total flooding, local application systems can protect a part of a larger space without flooding the entire area. Food preparation areas, such as deep-fat fryers and hoods and ducts, are one type of local application. Industrial processes, dip tanks, oil-filled transformers, pumps or motors, and access openings in containers of flammable liquids are other types of application. Because the carbon dioxide in local applications is dissipated into the larger volume, special attention is given to direction, duration of discharge, and ventilation. If the heat energy source and the fuels remain, there is the possibility of re-ignition after initial suppression.

Other types of CO_2 application involve hand hose lines or standpipes with a mobile supply. This use is another form of manual fire fighting. An evaluation for manual extinguishing effectiveness would be restricted to the CO_2 application. Detection, notification, agent application, and the final extinguishment are evaluated for the local building fire brigade. Size, experience, equipment, and training of the building occupants become major factors in the evaluation.

23.9   Clean Agent Systems

Clean agent systems describe a family of inert, non-corrosive, and electrically non-conductive gases that effectively extinguish fires by interrupting the chemical reaction and/or reducing the O_2 level. Unlike CO_2 systems, clean agent systems are designed to enable people to survive in the space when the agent is discharged.

Clean agent systems are particularly suitable for fires involving electrical equipment, flammable liquids and machinery spaces. Because clean agents do not leave a residue and are inert, they are often used in spaces that contain electronic equipment, legal documents and records, valuable clothing, and other materials that could be damaged by other types of extinguishing agent.

Clean agent systems are used for total flooding extinguishment in closed rooms and vaults where openings are small and the gas has an opportunity to remain for a time. Special design features must be incorporated to accommodate spaces where ventilation is greater or where equipment requires a period of time to shut down. Agent volume and storage requirements depend on the room volume and the design concentration of the agent.

Smoke, heat, or flame detectors usually actuate clean agent systems automatically. In some spaces, flammable vapor detection or detectors that sense other process abnormalities may be used. All clean agent systems must also have an independent means for manual actuation. In spaces where humans may be present for extended periods of time, such as electronic computer rooms, manual actuation may be the primary discharge actuation.

23.10   Dry Chemical Extinguishing Systems

Dry chemical extinguishing systems discharge a powder mixture onto flames. Portable hand fire extinguishers were the original major means of providing this type of extinguishment, and this continues to be the case. Applications can also utilize total flooding, local application, and hand line hose installations. In many ways, these systems use the same types of components as carbon dioxide systems. For our purposes, the evaluation process for the two systems would be similar, with adjustments for individual capabilities and limitations.

While dry chemical systems are a powder mixture, the term "dry powder" is usually reserved for the graphite and special compounds used to extinguish Class D fires. The dry chemical systems of this discussion are used primarily to extinguish Class B fires. They can also be used for Class A and Class C fires. Extinguishing agents may include chemicals such as sodium bicarbonate, potassium bicarbonate (also known as Purple K or PKP), and monoammonium phosphate (used in ABC portable fire extinguishers). The principal mechanism that contributes to a rapid-fire extinguishment is the interference of the dry chemical particles with the combustion chain reaction through the thermal decomposition of their chemical powders.

Some dry chemicals are slightly corrosive. Consequently, the scene should be cleaned up shortly after extinguishment. Also, dry chemicals have insulating properties that can cause electrical contacts and relays to become inoperative. Their use in these types of installation could cause substantial damage and repair expense.

Total flooding dry chemical extinguishing systems are functionally similar to those of carbon dioxide. Total flooding systems are used in enclosed spaces when re-ignition is not anticipated. A detector usually triggers actuation. Manual operation can also be used.

Fixed local application systems are similar to carbon dioxide systems. They discharge the agent directly onto the spot area being protected. These systems are commonly used for flammable liquid fires, such as dip tanks, electrical transformers, kitchen equipment, and storage vessels.

23.11   Water-spray Extinguishing Systems

Another type of fixed special hazard system is a water-spray extinguishing system. Water-spray systems are used primarily to extinguish fires in installations involving flammable liquid and gas storage tanks, electrical transformers, large electric motors, and industrial process piping. Water-spray systems are not used for building protection, but rather for special hazard installations to extinguish Class B fires or to protect these installations from exposure fires.

A water-spray system has fixed piping and specially designed nozzles that provide a defined pattern of water distribution, with specific water particle size, density, and velocity. Water sprays are deluge systems that do not have fusible elements. Water is discharged directly and forcefully on the surface being protected. Because of the hazards of the installations being protected, the water density is large and the entire surface area is covered. A variety of specific considerations, such as detection, actuation, and operating devices, water supply, drainage, and maintenance practices, are incorporated into the design. An analysis examines each of these influences on the reliability and design effectiveness.

23.12   Fine Water Mist Extinguishing Systems

Water mist systems are becoming more frequent in installations such as electrical equipment and computer rooms, museums and galleries, hydrocarbon fuels, and ships. Water mist is often considered as a replacement for gaseous agents and traditional sprinkler systems because of high efficiency, reduced water damage, and avoidance of toxicity.

Water mist systems use very fine water sprays (i.e. water mist) that have a large surface area to droplet size. This produces rapid conversion from water to vapor. The heat of vaporization transformation causes a large heat loss from the fire. In addition, the vapors displace oxygen and pre-wet unburned fuels, thus changing the combustion process.

The sprinkler installation, fire characteristics, and room features are important to performance. Water mist systems must be carefully designed to accommodate the needs of the type of fire. Sprinklers may have open nozzles or closed nozzles; pressure variations from low to high; continuous discharge or intermittent discharge. The room must be enclosed to contain the heat and water vapor and prevent spread beyond its boundaries. A carefully designed water spray system can be effective and economical to protect the special hazards for which they are selected.

23.13   Foam Extinguishing Systems

Another type of special hazard system for flammable liquid fires involves foam application. Foam is a mixture of water and a solution of specially formulated liquid agents which, when mixed, create gas-filled bubbles. The bubbles float on the flammable liquids and create a vapor layer that seals the surface, cools the material, and excludes air from the potential combustion process. Foam fire extinguishment is very effective for aircraft fuel fires or for flammable materials that spill and flow along surfaces.

Foam extinguishment can be delivered in three forms. One is a fixed system in an enclosed room. The room may be a tank or vault, or it may be an aircraft hangar. In a fixed system, the delivery system is essentially a deluge-type sprinkler system. The open heads allow a blanket of foam to be spread over the entire area.

The second is a local application system. Here the fixed deluge-type sprinkler equipment is positioned around the area to be protected and the foam is applied in a similar manner over the spot area. The third and most common application is by hand hose lines. Fire fighting foam is particularly effective for flammable liquid fires because the fire fighters can move against the flammable liquids and extinguish the fire in a progressive manner.

Foam is produced by mixing a predetermined amount of foam concentrate with the water that flows through the hose line or pipe. The foam may be mechanical (air) foam that is formed by pre-mixing or proportioning the solution with the water. Alternatively, the foam may be chemical foam that is created by a reaction between an alkaline solution and an acid solution. Foaming agents and equipment can generate expansions ranging from low to high. This involves bubble aggregation expansions from about 20:1 to 1000:1. Much of the technology involves the type of foaming agents, the mixing chamber and process, and the transportation of the equipment to the fire scene.

These special hazard systems can be designed for fixed, local application or portable hose lines. Their function is to extinguish a flammable liquid fire. Foam extinguishment can be very effective for these types of fire, especially when used as the extinguishing agent by fire fighting forces.

23.14 Explosion Suppression Systems

An explosion is an extremely rapid combustion that produces a rapid rise in temperature and pressure because of the speed and confinement of the reaction. The speed of the energy release and the resulting pressure rise become the explosion. The possibility of an explosion is present when concentrations of dusts, mists, gases, and flammable vapors are in a confined space. These explosive conditions can exist in a wide variety of industrial operations.

It is possible to design for an explosion by incorporating pressure-sensitive construction, such as walls that are designed to blow away cleanly with the rapid pressure rise. This releases the pressure and reduces the damage in other parts of the building. Alternatively, explosion suppression systems can be installed. These systems use extremely sensitive detection devices that actuate during the time lag between ignition and the development of the destructive pressure – a time lag measured in milliseconds. The detection device is electronically connected to the suppression agent discharge device. This device is an explosively actuated suppressor or a high discharge rate extinguisher. The suppression agent is discharged faster than the flame is propagated. The system confines and neutralizes the explosion.

23.15 Building Evaluations for Special Hazard Installations

Special hazard systems, also called spot systems, are designed and installed to protect a distinct and specific hazard. Most of the time, the hazards involve flammable liquids that are associated with industrial occupancies. Hood and duct systems in kitchens also are protected by these systems. However, industrial plants, restaurants, electrical installations, and aircraft hangars are all buildings, and one can evaluate their performance. Performance evaluations of special hazard systems must have a clear position within the complete analytical system.

The exact position of a special hazard system within the building evaluation depends primarily on its role. These systems may be viewed in three categories. The first is a total flooding system in which the agent covers an entire enclosed space. In this role the system is analogous to a sprinkler system. Although there are substantial differences, the function is to extinguish small fires and prevent extension of the fire throughout the space. If the system is successful, relatively little damage will occur.

When the system is unsuccessful, full room involvement occurs and succeeding barrier–space modules must be evaluated for fire propagation along other paths. It is common to have automatic sprinkler systems in addition to special hazard systems. If both systems exist in the same enclosed room, the two automatic suppression systems are evaluated separately.

Local application systems pose a different organizational problem because they protect specific zones within a larger room. In addition, the special hazard systems provide rapid initial extinguishment to a special problem. It is typically expected that a company fire brigade will provide additional extinguishing support in the case of insufficient agent capacity or a re-ignition. We incorporate this as a part of the prevention analysis.

In summary, special hazard installations are handled in one of two ways, depending on their function in the building. Total flooding systems are evaluated in a similar way to a sprinkler system because the entire floor area is covered. If these systems are to be augmented by an industrial building fire brigade, the effectiveness is evaluated and incorporated in the usual manner for a space. Subsequent reinforcement by the local fire department may be incorporated using the normal logic for a space evaluation.

23.16 Closure

Knowledge of a building unfolds logically as it is evaluated in an organized, structured manner. It is important to be able to uncouple and distinguish performance after EB from the conditions that lead to EB. This orderly process enables one to develop a clear understanding of the building's strengths and weaknesses.

Many times, the evaluation will not require a "prevent EB" analysis. Plans approval is one such situation; a risk identification and consequence evaluation is other. However, some buildings survive only on the success of fire prevention. Owners and occupants should be aware of these buildings. If post-EB protection cannot be improved, attention can be directed to the pre-EB protection.

Part III

The Analysis

The analytical framework is based on established techniques from other disciplines. It is cohesive, comprehensive, and has evolved to the point where confidence can be placed in its application.

Although the framework provides both form and function for fire safety analysis, it is not complete without numerical measures of performance. One must be able to quantify important numerical values in order to get a sense of proportion and to be able to compare differences. Engineering evaluations combine a framework for thinking and quantification for understanding, comparing, and deciding. One is sterile without the other. The Interactive Performance Information chart provides both structure and information for component and holistic analysis.

Unit Three consolidates the framework and the factors that influence performance. The analytical framework is based on the systems framework of Unit One and the operational performance of the fire and fire defenses of Unit Two. Examples for selected components augment the framework.

Fire Performance Analysis for Buildings, Second Edition. Robert W. Fitzgerald and Brian J. Meacham.
© 2017 John Wiley & Sons Ltd. Published 2017 by John Wiley & Sons Ltd.

24

Fire Performance: Framework for Analysis

24.1   Organizational Concepts

Fire behavior in buildings is complex. The fire itself is dynamic, changing in size and characteristics on a minute-by-minute basis. Building defenses and risks interact with the fire and are often influenced by other components. Additionally, they phase in and phase out of action in different sequences depending upon the building design and the fire.

Nevertheless, this complex process is analytically tractable. Figure 24.1 shows the relationship of the major components. The Interactive Performance Information (IPI) chart in Figure 24.2 provides the organization with which to order component actions independently or in combination. Time is the element that links all parts of this complex system.

24.2   Performance Evaluations

Deterministic analyses provide the basis for component quantification. One can establish a credible understanding of a building's dynamic fire behavior with state-of-the-art fire science and engineering. Although many calculations methods are still incomplete, they are sufficient to provide a base of understanding when they are combined with qualitative analyses. The analytical framework in combination with the IPI chart helps to bridge the gaps of knowledge and establish a holistic performance understanding.

Quantitative analysis and *qualitative analysis* provide a matched pair of techniques to quantify component behavior. Sometimes a quantitative analysis is supplemented by a qualitative analysis to substantiate numerical outcomes. At other times, a qualitative analysis will initiate an evaluation involving quantitative measures of behavior. In general, one will use both techniques during the same evaluation to gain a better understanding of performance.

Variation analysis provides a "what if" examination of differences in condition that may be reasonably expected in a building. A variation analysis enables one to identify reasonably different conditions that may influence the deterministic outcomes. This enables one to bracket ranges of performance to gain a more complete understanding of what to expect during a building fire.

Fire Performance Analysis for Buildings, Second Edition. Robert W. Fitzgerald and Brian J. Meacham.
© 2017 John Wiley & Sons Ltd. Published 2017 by John Wiley & Sons Ltd.

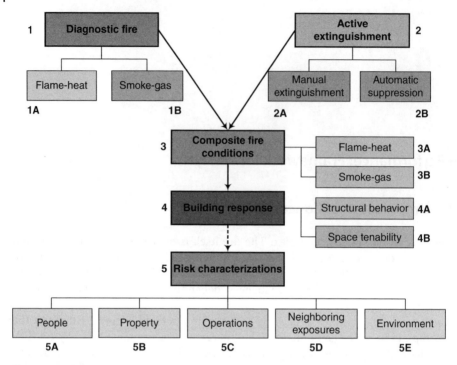

Figure 24.1 System components.

24.3   Analytical Framework

Analysis is based on component function and operating characteristics during a fire. *Unit Two: The Parts and How They Work* describes the components and their roles in the analytical process. *Unit Three: The Analytical Framework* identifies the analytical structure of the components and their integration within a holistic performance.

Unit Two provides the operational knowledge on which the analytical framework of Unit Three is based. Unit Three identifies the analytical structure and its use in component evaluations. Unit Three assumes that one understands component function, types, operating characteristics, and interrelationships between the fire and other components that were described in Unit Two.

The framework is structured with:

a) Event logic diagrams to identify constituent relationships within components.
b) Event trees to relate changes in component status to time.
c) Success (or fault) trees to structure the thought process for component analysis form the analytical framework.

The IPI chart organizes all of these tools into a time-related format.

Additional tools provide help to organize information and calculate outcomes. For example, networks are particularly valuable to identify a thought process. Additionally, networks can be used for quantitative analysis of events of certain analytical processes.

Working IPI template				Time			
	Section	Symbol	Event	5		10	
	1	**$\bar{I}$**	Diagnostic fire				
	A	$\bar{I}_F$	Flame-heat				
	B	$\bar{I}_S$	Smoke-gas				
	2		Active extinguishment				
	A	M	Fire department extinguishment				
	a	$M_{\text{part A}}$	Part A: Ignition to notification				
		MD	Detect fire				
		MN	Notify fire department				
	b	$M_{\text{part B}}$	Part B: Notification to arrival				
		MS	Dispatch responders				
		MR	Responders arrive				
	c	$M_{\text{part c}}$	Part C: Arrival to extinguishment				
		MA	Apply first water				
		MC	Control fire				
		ME	Extinguish fire				
	B	A	Automatic sprinkler suppression				
	a	AA	Sprinkler system is reliable				
	b	AC	Sprinkler system controls fire				
	3	**L**	Composite fire				
	A	L_F	Flame-heat				
	B	L_s	Smoke-gas				
	4	**R**	Building response				
	A	St	Structural frame				
	B	TS	Target space tenability				
	5	**L**	Risk characteristics				
	A	LS	People				
	B	RP	Property				
	C	RO	Operational continuity				
	D	RE	Exposure protection				
	E	RN	Environment				

(Left-margin groupings: Building performance → Fire response covers Sections 1–2; Building response covers Sections 3–4; Risk characteristics covers Section 5.)

Figure 24.2 Working Interactive Performance Information (IPI) template.

In general, a variety of techniques from other engineering disciplines can help to structure and quantify procedures.

Integrated networks of important events structure the thought process, quantification needs, and interactions relating to time. The IPI chart enables one to compare the state of one component with others that influence its behavior. The IPI chart becomes the primary tool to understand the dynamic interactions of different components. Event logic diagrams, event trees, success and fault trees, and network representations are all integrated in the IPI chart.

The IPI chart provides a simple and systematic way to document diagnostic fire conditions for a continuum of time in rooms of interest. This documentation enables global time to be related to local time. Examination of vertical columns enables one to relate diagnostic fire conditions and the status of all fire defenses at the same instant of time.

The IPI chart can accommodate any type of evaluation and building. Qualitative and quantitative documentation can be done conveniently. A series of charts can establish a clear set of documentation for any type of analysis. Supplemental videos, digital photographs, experimental data, calculation tables, and opinion enhance the information repository and the quality of assessment.

24.4   Fire, Risk, and Buildings

Diagnostic fire scenarios provide the foundation for performance analysis. The diagnostic fire identifies the species and characteristics of the products of combustion that relate to the evaluation of the defense and risk components. These are identified more specifically in the component evaluations.

The building is central to all performance evaluations. The fire and the fire defenses interact within the building. Risk characterizations are site-specific for the particular building being examined. In fact, risk may be defined in terms of *windows of time* the building will provide for people, property, and operational continuity with tenable conditions.

A window of time may be short or indefinitely long for different parts of an analysis. Building performance evaluations examine fire scenarios to describe the risk of different parts of the building to sustain functionality for people and operations.

Although the framework provides both form and function for fire safety analysis, it is not complete without numerical measures of performance. One must be able to quantify important numerical values in order to get a sense of proportion and to be able to compare differences. Thus, engineering evaluations combine a framework for thinking and quantification for understanding, comparing, and deciding. One is sterile without the other.

25

The Diagnostic Fire

25.1   Introduction

The diagnostic (design) fire serves as the "loading" against which the fire defenses act. The outcomes of load–resistance analyses establish time durations with which to characterize risks.

Neither professional engineers nor the regulatory community have yet established standardized fire characteristics for performance-based engineering. However, prescriptive regulations also do not describe fires in their codes of practice. Until standard fire conditions are established to describe expected performance, the engineer must identify the appropriate fires with which to test the building's fire defenses and characterize risks. This requires careful documentation of rationale for selections.

A diagnostic fire describes fire conditions that are used to examine the performance of each component of the fire safety system. Because the focus is on analysis, one does not consider factors of safety that may be associated with design situations. Eventually, codes and standards will handle these issues. Until that happens, the engineer of record and the authority having jurisdiction must agree on appropriate values that may include margins for unknown interactions and safety. Nevertheless, contemporary knowledge of fires enables one to gain far greater understanding of fire performance in buildings than routine regulatory compliance.

The diagnostic fire depicts a building fire. It is easier to describe a diagnostic fire for existing structures because one can examine prevailing conditions. It is more difficult to envision appropriate fire conditions for proposed new buildings, because one must anticipate individualized uses and routine building and interior design renovations.

Identification of a diagnostic fire scenario involves two decisions: the selection of appropriate rooms of origin; and the identification of the time of release of products of combustion against the fire defenses. Although these two roles are interactive we discuss them separately.

An initial overview, sometimes called "scoping," of the building gives an insight into functions, hazards, fire defenses, and risks. The process is a valuable professional skill that provides a good understanding of fire safety issues that must be addressed. As a part of this overview, top-down modular analyses help to recognize pockets of unusual hazards. Because barrier openings are a part of the modules, one may identify clusters of rooms through which fire can spread easily. The top-down analysis is rapid and a part of the initial overview. Computer programs can help to sort hazard clusters more easily.

Fire Performance Analysis for Buildings, Second Edition. Robert W. Fitzgerald and Brian J. Meacham.
© 2017 John Wiley & Sons Ltd. Published 2017 by John Wiley & Sons Ltd.

In addition to high hazard clusters, candidate rooms of origin may be selected to examine special needs such as:

- Room fire locations that cause greatest life safety problems.
- Rooms that create the greatest vulnerability for property protection or operational continuity.
- Room conditions or room clusters that may pose difficulties for manual fire fighting.
- Representative rooms for the building.
- Rooms that need special attention.

A top-down analysis is fast, informative, and gives a good understanding of fire safety problems in a building. Bottom-up analyses and computer fire modeling use similar factors. Top-down and bottom-up analyses complement each other.

25.2   Top-down Estimates

A top-down classification ranks a room's interior design for its relative ease to reach flashover (FO). The estimate uses five classification groups to segment the relative potential for an interior design to reach FO.

The interior design defines a composite interaction of room size and combustible contents. Usually, one gets an initial "feel" for the ease of reaching FO by viewing room size, shape, and ceiling height. Small rooms reach FO more easily than large rooms. Then, the size and distribution of combustible fuel groupings are examined. For example, FO occurs easily when fuel groups have the firepower to reach the ceiling. Interior finish is examined for its role in room fire development. Chapter 7 describes details for making top-down estimates.

The classification assumes stoichiometric burning. Room opening sizes and locations are not a part of the interior design classification even though we recognize their significance in the combustion process. The classification is based only on the room geometry and contents rather than the chance ventilation status of a door or window.

A top-down estimate associates a time to reach FO with the fire growth potential. Although default times are provided, the estimate requires a personal estimate because the inspector can recognize conditions that may increase or decrease default time durations. For example, fuel materials and configurations or ventilation conditions that significantly affect the time to FO can be incorporated into the estimate. Thus, one can adjust the time to FO to reflect specific observational conditions.

25.3   Modular Estimates

Any room can be a room of origin or a room beyond the room of origin. The room's interior design is the same in either situation. However, the type of barrier failure affects the ease and speed of fire development in rooms beyond the room of origin. The room and its barriers are modular because a single classification group can describe all outcomes for the room regardless of the actual room of origin.

A top-down modular estimate is done in three parts:

1) Classify the fire growth potential for the room's interior design. Treat this classification as if the room were a room of origin. Estimate a time to FO that seems appropriate for the room. While most interior designs would be expected within the default range, different values may be selected for special conditions.
2) Then, consider the room to be adjacent to a fully involved room. Because the fire must spread into the room being evaluated through a barrier, the type of barrier failure is important to the time to FO. Using the room's interior design classification, estimate a time to FO if one of the surrounding barriers exhibited a hot-spot ($\bar{T}$) failure.
3) Again considering the room to be adjacent to a fully involved room, estimate the time to FO if one of the surrounding barriers exhibited a massive ($\bar{D}$) failure.

Figure 25.1 provides a convenient way to describe these estimates. Each room module will have a single classification and three time durations to FO. One is as a room of origin. The other two represent fire spread expectations for different barrier failure types if the room were adjacent to a fully involved room.

The modular structure associated with Figure 25.1 is equally applicable to virtual barriers and to (real) physical barriers. Although a virtual barrier provides no resistance to flame-heat or smoke-gas movement, the modular organization enables one to estimate fire propagation through the different zones.

A variety of performance applications become feasible because top-down classifications can be done so quickly. For example, when all room classifications are shown on an architectural plan, patterns of fire growth hazards may be quickly visualized. Pockets of high (or low) hazard can be identified. Experience has shown that inspectors are able to identify "red," "yellow," or "green" rooms consistently and quickly to focus attention on building areas that can become important to a performance analysis.

	α	Default time (min)	Estimated time (min)	Default time (min)	Estimated time (min)	Default time (min)	Estimated time (min)
Very high	0.06	1–4	___	1–3	___	0–1	___
High	0.03	4–8	___	3–6	___	0–2	___
Moderate	0.003	8–12	___	6–9	___	1–2	___
Low	0.002	12–16	___	9–12	___	1–3	___
Very low	0.001	16–20	___	13–16	___	1–4	___
		Room as a room of origin		**Subsequent room with a $\bar{T}$ failure**		**Subsequent room with a $\bar{D}$ failure**	

Figure 25.1 Fire growth hazard (FGH) estimate template.

Although the procedures are associated with existing buildings, they can be adapted to proposed new buildings. The uncertainties in proposed buildings can be minimized with defined conditions and "what if" variability analyses. Nevertheless, top-down estimates provide a clear picture of building performance and associated risks. This information becomes an important component of preliminary screening and initial ordering for performance.

25.4   Bottom-up Scenario Analysis

The bottom-up scenario analysis selects an ignition location and examines expected fire behavior. A qualitative evaluation determines if "enough" factors can be combined at each realm to enable the fire to transition to the next higher realm. The major events from ignition to FO were highlighted in Chapter 7.

Figure 25.2 shows the event logic success tree that describes the events that must be completed for a fire to reach FO. The tree can be evaluated at successive time durations until all five events have been completed to identify the time to FO.

A bottom-up scenario analysis gives a clear understanding of the relative importance of different factors as a fire grows from ignition (IG) to established burning (EB) through FO. The examination identifies transitions for factors that phase in and out of importance. Chapter 7 identifies the factors that dominate behavior in each realm to enable one to match calculation methods with appropriate conditions.

25.5   Network Estimates

Figure 25.3(a) represents the event logic tree concepts in network format. A diagnostic fire considers only the conditions from EB to FO. Therefore, the events from the fire free status (FFS) to EB are considered a part of fire prevention and described in Chapters 23 and 34.

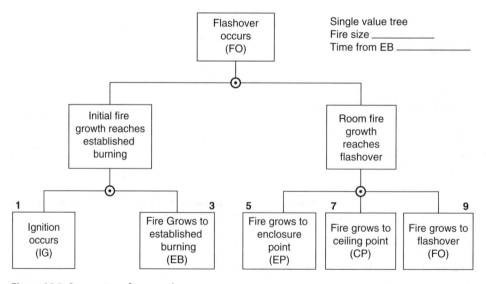

Figure 25.2 Success tree: fire growth.

A scenario analysis starts with a 25 kW "bucket of fire" (i.e. EB) that is located at the fire's origin. The network in Figure 25.3(c) guides examination of each successive event. The factors that influence each event are shown in Figure 25.4(a)–(e).

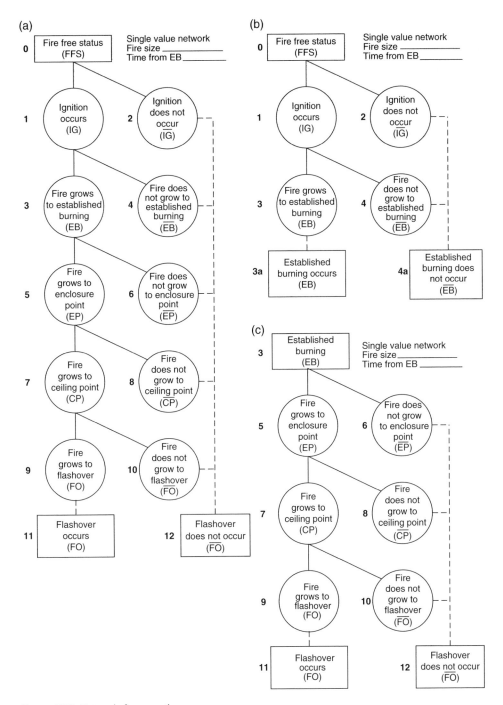

Figure 25.3 Network: fire growth.

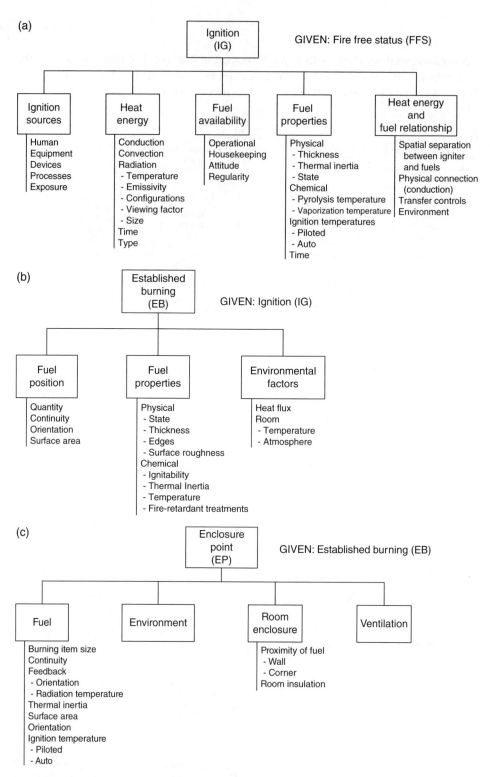

Figure 25.4 Evaluation factors: (a) ignition (IG); (b) established burning (EB); (c) enclosure point (EP); (d) ceiling point (CP); (e) flashover (FO).

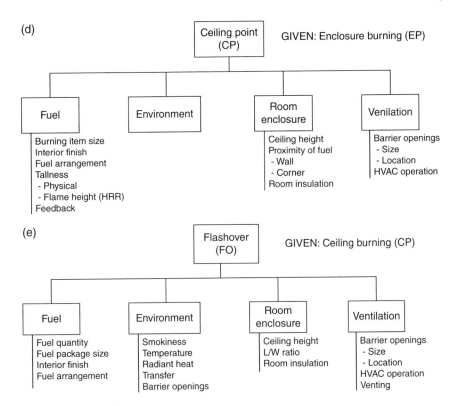

(d)

Ceiling point (CP)

GIVEN: Enclosure burning (EP)

Fuel
Burning item size
Interior finish
Fuel arrangement
Tallness
- Physical
- Flame height (HRR)
Feedback

Environment

Room enclosure
Ceiling height
Proximity of fuel
- Wall
- Corner
Room insulation

Venilation
Barrier openings
- Size
- Location
HVAC operation

(e)

Flashover (FO)

GIVEN: Ceiling burning (CP)

Fuel
Fuel quantity
Fuel package size
Interior finish
Fuel arrangement

Environment
Smokiness
Temperature
Radiant heat
Transfer
Barrier openings

Room enclosure
Ceiling height
L/W ratio
Room insulation

Ventilation
Barrier openings
- Size
- Location
HVAC operation
Venting

Figure 25.4 (Cont'd)

25.6   Scenario Applications

A bottom-up analysis is particularly suited to fire investigations. Potential locations of fire origins can be evaluated to establish what is needed for the fire to grow through its various realms. If an application examines only the cause and initial burning, Figure 25.3(b) would be separated. This is particularly valuable with fire prevention evaluations. If one were only interested in the fire growth potential, the truncated version of Figure 25.3(c) provides enough information.

During a fire investigation, the room can be examined to determine if the necessary features were present at the time of the fire. Specific laboratory tests can focus on materials, their configuration, or special conditions to provide better values for calculations and understanding of outcomes.

One can use bottom-up analyses to gain a greater understanding of computer fire model outputs. Often, computer models use approximations or default values to enhance calculation needs. The built-in calculation procedures may or may not be appropriate for the particular situation being studied. A bottom-up analysis provides a means to qualitatively examine calculation outputs to determine if the values are reasonable for the room. The results can compare similar conditions more easily.

The interaction of mathematical models, bottom-up scenario expectations, and top-down perceptions can provide greater confidence in estimates provided by each method. Conscious comparisons enhance observational powers and evaluation of modeling results. The analytical framework provides a means to "bridge the gap" between computational output and qualitative estimates.

25.7 Interactive Performance Information (IPI) Chart Applications

Top-down and bottom-up evaluations are useful and require little professional time to complete. They enable an engineer to organize a holistic analysis. When an appropriate room of origin is selected, more detailed quantitative fire analysis is normally justified.

Quantitative fire modeling provides numerical values of time durations for fire growth in the room of origin and the cluster of additional rooms that may become involved. Additionally, factors such as fire duration, heat release values, and products of combustion may be calculated. These can be compared with qualitative estimates to recognize when the two approaches do not generally agree.

The results of the diagnostic fire are then transferred to Sections 1A and 1B of the IPI chart. These values are then used as the "load" to examine fire defenses and associated risk characterizations.

26

Fire Detection

26.1   Introduction

One has confidence in a detection system when it has been designed and installed by quali-fied professionals in accordance with an accepted national standard. However, standards continually evolve, equipment improves, and original systems may become outdated. Also, standards allow one to tailor a detection system to design needs. After an initial design, commercial buildings frequently undergo interior changes to accommodate new tenants or needs and an existing detection system's performance could be compromised.

A performance analysis estimates the fire size when a hostile fire is first detected. Knowledge of diagnostic fire characteristics enables one to relate the time from estab-lished burning (EB) to the fire size at detection. Although the *fire detection* event initi-ates other fire defense operations and risks, any subsequent actions are not considered in a detection evaluation.

Both detection instruments and humans can detect a fire. An evaluation examines each independently. Detection instrument analysis is relatively straightforward. Quantitative methods are well advanced, although care must be exercised to relate out-comes accurately to actual building conditions. Qualitative techniques supplement the quantitative tools to give a better understanding of performance. A building evaluation normally combines both quantitative and qualitative analyses.

Human detection is more difficult to evaluate and involves greater uncertainty. Nevertheless, a methodical analysis of human detection, assuming an inoperative or no automatic detection system, gives a good understanding of the way the building works. In addition to evaluating human detection outcomes, one obtains a good introductory insight into occupant risks. Chapter 11 describes automatic detection systems and human fire detection. Chapter 26 examines performance analysis.

Fire Performance Analysis for Buildings, Second Edition. Robert W. Fitzgerald and Brian J. Meacham.
© 2017 John Wiley & Sons Ltd. Published 2017 by John Wiley & Sons Ltd.

PART ONE: AUTOMATIC DETECTION

26.2 Detection Analysis

A detection analysis involves the diagnostic fire, detector operation, and transport of products of combustion (POCs) from the fire to the detector. Figure 26.1 shows the major factors that affect performance.

Some of the factors in Figure 26.1 cannot be incorporated into contemporary quantitative methods. Qualitative techniques are inadequate for others. However, together, the strengths of one augment weaknesses of the other. Thus, we integrate both qualitative and quantitative methods to provide a better understanding of fire detection capabilities and determine the fire size at detection.

Figure 26.2 shows a single value event logic diagram for the success events that structure a time-related analysis. Figure 26.3 shows the corresponding evaluation network.

The network of Figure 26.3 provides a way to identify the time of detection for the Interactive Performance Information (IPI) chart. In theory, if the network were

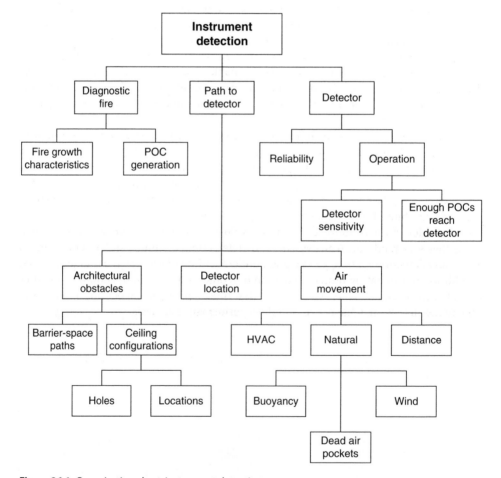

Figure 26.1 Organization chart: instrument detection.

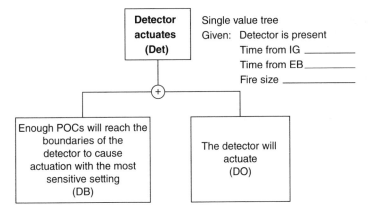

Figure 26.2 Success tree: instrument actuation.

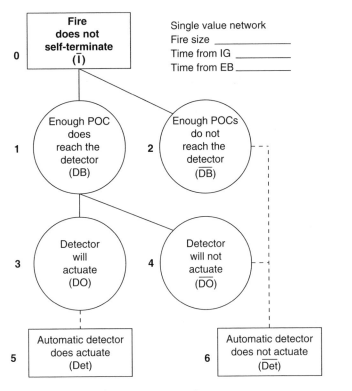

Figure 26.3 Network: instrument actuation.

evaluated at each minute after EB (or ignition, if appropriate) the result would be $\overline{\text{Det}}$ until the detector actuates. At that time the status changes to Det.

The events of Figure 26.3 are sequential. Event 1 is evaluated to determine if enough POCs can reach the detector to cause actuation of the most sensitive setting for the detector type. Then, detector sensitivity (Event 3) is examined to determine if the

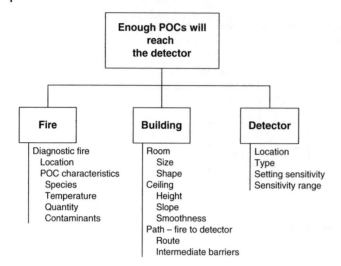

Figure 26.4 Organization chart: detector analysis.

detector will respond to the POCs that reach it at the time of being examined. If not, additional POCs must reach the detector (larger fire and more time) to cause actuation.

Both qualitative and quantitative analyses use the same conceptual process in Figures 26.1 and 26.3. A qualitative analysis envisions POC transport along its path from the fire to the detector. Features that influence POCs affecting detector response can be recognized to ensure that quantitative outcomes are reasonable. Quantitative analyses provide a sense of proportion for numerical values. Figure 26.4 identifies factors that have a major influence on the fire size at detection.

26.3 Detection Example

Fire detection initiates a building's fire performance analysis. Detection analysis uses the actual installation, including detector type, locations, the diagnostic fire, and the transport path from fire to detector. A performance analysis describes what to expect from a specific detection installation.

Although a compliance check of the applicable standard provides useful information, a performance analysis augments that knowledge with qualitative and quantitative examinations. Example 26.1 describes the process.

Example 26.1 Figure 26.5 shows the fourth floor plan for the model building. The detection system used minimum requirements for office occupancy to obtain approval from the authority having jurisdiction.

Fixed-temperature heat detectors with actuation temperature of 135°F (57.2°C) were selected. The *National Fire Alarm Code* (NFPA 72) identifies heat detector installation spacing based on detector listing. Maximum spacing (S) can be 50 ft (15.24 m) by 50 ft (15.24 m) on center, with the maximum distance to a wall of 0.5S. The maximum distance to an unprotected area (e.g. corner) is 0.7S.

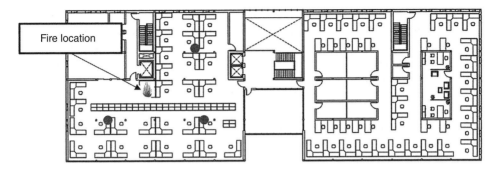

Figure 26.5 Example 26.1: floor layout.

The three detectors shown in Figure 26.5 (blue circles) provide minimum detection requirements for the open plan area in the west side of the building.

Assuming an ignition in this section of the building, select a fire size for detection.

Solution:
One can estimate the actuation time for heat detectors using algebraic methods or computational fluid dynamics (CFD) modeling. Algebraic calculations using a spreadsheet are faster. Computer models estimate growing fires better, although the setup time is greater. Thus, the value of improved information can be related to its professional cost.

We will assume an ignition location in a workstation nearly equidistant from the three detectors, as shown in Figure 26.5.

Quantitative Estimates. The response time index (RTI) ($m^{1/2}$ $s^{1/2}$) is a measure of how quickly a thermally activated material responds to a given gas temperature and velocity near the device. New heat detectors may have their RTI listed. If not, the RTI can be related to the time constant (τ_0) for a heat detector. The relationship is roughly estimated as RTI = $\tau_0 u_0^{1/2}$.

Values for τ_0 can be found in various resources, such as NFPA 72. For a heat detector with a listed 50 ft (15.24 m) spacing and an actuation temperature of 135°F (57.2°C), $\tau_0 = 44$, as measured at a reference velocity (u_0) of 1.5 m/s. The RTI for this device = $(44\ s)(1.5\ m/s)^{1/2} = 53.9\ m^{1/2}\ s^{1/2}$.

The heat detector response using Alpert's correlations is based on smooth, unconfined ceiling conditions. Equation (26.1) is appropriate for the open plan office space of Figure 26.5:

$$t_{actuation} = \left(\frac{RTI}{\sqrt{u_{jet}}}\right)\left\{\ln\left[\frac{\left(T_{jet}-T_a\right)}{\left(T_{jet}-T_{actuation}\right)}\right]\right\}$$

(26.1)

where:

RTI = detector response time index ($m^{1/2}$ $s^{1/2}$);
u_{jet} = ceiling jet velocity (m/s);
T_{jet} = ceiling jet temperature (°C);
T_a = ambient air temperature (°C);
$T_{actuation}$ = actuation temperature of detector (°C).

The ceiling jet temperature and ceiling jet velocity at the detector are needed to solve this equation. As a first-order approximation, Alpert's correlations for steady-state fires are typically used. These correlations are divided into two categories based on the ratio of the distance between the fuel and the detector, and the height of the detector above the top of the fuel item.

We select a constant heat release rate of 700 kW. H is the vertical distance between the fuel surface and heat detector (2.5m), and r is the horizontal distance from the centerline of the fire to the detector (9.7 m). This value for r/H of 0.95 is a valid representation for Equation (26.1) ($r/H > 0.18$).

The ceiling jet temperature when the ratio $r/H > 0.18$ is:

$$T_{jet} - T_a = 5.38 \frac{Q^{2/3} / H^{5/3}}{(r/H)^{2/3}} \tag{26.2}$$

The ceiling jet temperature when the ratio $r/H \leq 0.18$ is:

$$T_{jet} - T_a = \frac{Q^{2/3}}{H^{5/3}} \tag{26.3}$$

where:

Q = heat release rate of fuel (kW);
H = height of ceiling above top of fuel (m);
r = the radial distance from the plume centerline to the detector (m).

The ceiling jet velocity when the ratio $r/H > 0.15$ is:

$$u_{jet} = 0.197 \frac{(Q/H)^{1/3}}{(r/H)^{5/6}} \tag{26.4}$$

Ceiling jet velocity when the ratio $r/H \leq 0.15$ is:

$$u_{jet} = 0.947 \left(\frac{Q}{H} \right)^{1/3} \tag{26.5}$$

The detector actuates at 57.2°C assuming an RTI of 54 $m^{1/2} s^{1/2}$ and an ambient temperature of 20°C.

Equation (26.1) is valid for steady-state fires. The workstation would have a growing fire. Growing fires that follow a power law (e.g. t^2 fire) developed by Heskestad and Delichatsios, with modifications from Beyler, provide better approximations. The theory and calculation approaches are described in detail in the *NFPA Fire Protection Handbook*, the *SFPE Handbook of Fire Protection Engineering*, and Annex B of NFPA 72, *the National Fire Alarm Code*. These calculations give an actuation time of 450 seconds for the same conditions described earlier.

The third approach using a CFD model requires more professional time. Using the same conditions, this gives an actuation time of 400 seconds.

26.4   Detection Estimate

The calculation procedures described above provide the following information:

Alpert correlation: t = 490 s
HD & B calculation: t = 450 s
CFD model t = 400 s

The diagnostic fire and room conform somewhat closely to the conditions on which the quantitative methods are based. This improves confidence in the calculations.

Figure 26.3 uses quantitative concepts to guide the thought process for establishing the fire size (and time) at detection. Figure 26.4 identifies factors that influence performance. In this case, performance is defined as the fire size and time of detection. Because the qualitative process is based on quantitative modeling, one must adjust quantitative outcomes to reflect performance more confidently.

The calculations used the fuel characteristics described in Figure 9.7, although the furniture item is a part of a larger fire group. Therefore, additional fuels will become involved. Also, the fire group is located near a barrier. The combination of calculation results, fuels involved, room configuration, and general room conditions contributes to a selection of 360 seconds from EB and a fire size at detection of about 1200 kW.

The calculations used an RTI of 54 $m^{1/2}$ $s^{1/2}$. Detectors rated at 135°F (57.2°C) may have RTI values somewhat different from 54 $m^{1/2}$ $s^{1/2}$ and remain standard-compliant. The second event of Figure 26.3 provides a way to incorporate this type of variation into the performance estimate.

Detectors may be of different types and configurations. Each requires calculations that are suited to its specifications. Also, fires may occur at other locations and involve other fuel groups. POC transport is affected by building conditions. Regardless of the actual conditions of the building, combining quantitative measures with qualitative adjustments enables one to estimate performance situations with greater confidence.

Although we may have more confidence in a CFD analysis for this scenario, the professional costs of information must be balanced with the value of improved accuracy. The engineer must use qualitative judgment to augment computational results. Uncertainties are clearly understood. Chapter 35 discusses ways to manage uncertainty.

26.5   Detector Reliability

Evaluation procedures assume that the detector is operational when it is needed. However, evaluation of instrument reliability (Section 11.5) is a necessary part of a performance evaluation. Reliability for new detection systems is dependent on installation quality. Reliability for older systems is dependent on long-term maintenance practices.

Figure 26.6 shows the thought process for a reliability evaluation. Each event may be examined separately because they are independent. After an installation's acceptance test, only the maintenance quality need be evaluated. The performance of this event declines with age unless a programmed maintenance schedule is followed.

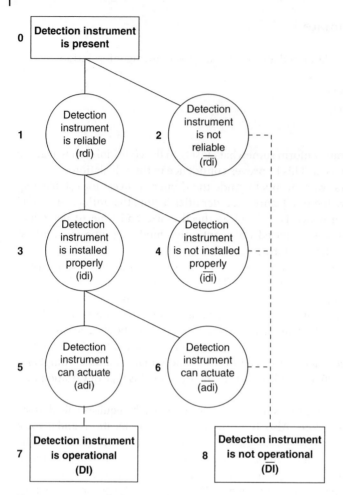

Figure 26.6 Network: detector reliability.

PART TWO: HUMAN DETECTION

26.6 Concepts in Human Detection Analysis

The concepts of human detection are similar to those of automatic detection except that an individual's location is movable. Humans are sensitive to the smoke, heat, light, and noise from a fire. Also, humans can be adept at distinguishing if the POCs are the result of an actual fire or just an abnormal condition. However, the time needed to make decisions and the reliability are less certain. Nevertheless, the exercise of evaluating the fire size at human detection provides an insight into the building and associated risks.

Figure 26.7 describes the factors involving POCs reaching an individual.

The ability of a human to discover a fire without a cue from an automatic alarm depends on the individual (who represents the detector), the fire, the building and how it works. Often, a building will be occupied for certain periods during the day and unoccupied for others. The building population may be large or small and involved in a variety of activities. Figure 26.8 identifies conditions that influence outcomes when the building relies on human detection rather than automatic detection.

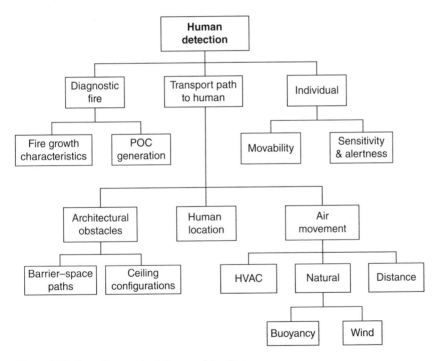

Figure 26.7 Organization chart: human detection.

26.7 Human Detection Analysis

A building analysis to determine the fire size at human detection can only be done with a qualitative analysis. The process involves selection of a base scenario to represent a reasonable, yet common situation. Then, a set of variability analyses is examined to

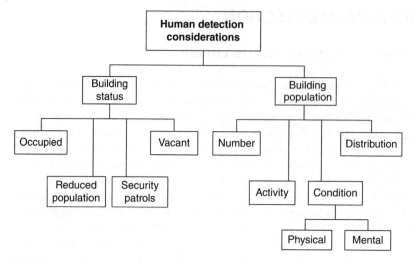

Figure 26.8 Organization chart: human detection.

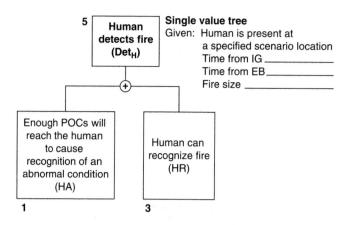

Figure 26.9 Success tree: human detection.

establish the range of outcomes that can be expected for other conditions that may arise during the year. Thus, the building's human detection capability can be defined by a series of "what if" and "when" situations.

As a basis for the selection of appropriate scenarios, the information gathering about people, fires, and building operations provides valuable knowledge about the building and how it works. This knowledge helps to structure risk characterization studies after a fire performance analysis has been completed.

Qualitative analyses involving human detection scenarios are normally evaluated quickly and easily. Time estimates for the events initially require more effort. As data for time durations to complete the events become available, confidence in outcomes increases and professional time needs decrease. The event logic diagram of Figure 26.9 and the evaluation network of Figure 26.10 express the thought process.

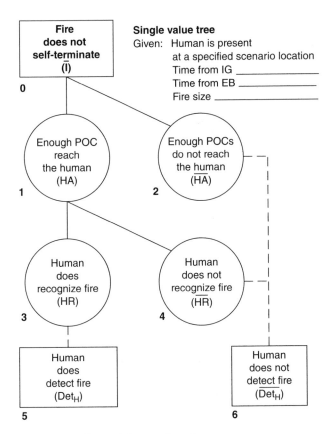

Figure 26.10 Network: human detection.

Figure 26.11 identifies factors that influence event evaluations. A base scenario analysis will provide a good understanding of the effect of the building and diagnostic fire on human detection. A set of variability analyses can identify reasonable expectations for differences in building status and the condition of the individual who represents human detection.

26.8 Closure

Fire detection initiates a holistic performance analysis. Methods of detection vary in buildings throughout the world. Some rely on humans only, some on detection and alarm systems, and some on both. Some systems are very effective. Others may be ineffective. The effect of detection on building performance can vary enormously.

Our confidence in a building's fire safety increases when we see detectors in a building. However, an evaluation examines three major functional aspects of detector performance. The first is the diagnostic fire. The original perception of the time–fire size relationship may have changed from the time of installation. Hazards can change over the years with different contents, materials, and arrangements. Is the diagnostic fire still appropriate for the building functions?

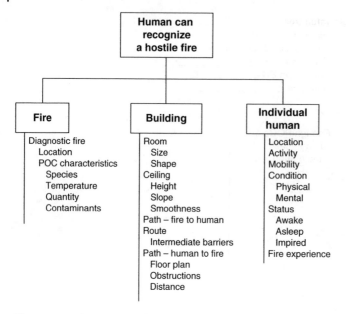

Figure 26.11 Organization chart: human detection.

The building is a second focus of attention. Statistically, commercial buildings undergo minor renovations every 7 years and major changes every 14 years. Thus, partition locations, room sizes, opening arrangements, heating, ventilation, and air conditioning operations, and functionality may differ from the initial design. Some architectural changes can affect fire size at detection.

The third part of a performance examination involves the detection system itself. Is the system still appropriate for the functions and hazards? Also, maintenance of the system can affect performance significantly. System reliability becomes more important as the building ages.

A detection analysis starts the performance evaluation. The selection of the fire size at detection is a central determination for the IPI chart. This value becomes the basis to establish time-related associations with all other parts of the entire system.

27

Fire Department Notification

27.1   Introduction

Fire department notification is a key link in the chain of events from ignition to extinguishment. This event transfers operational control for fire extinguishment from the building to the community.

Notification can be done by a human or by automated methods. Most automated methods involve people as part of the process. Therefore, this critical link depends on human actions and reliability somewhere in the process.

Figure 27.1 shows an event logic tree that describes the success for fire department notification.

The performance measure for fire department notification is the time duration between detection (MD) and notification of the fire department communication center (MN). Success or failure in manual extinguishment can be greatly influenced by the time to complete this event.

Individuals provide an essential link between the building and the local fire department. However, any component that involves human actions must deal with uncertainty. In spite of the difficulties in knowing the actions that will actually take place, an analysis can usually bracket time durations well enough to understand what to expect.

Chapter 12 describes the functions and operation of fire department notification methods. This chapter structures those methods into an analytical framework and illustrates ways of making estimates.

27.2   The Human Link in Notification

Fire department notification analysis starts at detection. Although notification can be done by human or automated methods, as noted in Figure 27.1, individuals provide the usual connection. Qualitative analyses examine this link for a site-specific building.

One develops an understanding of the building's role and the time for an individual to complete this event by analyzing a series of scenarios. A base scenario initially describes expected actions by a representative (hypothetical) individual who will notify the fire department. A base timeline combines time duration estimates for the major activities. Then, a group of "what if" alternative conditions are methodically introduced to

Fire Performance Analysis for Buildings, Second Edition. Robert W. Fitzgerald and Brian J. Meacham.
© 2017 John Wiley & Sons Ltd. Published 2017 by John Wiley & Sons Ltd.

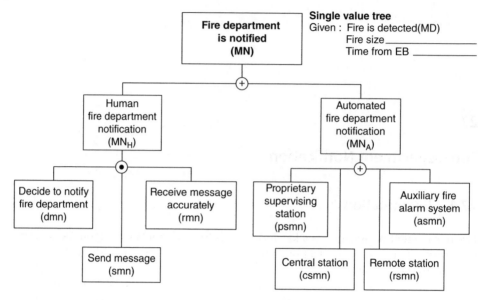

Figure 27.1 Success tree: fire department notification.

recognize the effect of different conditions that can change time durations. Qualitative analyses after the first scenario are normally very rapid.

The three essential parts of human notification in Figure 27.1 are recast into the evaluation network of Figure 27.2.

As with the cells of all columns in an Interactive Performance Information (IPI) chart, the events of Figure 27.2 are evaluated for a specific instant of time. Because Events 1, 3, and 5 are sequential, the status of each will change with each elapsed minute until the event has been completed.

27.3   Human Notification Analysis

The building, the fire, and the individual all influence outcomes for the events of Figure 27.2. One develops an understanding of the building by examining these events as they change status during a scenario analysis.

The process for evaluating a scenario is as follows:

1) Given: the diagnostic fire.
2) Given: fire has been detected, the fire size at detection, and the scenario that was used in the detection analysis.
3) Identify the representative (hypothetical) individual who will notify the fire department. Describe the individual's location, condition, and activities at the time of detection.
4) Select an appropriate IPI (column) time for analysis. Describe the individual's location, the building features, and the environmental conditions at that time. Will the individual decide to call the fire department? Note the time when the answer changes from no to yes.

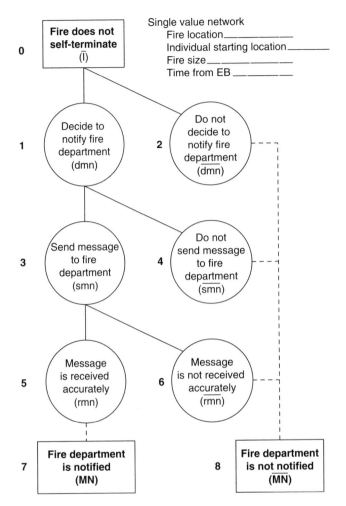

Figure 27.2 Network: fire department notification.

5) Given that the representative individual has decided to call the fire department (Step 4), examine the means to transmit the information. One knows the individual's location, the building features, and the environmental conditions at that time. Select appropriate IPI times and answer the following question: will the individual make initial connection with the fire service communication center? Note the time when the answer changes from no to yes.

6) Given that the individual made connection with the fire service communication center, was the message received accurately? When the answer changes from no to yes, note the time.

7) Add the time durations of Steps 4, 5, and 6 to determine the elapsed time between MD and MN (Event 7, Figure 27.2).

Building features or scenario conditions may accelerate or delay the representative individual's ability to complete all three events. In addition to estimating time durations, recognition of the influence of building features on the process improves one's understanding of performance.

27.3.1 The Role of Detection

Although detection and fire department notification are evaluated as separate components, the process is often continuous. An instrument or a human may detect a fire. Each mode has a different influence on notification.

Figure 27.3 shows the methods by which an individual who is the focus of scenario analysis becomes aware of a hostile fire. Each method can produce a different scenario or set of scenarios that become candidates for qualitative analysis. Each scenario places the individual in a slightly different role that can affect the time to a decision to notify the fire department.

27.3.2 Initial Scenario Analysis

An initial scenario on which to base qualitative analyses is usually related to how the building normally works. The diagnostic fire will have been established. Many of the features of building architecture and characteristics will have been identified during the detection analysis. Additional features that can affect the notification process may be incorporated.

The scenario focuses on a representative individual who will notify the fire service communication center. The initial scenario defines the individual's characteristics, such as location, activity, condition, fire experience, training, and instructions.

Figure 27.4 identifies factors that commonly affect notification scenarios. Greater understanding of the context of the scenario enables greater confidence in qualitative estimates.

The initial scenario develops a basic understanding of the effectiveness of the human fire department notification process. Often, the examination will uncover unexpected information during time duration estimates of important distance movement and decision-making activities. Evaluations envision the time a representative individual needs to complete the three major events of Figure 27.2. The analysis provides a

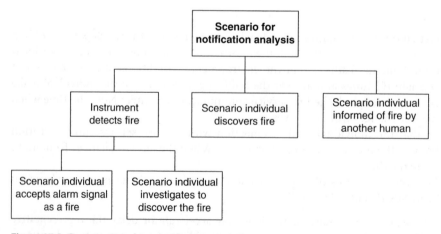

Figure 27.3 Organization chart: notification analysis.

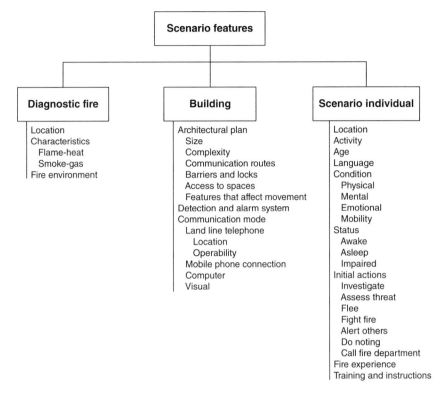

Figure 27.4 Organization chart: fire department notification.

realistic understanding of building and occupant effectiveness in notifying the local fire department of an emergency.

27.3.3 Information Augmentation

The information provided by an initial scenario analysis can be augmented with methodical examination of additional scenarios. The number and scope of additional conditions depend on the building and how it works. However, the added time to examine variations of the initial information provides sensitivity regarding what may be expected for differing conditions.

Qualitative scenario analyses provide insight into the strengths and weaknesses of the building and its occupants in fire emergencies. Although the focus of this examination provides time durations between MD and MN, the process often uncovers useful information for later use in risk characterization and management.

27.4 Human Notification

The analytical framework combined with the IPI chart events and time durations enables one to get a mental picture of the way a building will respond to a fire. Although the notification analysis of Figure 27.2 requires a qualitative analysis, the outcomes are time durations for the event and an understanding of this part of building performance. A methodical selection of scenario variations will provide a picture of time ranges.

27.5   Automated Notification Analysis

Figure 27.1 shows the four methods that provide automated notification. Although a network could be constructed for that branch of the event logic tree, we will not show it here.

Each of these automated methods of fire department notification has a similar process, although details differ. Figure 27.5 shows a network with which to analyze the system. The events are:

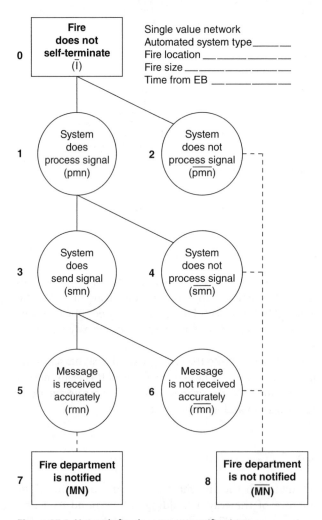

Figure 27.5 Network: fire department notification.

1) An installed fire detection system actuates (Det).
2) The signal is processed (i.e. the system "decides" to notify the fire department) (pmn).
3) The message is sent (smn).
4) The message is received accurately (rmn).

All of the automated methods rely on an installed detection system to detect the fire. After the detection event is completed, each automated method functions slightly differently. For example, a proprietary supervising system uses an individual to process the signal (Event 1) by requiring human decision and actions to decide to send the signal. However, a central station, remote station, and auxiliary fire alarm system all process the signal automatically.

The signal is sent automatically (Event 3) with an auxiliary fire alarm system. However, human decisions and actions are a part of the other automated systems. The final event of the fire service communication center receiving an accurate message is automated in all four methods.

A performance analysis examines the time duration to complete each of these three events. Although the procedures for each one differ, their speeds are usually quite fast.

A second performance component is the reliability of each system. Although the processes are automated, detailed installation examination gives a better understanding of potential sources of failure.

27.6   Closure

The time between detection (MD) and notification (MN) is one of the most difficult evaluations in a performance analysis. It involves much variation and uncertainty. Nevertheless, the information acquired from this examination provides some insight into the building's fire performance.

Although actual actions of individuals are uncertain, estimating time durations enables one to develop windows of time that can bracket "what if" scenarios. Fortunately, the efforts do not require substantial professional time commitments. The outcomes often provide information that becomes useful for risk management programs.

The IPI chart provides an organizational tool to order the information. Variations can be incorporated into a holistic analysis to recognize the effect of notification on other aspects of building performance. The IPI chart helps with variation analysis and organizes communication for others.

28

Fire Department Extinguishment

28.1 Introduction

Time is more important to the evaluation of manual extinguishment than to any other component except perhaps life safety by egress. Thus, estimating time durations for a set of relatively independent, although related, events becomes vital to understanding fire performance. The Interactive Performance Information (IPI) chart is an important tool for comparing fire conditions and fire suppression events.

Figure 28.1 shows an event logic diagram for successful manual extinguishment. The major events are sequential and, because the diagram is single value, their performance status changes as time progresses. Different events phase in and out of importance.

Figure 28.2 shows a representative timeline for the major events.

A building analysis relates event time estimates to diagnostic fire conditions. A hierarchical set of networks that correspond to the event logic diagram guides the analysis. An initial scenario analysis provides much knowledge of what to expect. Variation analyses using other scenarios augment this initial knowledge.

28.2 Framework for Analysis

The first level of decomposition in Figure 28.1 segments an analysis into three parts. Figure 28.3 shows a single value network that guides the thought process. Events 1, 3 and 5 must be completed sequentially before Event 7, fire department extinguishment (M).

Figure 28.4 shows the next level of decomposition. Detection (Event 1a) and notification (Event 3a) have been described in Chapters 26 and 27. Chapter 13 discusses fire department response. Chapters 14 and 15 describe extinguishment analysis. Collectively, all of the events are sequential and necessary to complete a manual extinguishment evaluation.

28.3 Notification to Arrival

The diagnostic fire, the building design, detection and fire department notification are under the management control of the owner and occupant. Fire department response from notification to first-in arrival is under the management control of the community.

Fire Performance Analysis for Buildings, Second Edition. Robert W. Fitzgerald and Brian J. Meacham.
© 2017 John Wiley & Sons Ltd. Published 2017 by John Wiley & Sons Ltd.

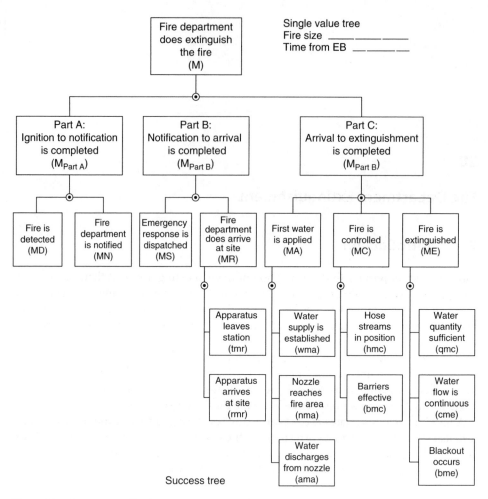

Figure 28.1 Success tree: fire department extinguishment.

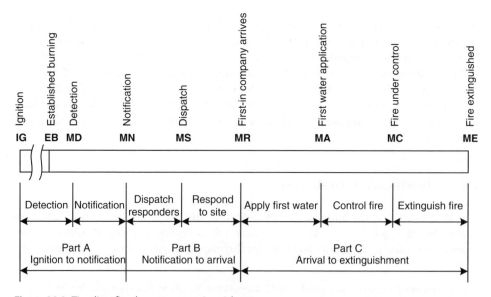

Figure 28.2 Timeline: fire department extinguishment.

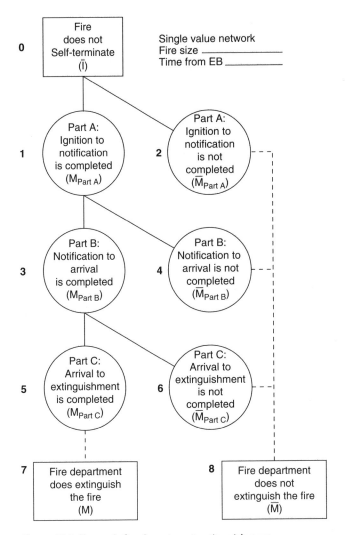

Figure 28.3 Network: fire department extinguishment.

Collectively, they form the basis of the analysis for *Part A: Ignition to notification* and *Part B: Notification to arrival*. The product of these parts is identification of the fire size at the time of first-in fire company arrival.

The analysis of fire size at detection and its associated time from established burning (EB) was described in Chapter 26. The time interval from detection to fire department notification was described in Chapter 27. These events trigger the local community response to a fire emergency. They are integral to fire department extinguishment evaluations because the time to complete these activities influences the fire size and conditions that the incident commander must manage.

28.4 Fire Department Response

Figure 28.4(b) describes the initial community actions to a fire emergency. Figure 28.5 defines sequencing activities between fire department notification and first-in arrival at the scene.

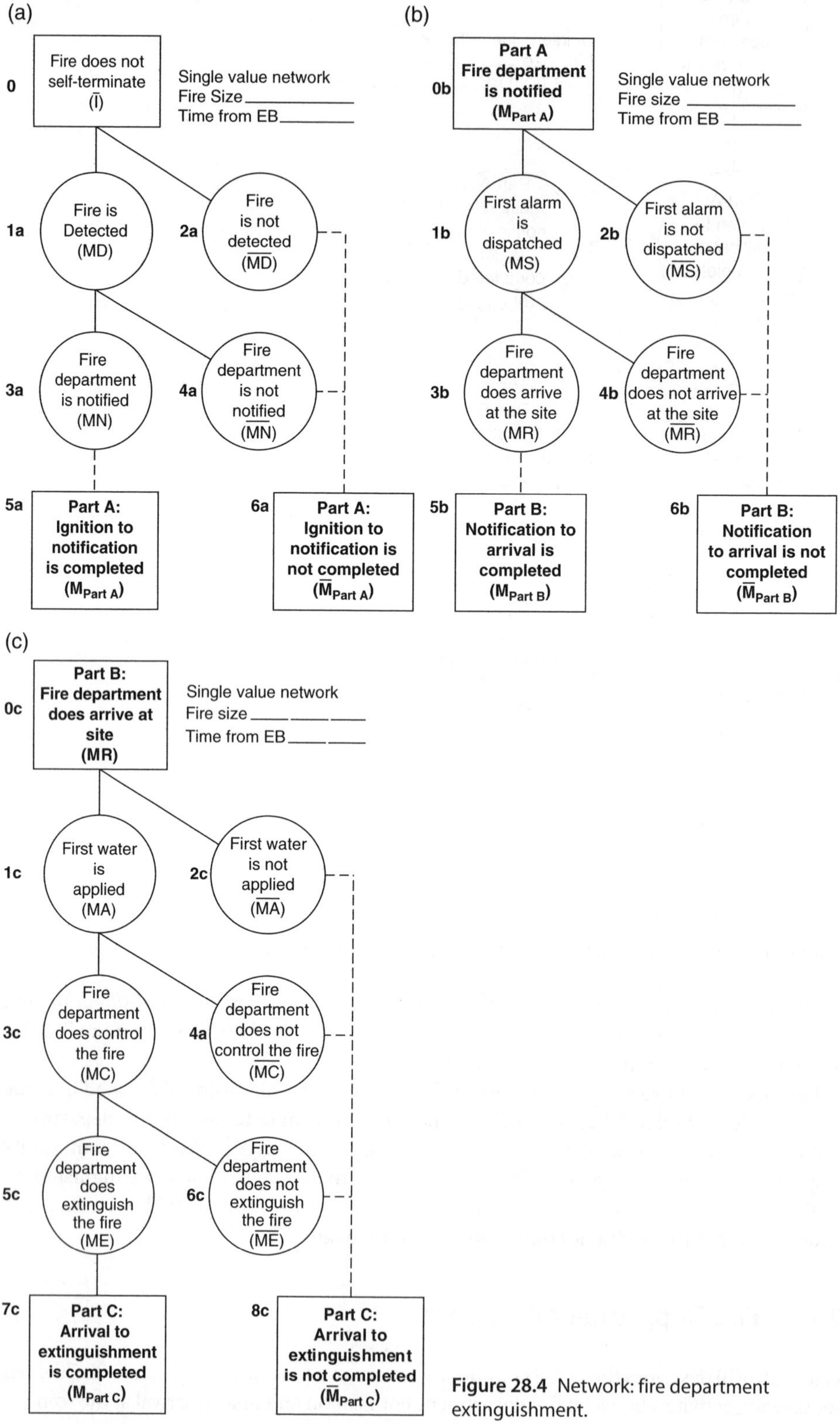

Figure 28.4 Network: fire department extinguishment.

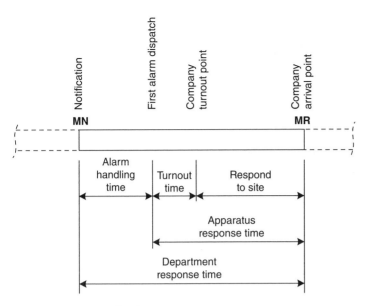

Figure 28.5 Timeline: fire department response.

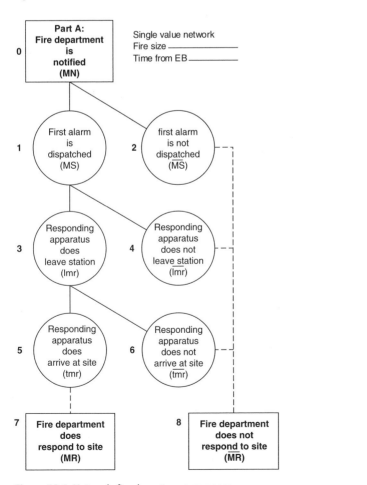

Figure 28.6 Network: fire department response.

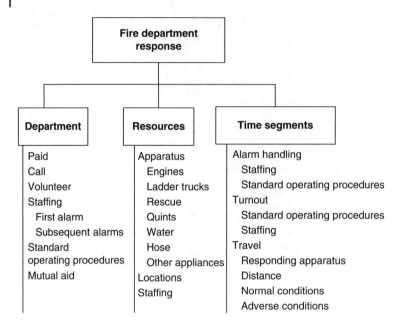

Figure 28.7 Organization chart: fire department response.

An analysis examines community procedures that affect each segment. Fortunately, the information is usually accessible and orderly. One usually has confidence in these time estimates. Figure 28.6 shows the single value network to guide thinking. Each of the events is conditional and sequential. Figure 28.7 identifies the information that influences response time and resources for a fire emergency.

28.5 Arrival to Extinguishment

A building performance analysis has the advantage of knowing the diagnostic fire location and conditions at the time of fire department arrival. An analysis also has information about occupant numbers, activity, and locations. The evaluation also knows the building and site information that influences fire fighting and fire ground operations. Although the "desktop" performance analysis has more information than an incident commander arriving at the scene, a performance evaluation gives an insight into the building's ability to work with the local fire department.

When the local fire department response arrives at the scene, fire ground operations evolve to improve the emergency. Fire suppression is one of the early actions and follows "routine" procedures that are adapted to the needs of each fire scene.

Part C: Arrival to extinguishment has three phases to an evaluation. Initially, Phase 1 provides an idea of the fire size when the first attack line can theoretically apply water (MA). A Phase 1 assessment puts the general situation into clearer focus. Phase 2 gives a sense of the likely boundaries for controlling the fire. It also identifies whether resources are adequate to stretch and provide water for additional attack lines. Overall, Phase 2 provides an opportunity for general assessment of conditions that may indicate advisability for offensive or defensive modes of fire fighting.

Phase 3 uses the information of Phase 2 to get a sense of where final extinguishment may occur. It also allows one to recognize needs for directing the attack and for protecting the exposures.

28.6   Phase 1 Analysis

Phase 1 determines the fire size at theoretical first water application. After arrival (MR), an analysis examines factors of finding the fire, identifying an attack launch point (ALP), and selecting a route to stretch attack lines. Figure 28.8 shows the network for evaluation.

Event 1 (wma) and Event 3 (nma) are evaluated independently. The order in which they are evaluated and the order that they are shown in Figure 28.8 are unimportant. However, because both Events 1 and 3 must be completed before discharging water, Event 5 must be the final event in the chain.

Because Figure 28.8 is single value, the IPI chart is needed to establish timelines for the events. Chapter 14 discusses IPI displays to develop sequencing of the events. Figure 28.9 shows the factors that influence the three events comprising interior water application.

28.7   Phase 2 Analysis

A Phase 1 analysis provides important information about how the building and site influence the interior fire attack. However, this analysis also gives a wealth of additional knowledge that can offer an insight into expectations of fire behavior and extinguishment decisions.

When it is obvious that the fire cannot be put out by the first interior attack line, additional resources are needed to complete extinguishment. Phase 2 provides an opportunity to "pause" and compare building and fire conditions with available resources to decide how to handle the problem.

An important decision relates to the ability to contain the fire within identifiable boundaries and control its behavior. The fire is described as "under control" when the incident commander believes that he or she understands what will happen until complete extinguishment.

Sometimes an aggressive attack can put out the fire relatively quickly. Sometimes a defensive stand is needed to prevent further fire extension and eventually put it out. The defensive stand may be used inside or outside the building, depending on the fire size and building conditions. Building barriers and hose streams are the means to contain the fire.

A decision to mount an aggressive attack or use a defensive mode (often in combination with offensive actions) until the fire goes out is largely judgmental. It is influenced by the building design and available fire fighting resources. An evaluation is based on specific building sections that identify a potential situation for controlling the fire. Figure 28.10 shows a network to guide estimation of the building area within which the fire scenario can be contained. Events 1 and 3 are independent, and the order of evaluation and network position is unimportant. Both actions are necessary to control the fire.

Figure 28.11 shows factors that influence the identification of the boundaries that can be used to contain and control the fire.

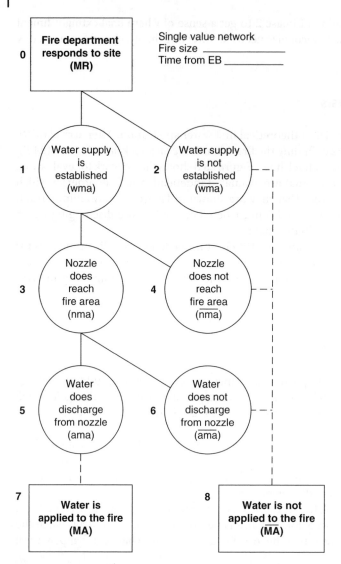

Figure 28.8 Network: first water application.

Figure 28.10 is a single value network. The fire size (i.e. conditions) for control may be different from that identified by Phase 1. The IPI chart can identify hose stream application and water flow as well as fire propagation to barrier boundaries that define control (MC).

28.8   Phase 3 Analysis

The time from control (MC) to extinguishment (ME) may be relatively short or very extended, depending on the composite fire size and conditions, building design, and fire department resources. Small fires go out quickly while large fires burn for long durations.

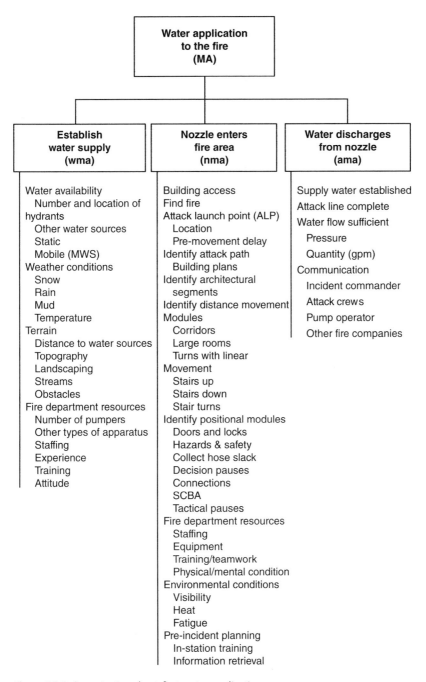

Figure 28.9 Organization chart: first water application.

A Phase 3 analysis examines conditions for extinguishing the fire without its growth to larger sizes. This depends on structural frame stability and the endurance of barriers to prevent fire extension. It also requires the continuous availability of enough water to put out the fire. The IPI chart enables one to order time durations.

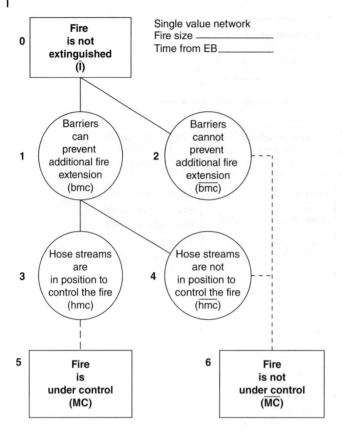

Figure 28.10 Network: fire department controls fire.

Figure 28.12 shows the network that guides the evaluation. The events are independent and the order of analysis is unimportant. The final event signifying that the fire is extinguished (blackout occurs) provides organizational completeness for the framework. Figure 28.13 shows factors that influence the events.

28.9 Putting It Together

The complete manual extinguishment analysis connects a logical path of events. The following example describes the evaluation process.

Example 28.1   Figure 28.14 shows the third floor and site plan for the model building. Figure 28.15 shows an IPI chart for the analysis of a fire in the internal office section of the third floor, west wing. Discuss the IPI chart that describes manual suppression analysis for this scenario.

Solution:
The IPI provides a foundation from which to discuss the dynamics of analysis. We will discuss the IPI entries and amplify the information in some cases.

Section 1A: Diagnostic Fire   The diagnostic fire starts in Room 3011 and spreads to the other rooms because of barrier openings. The room fires show two conditions. Cells

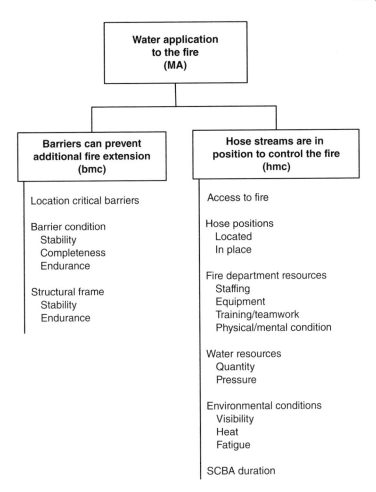

Figure 28.11 Organization chart: fire department controls fire.

with clear fill indicate a growing fire from EB to flashover (FO). The textured fill shows post-FO conditions to burnout.

Discussion   The room of origin and connecting rooms are internal to the building. For this diagnostic fire, we assume the corridor door to Room 3011 to be open. This provides sufficient ventilation for fire growth within the block of spaces. It also enables smoke to spread through the building.

The status of corridor barrier openings outside of the cluster of internal office spaces will affect fire conditions. If sufficient ventilation is available, the fire will behave as identified in the IPI chart. If all corridor doors are closed, as may be the case during non-working hours, only leakage ventilation will occur. This would cause slower fire growth in the office spaces and delay smoke movement. However, the ventilation-restricted fire growth period could also create backdraft conditions, as noted in Chapter 6. This causes additional safety concerns for fire fighters.

The goal for this scenario is to understand the way the building helps or hinders manual fire fighting. To create this base of understanding we will select severe, but reasonable, fire conditions provided by open corridor doors in the office suite. This

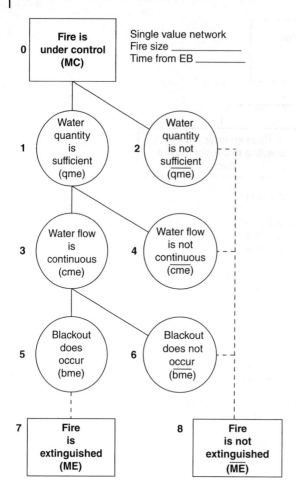

Figure 28.12 Network: fire department extinguishment.

status provides sufficient ventilation for fire growth and also allows smoke to spread through the building. After we gain an understanding of the building and fire department interactions from a complete manual suppression analysis, we can qualitatively extrapolate results to understand what one can expect for different "what if" scenarios.

Section 1B: Diagnostic Smoke   The diagnostic fire will develop smoke conditions. The IPI chart describes scenario conditions in the corridors.

Discussion   Chapter 9 described ways of estimating time and visibility conditions for smoke movement. The cells in Section 1B indicate when visibility is at 3 m. As the diagnostic fire spreads to additional fuel groups and rooms, smoke generation increases and visibility declines. Because the initial attack forces do not know the fire situation, caution and slower movement increase as visibility declines. The numbers in the IPI cells estimate visibility changes.

Section 2: Active Extinguishment   We focus attention only on the manual extinguishment component of Part A.

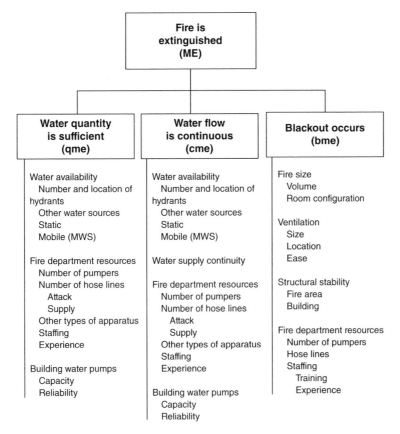

Figure 28.13 Organization chart: fire extinguishment.

Section 2Aa: M$_{Part\ A}$ Ignition to Notification   An analysis of the detection system indicates that MD occurs when the fire is at 50 kW. Considering the fire growth (heat release rate, $\dot{Q}$; not shown), the IPI shows this at about 2 minutes from EB. The IPI chart indicates an estimate of MN to be 5 minutes after MD. At MN, the diagnostic fire has not yet reached FO in the room of origin.

Discussion   The state of the art for automatic detection response provides some confidence in quantification. Integrating quantitative outcomes with qualitative observations improves confidence that the estimates represent reality. Chapter 26 describes quantitative and qualitative integration for detection.

Event MD is completed at the time of detector actuation. This building does not have direct fire department notification when a detector or pull station actuates. Therefore, humans must complete the MN event. Consequently, the time of day and population status will influence MN time estimates. For example, a fire during normal working hours should have a much shorter MN time than a fire during unoccupied conditions. If occupants become aware of a fire by an audible alarm, an analysis should determine if the building population knows whether the alarm is local or is transmitted automatically to the local fire department.

Although the fire size at detection (MD) can be established with confidence, human notification (MN) involves a substantial "window" of uncertainty. A prompt notification

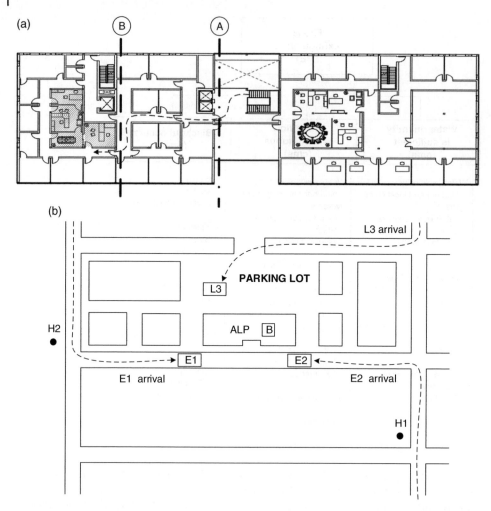

Figure 28.14 Example 28.1.

enables an earlier response, which will mean encountering a much smaller fire on arrival (MR). Delayed notification allows the fire to grow and this will require a very different fire ground strategy and tactics after arrival.

This scenario estimates a fire in the early evening (about 8:00 pm) when most of the population will have left the building except for a small cleaning crew. Although there is great uncertainty, a value must be selected for MN. As the complete picture of manual analysis unfolds using this MN estimate, one can identify the effect of other MN times on fire ground operations and damage estimates. Thus, "what if" variation analyses may be qualitatively extrapolated from this initial estimate to provide a relatively clear picture of the range of outcomes.

Section 2Ab: $M_{Part\,B}$ (notification to arrival) The IPI chart shows an estimate of 2 minutes for MS and an additional 4 minutes for first-in MR. The first alarm response of two engines and one ladder truck all arrive within 2 minutes of MR. The diagnostic fire size at MR shows Room 1311 fully involved and Room 1312 nearing FO.

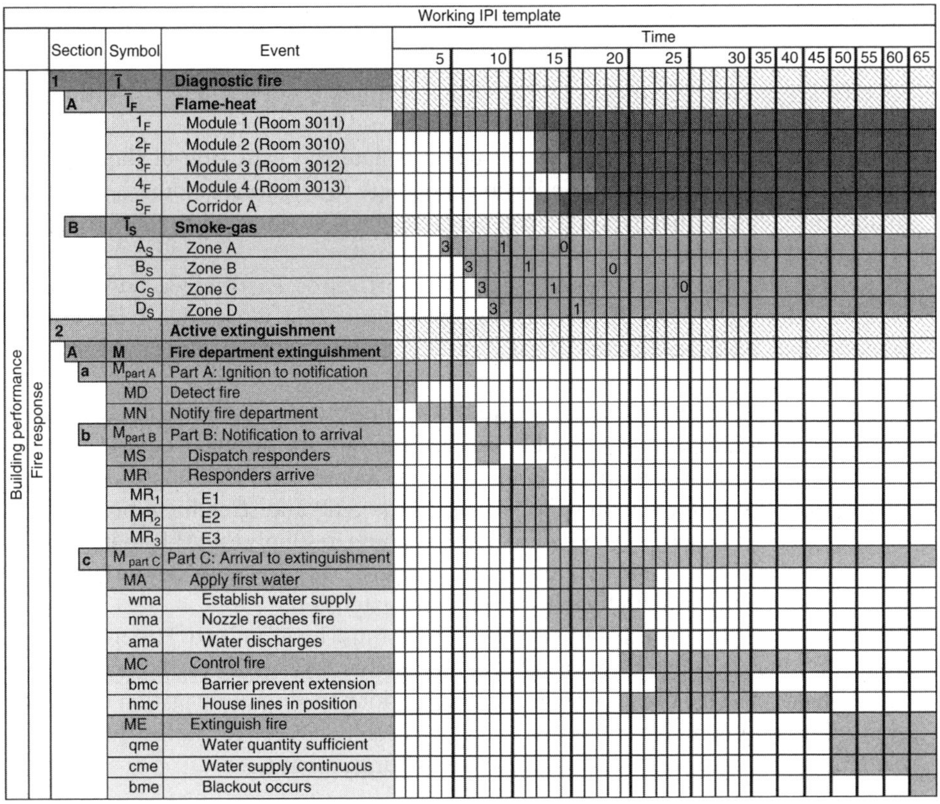

Section	Symbol	Event	Time (5–65)
1	**T**	**Diagnostic fire**	
A	T$_F$	**Flame-heat**	
	1$_F$	Module 1 (Room 3011)	
	2$_F$	Module 2 (Room 3010)	
	3$_F$	Module 3 (Room 3012)	
	4$_F$	Module 4 (Room 3013)	
	5$_F$	Corridor A	
B	T$_S$	**Smoke-gas**	
	A$_S$	Zone A	3 ... 1 ... 0
	B$_S$	Zone B	3 ... 1 ... 0
	C$_S$	Zone C	3 ... 1 ... 0
	D$_S$	Zone D	3 ... 1
2		**Active extinguishment**	
A	**M**	**Fire department extinguishment**	
a	M$_{Part A}$	Part A: Ignition to notification	
	MD	Detect fire	
	MN	Notify fire department	
b	M$_{Part B}$	Part B: Notification to arrival	
	MS	Dispatch responders	
	MR	Responders arrive	
	MR$_1$	E1	
	MR$_2$	E2	
	MR$_3$	E3	
c	M$_{Part C}$	Part C: Arrival to extinguishment	
	MA	Apply first water	
	wma	Establish water supply	
	nma	Nozzle reaches fire	
	ama	Water discharges	
	MC	Control fire	
	bmc	Barrier prevent extension	
	hmc	House lines in position	
	ME	Extinguish fire	
	qme	Water quantity sufficient	
	cme	Water supply continuous	
	bme	Blackout occurs	

(Left axis labels: Building performance / Fire response. Time scale across top: 5, 10, 15, 20, 25, 30, 35, 40, 45, 50, 55, 60, 65.)

Figure 28.15 Example 28.1: Interactive Performance Information (IPI) description.

Discussion The uncertainty that exists in estimating the time from MD to MN does not exist for M$_{Part B}$. One can estimate with reasonable confidence the time to dispatch first alarm response (MS) and time to arrival (MR) for all companies. The effort to gather the necessary information for the initial investigation may be used for other buildings in the same community.

Section 2Ac: M$_{Part C}$ (arrival to extinguishment) Fire ground suppression involves three phases of analysis. Phase 1 (MA) examines the time and associated fire conditions for the theoretical "fastest" time to apply first water. Although this examination may show that an offensive attack with a single hose is not feasible, the information needed for the analysis provides a good understanding of the building and site, the fire, and available fire fighting resources.

Phase 1 (MR–MA) The IPI chart indicates that the theoretically fastest time to apply first water (MA) is the sum of EB–MD (2 minutes) + MD–MN (5 minutes) + MN–MR (6 minutes) + MR–MA (9 minutes) = 22 minutes after EB. The diagnostic fire indicates three rooms fully involved, a fourth room nearing FO, flames into the corridor, and substantial heat and smoke filling the west wing corridors.

Discussion   A single hose line clearly cannot handle this fire condition. Nevertheless, the examination does provide the following information:

- An offensive attack resulting in a rapid extinguishment is not possible. Conditions are unsafe for fire fighters and back-up hose lines would be needed.
- A defensive perimeter (A) can be set up to isolate the west side of the building. A defensive perimeter (B) may be feasible if enough fire fighting resources can arrive in time to defend the barriers and prevent extension. We can calculate the arrival time of additional companies.
- Building access is available from the south and north ends. East and west sides cannot be used, although ventilation from the third floor west is possible from ground ladders. The type of window glazing should be examined to determine ease of ventilation.
- Fire fighter staffing must be augmented by additional multiple alarm companies. At least four attack hose lines will be needed for interior fire fighting. The time to place these additional lines in operation can be calculated by adding the response time for multiple alarms with supply water to the ALP and stretching interior hose lines. Chapter 14 describes ways to calculate the time to supply the ALP and to stretch attack lines from the ALP to the fire. The time to place the hose lines may be compared with diagnostic fire growth to get a sense of fire extension.
- Fire extension from the third to the fourth floor must be examined. Fire may extend from the interior suites to external offices and then externally to the upper floor. Additionally, modern fire experiences show that floor-to-floor propagation is possible even with code-complying fire endurance ratings (Chapter 18). Fires that involve more than one floor require substantial fire fighting resources.

Procedures described in Chapter 14 enable one to calculate the time to establish supply water to the ALP and the time to stretch interior attack hose lines. These activities are coordinated to determine the time between first arrival (MR) and first water application (MA). These procedures may be used to calculate the time to deliver water from each attack line. Expanding the IPI chart enables one to estimate the location and time at which each hose line can come online.

Phase 2 (MA–MC)   The IPI chart indicates that the fire can be under control at about 45 minutes after EB. This is based on the information provided by Phase 1 (MR–MA) and examination of multiple alarm resources, the building features that work with the fire department in controlling the fire, and calculated time sequencing to apply water to the fire.

Phase 3 (MC–ME)   The IPI chart indicates that extinguishment may be expected about 20 minutes after control (MC). The IPI chart indicates that the fire will be confined to the third floor and that Perimeter B will be successful in limiting the extent of the fire.

28.10   Discussion

The scenario for this fire occurred on a weekday evening at 8:00 pm. Most building occupants would probably have evacuated the building. This reduces complications for the incident commander. Nevertheless, some staffing would probably search endangered sections of the building to ensure that no occupants were trapped. Most of the response could be allocated to fire suppression activities. A fire during normal hours of operation would require fire fighters to direct and ensure occupant life safety. Although this would diminish personnel available to apply water to the fire, the Phase 1 (MR–MA) evaluation would be evaluated using the staff available to fight the fire.

This fire scenario indicates that major damage would occur to the building and its functions. A fire during occupied conditions may improve the $M_{Part\ A}$ somewhat, although not enough to reduce the damage significantly, because of the location of the room of origin. Rooms of origin in the exterior of the building would probably lose only an office if the doors were closed. A fire in the open office would be difficult to extinguish without extensive damage. A fire in the subterranean garage would be very difficult to fight. Significant smoke damage could be expected throughout the building unless smoke control measures were available.

The time durations for multiple alarm responses can vary significantly. Larger cities can usually augment first alarm needs quickly and with sufficient resources. Time durations and resource capability can be more difficult in smaller communities. Fortunately, the manual suppression analysis using the IPI chart can track all activities that affect outcomes and their time durations.

28.11   Closure

A building analysis compares the diagnostic fire and critical events for manual suppression in a routine, methodical manner. Specific building design information, site conditions, and fire department resources are known. Uncertainties are clearly recognized. The outcome identifies the limit of flame spread before the fire is extinguished.

The result of a single, carefully selected scenario can provide substantial information about the building's ability to work with the local fire department. Qualitative extrapolations can describe outcomes for "what if" variations. On occasions where additional IPI evaluations may be needed, analyses do not require significant professional time, because much of the information would have been obtained from the initial scenario.

29

Automatic Sprinkler Suppression

29.1   Introduction

When automatic sprinklers are discussed in the USA, the NFPA and FMGlobal standards usually come to mind. Certainly, these and their related standards provide completeness and excellence for design and maintenance. New installations developed in compliance with these standards give confidence for quality design. The same confidence exists for sprinkler systems designed to appropriate standards in other countries.

This book addresses analysis rather than design. Therefore, the focus is on:

a) Evaluation of an existing sprinkler system installation; or
b) Review of a proposed new sprinkler system design.

Sprinkler systems in existing buildings have been designed and installed for well over a century under a wide variety of conditions. Over the years, sprinkler standards have steadily evolved. However, the quality of existing installations can be quite variable. For example, existing systems may not have been updated to accommodate standards changes subsequent to the original installation. Also, a satisfactory initial design may have been compromised because changes in materials (i.e. fire growth conditions), building construction, or water availability provide different conditions from those on which the original design was based. Sometimes individuals with insufficient training may have designed an original system. Sometimes owners delay maintenance to finance other, more urgent needs.

A wide range of factors can affect the quality and effectiveness of a sprinkler system. Often, a system may be effective even if not in complete compliance. However, a system may also be inadequate. When an existing sprinkler system is important to manage risks, one must find a way to evaluate its potential effectiveness.

A performance analysis attempts to answer the questions:

- Will the sprinkler system control any fire in the building?
- What damage may be expected after the fire is extinguished?

Qualitative observations can provide an understanding of effectiveness for certain parts of a performance analysis. Other aspects require detective work to establish quantitative measures. Some information will remain unknown. Uncertainty clearly exists. However, a methodical approach to analysis enables one to systematically examine the

Fire Performance Analysis for Buildings, Second Edition. Robert W. Fitzgerald and Brian J. Meacham.
© 2017 John Wiley & Sons Ltd. Published 2017 by John Wiley & Sons Ltd.

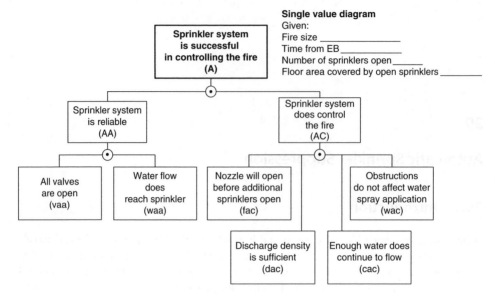

Figure 29.1 Success tree: sprinkler system control.

elements and gain an understanding of expectations. Often, gaps in knowledge can be bridged by logical extrapolation using the analytical framework and Interactive Performance Information (IPI) chart.

Figure 29.1 shows a single value event logic diagram that organizes the functional operations of sprinkler systems (Chapter 16). The corresponding evaluation networks are shown in Figure 29.2. Figure 29.2(a) describes the system reliability (i.e. initial agent application). Although these events are independent, operating conditions require valves to be open before water can flow to a fused sprinkler.

The four design effectiveness events are also independent, although the examination sequence normally follows the order in Figure 29.2(b).

Figure 29.2(c) combines reliability (AA) and design effectiveness (AC) to describe system performance.

29.2   Agent Application (AA)

Sprinkler systems are normally segmented into zones that cover a defined area. This provides economy and helps with practical maintenance practices. However, it also increases the number of valves and potential reliability features.

Agent application (AA), also defined as system reliability, is relevant to each sprinkler zone in the building. Figure 29.2(a) describes the two major events that comprise system reliability. Figure 29.3 identifies common factors that influence these events. An analysis identifies the good and bad factors that affect each of the building's sprinkler system zones.

These factors are generally associated with observational evaluations. Maintenance records give an indication of the attention to the system. System reliability is the only major part of a performance analysis that remains unaffected as the diagnostic fire conditions change.

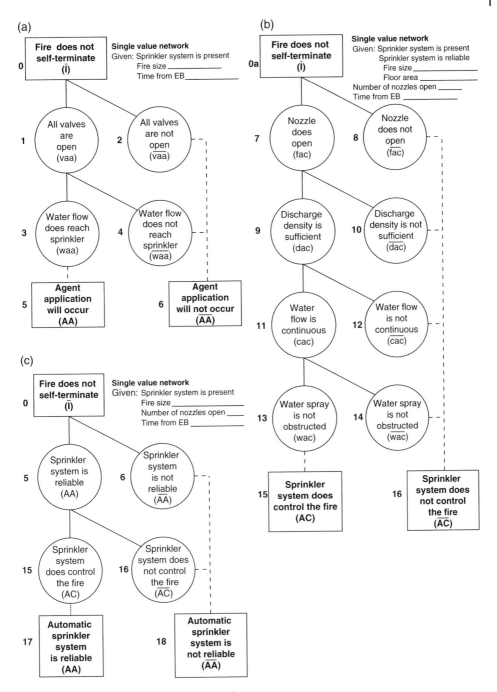

Figure 29.2 Network: sprinkler system control.

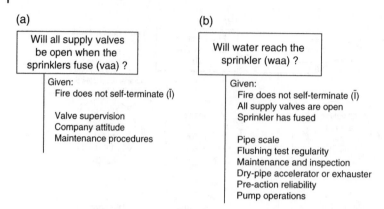

(a)

> Will all supply valves be open when the sprinklers fuse (vaa) ?

Given:
  Fire does not self-terminate ($\bar{I}$)

  Valve supervision
  Company attitude
  Maintenance procedures

(b)

> Will water reach the sprinkler (waa) ?

Given:
  Fire does not self-terminate ($\bar{I}$)
  All supply valves are open
  Sprinkler has fused

  Pipe scale
  Flushing test regularity
  Maintenance and inspection
  Dry-pipe accelerator or exhauster
  Pre-action reliability
  Pump operations

Figure 29.3 Organization chart: sprinkler agent application.

29.3 Design Effectiveness (AC)

Figure 29.2(b) shows the four major events that examine the sprinkler's ability to control a fire. Although each of these events is examined independently, events dac, cac, and wac are conditional on the sprinklers fusing (fac).

The analysis first examines the fire size at first sprinkler actuation. Then, it methodically examines successively larger areas (numbers of sprinklers fusing) until the fire is controlled or the fire overwhelms the system.

29.3.1 First Sprinkler Fusing (fac)

The initial event in a design effectiveness (AC) analysis determines the fire size at first sprinkler actuation. To give a sense of proportion, for ceiling heights less than 3.5 m (11.5 ft) the fire is usually in Realm 4 (enclosure point– ceiling point) with flame heights of about 1.5–2 m (4.5– 6.5 ft).

29.3.2 Multiple Sprinkler Fusing (fac)

A single sprinkler extinguishes many fires. Less than four sprinklers extinguish most fires in compartmented buildings because barriers assist in the process. Evaluations involving large open spaces and buildings with high ceilings require more attention.

Although one can calculate the fire size at first sprinkler actuation, observations and judgment are needed to estimate the additional sprinklers that will fuse. The initial water application upsets the natural fire and turbulence occurs. Mathematical relationships are difficult to model.

An evaluation for multiple sprinkler fusing examines the room and the sprinkler system to envision incrementally increasing areas in which the fire may extend. This qualitative analysis estimates if the sprinklers will fuse before the fire extends beyond the incremental coverage area. The water density (dac) and water continuity (cac) are examined as a part of the estimate for each area considered.

Figure 29.4 shows the factors that influence the actuation of one or multiple sprinklers.

```
┌─────────────────────────────────────┐   Given:
│  Nozzles(s) open before the fire grows │     Fire size _____
│   Larger than the size under study     │     Floor area _____
│              (fac)                     │     Number of nozzles
└─────────────────────────────────────┘        being evaluated _____
      │                                        Time from EB _____
      │ Fire
      │   Type
      │   HRR
      │   Growth speed
      │ Link temperature
      │ Response Time Index (RTI) characteristics
      │ Sprinkler link protected from heat of the fire
      │ Sprinkler link corroded
      │ Sprinkler link bagged, taped, painted, etc.
      │ Sprinkler skipping
      │ Distance from fire to sprinkler (e.g. ceiling height)
```

Figure 29.4 Organization chart: sprinkler fuses.

```
┌─────────────────────────────────────┐   Given:
│                                       │     Fire size _____
│  Water discharge density is sufficient│     Time from EB _____
│              (dac)                    │     Number of nozzles open ___
│                                       │
└─────────────────────────────────────┘
      │
      │ Fire characteristics
      │   Rate of heat release ($Q$)
      │   Fire plume momentum
      │   Speed of fire growth
      │   Fire size
      │
      │ Water density needed to control the fire
      │   Actual Delivered Density (ADD)
      │   Required Delivered Density (RDD)
      │
      │ Water density available
      │   Quantity
      │   Pressure
      │
      │ Water droplet size and characteristics
      │
      │ Room enclosure characteristics
      │   Size
      │   Ceiling height
      │   Ventilation
      │   Shafts
```

Figure 29.5 Organization chart: sprinkler discharge density.

29.3.3 Discharge Density (dac)

The discharge density for one or more sprinklers can be calculated from the hydraulic data of the sprinkler system. The objective is to determine if the water spray can control the fire without further extension. Figure 29.5 shows the factors that influence the discharge density evaluation.

29.3.4 Water Continuity (cac)

The incremental areas enable one to estimate the ability of the sprinkler system to deliver enough water to provide the density expected in the dac evaluation. One might

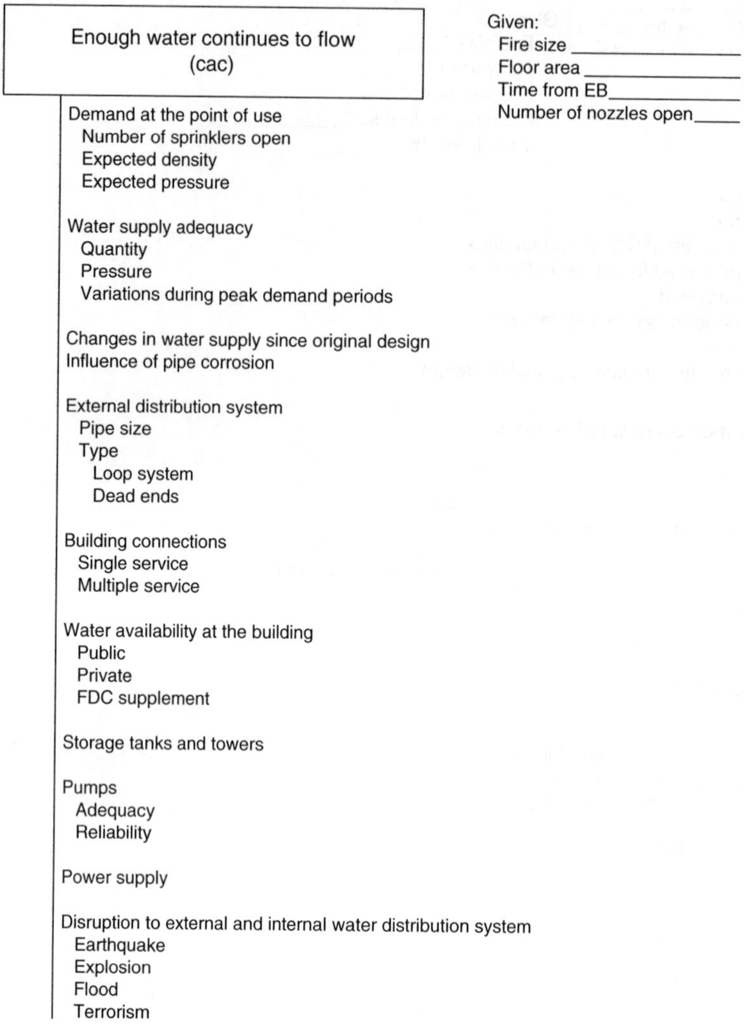

```
┌─────────────────────────────────────────┐
│     Enough water continues to flow       │
│                (cac)                      │
└─────────────────────────────────────────┘
```

Given:
Fire size _____
Floor area _____
Time from EB_____
Number of nozzles open_____

Demand at the point of use
  Number of sprinklers open
  Expected density
  Expected pressure

Water supply adequacy
  Quantity
  Pressure
  Variations during peak demand periods

Changes in water supply since original design
Influence of pipe corrosion

External distribution system
  Pipe size
  Type
    Loop system
    Dead ends

Building connections
  Single service
  Multiple service

Water availability at the building
  Public
  Private
  FDC supplement

Storage tanks and towers

Pumps
  Adequacy
  Reliability

Power supply

Disruption to external and internal water distribution system
  Earthquake
  Explosion
  Flood
  Terrorism

Figure 29.6 Organization chart: sprinkler water supply.

reasonably assume that a standards-complying design would easily provide the water for any area within the design zone. Although that may be a reasonable assumption for standards complying designs, many designs exist that are less than standards-complying. Calculations and flow tests may provide useful information with which to make this assessment. Figure 29.6 shows the factors that influence this event.

29.3.5 Obstructions (wac)

Although measurements and calculations guide dac and cac evaluations, observations become the means for evaluating the influence of obstructions within the incremental areas. The objective is to recognize the presence of obstructions to water spray distribution patterns. Water spray interruptions are common, particularly in older

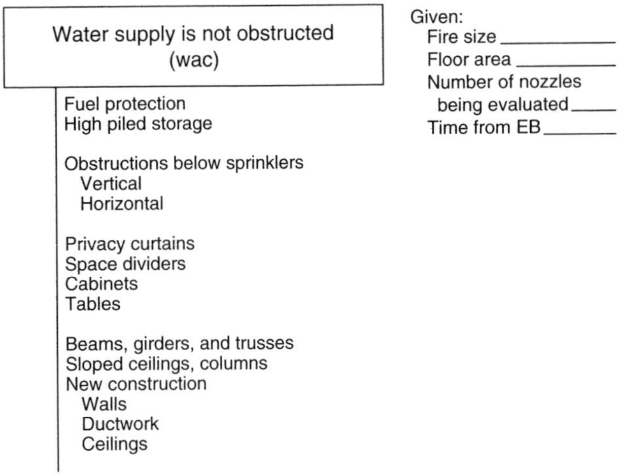

```
┌─────────────────────────────────────┐   Given:
│   Water supply is not obstructed     │     Fire size _____
│              (wac)                   │     Floor area _____
└─────────────────────────────────────┘     Number of nozzles
    │                                          being evaluated_____
    ├── Fuel protection                      Time from EB_____
    │   High piled storage
    │
    ├── Obstructions below sprinklers
    │     Vertical
    │     Horizontal
    │
    ├── Privacy curtains
    │   Space dividers
    │   Cabinets
    │   Tables
    │
    └── Beams, girders, and trusses
        Sloped ceilings, columns
        New construction
          Walls
          Ductwork
          Ceilings
```

Figure 29.7 Organization chart: sprinkler obstructions.

buildings. Statistically, buildings undergo minor renovations every 7 years and major alterations every 14 years. Modern buildings normally design the floor system to carry partitions in any location. This permits a new tenant to move partitions to accommodate different spatial needs. One commonly recognizes that inattention to fire protection systems during building renovations provides different conditions than were envisioned at the time of construction. Occupants may also cause problems with water spray discharge by arranging storage or other contents in a way that shields potential fire growth.

Sometimes, building or contents obstructions cause only a temporary problem that enables the fire to grow to larger sizes before extinguishment. The incremental analysis enables one to recognize when obstructions are relatively inconsequential and when they become important. Figure 29.7 identifies the major factors that influence observational obstruction evaluations.

29.4   Automatic Sprinkler Suppression (A)

A methodical sprinkler system evaluation gives an insight into strengths and weaknesses in the specific building being investigated. Most sprinkler systems perform well, even when a few flaws are present. However, some systems have critical weaknesses that can lead to failure during an emergency. Thus, the observation that a building is sprinklered is necessary, but not sufficient for a performance analysis. Fortunately, many common flaws can be recognized by observation. Some may be hidden. Although uncertainty is routine, the analytical framework provides a way to organize thinking and bridge gaps in knowledge or understanding.

29.5   Automatic Sprinkler System Analysis

The framework for sprinkler performance analysis is in harmony with state-of-the-art sprinkler technology. The qualitative judgments and quantitative estimates used in evaluations are compatible with sprinkler standards and practices.

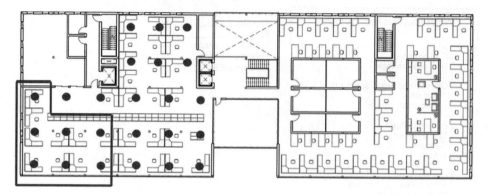

Figure 29.8 Model building: partial sprinkler layout.

Before we discuss sprinkler performance analysis, let us identify sprinkler system design requirements for the model building. Design standards permit some decision flexibility to accommodate specific building needs. Some decisions affect performance expectations more than others. Our design example will be standards-compliant, although decisions will be based on minimum costs and requirements rather than high performance.

Figure 29.8 shows the model building's fourth floor plan and a partial sprinkler layout for the open office. Assume the sprinkler system was designed for an office occupancy that used minimum requirements to obtain approval of the authority having jurisdiction (AHJ). Information for this design is described below.

The sprinkler system design (USA standards) is based on a light hazard (business) occupancy. A single system can be used because the floor area [14 920 ft^2 (1390 m^2)] is less than the allowed maximum [52 000 ft^2 (4831 m^2)]. When the system is designed hydraulically, the maximum area for "standard coverage" sprinkler heads is 225 ft^2 (20.9 m^2) with a maximum distance of 15 ft (4.5 m) between heads. The hydraulically most remote area supplies the seven operating heads shown in Figure 29.8.

The open plan area on the west side of the building would require 24 heads, roughly laid out as shown in Figure 29.8 (red circles). A design area of 1500 ft^2 (139.4 m^2) requires a minimum density of 0.1 gallons/minute (gpm) per ft^2 [4.07 liters/minute (lpm) per m^2]. The calculated flow rate for the design area is (7)(22.5 gpm) = 157.5(596 lpm). The minimum flow rate over the design area is 0.1 gpm/ft^2 × 1500 ft^2 = 150 gpm. The larger value governs and the minimum design flow rate is 157.5 gpm (596 lpm).

Some codes require additional flow to supply hose streams. Light hazard occupancies in the USA require 100 gpm (379 lpm). The total required flow rate becomes 157.5 + 100 = 257.5 gpm (975 lpm). This total supply of 257.5 gpm (975 lpm) is required for at least 30 minutes.

The required pressure is based on the K-Factor of the head, friction loss in the piping, elevation, and standpipe system requirements. We assume the sprinkler head for this system uses a K-Factor of 2.8 (40.3 SI), minimum operating pressure of 7 psi (0.5 bar), and a maximum working pressure of 175 psi (34.5 bar). The required pressure at the head is calculated as $P = Q$ (flow rate)/K^2. This requires $P = (22.5/2.8)^2 = 64.5$ psi (4.45 bar).

The operating temperatures for the sprinklers can range from 135 to 360°F (57–182°C). Operating temperatures for glass bulb/fusible link sprinklers are typically selected to handle expected ceiling temperatures and the desired response. For a maximum ceiling temperature of 100°F (37.8°C) this sprinkler has two temperature options: 135°F (57°C) or 155°F (68°C). We select 135°F (57°C) for this system.

29.5.1   Role of Performance Analysis

National standards that guide sprinkler system design are based on well over a century of experiential, statistical, and research information. The standards are thoughtfully organized to guide successful fire control for buildings. Regular updated editions are published to incorporate improvements and better designs.

Technical information is continually evolving to improve understanding of the enormously complex interactions of fire, water, and equipment, as well as predictions of expected outcomes. However, we should recognize that the role of a performance analysis for an existing or proposed new building is different from that of design for standards compliance.

A performance analysis augments standard design behavior expectations with additional quantitative knowledge and qualitative judgments. The composite information gained from integrating quantitative information with qualitative judgments enriches performance understanding of expected outcomes.

Applications using observation and judgment can be done quickly. They enable one to recognize abnormalities that could cause sprinkler system performance deficiencies. The organization described above provides a framework for thinking and an inventory of factors that influence performance. Practice in making observations and judgments and relating them to performance provides a way to continually improve understanding. Chapter 35 discusses ways to deal with imprecision and uncertainty and Chapter 36 provides ways to communicate comparisons of different conditions.

Performance analysis is different from regulatory design. For example, we may obtain AHJ approval for the design identified earlier. But how will the sprinkler system actually perform in this building? A performance analysis might indicate that the sprinkler system should work very well for certain locations within the building, but be questionable for other locations. A performance analysis attempts to understand where the sprinkler system will work well, where it may not be as effective, and reasons for the conclusions.

29.5.2   Organizing Performance Analysis

A performance analysis separates design effectiveness (AC) from the system's agent application (AA) (i.e. reliability). The major focus for AC performance is identified by the events of Figure 29.2(b). Reliability (AA) is identified by the events of Figure 29.2(a).

The process selects a specific fire scenario and envisions the interaction of the fire and the sprinkler system for successive increases in floor area of fire involvement. Figure 29.9 shows a continuous value network (CVN) to guide the thought process for judgments. The outcome for each event is judged by envisioning fire conditions within a coverage area and asking the question: "Given the fire was not controlled within Area A_n, will it be controlled within Area A_{n+1}?"

Row 2Bb (sprinkler control, AC) of the IPI chart shows the outcomes for Figure 29.9. The decision concerning whether the fire is controlled within any specific area AC_n is

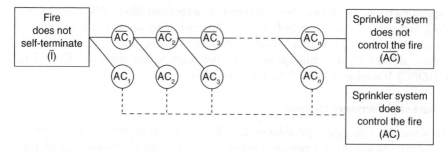

$\overline{AC}_n$ = Sprinkler system does not control the fire within area A_n.
AC_n = Sprinkler system does control the fire within area A_n.

Figure 29.9 Continuous value network (CVN): sprinkler control.

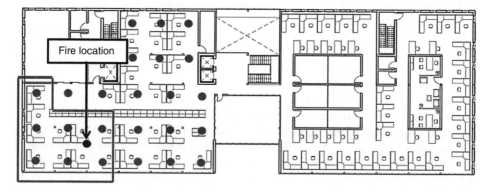

Figure 29.10 Example 29.1: sprinkler layout.

determined from evaluating the events of Figure 29.2(b). The outcomes for these events are shown in the IPI cells associated with each fire size. The floor areas of involvement can be segmented with virtual barriers, as discussed in Chapter 7.

Example 29.1 describes performance analysis concepts for design effectiveness (AC). Initially, we will examine a fire at the center of the four sprinklers, as shown in Figure 29.10. An understanding of the fire performance for this condition will enable one to extrapolate qualitative differences in condition for other locations in the building.

Example 29.1 Assume the sprinkler system described earlier had been designed and received AHJ approval when the building was initially constructed. You have been asked to develop a risk management program for a new owner. A part of the risk management program (see Chapter 37) examines the building's sprinkler performance. How would you evaluate the performance of this building's existing sprinkler system?

Solution:
Assume that you have the design information described in Section 29.5. You examine a potential fire to gain an initial basis for understanding performance. Here, we select an ignition location equidistant from four sprinklers, as shown in Figure 29.10.

The CVN of Figure 29.9 guides the process. Examining the events of Figure 29.2(b) at each selected fire size provides the basis for performance understanding.

Sprinklers fuse (fac)   The initial focus examines the fire scenario and sprinkler behavior. Calculation methods can give a sense of fire size when the first sprinkler actuates. The information from this analysis helps one to make qualitative judgments of sprinkler operations beyond the first sprinkler actuation.

The number of sprinklers needed to control the fire is a major concern. If the fire spreads faster than sprinklers actuate, too many sprinklers will open, and water flow and suppression quality are reduced.

First sprinkler actuation can be calculated in the same way as heat detector response described in Chapter 26. Relatively simple spreadsheet analyses can be made for steady-state fires using Alpert's correlations, or for growing fires using the power law approach of Heskestad and Delichatsios with modifications by Beyler (HD&B). A more complete analysis can be made with computational fluid dynamics (CFD) modeling. The value of the information must be related to its cost. Normally, sufficient information can be developed from the more simple analyses. For illustrative purposes, the results of these calculations for a fire in the center of the four sprinklers, as illustrated in Figure 29.9, is as follows: (response time index, RTI = 100, actuation temperature = 57°C)

Alpert's steady-state correlation: $t = 200$ s
HD&B Quasi-steady state: $t = 350$ s
CFD Analysis: $t = 360$ s

A sense of sprinkler actuation can be obtained examining the parameters for steady-state fires from equation (29.1):

$$t_{actution} = \left(\frac{RTI}{\sqrt{u_{jet}}} \right) \left\{ \ln \frac{\left(T_{jet} - T_a \right)}{\left(T_{jet} - T_{actuation} \right)} \right\} \tag{29.1}$$

A sprinkler's RTI is a measure of how quickly the thermally activated material responds given gas temperature and velocity near the head. Standard response sprinklers have an RTI of at least 80 $m^{1/2}\,s^{1/2}$ but it can be as high as 400 $m^{1/2}\,s^{1/2}$. To illustrate the effect of RTI on a steady-state fire, assume in equation (29.1) that the ambient temperature, $T_a = 20$°C, the ceiling jet temperature, $T_{jet} = 100$°C, and the ceiling jet velocity, $u_{jet} = 0.8$ m/s. Using these values in equation (29.1) gives an actuation time of $t_{actuation} = (0.69) \times RTI$. For steady-state fire conditions and standard response sprinklers, the actuation time will vary from 56 seconds for an RTI of 80 $m^{1/2}\,s^{1/2}$ to 276 seconds for an RTI of 400 $m^{1/2}\,s^{1/2}$.

The location of the fire relative to the sprinklers is also a part of this analysis. In Figure 29.9, the sprinklers are spaced 15 ft (4.5 m) on center. The radial distance from the fuel to the sprinklers (r) is 10.5 ft (3.2 m). If we assume the fire height is 0.5 m and the ceiling is 2.5 m, then $H = 2.0$ m. The ratio r/H is 3.2 m/2.0 m = 1.6. Relationships are given for T_{jet} when $r/H > 0.18$ and for u_{jet} when $r/H > 0.15$. A spreadsheet analysis shows that for heat release rate (HRR) = 500 kW, $T_{jet} = 98$°C and $u_{jet} = 0.82$ m/s. These are close to the values used earlier. Thus, sprinkler actuation temperature, RTI, ceiling height, and spacing can be used to estimate the fire size at first sprinkler actuation.

A family of curves that relate fire size to the most common parametric values for room fires could be prepared for digital access. Using calculated results with practice in observation and judgment helps to develop a sense of proportion for fire size and initial sprinkler response.

For this scenario, we select an HRR, $\dot{Q}$, of about 600 kW for initial sprinkler actuation.

In addition to routine examination such as this, one should examine the sprinklers for other factors that can delay actuation. These are identified in Figure 29.4.

Discharge density (dac) Depending on the type of fire, its location, and the sprinkler pattern, one sprinkler will often discharge enough water to cool the fire, wet unburned fuel, and prevent its continued growth. When the fire is equidistant from several sprinklers, four sprinklers may "surround and contain" fire spread. However, when water application and discharge density are not able to control the fire, additional sprinklers may open. As more sprinklers open, the water discharge density decreases. If the fire continues to spread, the sprinkler system may exceed its capability and become ineffective.

The discharge density (dac) event estimates whether the water discharge is able to control the fire within the area of analysis without extension to a larger area. This deterministic assessment is very difficult to make because at this time there is no way to predict the actual amount of water that will be delivered to a specific location. Nevertheless, the event is essential to sprinkler success, and we must find a way to distinguish between success and failure for successive fire sizes. Fortunately, several concepts help with these estimates.

One observation involves the sprinkler type. Although this problem notes that the selection is a "standard coverage" sprinkler, one should ensure that the sprinkler matches the need. It is not uncommon for uninformed maintenance workers to use inappropriate heads for replacements. Examining the installed sprinklers can eliminate this concern.

The water pressure, its quantity, droplet sizes, and size of the wetting area are all interrelated. Unfortunately, precise values cannot be calculated. Nevertheless, one can get a sense of extinguishment capability with water pressure and quantity values. Therefore, a check of water pressure is useful information for this evaluation.

Another concept that provides useful guidance for this evaluation uses required delivered density (RDD) and actual delivered density (ADD). Although this research has been used with early suppression fast response (ESFR) sprinklers, the concept is useful to envision sprinkler actions. Contemporary research is attempting to extend the ESFR concepts to ordinary hazard fires.

The ADD of water is a measure of the water flux that penetrates the fire plume and is delivered to the top surface of a burning fuel package. The ADD depends on water droplet size, spray pattern, discharge rate, and fire size. While ADD is not measured for all sprinklers, it is an indication of whether one or more sprinklers can control the spread of a fire. It also gives an indication as to when a fire might grow beyond the capacity of a sprinkler system. Key issues when considering the impact of fire size and ADD are type of head (e.g. standard, large drop, etc.), actuation time of sprinklers, size of fire, and speed of fire growth.

Standard sprinklers with normal (small) drop size control relatively small fires very well. However, standard sprinklers do not control fast-developing fires with high HRRs as completely. Because the design area requirement anticipates multiple sprinkler operation, most fires can be controlled. However, when the fire is intense and water cannot penetrate the plume, the fire can spread. If the fire actuates more than the number of sprinklers considered in the design area, the water quantity may be insufficient to control the fire, and the fire could overwhelm the sprinkler system.

The ADD concept is important for standard sprinkler ratings. However, it is more important for extended coverage sprinklers because spacing distances are increased, which influences the time to sprinkler actuation and associated fire sizes. A performance analysis examines sprinkler actuation tendencies with the size and speed of fire growth. The number of sprinklers in operation affects the water density application and its ability to control the fire. Figure 29.5 identifies factors that affect (dac).

Continuous water flow (cac)   This event examines the duration of the water supply with sprinkler operations. Some installations have "unlimited" water supply. In others, water flow is possible for only limited time durations. The source and ability of the water to be delivered for the number of sprinklers that fuse and for the fire duration are a part of all sprinkler evaluations. Figure 29.6 lists factors that affect cac.

Water spray obstructions (wac)   Observation of water spray obstructions is the easiest to evaluate of the four events that affect sprinkler control (AC). The obstructions may be caused by the building or from shielding the fire by contents. Visual examination for products of combustion shielding is very important from a fire spread viewpoint. Figure 29.6 identifies factors that affect water contact with a fire.

29.5.3   Performance Evaluation

We will initially describe a performance evaluation for the scenario of Example 29.1 Then, we will extend the discussion to other areas in the building.

For the initial scenario, the basis for a performance evaluation is the opinion regarding whether the fire can grow faster than the sprinkler system is able to control its movement. We estimate that a fire that is equidistant from four sprinklers will reach about 600 kW before the first sprinkler actuates.

Selecting a diagnostic fire is an important part of a performance analysis. This would have been done earlier as a part of the diagnostic fire (Section 1A of the IPI chart). Figure 29.11 shows several diagnostic fires that are feasible for office arrangements. The diagnostic fire relationships are based on furniture calorimeter measurements and interactions of fuel groups and fuel packages described in Chapter 6. Diagnostic fire D may represent traditional office fuel packages of wood and metal. Diagnostic fire A would represent a type of modern workstation, as illustrated in Chapter 6.

In this scenario the open office has no physical barriers. However, one can associate the diagnostic fire with floor areas using virtual barriers to represent the analytical segments of Figure 29.9.

If the office diagnostic fire were similar to fires C and D in Figure 29.11, the sprinkler system described by the design criteria would probably control the fire. On the other hand, if the office fuel packages exhibited fire growth such as that of fire A, a fire that is not controlled very early would probably spread faster than the sprinklers could operate. If the fuel package were located directly under a sprinkler actuation with a smaller fire size, control would be easier. If extended coverage sprinklers were used rather than standard sprinklers, one would expect a larger fire at actuation and less chance of controlling the fire when it is small.

If the original sprinkler design were used with compartmented offices, as in Floors 2 and 3, the physical barriers would constrain movement. Steam production would

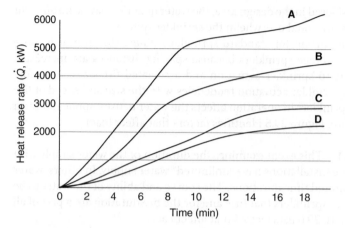

Figure 29.11 Illustration: Diagnostic Fire Types and Sprinkler Response.

reduce oxygen in the space, and wetting would be easier. One would expect this sprinkler system to be successful in small and moderate-sized offices regardless of the combustibility of the furnishings.

Floor 1 has a restaurant and a retail store. The restaurant cooking area may need hood and duct suppression in addition to the sprinkler system. The retail store may have different diagnostic fire conditions as well as greater potential for water spray obstructions. One should ensure that the sprinkler design is appropriate for mixed-occupancy buildings such as this.

A basement sprinkler system for this building requires special attention. Modern automobiles display very different diagnostic fire characteristics. Large, fast fires involving several vehicles are possible. Unheated areas must be investigated. Compartmentation is not feasible. One should ensure that the office sprinkler design was not inadvertently used for the garage.

29.6 Sprinkler Reliability

A deterministic performance analysis documents the factors that can adversely affect water discharge from a fused sprinkler. Factors that affect agent application are discussed in Chapter 16 and Section 29.2.

Uncertainty is a part of an agent application (AA) evaluation to define reliability. Most factors are difficult to identify in terms of whether or not the sprinkler system will deliver water when needed. Nevertheless, a thoughtful examination of factors that affect reliability is valuable to distinguish between good and bad performance. Chapter 35 discusses ways to calibrate the "shades of gray" that are a common part of many performance evaluations.

29.7 Closure

A performance analysis for sprinkler systems augments code and standards design procedures. Its function is to examine an existing system or a proposed new system to recognize any conditions that could compromise effective sprinkler suppression.

Sprinkler systems employ a "one size fits almost all" approach for the hazard classification used for the design. Over the years, many things can happen to the original design decisions. For example, the diagnostic fire can change because of different materials that enter the market. Large, fast fires with modern contents materials can easily appear. Differences in conditions may arise because of changes in operation or use. For example, a manufacturing facility may change its product packaging material from metal or paper to plastic. Storage may change from low pallet packages to high rack storage. These changes will alter both the diagnostic fire and the water distribution effectiveness.

The interior plans of commercial buildings change regularly over the years. Often, the change may involve only a few rooms or perhaps an entire floor. However, fire performance can be affected by changes in spatial configuration (i.e. partition locations), installation of partial barriers, ductwork, or furnishings and storage methods.

A performance analysis for sprinkler systems is largely qualitative. Quantitative methods are used when they are available. Fortunately, continued research is creating information that enables gradual transition from qualitative judgment to quantitative proportions. Nevertheless, qualitative observation and judgment can enrich our understanding of sprinkler system performance. This enables one to recognize areas in which the sprinkler system may not perform as well as might have been expected. When these locations are identified, the change in risk can be managed.

30

The Composite Fire

30.1 Introduction

The composite fire combines the diagnostic fire with automatic and manual extinguishment. Figure 30.1 shows the general relationships. Time links all components.

The diagnostic fire is the composite fire until first water application disrupts the natural fire behavior. After first water application, the fire becomes turbulent and changes character. Water application absorbs heat energy, wets unburned fuels, and steam conversion displaces some air. All of these actions help to diminish the fire.

Often, the fire can be extinguished relatively quickly, and the flame-heat and smoke-gas products of combustion stop. Sometimes, the fire may continue to grow, even after initial water application. Eventually, the fire will be controlled and extinguished. At that time, the fire ceases to generate any more products of combustion.

30.2 Event Logic Description

The time of first water application by a sprinkler system or the local fire department marks the end of a mathematically based diagnostic fire. Initial water application disrupts natural fire behavior and the composite fire must incorporate changes caused by extinguishment actions. Fire dynamics cannot quantify fire behavior for these conditions and qualitative methods must describe the composite fire environment. Fortunately, analyses for sprinkler suppression and fire department extinguishment can provide information to estimate limits for expected outcomes and make credible qualitative estimates.

Figure 30.2 shows an event logic diagram for the success of active extinguishment. When a sprinkler system is present in the room of fire origin, the automatic suppression component starts the process. The fire department acts as "mutual aid" to complete extinguishment, stop water application, and check for hot spots. When a sprinkler system has not been installed, the local fire department must initiate and complete active extinguishment.

Fire Performance Analysis for Buildings, Second Edition. Robert W. Fitzgerald and Brian J. Meacham.
© 2017 John Wiley & Sons Ltd. Published 2017 by John Wiley & Sons Ltd.

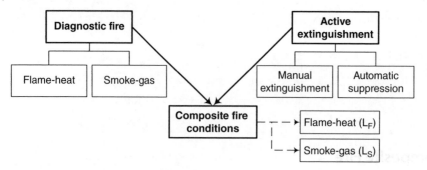

Figure 30.1 Organization chart: composite fire.

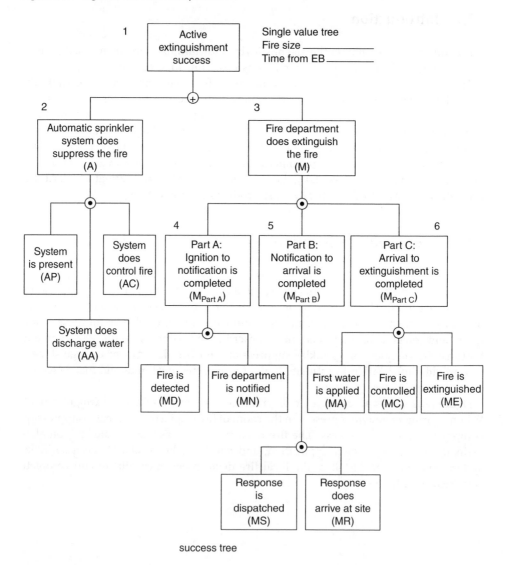

Figure 30.2 Success tree: active extinguishment.

30.3   Network Description

Figure 30.3 shows the network diagram that guides analysis. Only the first level of the event logic diagram is shown here. Additional levels for automatic sprinkler suppression, as well as guidelines for analysis are given in Chapter 29. Chapter 28 shows complete networks and application guidelines for local fire department extinguishment.

The normal procedure for analysis examines each active extinguishment component as an independent event for the fire condition being evaluated. This provides distinct advantages in understanding the fire performance in buildings. Because sprinkler suppression (Event 2) and manual suppression (Event 4) are independent, their order in Figure 30.3 is irrelevant. Also, the Interactive Performance Information (IPI) chart Sections 2A and 2B document performance assuming no interaction between the components.

In situations when the sprinkler system operates first and the local fire department augments sprinkler suppression, conditionality is introduced. Manual extinguishment evaluations then become based on composite conditions in which the sprinkler system has changed the fire conditions at the time of first water manual suppression (MA). While Figure 30.3 can remain accurate, the interpretation of the events becomes "M given that A is not successful $(M | \bar{A})$."

The analytical framework can accommodate any special situation such as this. The composite fire (IPI Section 3A) provides a location to document this situation. Usually,

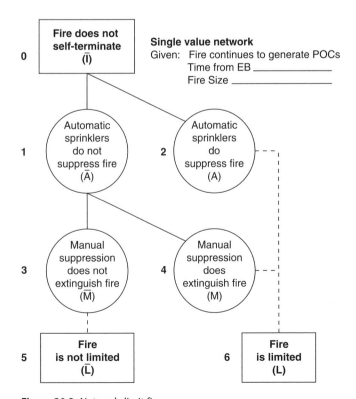

Figure 30.3 Network: limit fire.

a separate IPI sheet should be prepared to show these actions as an alternative. The process rapidly creates new IPI sheets and can easily document the performance.

30.4   Summary

Although each scenario in a building has a potentially different limit (L), an aspect of the art of engineering involves selection of appropriate scenarios on which to base the building analysis. Using a good scenario as a basis for comparison, one can extrapolate for other scenarios to recognize atypical situations. If necessary, additional detailed scenarios may be examined to establish a clear understanding or specific details that affect building performance.

The composite fire in IPI Section 3A provides an opportunity to document different situations or other alternatives. The smoke section in IPI Section 3B enables one to document smoke spread. In both cases, values before first water application may use a combination of quantitative and qualitative methods.

31

Structural Performance

31.1   Introduction

Traditional building codes consider structural fire protection as both barrier integrity and structural stability. The criteria for selection is the fire endurance provided by the standard ASTM E-119 or ISO 834 fire test. Although the objective of the test is to compare the fire endurance of different structural assemblies under the same fire conditions, the results are also used to organize the code's fire safety requirements. Hourly ratings encourage "substantial" construction for large or fire-sensitive buildings.

Performance-based compartment analysis separates the barrier integrity function from the structural frame's collapse or deformation. Figure 31.1 shows an organization tree to describe the activities. The fire safety engineer is responsible for describing the diagnostic fire. This includes multi-room fires caused by fire propagation through barriers due to openings. It also includes analysis of heat transmission through the barriers that may cause ignition in the adjacent room.

The structural engineer estimates the deformations and potential structural collapse conditions. Structural collapse may be local for members or room size assemblies. Structural collapse may also involve a more global analysis to recognize the potential for larger building collapse conditions. The fire safety engineer is responsible for identifying the diagnostic fire for structural analysis, as noted in Figure 31.1

31.2   Interactive Performance Information (IPI) Documentation

The IPI chart provides an organizational framework for ordering dynamic information. IPI Section 1A defines the diagnostic fire for the room of origin and the cluster of additional rooms beyond. Figure 31.1 Section A identifies $(\overline{T})$ and $(\overline{D})$ failure conditions that create multi-room fire scenarios. In addition, the heat from a fire may transfer through barriers and ignite kindling fuels on the unexposed side. This scenario, particularly for floor-to-floor vertical propagation, seems to be more common than is generally assumed. The fire safety engineer is responsible for information in IPI Section 1A.

Section 4A describes the structural frame performance. This is the responsibility of the structural engineer. This analysis is based on the diagnostic fire characteristics provided by the fire safety engineer. We note that fire in a room normally affects the

Fire Performance Analysis for Buildings, Second Edition. Robert W. Fitzgerald and Brian J. Meacham.
© 2017 John Wiley & Sons Ltd. Published 2017 by John Wiley & Sons Ltd.

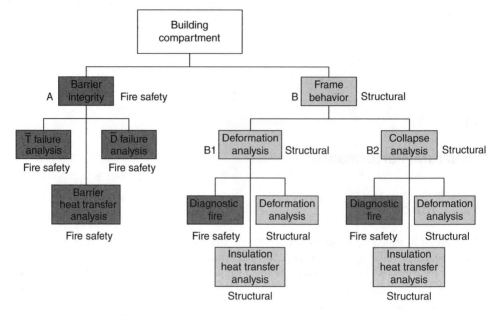

Figure 31.1 Responsibility divisions: fire safety engineer and structural engineer.

framing for the rooms above. Therefore, one must be careful to associate structural frame performance with the appropriate rooms in Part 4A.

The diagnostic fire identifies not only the time–temperature relationships for the fire compartment but also the compartments of interest. In addition, specialized compartment locations that are particularly important to structural performance may be selected for attention.

IPI section 4A is designated for structural performance information. One may use this IPI section in the following ways:

1) *No information.* The cells of the IPI chart may be left blank. This signifies that the structural frame was not examined for performance expectations. There are many reasons why structural performance might not be of interest or possible for an evaluation. However, neglecting this part of a holistic analysis must be noted and the reasons for the omission also noted.
2) *Non-analytical comments.* Sometimes the structural frame will be clearly stronger than any fire threat. Rather than developing a detailed structural examination, the professional may recognize that collapse or excessive deformation will not be a concern to the building performance. A brief comment in the cells of the IPI chart will note that structural performance will be adequate without calculations.
3) *Numerical estimates.* A structural analysis based on the concepts described in Chapter 19 will provide load capacity related to time durations of the diagnostic fire. These may be based on either concern about excessive deformations or actual collapse.

4) *Visual information.* The versatility of spreadsheets to incorporate digital pictures and other visual information expands the application scope. Additional pages devoted to IPI Section 4A can document digital images of actual conditions or personal notes. These may be used with the analysis or as ancillary documentation.

Structural frame information is organized to meet the needs of the analysis. The IPI chart provides a structure to record any information efficiently.

31.3 IPI Numerical Estimates

A structural analysis may examine a variety of conditions of interest to risk management. Figure 31.1 shows that one may analyze the frame with regard to fire reconstruction after the fire is extinguished. Any defined structural performance needs may be examined.

Figure 31.2 shows representative structural performance relationships described in Chapter 19. This type of graph may be constructed for any structural element (e.g. beam, column, truss, etc.), assembly, or framing system. They may be adapted for all normal construction materials. Although the criteria and behavior of wood, steel, reinforced concrete, and prestressed concrete will differ, performance curves can represent the specific situations that may be analyzed.

The cells of the IPI chart can document the performance needs. For example, the representative performance diagram in Figure 31.2 shows collapse loads and loads that would produce first yield (i.e. permanent deformation). These values would be of particular interest for members of structural steel and reinforced concrete. The dead load would be known with confidence, and the applied live load may be determined from Figure 31.2. Analogous curves may be constructed for appropriate failure criteria for wood and prestressed concrete.

The IPI chart may record the live load capacity for conditions of first yield (point "a") or collapse (point "b"). Additionally any comparisons with temporary or fixed live loads, or different loading conditions may be used. Additional sheets enable one to examine a variety of conditions to develop a clear understanding of structural performance.

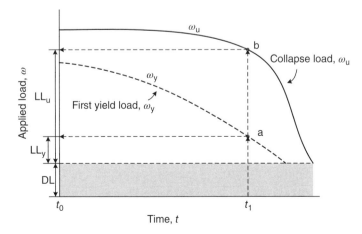

Figure 31.2 Representative structural performance.

Local times may be coordinated with the global time from established burning in a room of origin according to the IPI chart. Thus, one will be able to recognize the structural expectations at any time during the fire.

31.4 Summary

A performance-based structural analysis can describe the time at which the diagnostic fire will cause non-collapse deformations sufficient to require reconstruction. An analysis can also identify the time at which a diagnostic fire will cause structural collapse.

Collapse loads require interpretation. For example, should the analysis be based on code-specified live loads, or some lesser amount? Collapse potential can be estimated for a variety of situations. In some cases, live loads may not be a concern and collapse may be based on dead loads only. On the other hand, some immovable live loads will require full loading conditions. Fortunately, almost any type of condition can be examined.

Collapse may be examined for local framing directly affected by the diagnostic fire. Collapse may also be examined in terms of general collapse. Often, loads may be redistributed to delay or avoid failure. An engineer may be able to provide some guidance regarding holistic performance through qualitative estimates of structural support conditions.

32

Target Space Smoke Analysis

32.1 Introduction

Target space analyses are essential to understanding risk characterizations for people, property, and operational continuity. The window of time during which a target space remains tenable is a measure of performance. The Interactive Performance Information (IPI) chart Section 4B records this information.

Although the number of target rooms and scenarios often appears to be large, quantitative analysis of a few carefully selected spaces usually gives a clear understanding of performance. The quantitative analysis gives confidence regarding qualitative extrapolations for other spaces. Also, it is possible to substitute zones (i.e. multiple rooms) for individual rooms in some situations. The goal is to identify clearly the specific spaces for which risks must be characterized, time durations for tenability, and environmental conditions that affect risk and tenability.

Figure 32.1 shows the event logic success tree for the critical events of target space analysis. Because this single value tree identifies event status changes as time progresses, outcomes can be recorded on the IPI chart for each instant of time.

Chapter 20 discusses target rooms, tenability considerations, and an overview of the process. This chapter describes the analytical framework. After tenability is defined and smoke paths from the room of origin to the target room are identified, an analysis examines three questions:

1) At what time will enough smoke to cause untenable conditions reach the target room boundaries? This provides an understanding of smoke generation and the features that affect movement from fire origin to target space.
2) Will enough smoke enter and remain in the target room to make the room untenable? This examines the effectiveness of barrier protection, venting, and mechanical pressurization.
3) What is the window of time during which the target room will remain untenable? The responses to selective single value analyses become the entries for the IPI chart, Section 4B. The IPI chart shows both the dynamic changes over time and the duration during which the target room remains tenable.

Fire Performance Analysis for Buildings, Second Edition. Robert W. Fitzgerald and Brian J. Meacham.
© 2017 John Wiley & Sons Ltd. Published 2017 by John Wiley & Sons Ltd.

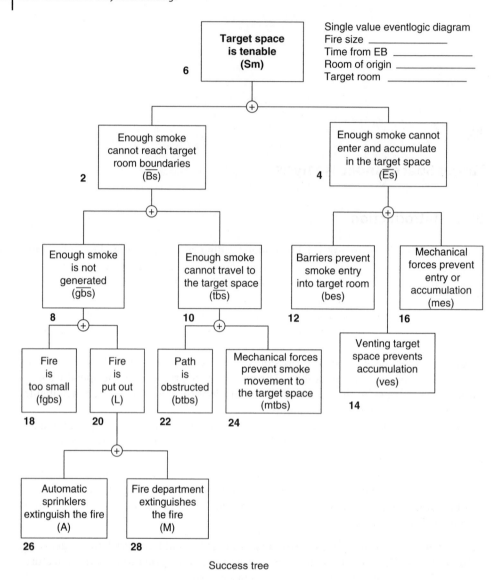

Single value eventlogic diagram
Fire size _____
Time from EB _____
Room of origin _____
Target room _____

Success tree

Figure 32.1 Success tree: target space tenability.

32.2 Success or Failure?

Event logic diagrams and their associated network diagrams guide building evaluations. The earlier networks in the analytical system have had an orderly structure where the logic of "failure" and "success" events formed a consistent pattern with outcomes. Unfortunately, the description of events for smoke analysis forces a slight change in the methodical pattern. Therefore, one must read the analytical network with greater care to ensure that the event evaluations are ordered properly.

The analytical logic to evaluate target room success (i.e. maintain tenable conditions in the target space) requires OR gates, as shown in Figure 32.1. The IPI chart is based on a set of success events for all components. Event descriptions for smoke tenability are difficult to formulate (i.e. describe) in a manner that is consistent with other components. In particular, the descriptor overbar (" ¯¯ ") is intended here to mean "success" rather than "not." The descriptors are more compatible with evaluation measures and an analytical thought process. Consequently, one must follow the logic rather than a "preconceived" indication of overbar location.

The difficulty is overcome with the working network diagrams of Figures 32.2(a)–(c). The numbered events of the event logic diagram are coordinated with the network diagrams. These network diagrams guide the thought process and allow one to evaluate either success or failure at every event. Thus, modern quantitative methods are easier to use in an evaluation.

Figures 32.1 and 32.2 are single value. The results are analogous to photographs taken at each instant in time. The IPI chart establishes time sequencing changes by the horizontal rows that represent event tree success events. Therefore, at each instant in time, the status of each event in Figure 32.2 changes. The IPI chart enables one to systematically order all status changes.

32.3 Target Room Performance Bounds

The times at which the best and worst conditions for target room tenability can occur give an understanding, and perhaps a cost-effectiveness measure, of fire defenses. These bounding conditions are determined by examining:

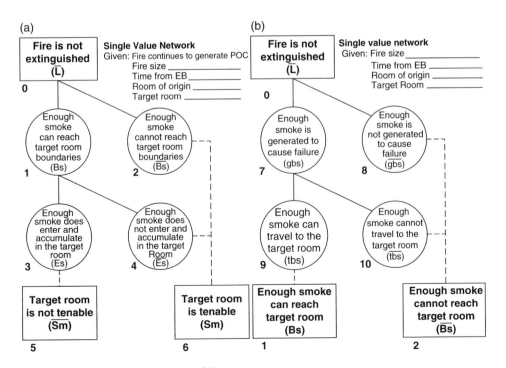

Figure 32.2 Network: target space tenability.

(c)

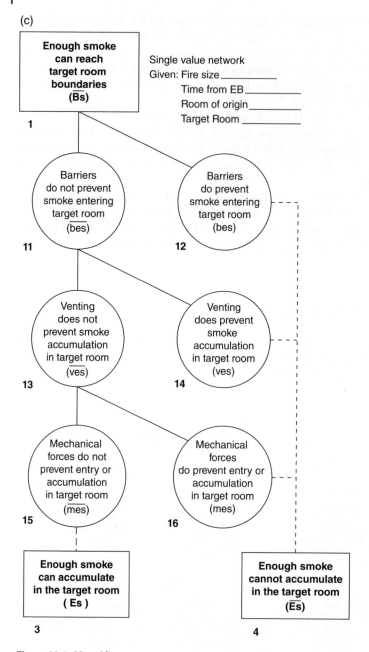

Single value network

Given: Fire size _____

Time from EB _____

Room of origin _____

Target Room _____

Figure 32.2 (Cont'd)

1) The minimum time for a target room to become untenable using the diagnostic fire alone with an unobstructed products of combustion (POCs) travel route.
2) The target room tenability time when the fire is extinguished (L) by either automatic sprinklers (A) or the local fire department (M). Extinguishment stops the generation of POCs.

The first condition gives an insight into building performance for a fire having no suppression intervention and no way of controlling the smoke. This gives an insight into sensitive locations that affect smoke movement as well as a worst-case scenario for risk characterizations. It also gives a base with which to compare the effectiveness of smoke control measures that may be installed.

The second bound enables one to recognize the effect of extinguishment on target space tenability. Smoke generation stops when the fire is extinguished. Timely fire extinguishment becomes an option for fire risk management programs. Again, performance comparisons can be made to measure cost and effectiveness of different risk management alternatives.

Interactive Performance Information chart comparisons provide a systematic and relatively easy way to compare alternatives with the best and worst performance conditions. After the two bounding conditions have been examined, alternatives may be analyzed to determine the effect of actual or proposed measures on smoke tenability in the target spaces.

33

Life Safety Analysis

33.1 Introduction

A performance analysis for life safety does not predict the likelihood of death or injury. Instead, it recognizes scenarios when people and untenable conditions can occupy the same space at the same time.

Time and windows of time are the fundamental determinants for occupant safety.

A building is defined by a fixed array of space–barrier modules whose fire environment can change dynamically. One part of a performance analysis identifies the windows of time during which important modules, called target spaces, remain tenable. A life safety analysis examines scenarios in which an individual may encounter untenable conditions in the building.

Occupant actions – or inactions – have a freedom of movement within the building's array of barrier–space modules. An analysis establishes the maximum pre-movement delay time available to occupants before they can expect to encounter untenable conditions during a scenario. Analyses provide a basic understanding of how the building's performance affects life safety.

33.2 The Exposed

Life safety involves four occupant categories:

- Occupants who attempt to leave the building during a fire emergency.
- Occupants who are restrained, incapacitated, or unable to leave the building.
- Occupants who decide to remain in the building.
- Fire fighters conducting emergency operations.

Occupant rescue or fire fighters directing occupant movement to a safe location is not considered in a performance analysis. Nor does an analysis consider emergency evacuation through windows. Although Chapter 22 identifies these actions as part of a complete picture, life safety analyses consider only movement through interior building spaces or remaining in the building.

A performance analysis also does not make decisions about what individuals may or may not do during a fire. However, performance analyses can describe windows of time

Fire Performance Analysis for Buildings, Second Edition. Robert W. Fitzgerald and Brian J. Meacham.
© 2017 John Wiley & Sons Ltd. Published 2017 by John Wiley & Sons Ltd.

for "if ... then ..." occupant scenarios. These scenarios provide an understanding of how the building's design affects life safety.

33.3   The Exposure

A building fire can cause death or injury by:

- *Local or general building collapse.* This can be estimated from structural analysis for fire conditions.
- *Flame-heat burns.* Burns occur when an occupant is near the fire. These burns usually occur during the early stages of a fire or during fire fighting.
- *Smoke-gas inhalation.* Exposure to smoke-gas may cause an occupant to die or become incapacitated. Later, the individual may also experience severe burns if the fire propagates to the location.

Although conditions that cause these outcomes may be recognized during an analysis, we note their potential from observations rather than dedicated evaluations.

Chapter 21 describes the effect of products of combustion on humans. Although physiological effects may be recognized, human vulnerability cannot be defined in terms of measurable products of combustion during a fire. We recognize that humans can endure exposure to untenable smoke-gas conditions for a limited period of time. Humans may move through untenable spaces or they may seek another route. Therefore, untenable conditions are not necessarily synonymous with death or injury.

Quantifiable conditions to define tenability must be selected to provide a measure to describe building performance. The most practical tenability measure is a distance that individuals can see through smoke under normal lighting conditions. This definition of tenability can be used with the current state of fire technology. It provides a measure that can be understood by laypersons and professionals alike.

33.4   The Window of Time

A building analysis identifies windows of time that target spaces remain tenable. This describes an available safe egress time (ASET) for the target space. ASET may also be used for sequential barrier–space modules within an egress path.

The definition can be extended to describe an available tenability duration for individuals who remain in a building with a "defend-in-place" concept.

In theory, windows of time may be used for every space in a building. In practice, selected spaces may be associated with the scenarios that are used to describe life safety. Initially, we will consider occupant egress during a fire. Later we extend the concept to defend in place analysis.

The following procedure describes a life safety (by egress) analysis:

1) Identify the diagnostic fire scenario.
2) Evaluate the composite fire that combines the diagnostic fire and active suppression.
3) Select a scenario to analyze occupants leaving the building. This requires a path that includes the room of occupant origin, a destination, and occupant characteristics.

4) Identify the combination of barrier–space modules that describe the egress path being evaluated, and establish a tenability window of time for these spaces (Chapter 32).
5) Determine the maximum pre-movement time that is available to an individual in order to leave the building without encountering untenable conditions along selected routes.

The Interactive Performance Information (IPI) chart helps to order time durations and recognize potential conditions for entrapment of people. The ease of transferring spreadsheet information allows one to use a different IPI worksheet for each scenario.

Carefully selected "if... then..." variability analyses give an insight into time durations, status of opening protectives, and route selections that affect life safety. Identification of maximum pre-movement times for the scenarios gives an insight into the building's performance for life safety.

33.5   Pre-movement Time for Egress

Figure 33.1 shows the event logic diagram that relates pre-movement events. The diagram is single value, and event completions are sequential. The time intervals from ignition to fire detection (OD), to alerting the occupant (OA), and for the occupant to make a decision and actually leave the room (OT) comprise pre-movement.

A performance analysis does not make decisions about what an occupant should do. One can speculate about different behaviors. A chain of events that describes human decisions is described below. Statistical data on occupant decision-making are useful for risk management. Behaviors can be categorized using variability analyses.

Although a performance analysis does not predict occupant actions, it can provide information that is useful to characterize occupant risks for different decisions.

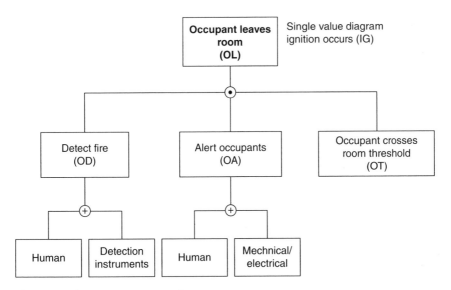

Figure 33.1 Success tree: occupant leaves room.

A performance analysis identifies the maximum pre-movement delay that the building can provide useful information about experiencing untenable conditions during egress. The available pre-movement delay becomes a performance measure for the building's life safety characterization.

Figure 33.2 shows the analytical network to examine pre-movement activities. The network is single value and events are conditional on completion of the prior event. Time durations are provided by the horizontal rows of the IPI chart that examines single value outcomes (vertical columns) for sequential instants of time.

33.5.1 Fire Detection (OD)

We distinguish between detection as a part of fire department notification (MD) and detection as a part of occupant life safety (OD). Normally, fire detection is a single event

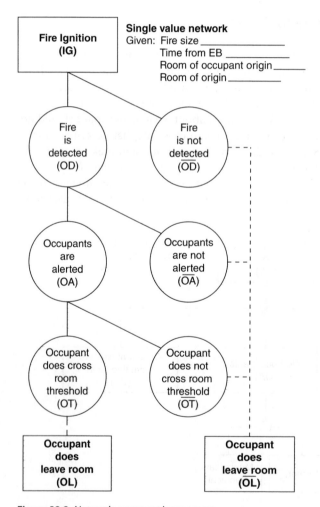

Figure 33.2 Network: occupant leaves room.

that is the same for each activity. Even though the detection event may be the same, we focus on its role for the component being evaluated.

Chapter 26 discusses detection analysis. The factors of Figures 26.1 and 26.6 are valid for detection relating to occupant life safety (OD). They also guide estimates for appropriate time durations.

33.5.2 Alert Occupants (OA)

An initial audible alarm after detection is a common way to alert occupants. However, occupant alerting can be very different for some buildings. Building design and management decisions can cause intentional delay from detection (OD) to alerting occupants (OA). A wide variety of building situations exist, and each installation has a unique process that must be examined.

Figure 33.3 associates common factors that affect events. Tracing the specific building's installation provides a basis for estimating the time between detection (OD) and occupant alerting (OA). When humans are the linkage, the scenarios specifically distinguish characteristics of the alerting individual from those of occupants being alerted.

33.5.3 Occupants Start Egress (OT)

The final event in the sequence involves occupant decisions to leave the room and begin the egress journey. Although conventional opinion often assumes that an individual will quickly begin egress, this is often not the case. Delay between occupant alert (OA) and

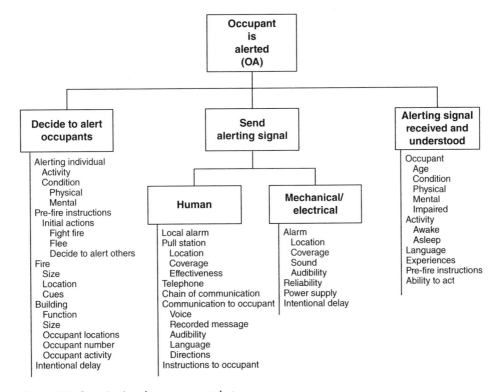

Figure 33.3 Organization chart: occupant alert.

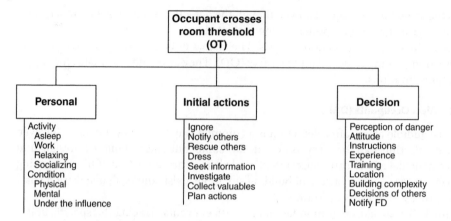

Figure 33.4 Organization chart: occupant leaves room.

the occupant deciding to leave the room to begin egress (OT) can be significant. Figure 33.4 organizes factors that influence the time to complete this event. Variability analyses can characterize occupant conditions that affect time delays for movement.

33.6 Occupant Life Safety (LS)

Potential egress routes are the usual ways out of a building by ambulatory means. These include normal egress components as well as the use of elevators and escalators. Any path from an occupant room to the building discharge may be considered. However, a performance analysis does not include evacuation through windows, roofs, and fire department-assisted rescue.

An egress analysis initially identifies the chain of barrier–space modules from the occupant room to the exit discharge. The continuum of time that describes an individual's movement is incorporated in the IPI chart Section 5Ab, because horizontal rows represent time progressions of an event tree. The window of time for tenability of these target spaces would be shown with the smoke evaluation recorded in the IPI chart Section 4B. One compares the time an individual may use each target space with its window of tenability.

The IPI chart Section 4B shows the window of time that each target space along the egress path remains tenable. Therefore, a simple comparison of tenability and the individual's location provides the answer for the events of Figure 33.5. The pre-movement delay time would have been recorded in the IPI chart Section 5A. Examination of these IPI sections enables one to determine the maximum pre-movement time the building can give an individual before he or she encounters untenable conditions along the egress path.

33.7 Discussion

A performance analysis for life safety by egress describes the maximum pre-movement time the building provides for an occupant. An occupant can expect to encounter tenable conditions at all points along the egress route when he or she leaves the room

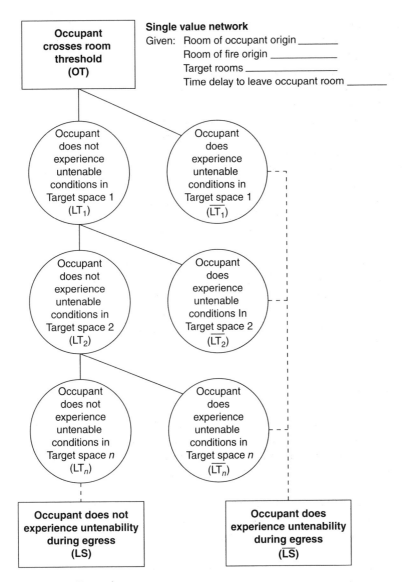

Single value network

Given: Room of occupant origin _____
Room of fire origin _____
Target rooms _____
Time delay to leave occupant room _____

Figure 33.5 Network: occupant egress.

before the pre-movement time is exceeded. However, the matching of untenable conditions and occupant location depends on the speed of occupant movement. Consequently, building performance for egress is linked to movement speed.

In theory, the life safety by egress performance concept is straightforward. In practice, the number of variables includes the fire location, diagnostic fire conditions, fire defenses, and the location and movement characteristics of the individual. Although the number of variables can appear overwhelming, a methodical analysis can provide enough information to be able to characterize the building for risk.

The analysis for a single representative occupant in a specific room provides much information about the building's life safety behavior. The procedure is well suited for occupancies such as hotels, apartments, and many offices where individual decisions are more common. Although analytical concepts are the same, large assembly spaces require additional crowd movement planning.

Time is incorporated into the analysis with the speed of occupant travel. Building occupants have a wide range of ages, physical and mental abilities, and impairments. Acceptable performance for one individual may not be adequate for another.

We base time durations on scenarios that depict representative individuals. The scenario identifies movement speed and agility to traverse architectural segments, such as corridors, stairs, and doors. Speeds may vary to reflect different architectural segments along the travel route. Characteristics define occupant mobility, physical impairments, and architectural obstacles. For example, an athletic, unimpaired individual exhibits one type of movement. An elderly, impaired individual unable to move easily along corridors and stairs exhibits another. Blind, deaf, young, elderly, and frail individuals are other categories. Different building performances are needed to represent each characteristic group.

An individual usually has choices of escape routes both initially and as the egress evolves. A performance analysis examines each specific route of interest. "If... then ..." variability analyses enable one to obtain an understanding of building performance to describe risk characterizations for different individuals and origin locations. The understanding that evolves provides knowledge to document rationale and develop better risk management plans.

33.8   Defend in Place

A defend-in-place strategy is often intentional. Occupants who may be incarcerated or restrained by illness or immobility do not have free choice to leave their rooms in a fire. Other occupants may remain in their room of origin by choice. Still others may need emergency refuge because they cannot move through a planned escape route. A performance analysis evaluates these conditions in the same way.

Conceptually, analysis of a room to defend occupants in place is relatively straightforward. The room becomes a target space and is evaluated with the techniques of Chapter 32. If room ventilation can dilute the smoke-gas concentration, the problem is reduced.

The difficulty arises in defining tenability. Visibility is measurable and able to be calculated. Toxicity and its cumulative effect are not measurable for smoke travel, and the delivery cannot be calculated. The cocktail effect of gas combinations is unknown, but recognized. However, we do know that the condition exists and is not good for humans.

33.9   Closure

A life safety analysis matches two parts. One is the occupant who may remain in a room or move through sequential barrier–space modules in an attempt to leave the building. The second is the time-related environmental changes in selected barrier–space modules. The analysis first analyzes the windows of time during which the target modules remain tenable. Then, an analysis attempts to determine time durations during which people may experience untenable conditions.

During egress, travel time may be estimated for the physical condition of the individual moving within the expected environment. The sequential time intervals make a chain through the scenario route. The pre-movement time delay between ignition (IG) and the start of travel movement has an influence on the likelihood that an individual will encounter untenable conditions during the egress travel.

An individual may leave the building without encountering untenable conditions if pre-movement time is small. When the pre-movement time is long, the escape routes may become blocked. Sometimes, escape routes can be blocked even when an occupant decides to leave immediately after alert (OA). This is because built-in delays in detection (OD) and occupant alert (OA) may be sufficient to cause escape route target modules to become untenable before occupant movement starts.

Scenarios can portray occupant decisions on the expected tenability conditions during travel through the egress modules. Different occupant movement scenarios can be portrayed with different sheets of the IPI chart to gain a better understanding of the influence of pre-movement time on successful egress. The influence of more inflexible events of detection (OD) and occupant alert (OA) is more easily recognized when compared with the available time for an occupant to react.

34

Prevent Established Burning

34.1 Introduction

Fire prevention is an important part of the complete fire safety system. Many buildings survive on their ability to avoid a hostile fire. Some owners recognize this condition. Others are unaware of the role that fire prevention plays in their survival. Regardless of owner or management awareness, we consider fire prevention to be significant, particularly in applications of risk management.

For an engineering analysis, the term "fire prevention" means both "prevent ignition" (IG) and "prevent established burning" (EB). The role of each is different, and the ability to affect outcomes is managed in different ways. The framework allows one to examine details and develop an understanding of specific features that are sensitive to success.

We recognize that an occupant extinguishes many ignitions – perhaps the vast majority. Of course, the occupant must be near the fire to be successful. And the fire must be small, because an untrained occupant cannot extinguish a fire more than about 0.5 m (1.5 ft) in size, about 50 kW. For these reasons we select a fire size of EB as the limit of occupant extinguishment effectiveness.

A fire size for EB that is about knee high (or waste basket size or 25 kW) seems a reasonable demarcation between occupant extinguishment and building performance. This size is good for a performance analysis because outcomes can be predicted with greater confidence. It is also suitable for prevention analysis because many occupant-related features can be examined for a fire of this size.

Initial occupant activities can be very important to fire safety, and incorporating these actions into the prevention program is both practical and reasonable. Also, the building analysis is less complicated when we do not commingle expectations for untrained individuals with automatic sprinkler functions and professional fire department operations.

Fire prevention is an important part of a complete building analysis. When we pay close attention to the details, the planning becomes site-specific and effective. We separate the process into two components: (1) traditional ignition prevention; and (2) activities from appearance of the first fragile flames to the fire size identified with EB. Figure 34.1 shows the event logic diagram for a prevention analysis. Figure 34.2 shows the top-level evaluation network that guides an evaluation.

Fire Performance Analysis for Buildings, Second Edition. Robert W. Fitzgerald and Brian J. Meacham.
© 2017 John Wiley & Sons Ltd. Published 2017 by John Wiley & Sons Ltd.

Fire prevention and initial actions after ignition occurs can be very effective in industrial occupancies and some commercial occupancies. Special attention to the early actions of occupants can be tailored to the operational activities of the building. Although this book focuses primarily on analysis, the weak link recognition can customize solutions for effective prevention and initial action plans.

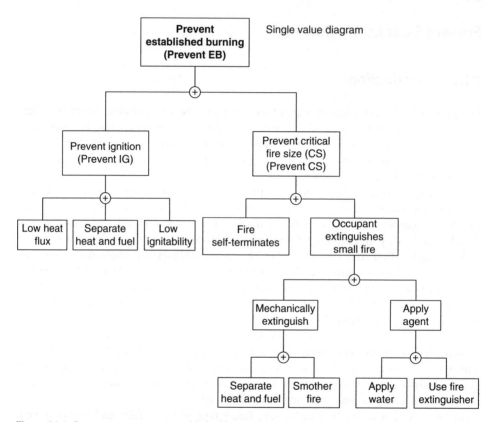

Figure 34.1 Success tree: prevent established burning.

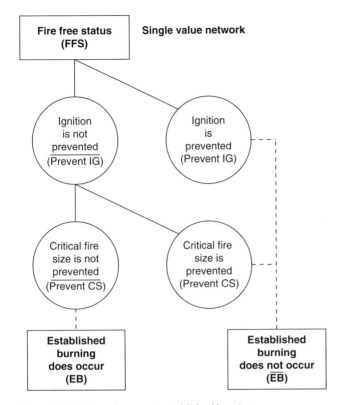

Figure 34.2 Network: prevent established burning.

PART ONE: ESTABLISHED BURNING PREVENTION

34.2 Ignition Potential

Chapters 6 and 7 describe realms of fire growth for bottom-up analyses. Realm 1 for fire growth describes conditions from a fire free status to ignition, which is the appearance of the first fragile flame. Success in fire prevention is the avoidance of ignition.

An analysis estimates whether an ignition will occur for selected locations within the building. Commonly, evaluations use qualitative judgment based on observations of conditions that could lead to ignition. A "grade" may reflect when sufficient conditions for ignition are present or could easily be moved to a proximity that can cause an ignition. Figure 34.3 shows the factors that influence ignition conditions.

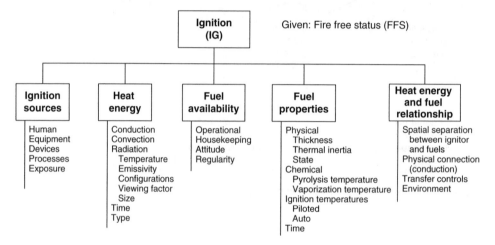

Figure 34.3 Organization chart: prevent ignition.

Deterministic evaluations use "pass" or "fail" as the assessment grade. However, it is possible to borrow concepts from Chapter 8 to grade the relative likelihood of ignition with the descriptive terms shown in Figure 34.4. The concepts in Unit Four look at additional ways of handling uncertainty.

A grade may be based on heat source locations and kindling fuel distributions. For example, a useful evaluation technique identifies heat source locations on space utilization plans. Kindling fuels can be located on another set of plans. In the initial stages, the two sets can be superimposed to highlight potential problem areas. Heat source or kindling fuel mobility can then be examined to note additional potentially unwanted ignition locations. Figure 34.5 illustrates a superimposed set of plans that can facilitate an understanding of operational ignition conditions.

34.3 Established Burning Evaluation

Given ignition, a small fire can grow or self-terminate. If it is able to grow, an occupant may or may not extinguish it. The events in Figure 34.1 show the logic. We can see that when conditions for continued growth are not easily recognized, fire development may

Figure 34.4 Ignition potential classifications.

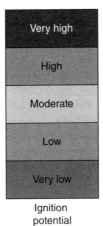

Ignition
potential

take additional time to evolve. This may provide an occupant with additional time to put out the fire.

34.3.1 Fire Self-termination

Many small ignitions self-terminate because there are not enough critical factors to help the fire grow. An awareness of the influence of the synergistic relationships can provide an awareness of initial fire growth. Figure 34.6 shows the factors that influence the growth or diminishment of initial fire growth from IG to EB.

If desired, this analysis, similar to other qualitative estimates, can be "graded." Figure 34.4 is also useful for ranking conditions for fire growth.

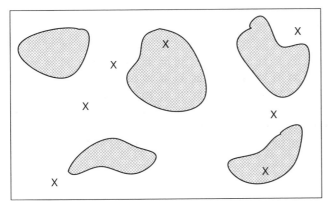

X = heat sources

⌒ = potential kindling fuel locations

Figure 34.5 Recognizing ignition potential.

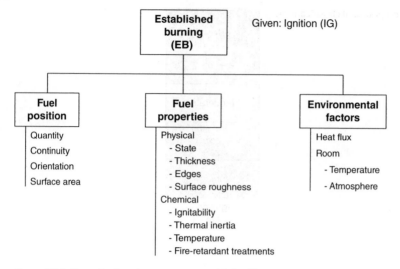

Figure 34.6 Organization chart: prevent established burning.

34.3.2 Occupant Extinguishment

Occupant extinguishment of small fires can be very effective – so effective, in fact, that most ignitions go unreported. Even in organizations that want to maintain records, unwanted ignitions frequently go unreported because they are small incidents and people don't want to go through the reporting process. When one wants to use statistics for comparison, the term "ignition" is often interpreted to mean "established burning."

When one wants to evaluate occupant actions in a more disciplined manner, the network shown in Figure 34.7 can be used to examine conditions more carefully. Each event is important to the process, and detailed examination identifies potential weaknesses in the process.

Figure 34.8 identifies the factors that most frequently affect the outcomes of each event.

34.4 Scenario Selection

Occupant extinguishment depends on the availability and inclination of a building occupant to put out a nascent fire. Therefore, scenarios must be selected to reflect the status of different building conditions. These include the building in normal operation, the building occupied with only a small workforce, such as a cleaning crew, and the building unoccupied.

Figure 34.8 identifies factors that influence occupant extinguishment. Appropriate scenarios should be chosen that reflect the status of the building during a complete annual life cycle.

34.5 Prevent EB: Discussion

Conscious, disciplined integration of fire prevention augments opportunities in risk management applications. The building performance after EB will be the same, whatever the cause of ignition. A clear analytical separation avoids obscuring building

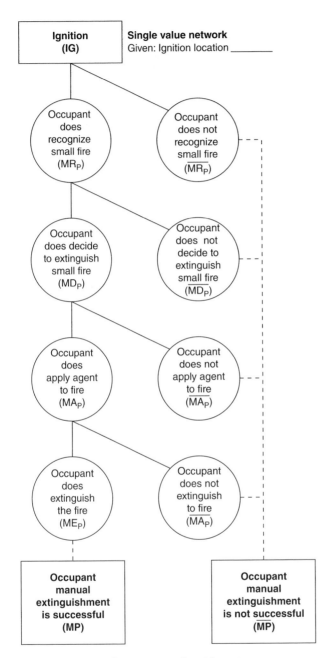

Figure 34.7 Network: occupant extinguishment.

performance with fire prevention effectiveness. This creates a better understanding of failure and the ways to improve performance. This separation allows a clearer communication with others regarding the roles and outcomes of failure.

One or a combination of the following ways of describing fire prevention can help to communicate composite building performance:

(a)

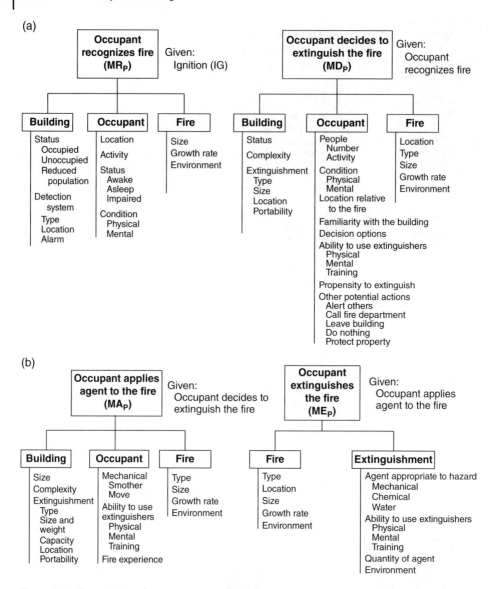

Figure 34.8 Organization chart: occupant extinguishment.

1) Start with a fire free status and use a statistical frequency of ignition. This type of analysis is statistically based and indicates the frequency of ignition in a class of buildings similar to the one being evaluated. This information may be useful to inform others about the regularity of fires for comparative purposes. One must interpret the statistics to determine if the term "ignition" really means "established burning." Documentation procedures often unintentionally encourage reporting only those ignitions that cannot be ignored. Fires the size of EB may be the actual incident being reported.

2) Consider selected scenarios in order to tell a story to the building management. This procedure starts with an ignition [i.e. $P(IG) = 1.0$] in a specified location. It then

examines initial fire growth and occupant manual suppression activities. This technique is useful to evaluate the effectiveness of different fire prevention programs under consideration.

3) Compare the relative potential for ignition at various locations with different conditions of heat energy, kindling fuels, and heat/fuel separations. Observations using qualitative estimates give an insight into the effectiveness of ignition prevention. Ranking ignition conditions provides a relative index that relates critical factors and combinations to organize possible changes in fire prevention recommendations.

4) Rank the importance of ignition locations to building performance. This technique assumes that ignition will occur sometime during the life of the building. The analytic hierarchy process (AHP) is a technique for making multi-level decisions and translating judgments into quantified priorities. This technique can "weight" or "compare" a site-specific building with statistical data. This technique is more suitable for insurance industry applications than for individual analyses. The technique is noted here to highlight the fact that there a variety of procedures that can use a systematic framework for analysis of fire safety decisions. Detailed AHP procedures can be found in the literature.

Each of these options has an application for corporations or the insurance industry. A detailed fire prevention program could get a clearer picture of effectiveness by comparing alternatives with cost and value.

PART TWO: SPECIAL HAZARDS PROTECTION

34.6 The Role of Special Hazards Suppression

Special hazard suppression equipment is installed to protect a small (spot) area within a larger space. Usually these hazards involve the potential for rapid flame spread, such as Class B fires in industrial buildings or food preparation areas in restaurants. Applications such as flammable liquid storage, dip tanks, oil-filled transformers, and deep-fat fryers for cooking are examples of hazards that require special attention.

It is common to have a general area sprinkler system for the entire building, in addition to the special hazards suppression systems that protect local problem areas of an operation. However, the framework separates the role of automatic special hazards protection from that of the general area automatic sprinkler system. In industrial or commercial occupancies where this type of automatic protection equipment is installed, each component is evaluated independently.

Special hazards equipment includes the following:

- Carbon dioxide system
- Dry chemical system
- Water-spray system
- Foam extinguishing systems
- Explosion protection system.

An evaluation of the reliability and design effectiveness of special hazards automatic suppression equipment first considers characteristics of the flammable materials and combustion products and the speed and extent of fire propagation. After these potential fire characteristics are understood, we evaluate the reliability and design effectiveness of the suppression system for:

- Agent selection
- Detection system
- Release system
- Storage and delivery system.

The examination determines whether the agent is appropriate to the hazard. One may discover a special hazard being protected by an agent that is inappropriate for the fire. Detection and release systems must account for the expected fire propagation speed and the released combustion products.

Special hazards equipment is frequently located in corrosive environments, and long-term maintenance is important in evaluating system reliability.

Although not a part of performance analysis, the cost of clean-up and salvage from fire suppression is considered in an evaluation. Damage to neighboring materials susceptible to the effects of the extinguishing agent can be significant.

34.7 Framework for Analysis

The operation of special hazards equipment is considered a part of fire prevention. This position is a convenience to enable performance evaluations for most buildings to be analyzed with a routine, consistent procedure. Nevertheless, special hazards suppression is very important for conditions in which it is needed.

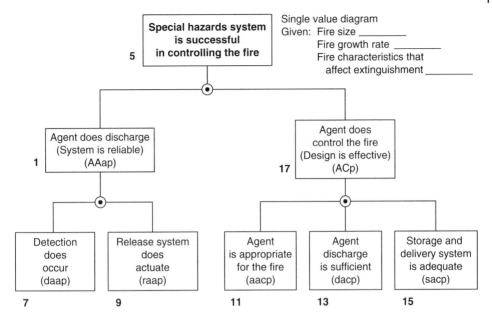

Figure 34.9 Success tree: special hazards extinguishment.

The framework for analysis is similar to that of a sprinkler system. That is, the system reliability (AA_P) is evaluated independently of the design effectiveness, given that the system does discharge agent.

Figure 34.9 shows the single value event logic diagram that organizes an analysis, while Figure 34.10 shows the network for the analysis. The event numbers indicate the correspondence between the event logic diagram and the analytical network.

34.8   Special Hazards Analysis

The network shown in Figure 34.10 guides analysis of a special hazards installation. Figure 34.11 identifies the guidelines for an evaluation.

34.9   Protection Combinations

The location of a special hazards installation depends on the hazard being protected. Normally, these hazards are located within larger spaces that are beyond the range of the special hazards installation. The building may have other types of active extinguishment features that often, but not always, include an automatic sprinkler system. The framework for analysis remains consistent in its use for building analysis.

Virtual barriers segment large open spaces that have spot type protection within their bounds. The virtual barriers segment open spaces with imaginary barriers that have no resistance to flame-heat or smoke-gas. However, they do allow one to examine different zones in a large space with the same barrier–space modular analysis. The virtual barriers may be placed to establish useful zones for analysis.

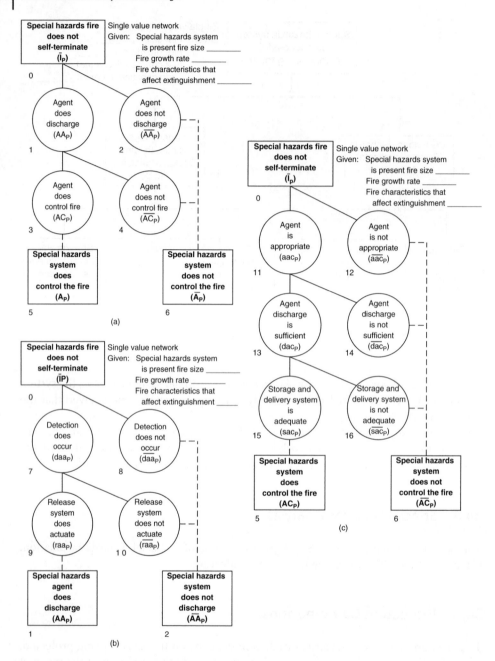

Figure 34.10 Network: special hazards extinguishment.

Figure 34.12 shows a network that combines automatic special hazards suppression with an automatic sprinkler system. The analysis examines the suppression for successively larger fire sizes involving diagnostic fire and composite fire conditions. After initial device actuation, qualitative analysis must be used to estimate the synergistic actions. Estimates are based on the installation features that are expected to be in operation for each potential fire size.

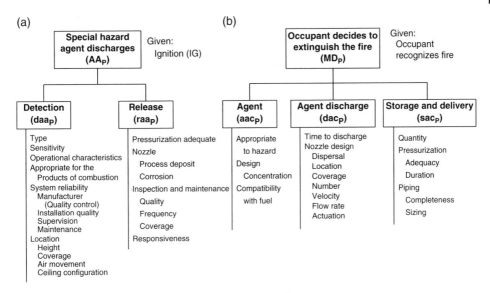

Figure 34.11 Organization chart: special hazards extinguishment.

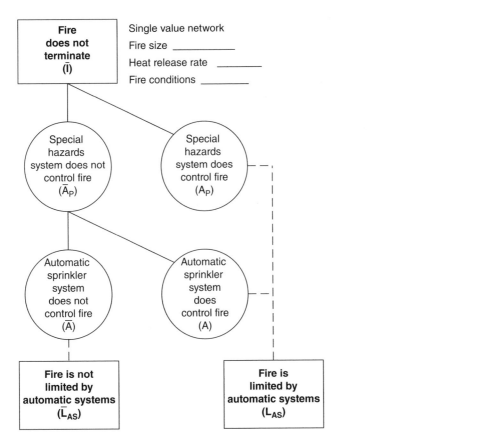

Figure 34.12 Network: automatic systems analysis.

The events of Figure 34.12 may be evaluated using the decomposed levels of events described earlier. The evaluation can also include the building fire brigade (if appropriate) and the local fire department with the event analyses described earlier. The main caution is to keep the network logic consistent.

34.10 Closure

Fire prevention analysis is important. This analysis may or may not be a part of a building evaluation, depending on the purpose of the evaluation. However, risk management for existing buildings commonly incorporates "prevent ignition" into the complete program.

"Prevent EB" has two parts. The first is the prevent ignition (IG) analysis that is a normal part of risk analysis. The second is prevent EB. The start of prevent EB analysis is an ignition. This analysis examines the ease and likelihood that the ignition will be able to grow into the critical fire size that defines EB. In addition, the analysis evaluates the ability of an occupant to extinguish the fire before it reaches the size of EB.

Occupant extinguishment may be a significant part of a risk management program – or it may be irrelevant. We determine the importance of the occupant in the system by evaluating a series of scenarios with examination of details of the process. Industrial and commercial enterprises are often different from residential occupancies. Detailed scenario analysis with qualitative judgmental event outcomes gives an insight into performance. Certainly, these scenario analyses enable one to identify conditions where occupant actions can be effective and where they will be completely ineffective. These prevent EB analyses become useful for developing risk management programs.

Special hazards systems are installed to control unusual fire operations within larger activities. The "spot" areas often use special hazards systems to address unique fast fire conditions. Although these systems are sometimes described as "automatic" fire suppression, their functions are different from automatic sprinkler systems. Consequently, we separate the two because they address different types of concerns.

For convenience, we associate special hazards systems with fire prevention. This may not always be appropriate. However, it serves the purpose of separating special hazards conditions from normal building fires. Although evaluations for each component and package of components use a structured framework, one can incorporate additional conditions if necessary.

Part IV

Managing Uncertainty

A performance evaluation is based on deterministic analysis for a specific building. Fire safety is an emerging engineering discipline. Some of the components are relatively well-developed and reaching early maturity while others are moving from infancy to adolescence. Consequently, the state-of-the-art for quantification is uneven. This is a typical evolution that all engineering disciplines have experienced.

Uncertainty is a normal expectation for quantitative evaluations. However, the site-specific uncertainty can be narrowed substantially because the framework for analysis and the associated IPI chart force the engineer to describe performance within recognized conditions. This enables one to identify the range of variation that may be expected. One can select appropriate values to create an understanding of performance. Often, the most pessimistic, most optimistic, and most reasonable values may be selected for performance evaluations.

Unit Four describes ways of dealing with uncertainty to provide credible engineering analysis and clear understanding of building performance. Chapter 35 describes the important concepts and ways to discriminate within the window of uncertainty. Chapter 36 discusses ways for understanding the dynamic behavior of uncertainty and communicating expectations to other professionals and to the general public.

Understanding fire performance becomes the basis for managing risk characterizations. The general process of risk assessment and management is an important application of an engineering performance analysis. Chapter 37 organizes procedures for methodical and logical performance-based risk management.

The method described in this book has always been deterministic for site-specific buildings. During the 1970s and 1980s the state-of-the-art for component behavior was much less developed that at the present time. Consequently, performance descriptors were based on degree of belief estimates founded on the way components were expected to behave. This reliance on probability expectations caused confusion and questions about the validity of a performance analysis. Nevertheless, the use of probabilistic methods with systems analysis enabled the framework to mature more rapidly than would otherwise have happened. Chapter 38 discusses the theoretical foundations from which this analytical framework evolved.

Fire Performance Analysis for Buildings, Second Edition. Robert W. Fitzgerald and Brian J. Meacham.
© 2017 John Wiley & Sons Ltd. Published 2017 by John Wiley & Sons Ltd.

35

Understanding Uncertainty

35.1   Introduction

Fire safety evaluations are based on deterministic quantification procedures. Although fire safety is not yet a mature engineering discipline, the analytical framework can integrate quantitative methods with qualitative understanding to produce credible performance descriptions. Contemporary state-of-the-art engineering practice accepts that uncertainty is a normal part of evaluations and decisions to achieve acceptable risk.

This chapter describes techniques to deal with uncertainty in site-specific applications. Although the management of uncertainty can be a subject in its own right, we will describe ways to incorporate uncertainty to improve fire performance understanding.

We start by discussing the window of uncertainty in the context of time and knowledge. The analytical framework is used to structure thinking. We illustrate techniques with examples that show practical methods for incorporating uncertainty into performance. The way they are used depends on the needs of the application.

Although one expects some variation in numerical agreement when different computational models evaluate the same condition, the illustrations have been coordinated to demonstrate coherence. Our goal is to provide a way to deal with the uncertainty of deterministic applications.

35.2   Window of Uncertainty

The Interactive Performance Information (IPI) chart helps to organize, analyze, record, and visualize performance. Vertical columns allow us to "stop the world" and examine the status of associated components for specific events or instants in time. Horizontal rows enable one to track the dynamic changes of a single component over time.

We begin to manage uncertainty by introducing the "window of uncertainty" in a horizontal row of the IPI chart. Then, we extend this concept to other related topics. Chapter 36 expands these techniques to communicate performance and risk.

Time is the factor that links all components, and deterministic evaluations examine critical events at specific time durations. For example, assume a building analysis requires the fire department's first-in arrival time. Suppose the analysis uses a "most appropriate" time of 12 minutes after established burning (EB). One can recognize that this value might fluctuate a little on either side of 12 minutes. Assume additionally that

Fire Performance Analysis for Buildings, Second Edition. Robert W. Fitzgerald and Brian J. Meacham.
© 2017 John Wiley & Sons Ltd. Published 2017 by John Wiley & Sons Ltd.

Component timeline:

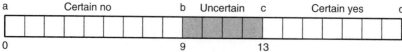

Figure 35.1 Window of uncertainty.

the engineer is "certain" that arrival will not occur before 9 minutes or after 13 minutes. The 4-minute interval from 9 to 13 minutes describes a window of uncertainty for the event.

Every component in the analytical framework has three regions along its timeline. The first is a time period during which the component will certainly not operate. Another is a time after which the component will certainly have acted or is no longer relevant. Between these two regions is a window of uncertainty in which one is unsure when the component will actually act. Figure 35.1 illustrates these regions for a row of an IPI chart.

The window of uncertainty gives a sense of possible times during which a component may be active. Some uncertainty will be associated with all calculations and estimates within the window.

35.3 Calibrating Uncertainty

The window of uncertainty shown in Figure 35.1 can be calibrated. For example, the cells can use verbal terms (e.g. "very unlikely," "a good chance," "a high probability," etc.). Alternatively, the performance in those cells can be given numerical measures (e.g. a probability of 10%, 60%, 80%, etc.). Both descriptions are of value in understanding and communicating fire performance. Different application needs use them to enrich performance evaluations and communication.

Probability of success estimates for a component's performance at each instant in time is the most useful calibration of uncertainty. Although the word "probability" typically implies a statistical association, here it is defined as a *degree of belief*. In this context the term reflects the engineer's judgment and opinion of the likely success of an action at that instant in time.

A degree of belief encodes a state of knowledge. Assessments are based on acquired knowledge with the site-specific operation of the building. The acquired knowledge includes observed facts, calculated values, computer models, physical relationships, experimental information, failure analyses, and personal experience. Assessments use the full spectrum of available information that seems relevant to the problem. Uncertainties can be clearly recognized and their documentation is straightforward.

Degree of belief estimates are based on the deterministic analysis that is used with the analytical framework to evaluate a performance. The framework and IPI cell establish specific, unambiguous scenario bounds for each assessment. The process requires one to understand events and interactive conditions as well as the nature of the uncertainty. The framework ensures that uncertainty gaps are neither so large that the engineer cannot comprehend the situation nor so small that an evaluation becomes tedious. The process integrates qualitative and quantitative procedures. Degree-of-belief estimates use theory and observation to compensate for shortcomings in computer models, calculation procedures, and qualitative assessments.

Event success is expressed as a probability having values between 0.0 and 1.0. Both extremes show certainty. A probability of success of 0.0 means that the individual believes there is a certainty that the successful action *will not* occur. A value of 1.0 indicates certainty that success *will* occur. When the expected outcome of an event falls between these extremes, performance is expressed as a value $0.0 < P(\text{success}) > 1.0$.

Degree-of-belief assessments describe an expected performance for the site-specific application. For example, one should not interpret degree-of-belief probability values of 0.4 or 0.9 literally as a statistical frequency. Rather, one should interpret the numbers as a calibration between 0% and 100% of the belief that the component will be successful *within the circumstances of the scenario*. Its value is based on all information that is available for the assessment.

Let us pause for a moment to consider the meaning of "certainty." An absolute certainty means that there is no possibility. For example, there is an absolute certainty that the probability of success of automatic suppression is $P(A) = 0.0$ for a room having no automatic sprinkler system.

However, a functional certainty would be used when the room has a sprinkler system but one is unsure of an outcome at a specific time. For example, specific conditions such as the fire size when a sprinkler will fuse or the maximum suppression limits for a sprinkler system are difficult to identify with accuracy. In cases where the certainty limit is difficult to decide with precision, we use functional certainty. Functional certainty is commonly viewed to mean "about 1 in 20" for performance analyses. Therefore, functional certainty values of 0.0 and 1.0 are taken to mean "nearly 0 or 1." One may recognize the bounds for a window of uncertainty to be near those values.

35.4 Degree-of-Belief Estimations

Degree-of-belief probability estimates are based on subjective judgment for the specific situation being evaluated. The quantitative output is examined in the context of available qualitative information. An analysis identifies scenario conditions by isolating an IPI cell at the intersection of a continuous value row and a single value column. The boundary conditions for the evaluation involve all factors in the column that contribute to the probabilistic evaluation.

The use of degree-of-belief probabilistic values for identified scenario conditions enables all cells within the window of uncertainty to be evaluated with the same criteria. The limits of the window of uncertainty are established when one is certain that the event will not occur [$P(\text{event}) = 0.0$] and that the event will occur [$P(\text{event}) = 1.0$]. The performance estimate for any specific cell within this range is $0.0 < P(\text{event}) < 1.0$.

The next few sections illustrate mechanics of the process. The concepts are valid for all fire performance components. Each evaluation increases understanding of the scenario's performance. A probabilistic value represents the individual's degree of belief in the success of component performance at the time and for the conditions being considered. Conditions may include both the targeted component and the influence of other components that affect its behavior at the specific time being examined. The logic for value selection becomes a basis for narrative documentation.

35.5   The Role of the Analytical Framework

The main function of networks is to guide a deterministic thought process that incorporates dynamic interactions. However, the networks are also structured to accommodate mathematical operations involving probabilistic estimates. The calculus ensures mathematical consistency regardless of the manner in which probabilistic values are determined. Thus, whether classical frequency methods, Bayesian procedures, or degree-of-belief philosophies are the basis of probabilistic values, the mathematical operations guided by the network structure remain valid.

The analytical framework structures a performance analysis by integrating single value network (SVN) evaluations with continuous value network (CVN) changes. The complete framework is carefully constructed to provide rigorous, logical analysis.

Component analyses remain deterministic. This enables one to use state-of-the-art quantification to calculate behavior in evaluations. Although event performance is expressed as a probabilistic degree of belief, the analyses remain deterministic.

Normally, engineers think in terms of success, as in "How can I make it work better?" Risk managers often think in terms of failure, as in "How much can I lose?" From a mathematical viewpoint, success and failure are two sides of the same coin as

$$P(\text{success}) + P(\text{failure}) = 1.0 \qquad (35.1)$$

The IPI chart is structured to display components in terms of success. Because evaluation networks show both success and failure, one can evaluate any specific event for the best available information. However, when transferring results to the IPI chart, one must be meticulous about ensuring that (success) values are consistent.

35.6   Sprinkler Analysis Networks

The next few sections illustrate applications using degree-of-belief probabilistic estimates with the framework structure. Although we use automatic sprinkler performance to illustrate the process, the techniques are appropriate for all components of the complete building–fire safety system.

Figure 35.2 reproduces the SVN and CVN that were described in Chapter 29. In addition to guiding the thought process and deterministic evaluations of sprinkler system analysis, the networks are also structured to organize calculations using probabilistic estimates.

Sprinkler reliability (AA) and system control (AC) are independent events that must be combined to describe sprinkler system performance, P(A). We will first look at the evaluation of sprinkler control (AC) for a building, assuming that the system is reliable. Then, reliability (AA) will be combined with sprinkler control of the fire (AC) to determine automatic suppression performance (A).

The following sections illustrate a way to incorporate performance evaluations into the framework for analysis. One goal is to understand performance and discriminate between different conditions in a technically defensible way. A second goal is to enhance communication with knowledgeable colleagues as well as non-technical decision-makers.

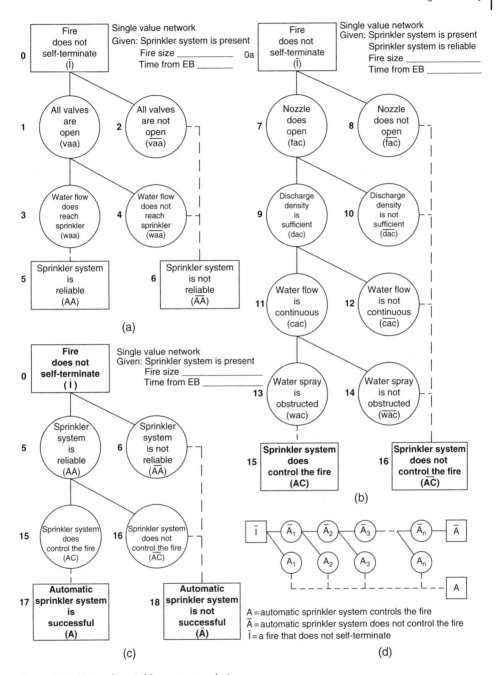

Figure 35.2 Network: sprinkler system analysis.

35.7 Sprinkler Control (AC)

An evaluation estimates fire conditions at selected sizes after initial water application. For these conditions, two principal actions are examined. The first looks at fire and water interaction. This considers speed of fire development with characteristics of sprinkler fusing (fac) and the effectiveness of water application (wac) in controlling the fire. The second part looks at the effect of water supply (cac) and discharge density (dac) with the number of open sprinklers.

These parts are considered during an initial assessment. An evaluation integrates deterministic information relating to observed conditions and experiential knowledge. All relevant information is used in estimating a degree of belief for operational performance.

Example 35.1 A new owner for the model building wants to develop a risk management program for important operations. As part of the information acquisition, the engineer is evaluating sprinkler system performance. Examining the sprinkler installation on the fourth floor west office space provides an initial understanding. Expectations for other parts of the building can be extrapolated from this information base.

The following estimates are based on the fourth floor sprinkler analysis in Example 29.1. Information from that analysis with diagnostic fire B (see Figure 29.11) has been used for the following degree-of-belief performance estimates.

The final row, sprinkler system controls the fire (AC), is a calculation for the degree-of-belief probabilities estimated in the cells. The calculations are described in Section 35.9.

35.8 Pause to Organize Thoughts

The initial goal is to understand how the sprinkler system will perform in this specific fire scenario. Then, we extrapolate that knowledge to highlight, with qualitative analysis, any other parts of the building in which sprinkler performance may be inadequate. Finally, we use our understanding to write a narrative that describes sprinkler performance for this building in such a way that one can recognize problem areas. If desired, one can develop a risk management program to change the risk characterizations for those problem areas.

Deterministic methods enable one to estimate the approximate fire size at which the first sprinkler fuses. A water supply analysis gives information to estimate the water discharge density for the number of open sprinklers. This knowledge, combined with additional theory, knowledge about sprinklers, diagnostic fire growth, and scenario observations, provides a basis for estimating performance within the window of uncertainty.

The fire and initial water spray interaction causes turbulent fire conditions that mean the natural (diagnostic) fire conditions are no longer accurate. Nevertheless, the diagnostic fire estimates do give information that enables one to anticipate burning conditions after initial fire growth is upset, but not yet extinguished.

The state of the art of fire science and engineering has not yet developed calculation procedures to evaluate sprinkler operations beyond the initial actuation. However, sprinkler system observations provide additional background about what to expect in these situations. Performance understanding helps to bridge the gap in the knowledge base to calibrate expectations.

The IPI chart is a central tool in establishing dynamic performance understanding. In addition to being a repository for information, the spreadsheet capabilities enable one to make calculations, draw graphs, compare different component events, and develop other useful knowledge. However, we will leave IPI applications until other calculation procedures have been discussed.

Time is the factor that links all parts of the fire safety system. Fire spread and products of combustion generation and movement are other common factors in evaluations. Time and the fire enable one to establish additional parameters such as floor area of fire involvement, space and barrier relationships at any time, and other factors that relate time and the fire. Therefore, fire and time can be related to other conditions that enable one to understand performance better. Although a variety of performance descriptors may be used, they all relate in some way to time.

Figure 35.2 shows the analytical framework for uncertainty evaluations. Although we will discuss the techniques separately, numerical values will be coordinated to provide consistency and show interrelationships.

In practice, some numerical variation is natural when a single individual uses the techniques at different times. Some variation is also expected when different individuals use the same technique for the same assessment. However, because of framework cohesiveness and boundary conditions for analysis, most variation is within a relatively narrow range. Occasionally, major differences can occur among individuals. These differences are almost always due to very different perceptions of the same situation or physical behavior of components. Discussion of the scenario bounds and logic for the assessments usually resolves the differences.

35.9 Calculating Single Value Outcomes

Throughout this book the organization of SVN networks, CVN networks, and the IPI chart have been carefully constructed to comply with systems theory and practice. This enables one to use the networks as a framework for calculations as well as a guide for analysis. Chapter 38 provides background to the organizational structure.

All event logic and network diagrams have systemic coherence. Their structure adheres to logical rigor. When probabilistic estimates for individual components have been selected, one can calculate outcomes for hierarchical combinations.

A single value analysis evaluates sprinkler performance for a specific fire size. Figure 35.2(b) shows the SVN that is used to evaluate the likelihood that the sprinkler system will control the fire without extension to a larger fire size.

Example 35.2   Table 35.1 shows selected fire size estimates for the situation of Example 35.1. Use the SVN of Figure 35.2(b) to calculate the probability that the sprinkler system will control a fire within an area of 60 ft^2 (5.6 m^2) without extension to a larger size. Values for a 3 MW fire from Table 35.1 are:

$P(\text{fac}) = 0.97$
$P(\text{dac}) = 0.96$
$P(\text{cac}) = 0.98$
$P(\text{wac}) = 1.0$

Calculate the probability that the sprinkler system will control the fire to an area of 60 ft^2 (5.6 m^2) without extension to a larger size.

Table 35.1 Example 29.1: degree-of-belief performance estimates.

	600 kW 4 ft^2	600 kW 4 ft^2	1 MW 10 ft^2	3 MW 60 ft^2	Fire size/area 5 MW 300 ft^2	8 MW 600 ft^2	12 MW 1500 ft^2	18 MW 4000 ft^2
fac	0.0	0.6	0.9	0.97	1.0	1.0	1.0	1.0
wac	0.0	0.95	0.98	1.0	1.0	1.0	1.0	1.0
dac	0.0	1.0	1.0	0.96	0.92	0.75	0.4	0.0
cac	0.0	1.0	1.0	0.98	0.97	0.9	0.8	0.1
AC	0.0	0.57	0.88	0.91	0.89	0.68	0.32	0.0

Solution:
These values were selected to illustrate the calculation process for SVNs.

The networks structure calculations to comply with the calculus of probabilistic operations. Figure 35.3 shows the SVN to calculate the value for AC and $\overline{\text{AC}}$. The events use the symbolic notation for convenience.

Because all of the events are described in binary terms, the sum of all probabilities at each level must add to 1.0. This includes the results of $P(\text{AC}) + P(\overline{\text{AC}}) = 1.0$.

When probability values are shown on the branches separating events, as in Figure 35.3, only two rules are needed to calculate the outcome of any event:

1) Multiply the probabilities along a continuous path to determine the outcome for the terminal event of the path.
2) Add like outcomes.

Thus, the outcome of Event 15 may be calculated as:

$$P(\text{AC}) = P(\text{fac}) \times P(\text{dac}) \times P(\text{cac}) \times P(\text{wac}) = 0.912$$

The outcome of Event 16 is calculated to be:

$$P(\overline{\text{AC}}) = \text{the sum of all like outcomes}\left(\text{i.e. AC failure}\right)$$

$$P(\overline{\text{AC}}) = 0.03 + 0.039 + 0.019 + 0.0 = 0.088$$

All SVNs structure calculations in this way. The results provide a single coordinate for a performance graph.

35.10 Graphing Results

Figure 35.4 shows three graphical descriptors that are useful for understanding performance and communicating outcomes. Evaluator A is useful for general thinking and discussion. Evaluator B is best suited to individual component descriptions. Evaluator C combines components to portray integrated performance.

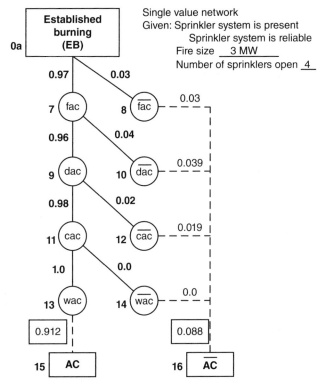

Single value network
Given: Sprinkler system is present
         Sprinkler system is reliable
       Fire size ___3 MW___
       Number of sprinklers open _4_

AC = automatic sprinkler system control the fire

$\overline{AC}$ = automatic sprinkler system does not control the fire

Figure 35.3 Example 35.2: sprinkler control calculation.

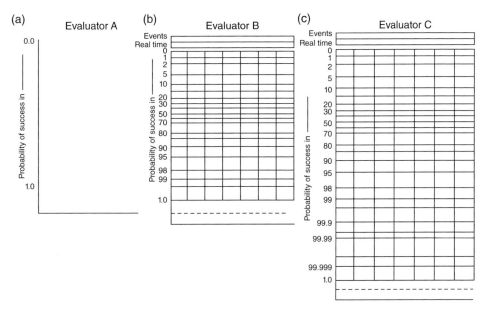

Figure 35.4 Performance analysis graphical descriptors.

These graphical evaluators are used to record performance, portray trends or behavior, and compare alternatives. They are not used for common graphical applications such as calculating slopes or areas.

The ordinate and abscissa scales are unorthodox. For example, the ordinate shows probability values. While values from 0.0 to 1.0 could be used in the usual manner, we change convention to show positive values in the downward direction. Although either direction may be used to describe mathematical relationships, experiments indicate that when the downward direction is labeled as positive, fewer errors occur and better communication takes place.

From a historical perspective, ordinate directions initially distinguished between failure analysis (for risk management) and success (for engineering design). Figure 35.5 shows two graphs, one depicting failure and the other success. Only the ordinate directions are different, because failure is the complement of success.

The analytical framework in this book maintains rigorous cohesiveness and attention to detail. All evaluations, calculations, and graphical descriptors are coordinated. Thus, in Figure 35.5(a), all active fire defenses are shown in terms of success as moving from "certainly no" [i.e. $P(event) = 0.0$] through the window of uncertainty to "certainly yes" [i.e. $P(event) = 1.0$]. This depicts an active fire defense component remining inactive until actuation occurs at a time greater than EB. Beyond this point, the component shows a successful operation. Eventually, it becomes irrelevant.

All passive fire defenses (e.g. barriers, structural frame, target space smoke tenability) start with complete success, after which performance decreases until failure occurs. Inadvertent errors are reduced with this unusual positive scale direction.

The ordinates in Figure 35.4 have another unorthodox feature. Evaluator A shows only the terminal events of 0.0 and 1.0, with intermediate values not shown. This evaluator is used for thinking and communication. One can avoid distractions of numerical values in describing or comparing expectations, behavior, and trends.

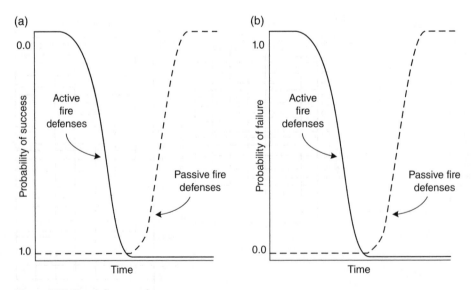

Figure 35.5 Fire defense performance.

Evaluators B and C use a modified scale of probability. Some fire event probabilities are often very large or very small within certain ranges. The scales of Evaluators B and C enable one to distinguish very large and very small changes more easily. Evaluator B is generally used for individual components while Evaluator C distinguishes larger differences which occur when components are combined.

Evaluators B and C modify the traditional probability scale by including 0.0 and 1.0 to show certainty. In addition, the bottom of a traditional probability scale paper was relocated to the top to distinguish larger numerical values near both extremes. Because these scales are unconventional, usual mathematical operations involving graphs are not accurate. However, numerical comparisons, trends, and information are easily portrayed.

The abscissa is based on time. However, time is related to fire growth, area of fire involvement, and sequential room fire propagation. Thus, the abscissa can show any measure or scale for the fire. The IPI chart provides the interrelations, if needed.

Example 35.4 Show the results of Example 35.2 on an Evaluator B.

Solution:
Figure 35.6 graphs values of $P(AC)$ on Evaluator B.

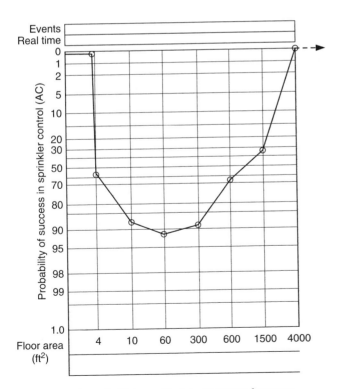

Figure 35.6 Example 35.4: graphical descriptor performance.

Discussion:

The graph in Figure 35.6 illustrates common sprinkler performance for large areas in which the first few sprinklers may not control the fire. Fire suppression improves as additional sprinklers come online. However, when too many sprinklers fuse, the water supply becomes inadequate. When this happens, the suppression capabilities of the system decline. If the fire has not been controlled before fire control occurs, the probability of success (AC) diminishes until eventually reaching $P(AC) = 0.0$.

Discrete evaluations for successive areas of fire involvement give a useful picture of sprinkler performance. The bounds of the window of uncertainty identify the fire size at first water application and the maximum size that can be controlled by the water supply. Within the window of uncertainty, one examines the interaction of fire conditions and water application to estimate the likelihood that the sprinkler system can control the fire.

One may have an opinion that failure should not occur if the sprinkler system were designed in accordance with recognized standards. However, we observe that over time, building conditions often change. For example, minor alterations in commercial buildings statistically occur about every 7 years. Major modifications occur about every 14 years. The sprinkler system is often overlooked when renovations occur.

Interior designs using different materials may create a faster or more intense fire. Relocated partitions or ducts can affect water spray patterns. Increased building development in the area can affect the water supply. Also, inadequate attention to system maintenance by the building management can reduce quality of response. And, the system may have been inadequate at the time of installation.

Performance analysis of buildings does not assume an existing sprinkler system will perform successfully. However, observations and available technology enable one to have a basic understanding of what to expect. Combining quantitative and qualitative techniques enables one to evaluate the fire and the sprinkler system. Performance evaluations of new designs, even those in compliance with design and acceptance standards, may also be examined for diagnostic fire and system interactions.

35.11 Cumulative Evaluations

Graphical displays of single value (SVN) outcomes and continuous value (CVN) outcomes provide slightly different perspectives and, when used together, give a better understanding of performance. SVNs and CVNs are interrelated and use the same factors for analysis. The IPI chart can be used to construct performance graphs. The IPI can also relate the SVN and CVN and calculate cell values for any given graph.

Continuous value networks use conditional probabilities to structure evaluations and order the mathematics to construct continuous value graphs. The CVN in Figure 35.7 describes the thought process for analyzing sequential events of sprinkler system performance.

In this CVN, an initial probability of success, $P(AC_1)$, is estimated. Normally, the value of $P(AC_1) = 0.0$ is used to identify the fire size when the first sprinkler fuses. Each sequential fire size is evaluated by asking the question, "Given that the fire is not controlled within size AC_m, what is the probability it will be controlled at AC_{n+1}?" Because the outcomes are binary, $P(AC_n) + P(\overline{AC_n}) = 1.0$.

This thought process enables one to envision outcomes for sequentially larger fire sizes. Normally, sprinkler control improves for larger fire sizes as additional sprinklers actuate. Eventually, one will recognize that the sprinkler system will certainly control a

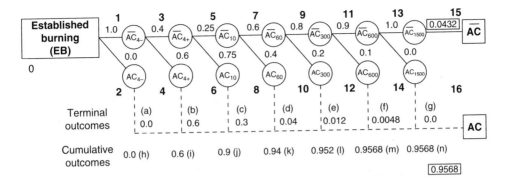

AC = Automatic sprinkler system controls the fire
$\overline{AC}$ = Automatic sprinkler system does not control the fire

Figure 35.7 Continuous value network: sprinkler performance.

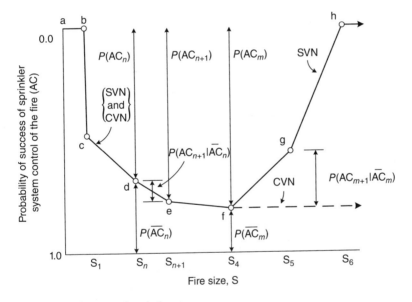

Figure 35.8 Outcome descriptions.

fire of the size being evaluated. $P(AC_n) = 1.0$ for that condition. One need not evaluate sizes beyond that fire size.

However, an evaluation may also indicate that the fire cannot be controlled beyond a particular fire size. At that time, the sprinkler system becomes overwhelmed and the fire can continue to grow beyond that point.

We identify a *sprinkler control point* as the fire size beyond which improvements in sprinkler control are no longer possible. This is illustrated as coordinate *f* in Figure 35.8.

The probability of sprinkler control steadily improves up to the control fire size. Figure 35.8 shows that sprinkler control, $P(AC_{n+1})$, follows the relationship:

$$P\left(AC_{n+1}\right) = P\left(AC_n\right) + P\left(AC_{n+1} \mid \overline{AC_n}\right). \tag{35.2}$$

Equation (35.2) is valid up to the maximum sprinkler control size. Beyond that fire size, the CVN and SVN separate. The CVN conditional probability remains as $P(AC_{n+1} \mid \overline{AC_n}) = 0.0$ beyond the sprinkler control point and the cumulative curve remains horizontal.

The SVN value becomes negative beyond the sprinkler control point. Negative conditional probability values in a continuous process are very difficult to observe and estimate. And probabilities can only have positive values. Therefore, cumulative values are shown as a horizontal line beyond the sprinkler control point.

Example 35.5 illustrates a cumulative value evaluation for the sprinkler system of Example 35.1. Because the example attempts to illustrate the thought process, values have been selected to give approximately consistent results with those of Example 35.3. In practice, some larger variation should be expected because one is using subjective evaluations with two different thought processes. Because the conditions and performance factors are the same, results should be roughly comparable.

Example 35.5 Construct a cumulative curve to describe the performance of the sprinkler system in Examples 29.1 and 35.1.

Solution:
The thought process for evaluating the scenario and estimating performance is shown below. The values are incorporated into Figure 35.9.

Given EB, the fire size at which the first sprinkler will fuse is 4 ft², $P(AC_{4-}) = 0.0$
Position Event 1: Given that the fire was not controlled at 4⁻ ft², the probability that the sprinkler system will control the fire when the first sprinkler actuates is $P(AC_{4+}) = 0.6$.
Position Event 3: Given that the fire was not controlled at (4⁺ ft²), the probability that the sprinkler system will control a fire of 10 ft² is $P(AC_{10}) = 0.67$.
Position Event 5: Given that the fire was not controlled at (10 ft²), the probability that the sprinkler system will control a fire of 60 ft² is $P(AC_{60}) = 0.3$.
Position Event 7: Given that the fire was not controlled at (60 ft²), the probability that the sprinkler system will control a fire of 300 ft² is $P(AC_{300}) = 0.1$.

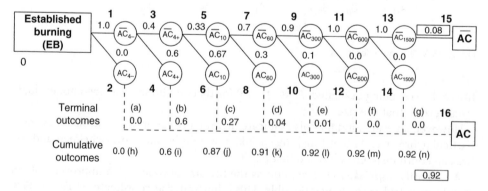

Figure 35.9 Example 35.5: continuous value network calculations.

Position Event 9: Given that the fire was not controlled at (300 ft^2), the probability that the sprinkler system will control a fire of 600 ft^2 is $P(AC_{600}) = 0.0$.

Position Event 11: Given that the fire was not controlled at (600 ft^2), the probability that the sprinkler system can control a larger fire is $P(AC) = 0.0$.

These values are shown on the CVN of Figure 35.9. The conditional probability values are shown as the terminal outcomes, values (a)–(g). The cumulative outcomes are shown as values (h)–(n). Figure 35.10 shows the associated CVN graph.

Up to a fire size of 600 ft^2 the values compare with those in Example 35.3 and Figure 35.6. This merely illustrates variation between the different approaches. In practice, an even larger numerical difference might be expected. Nevertheless the different ways of looking at the same sprinkler system should be comparable in terms of recognizing the same deficiencies.

Discussion:

This example illustrates that one can construct a sprinkler performance curve in two different ways. The CVN approach is faster, although not quite as "careful" as the SVN approach. In theory, if both were applied in the same way, the results would be identical. However, subjective decisions are not precise, and some practical differences are to be expected. Nevertheless, because evaluations consider the same factors, results should be comparable.

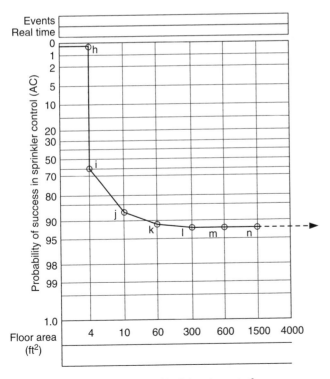

Figure 35.10 Example 35.5: graphical descriptor performance.

35.12   Sprinkler Reliability (AA)

Water must be able to discharge from the first fused sprinkler before the sprinkler system can begin to control the fire. A sprinkler reliability analysis estimates the likelihood that agent application (AA) will occur. Separating the reliability of the system (AA) from the suppression capabilities (AC) enables one to examine complete system performance in an orderly manner.

Reliability (AA) need only be evaluated one time for each sprinkler zone. The zones are usually similar enough that a single evaluation is often suitable for all analyses.

Examination of factors that influence reliability is an important part of a complete analysis. Failures in fully sprinklered buildings can often be attributed to parts of the system that fail to deliver water to the sprinklers. Human negligence is often a cause of reliability failures.

Figure 35.2(a) separates reliability analysis into two parts. The first (vaa) estimates the likelihood that all sprinkler control valves will be open at the time that water is needed. The second (waa) estimates the probability that water can reach the sprinkler, given that the valves are open. Figures 29.3(a) and (b) identify the major factors that influence these events.

Although many factors that affect reliability are well known, non-proprietary ways to establish numerical estimates do not yet exist. However, an examination of the site-specific factors that influence system reliability can often uncover weaknesses that might otherwise go undetected. These can be documented for future action.

A numerical measure for the quality of this part of the system is useful for building performance descriptions. We recommend that the probabilistic measure of reliability be considered as a subjective "grade" for quality. The grade would incorporate:

- Observed factors that reduce sprinkler reliability.
- Maintenance records.
- Company attitude and practices toward maintaining the sprinkler system.

When we assign a grade to the system reliability (AA), the probabilistic measure conveys valuable information about an engineer's belief in the sprinkler system's ability to function when a fire starts.

To illustrate its use, assume that the engineer for the previous example estimates a degree of belief (i.e. "grade") for sprinkler reliability of $P(AA) = 0.94$. This may indicate some observed shortcomings in the maintenance or the water delivery features of the system.

Example 35.6   An engineer examined the sprinkler system, valves, pumps, maintenance records, and practices. Assume that a subjective probability was assigned as $P(AA) = 0.94$. This numerical value was based on the following:

- $P(vaa) = 0.96$
  - Valves were open at the time of inspection. No measures were in place to ensure the valve would be open at all times (except routine maintenance).
  - Maintenance records were not retained.
  - Personnel were not aware of expected sprinkler maintenance requirements.
  - Management does not incorporate fire protection systems maintenance into corporate thinking.

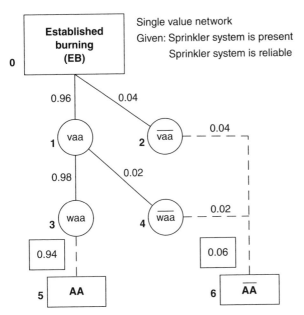

Figure 35.11 Example 35.6: sprinkler reliability.

- $P(\text{waa}) = 0.98$
  - Wet pipe system with no observed deficiencies.
Grade the reliability for the sprinkler system in Example 35.1.

Solution:
Figure 35.11 shows the network of Figure 35.2(a) with the "grades" incorporated. This produces a value for reliability of $P(\text{AA}) = 0.94$.

35.13 Sprinkler System Performance (A)

A sprinkler system performance evaluation combines the system's ability to control the fire (AC) and the system reliability (AA). Fire control changes as fire size (or a related parameter, such as floor area or time) increases. It continues until either fire control is successful, $P(\text{AC}_n) = 1.0$, or until the sprinkler system can no longer successfully control the fire, $P(\text{AC}_n) = 0.0$. System reliability (AA) is a constant for the zone being evaluated.

Figure 35.2 shows the hierarchical networks for single value automatic sprinkler performance evaluations. Figure 35.12 illustrates calculations for AA and AC at a fire size of 600 kW. Table 35.2 shows values for sprinkler performance [$P(\text{A})$] using a reliability of $P(\text{AA}) = 0.94$ and the $P(\text{AC})$ values in Table 35.1.

Figure 35.13(a) shows the absolute values for sprinkler performance $P(\text{A})$ that incorporates reliability $P(\text{AA}) = 0.94$ with values for sprinkler fire control, $P(\text{AC})$. Figure 35.13(b) shows the cumulative performance curve for the values.

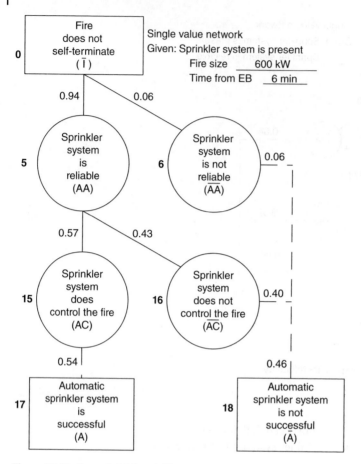

Figure 35.12 Example 35.6: sprinkler system performance calculations.

Table 35.2 Example 29.1: reliability-adjusted performance estimates.

	Fire size/area							
	600 kW 4 ft^2	600 kW 4 ft^2	1 MW 10 ft^2	3 MW 60 ft^2	5 MW 300 ft^2	8 MW 600 ft^2	12 MW 1500 ft^2	18 MW 4000 ft^2
$P(AC)$	0.0	0.57	0.88	0.91	0.89	0.68	0.32	0.0
$P(AA)$	0.94	0.94	0.94	0.94	0.94	0.94	0.94	0.94
$P(A)$	0.0	0.54	0.83	0.86	0.84	0.64	0.30	0.0

35.14 Control and Extinguishment

Professionals who are experienced with sprinkler system operations commonly expect sprinkler systems to completely extinguish a fire in about half of the operations. For those cases in which the fire is not completely extinguished, the sprinkler system alters

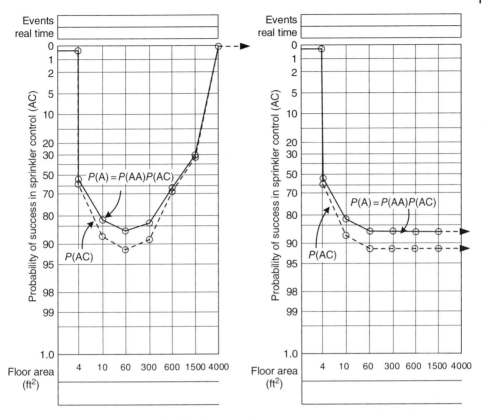

Figure 35.13 Example 35.6: graphical descriptor performance.

fire growth significantly. The water discharge pre-wets unburned fuel; steam generation starves the fire of oxygen; and water cools the fire and causes it to diminish. All in all, a successful sprinkler operation weakens the fire so greatly that complete extinguishment can take place by an occupant using a portable fire extinguisher or the local fire department completing the task.

Although we use the term "control," the concept means "extinguish" most of the time when the incidental help is available. Of course, when the sprinkler system does not control a fire and it extends and becomes quickly out of control, the fire department must then contend with a significant fire.

An occupant must have fire extinguishers available and some inclination to use them. And the fire department must know that a sprinkler operation is in progress. A performance analysis examines these details to examine the manner and capability of humans to complete the task.

Additional topics such as combining of fire department extinguishment with sprinkler operations will be identified in Section 35.19. However, within the discussion of the sprinkler extinguishment process, a performance analysis examines water flow alarm transmission to the community fire alarm center or to a supervisory service. Early fire department response can reduce fire damage from continued burning and damage from continuing water flow after control. Details are examined in a performance analysis because not all buildings use this alarm transmission service.

35.15   Sprinkler Performance for a Building

The examples have thus far centered on a single scenario in the open office. In this example, the first sprinkler will fuse at about 600 kW. This size may vary somewhat in other rooms because observed conditions may affect heat flow to the sprinklers. However, with functional certainty, one may estimate fire size in other rooms at first sprinkler opening.

The knowledge gained from an initial detailed analysis may be extrapolated to consider other scenarios. For example, the sprinkler system in smaller, enclosed office rooms should extinguish a fire without extension if at least one sprinkler is in the room. The small volume, barrier ability to contain fire and water within the small space, effect of steam generation, and wetting of unburned fuel all lead one to expect extinguishment success [$P(AC) = 1.0$] in those spaces. Water density and continuation will certainly be adequate for the smaller rooms. Reliability (AA) would not change.

Other spaces in a building may require more careful scrutiny. For example, in the model building the restaurant area is large and open. Fast fires uninterrupted by barrier delays are more likely. The retail stores have many of the same concerns in addition to a stronger diagnostic fire. Garage conditions for sprinkler operation may be very different from heated, above-ground spaces. The space is open and modern vehicles often have large heat release characteristics.

The occupancies on the ground floor of the model building have a higher hazard classification than office occupancy. Was the sprinkler system initially designed for these hazard classifications? Sometimes, a design may be based on the dominant classification. Buildings of this type regularly change tenants and hazards. Even if the hazard classification were correct at initial design and acceptance, is the expected sprinkler performance appropriate today in these spaces? Is another detailed performance evaluation necessary to answer this question, or is extrapolation good enough?

The underground garage requires special attention. At the time of initial construction, a number of different alternatives were possible. These include:

1) No sprinkler system was installed.
2) The system may have been compromised. For example, specifications for the office occupancy sprinkler might have been used, rather than a system design for the garage. If the garage is unheated and the building is in a location where temperatures are not below freezing, the mistake may not be recognized. If cold weather occurs several months a year, an unprotected wet-pipe system would be quickly recognized when pipes freeze. At that time the sprinkler system may have been disconnected for this space and no operable system would exist.
3) Low temperatures were recognized for this space and a dry-pipe sprinkler system was designed and installed.
4) The garage may be heated.

The initial sprinkler analysis provides a basic understanding of sprinkler system performance. The understanding that evolves regarding the system can be extrapolated to determine spaces that will be as good as or better than the initial space evaluation. These usually require no special analysis. On the other hand, spaces in which performance can be very different may be investigated more carefully to determine if sprinkler control is a problem.

35.16   Visual Thinking

Diagrams that describe component behavior provide a visual perception of performance and alternative comparisons. We pause briefly to illustrate ways in which performance curves can help to understand holistic building performance better and to communicate effectively with others. Chapter 36 provides additional discussion of this topic.

For convenience, performance curves are described by their function. Thus, the A curve describes the cumulative probability that the automatic sprinkler will control the fire. The M curve describes the cumulative probability that the local fire department will extinguish the fire manually. Each component's cumulative performance curve is described by a distinctive symbolic name.

Sub-component performance can also be illustrated with graphical descriptors. For example, the AC curve describes the cumulative probability of fire control by automatic sprinklers. The Det curve depicts the cumulative probability that the detector will actuate for a fire. Thus, all components and sub-components can be represented by a unique curve that describes their behavior in the site-specific building over time or other related measures.

We use degree-of-belief calibrations to describe performance for conditions in which deterministic methods are still inadequate. Uncertainty is reduced when we use judgmental estimates to augment deterministic knowledge. Often, this thinking helps to sort out important alternatives. Graphical descriptors augment the techniques to give a more complete picture of the thinking.

Figure 35.14 uses Evaluator A to order thinking and compare alternatives. The ordinate shows the end values of probability of success, and intermediate values are not specified. The abscissa may be used for a variety of measures that relate to fire size. Here we will use floor area of fire involvement.

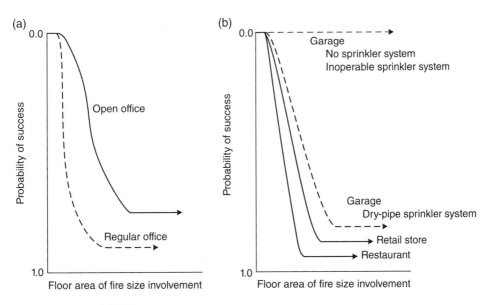

Figure 35.14 Visual thinking.

The curves are constructed to describe sprinkler system suppression behavior for specific spaces. Earlier knowledge from the sprinkler system analysis provides an insight into the expected performance. Additional observed information augments or changes estimates. Estimates are rapid and provide an insight into the way the engineer understands the conditions.

Figure 35.14(a) shows the open office that was evaluated in earlier examples. On the other hand, for regular offices the A curve notes that the sprinkler system will have no difficulty in controlling a fire. The office A curve is very good. The grade given to the system reliability is only reason P(success) does not reach certainty.

Figure 35.14(b) portrays qualitative comparisons of sprinkler system performance in the noted spaces. Performance would express calibrations based on observations and previous information from the open office analysis. Two performance curves are shown for the garage to illustrate differences.

35.17   The IPI Chart

The IPI chart is a platform for thinking, recording, and calculating. Figure 35.15 shows Worksheet 1, which describes the base analysis for the sprinkler system. Additional worksheets can describe information for other rooms, alternatives, and conditions. One can have a broad range of information available from which to develop the understanding for a variety of alternatives and conditions. IPI worksheets are particularly suited to comparing variable "what if" scenarios or adaptations. Spreadsheet capabilities of the IPI chart enable graphs to be constructed from the data or calculations or qualitative estimates.

35.18   The Narrative

A narrative describes specific component performance or building performance or risk. A narrative may be written for a building owner, authority having jurisdiction, architect, fire department, or any other individual or authority that has an interest in decisions concerning the building.

35.19   Sprinklers and the Fire Department

The sprinkler system and the fire department can interact in several ways that are beneficial to building performance. The more common interactions include the following:

a) The fire department is notified by a sprinkler alarm system. The fire department responds and extinguishes the remainder of the fire, shuts off water, and reduces water damage.

Section	Event	Time																		
		1	2	3	4	5	6	7	8	9	10	11	12	13	14	15	16	17	18	19
2b	A						0.0	0.54		0.83			0.86	—							→
	AA	0.94	—																		→
	AC						0.0	0.57		0.88			0.91			0.89		0.68			
	Floor area						4	5		10			60			300		1500			

Figure 35.15 Interactive Performance Information (IPI) chart: automatic sprinkler performance.

b) The fire department augments sprinkler water supply by pumping into the sprinkler standpipe. This enables the sprinkler system to have better density and water continuation for multiple sprinkler actuations beyond the initial area limitations.

c) The fire department uses existing sprinkler plumbing as extensions to deliver water to remote, inaccessible locations.

d) The fire department uses hand lines while the sprinkler is in operation. Standpipe and water capability must be evaluated to determine the water delivered to each component.

35.20   Other Components

Calculations, performance curve construction, and narratives to explain behavior may be prepared for all components and sub-components of the IPI chart. The analytical framework structure was formulated to do the following:

1) Track an engineering thought process to evaluate each component as its status changes during the progression of time.

2) Integrate dynamic component interactions into the evaluation of performance.

3) Use deterministic methods for numerical measures of performance. Where the state of the art for calculations is inadequate, use the framework to bound conditions to estimate degree-of-belief performance measures.

4) Use the hierarchy of single value and cumulative value networks to calculate discrete and cumulative values for event performance.

5) Construct performance curves from the calculations.

6) Estimate performance curves to sort thinking and select among alternatives.

7) Use the IPI chart to organize the framework, store associated information, make calculations, and construct performance curves.

Calculations and the associated performance curves follow the pattern described for the automatic sprinkler system. The framework uses a rigorous organization that complies with theory and accepted practices of the disciplines from which they have been adapted.

The description of sprinkler system analysis in this chapter serves as a model for evaluation of all framework components. All components have three segments that define the window of uncertainty. Within the uncertainty range, performance is described using degree-of-belief probability measures. The framework is a template for all calculations. Decomposition hierarchy must be maintained. Performance curves may be constructed from the probabilistic measures.

35.21   Summary

The full range of component performance may be viewed from the IPI chart. All components are represented on specific rows of the chart. Each component has three segments that describe regions where:

- the component certainly will not operate;
- the component certainly will operate (or is irrelevant);
- operation and performance are uncertain.

Performance is evaluated by examining specific times, events, or other related measures of fire size or condition. At each discrete instant in time or condition all relevant information is used to estimate a probability of success. The probability measure expresses an individual's degree of belief of the event success. The degree of belief is based on available evidence relating to the component, fire conditions, and other building actions that affect the component being evaluated for that instant in time.

A degree-of-belief estimate has relatively clear, narrow bounds for evaluation. One should ensure that the comfort level in making the judgment is good. When one is unsure because the space between bounds for an evaluation is too great, the conditions should be adjusted. The "1 in 20" rule is appropriate to give reasonable demarcation between "certainty" and uncertainty. Similar flexibility is appropriate for other estimates. Understanding performance is more important than numerical precision. Nevertheless, sloppy evaluations should be avoided.

Normally, all major events require evaluation of their sub-components. The framework identifies the components and structures the calculation procedures. One need only ensure that the event conditionality or independence is incorporated into the evaluations. While this concern is important, rarely are the conditions difficult to determine.

Framework networks, event evaluations, calculation of results, and graphing of performance form an interrelated package. Although we express performance in probabilistic terms, the analysis is deterministic. It uses all of the information that seems appropriate, much of which is obtained during the analysis that establish the window of uncertainty.

36

Visual Thinking

36.1 Introduction

Fire performance in buildings is dynamic. The fire changes constantly, and fire defenses move in and out of operation. Risk characterizations change with the composite fire.

There are many parts that interact, and all parts are time-related.

Performance evaluations of building fires are complicated. An engineer must often talk with others about what is likely to happen. Because most individuals easily comprehend graphical descriptions of performance, we introduce a few useful concepts for visual thinking.

Visual perceptions of performance are powerful tools of communication. Pictures can give understanding that transcends any professional background. Visual descriptions can be even more valuable during the engineering analysis to discuss performance expectations with colleagues as well as to enhance personal understanding. The ability to sketch one's perception of how a situation will change over time gives an insight into performance. A preliminary estimate can be later refined with a detailed engineering analysis.

This chapter describes ways in which to sketch expected fire performance. Freehand sketches that depict expected component behavior during the fire enrich performance thinking and communication.

36.2 A Case Study

Before discussing other aspects of visual communication, let us pause briefly to describe a building in which an engineering performance analysis had been done 8 months before the fire occurred. The objective of this analysis was to understand the building's existing fire performance and describe how alternative proposals would change the fire risk. The presentation to management attempted to enable decision-makers to understand the problem and the implications of different alternatives. Management would then decide what to do.

Military Records Center

The Military Records Center in Overland, Missouri, was a six-story reinforced concrete structure that stored military service records of members of the United States Army, Navy, Air Force, and Marines and provided information processing involving those

Fire Performance Analysis for Buildings, Second Edition. Robert W. Fitzgerald and Brian J. Meacham.
© 2017 John Wiley & Sons Ltd. Published 2017 by John Wiley & Sons Ltd.

records. The value of the information housed in the building was enormous, and its loss would be a financial disaster. Consequently, the government had a great interest in avoiding loss.

The building was 726 ft long by 286 ft wide (221 m × 87 m). This is approximately the area of four contiguous football fields. The floor-to-ceiling height was about 12 ft (3.7 m). At the time of the fire, 60% of the space housed records storage and 40% was used for office operations. Two thousand occupants worked in the building during the day, and six occupants were on duty at night.

The building was built in 1956. At that time, most records were stored in metal file cabinets. In 1960 the General Services Administration (GSA) was given responsibility for the building. This change resulted in substantial expansion of activities and a rapid increase in records storage. Most of the added records were placed in cardboard boxes on open shelving that extended from floor to ceiling in large open spaces.

Although the building had been evaluated for code compliance on two earlier occasions, the GSA conducted an engineering performance evaluation in October 1972. The four-page report noted the relative ease and speed of potential fire growth and the lack of effective fire defenses.

The presentation to management used L curves to describe the building's performance, as it existed, and to compare proposed changes. An L curve is a graphical descriptor of the limit to which a composite fire (flame-heat) will extend until extinguishment. An L curve is a significant performance descriptor because all risk factors can be related to the composite fire behavior.

At the presentation, management quickly understood the magnitude of the fire problem and was able to make an informed decision among alternative proposals to reduce the risk.

Unfortunately, the fire occurred before the fire defense changes could be implemented. However, the graphical descriptors proved valuable for conveying much information to non-engineering decision-makers within a very short presentation time. The fire reconstruction showed that the engineering degree-of-belief estimates were extremely accurate descriptors of the scenario.

Figure 36.1 shows that under existing conditions (shown by the solid lines) there was a high probability of extensive loss during normal operating hours and of disastrous damage during non-working hours. If the cardboard boxes containing files were rotated 90° to prevent paper exfoliation, the speed of the fire would be slowed, thus allowing a more timely intervention by the fire department. This consideration was included for illustrative purposes. Changing box orientation showed only marginal risk improvement, and work operations would be negatively impacted.

Two additional alternatives that examined both working and non-working hours of operation were portrayed in the graph. The most effective design alternative from a fire damage viewpoint was to store the records in metal file cabinets. This changed fire development conditions and speed of propagation. However, the cost was greater than the installation of sprinklers, which was also shown. This graphical story of potential damage under existing conditions and the cost-effectiveness of changes under the alternatives enabled management to make an informed decision. It decided to install a sprinkler system. Unfortunately, the fire occurred before the decision was implemented.

There are several significant aspects of this case study. One is that the engineers looked at fire performance both for functional components and also as an integrated

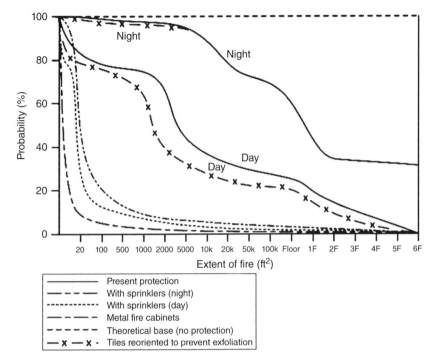

Figure 36.1 Military Records Center: decision alternatives.

(holistic) system. This enabled the team to understand time-related interactions and focus on the issues that affected performance. Evaluations were expressed as a degree-of-belief judgment using the technical knowledge base of 1970. Although the fire occurred on the sixth (top) floor, the descriptions in the report written 8 months earlier were very accurate. The rationale for the conclusions was provided in the four-page report.

The L curve became a powerful communication tool. This tool enabled management to comprehend the problem, envision the cost and effectiveness of alternatives, and confidently make informed decisions about a solution.

We pause to discuss a few graphical features that were introduced in Chapter 35. The ordinate used a uniform scale from a probability of 0% to a probability of 100%. Here, positive is in the upward direction because it described failure (i.e. "How much can I lose?"). The scale helped management to get a sense of proportion regarding the information. If numerical details were significant to the presentation, a different scale could have been used.

The abscissa of Figure 36.1 is not uniform. The first three-quarters of the scale depicts a single floor, the scale of which is also non-uniform. The remainder shows the five additional floors. Although the portrayal may be unconventional, the meaning is clear. Much information can be conveyed within a very short presentation time with visual comparisons. The ability to adjust scales to convey information effectively to the target audience is useful.

36.3   A Way of Thinking

We sketch performance curves to tell a story. The story tells what we think will happen during a fire. Sometimes the curves are intended to help decision-makers to understand relative differences among the alternatives. Functional, operating, and capital costs can be related to each.

Sometimes curves are used to guide discussion with colleagues on operating details or locations of selected parts of the fire safety system. Performance sketches can describe expectations for expected behavior as well as the effect on the holistic building performance. Differences of opinion can focus on the factors that affect performance. Interactive discussions enhance understanding.

Performance descriptors help an individual to think professionally. For example, an engineer may wish to develop an understanding about the effect of alternative locations or equipment, or even whether or not to use certain types of fire protection. There are many factors to consider in a design decision. Performance sketches can help to organize complex, often conflicting considerations in the selection process.

Any event can be described by a performance curve. The curves help to organize thinking. They also show relative outcomes more explicitly. This chapter discusses these concepts to create an awareness of the power of graphical thinking and communicating.

36.4   The Interactive Performance Information (IPI) Chart Relation

The IPI chart integrates the complete fire and building system in a hierarchical manner. Appendix A provides detailed IPI charts for the decomposition of each component.

One may sketch a performance curve for any component listed in Appendix A. Changes in behavior are described in the cells of horizontal rows of the IPI chart. Effectively, the performance changes of any IPI row may be represented by a graph. When we use degree-of-belief probabilistic calibration, we are describing our belief about how the component will behave during the fire scenario under discussion.

Each event in the IPI chart is given a distinctive symbol. Although one can sketch a performance curve for any component, the dominant curves are as follows:

- *I curve* – describes the diagnostic fire. In fact, the I curve represents the probability of self-termination of a fire. This is its success in fire self-termination. Its complement (i.e. $\bar{I}$, a fire that does not self-terminate) is the diagnostic fire that is used to evaluate fire defenses and risk characterizations. The importance of the I curve has diminished in recent years because we use a diagnostic fire (actually $\bar{I}$) more explicitly. However, we include the component for theoretical completeness.
- *A curve* – describes automatic sprinkler performance success for the scenario.
- *M curve* – describes local fire department manual extinguishment success for the scenario.
- $\bar{T}$ and $\bar{D}$ *curves* – describe barrier failure performance for the fuel and construction conditions of the scenario.
- *L curve* – the limit of flame-heat movement. This combines the diagnostic fire with fire defenses that limit fire propagation.

Other curves, such as smoke tenability, detection, occupant alert, fire department notification, can be sketched. Any IPI row can be expressed by a performance curve if it is useful for evaluation or communication.

36.5 Performance Evaluators

A performance curve "tells a story" of what to expect during the fire scenario. This allows one to examine the behavior of any component in isolation as well as to show related components in order to recognize interactions.

Figure 36.2 shows three types of graph that have been useful for visual applications. These forms evolved over many years through experiences with engineering students and practicing professionals.

Although one may adapt evaluator scales to meet personal preferences, the following observations have emerged:

- *Evaluator C* is most useful for comparing calculation-based outcomes that often use holistic, cumulative value evaluations. The ordinate scale is based on standard probability paper that has been modified by relocating the bottom section to the top. This enables one to compare extremely large or small values. For example, the expanded scales allow one to discriminate more easily values such as $0.99 = 99/100$, $0.999 = 999/1000$, or $0.9999 = 9999/10000$. This unusual scale is of value in visual comparisons.
- *Evaluator B* is more useful with specific (single) components. This scale enables qualitative and quantitative predictions to be compared more easily. The scale is changed because the awareness of proportion improves with fewer significant figures.

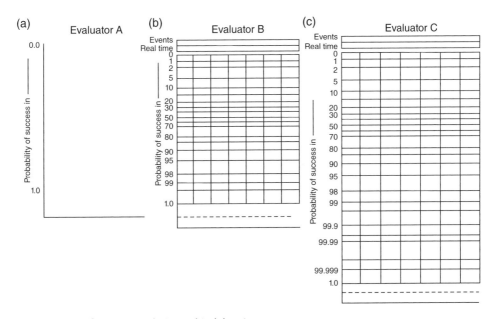

Figure 36.2 Performance analysis graphical descriptors.

- *Evaluator A* is a thinking tool to describe qualitative estimates. A numerical scale is often unnecessary to describe general performance characteristics and trends, and relative orders of magnitude are of interest. If numerical values are useful to give a sense of proportion, only a few intermediate values at general locations used in Evaluator B will suffice.

Selecting the origin at the top with positive in a downward direction is easy to adapt and seems to enhance communication.

The uniform ordinate scale of Figure 36.1 does not discriminate well. Differences such as 1/100 and 1/1000 cannot be detected visually, even though this type of distinction is useful for many applications. The variable scale, loosely related to probability, enhances understanding even with its theoretical flaws.

The abscissa describes component performance conditions. Although everything relates to time, one may use fire size, floor area of involvement, sequential barrier–space modules, or any measure of interest. Time may be either local or global, depending on the intent of communication.

Relationships between time, fire growth, floor area or the other measures are non-linear. Two bars are shown at the top of the graph. These spaces can show related units of measure.

These slight changes to the format of Figure 36.1 have evolved over the years because evaluation procedures have become more rigorous, systematic, and complete. As the transition from qualitative assessments to quantitative procedures continues, communication is greatly enhanced by tools that can be adapted to contemporary needs and discarded when they become obsolete.

36.6   Reading Performance Curves

The following examples illustrate ways to use performance curves. We vary the applications to illustrate different means of communicating information. The curves are used to tell a story that is based on important details that influence performance. You are asked to relate stories and curves.

36.6.1   Detection

Detection may be by instrument or by humans or both. Example 36.1 shows several performance curves that may be used for discussions with colleagues.

Example 36.1   Figure 36.3 shows several offices on the third floor west wing of the model building. Several alternative detection conditions have been discussed. Figure 36.4 illustrates performance for the considerations. Match the performance curve with the scenarios:

- *Scenario 1* – workstation fire in Office 3011; door between Offices 3011 and 3010 is open; transom wall extends 12 in (300 mm) below ceiling to door soffit; 135 °F (57.2 °C) heat detector in Room 3010 on ceiling 6 in (150 mm) from wall.
- *Scenario 2* – same as Scenario 1 except that the detector is on the transom wall 6 in (150 mm) above the door.
- *Scenario 3* – same as Scenario 1 except that the door between Offices 3011 and 3010 is closed.

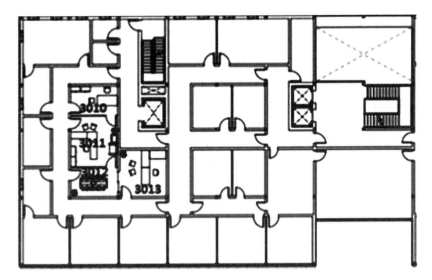

Figure 36.3 Partial Floor 3.

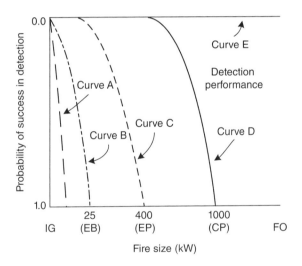

Figure 36.4 Example 36.1: qualitative (Det) performance descriptors.

- *Scenario 4* – same as Scenario 1 except that a photoelectric smoke detector replaces the heat detector.
- *Scenario 5* – same as Scenario 1 except that a photoelectric smoke detector is placed in Office 3011 rather than Office 3010.

Solution:

The best matches are as follows:

Scenario 1: curve C

Scenario 2: curve D

Scenario 3: curve E
Scenario 4: curve B
Scenario 5: curve A

36.6.2 Fire Department Notification

The component having the most significant window of uncertainty is fire department notification (MN). For this reason, scenario definitions for the building and occupants are important in understanding this important link of building performance.

Fire department notification is a part of manual suppression that is under the management control of the building owner. Example 36.2 shows several scenarios that may be used to discuss performance with a developer.

Example 36.2   As part of a manual suppression (M) analysis of the model office building, the engineer studied the building's fire department notification (MN). Although the building does have an automatic detection system that sounds a local alarm upon actuation, there is no automatic transmission to the fire department.

Use Evaluator A to describe different realistic MN scenarios that affect building performance.

Solution:
Figure 36.5 shows the way of thinking that will become the basis for evaluating the MN component of an M Curve.

Discussion:
A performance analysis examines the building over an entire year. The scenarios for MN performance are based on successful operation of an automatic detection system. The curves describe only the local time for the MN component and are based on successful completion of MD.

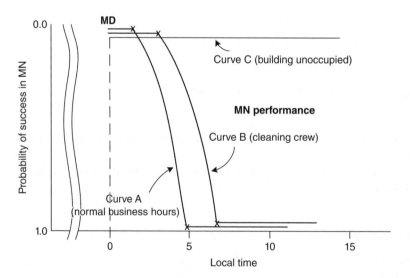

Figure 36.5 Example 36.2: qualitative (MN) performance descriptors.

Curve A   A fire alarm sounds during normal business hours. The scenario is based on an occupant deciding to call the fire department and reporting the fire on a landline or a cell phone. The shortest time to complete the event is 30 seconds after initial detection. The other bound for the window of uncertainty is estimated at 5 minutes.

The curve A estimate is based on a local alarm and no individual having responsibility for notifying the local fire department. Occupants often assume that the fire department is automatically notified when a local alarm sounds. This may or may not be correct, and precious time can be lost until someone investigates the situation to establish the actual existence of a fire or makes an independent decision to call the fire department. Although this "best case" has uncertainties, the assumption that "someone" will call the fire department (eventually) seems very good.

Curve B   A fire alarm sounds outside of normal working hours, but during routine cleaning operations. The cleaning crew comprises five individuals and a supervisor. In the absence of emergency instructions, it is assumed that "someone" will investigate before actually making a call. Uncertainty is greater, and the window of uncertainty is estimated at between 2 and 7 minutes after the initial alarm sounds.

Curve C   The fire alarm sounds when the building is unoccupied. The situation of an individual passing in the street and notifying the fire department is not considered in a performance study. Therefore, no notification will occur until the fire is very large.

Reliability is not included in this initial estimate. However, MN performance is a weak link in the M curve chain of events.

36.6.3 Sprinkler Control

Sprinkler systems that comply with standards can exhibit different performance expectations, depending on building conditions. Example 36.3 may guide performance expectations with an architect developing a building modification.

Example 36.3   An architect is preparing schematic drawings for a building modification. The fire safety engineer examines the schematics and sketches a qualitative opinion of how sprinkler system performance will be affected by the proposed plans. Describe the performance depicted by the sprinkler performance curves of Figure 36.6.

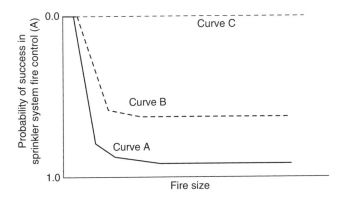

Figure 36.6 Example 36.3: qualitative (A) performance descriptors.

Discussion:

Curve A describes the expected performance for an open office as it was originally designed. The probability of sprinkler success for original conditions of moderate initial fire growth speed and intensity (usual work station materials and configurations) is excellent.

Curve B describes sprinkler system expectations if workstations are changed to materials and configurations that produce a very fast and high heat release condition. The initial sprinklers have a chance to control the fire when it is small. However, if the sprinklers cannot handle the fire when it is small, the speed of fire growth will "outrun" the ability of the sprinklers to fuse in time to control the fire. When too many sprinklers do fuse, the water supply will be inadequate.

Curve C shows a condition when new partitions are erected without regard to sprinkler location. When sprinklers are shielded, water cannot reach the fire and the office space is no longer protected by the sprinkler system.

36.6.4 Fire Extinguishment

Relatively few owners and architects realize the ways in which a building design influences fire department extinguishment. Example 36.4 compares three buildings that lead to different levels of damage for a fire.

Example 36.4 Figure 36.7 shows M curves for three buildings in the same geographical area. They have similar occupancy, size, and construction types. The same fire department and community resources apply to each building.

You have not yet seen the written report or spoken with the inspector. Using only this diagram, what do you think the inspector is trying to say about these buildings with regard to fire department suppression?

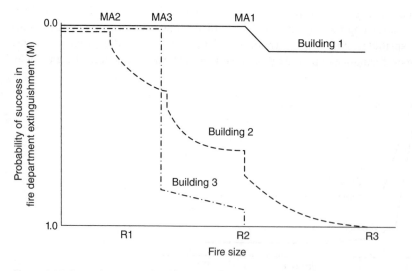

Figure 36.7 Example 36.4: qualitative (M) performance descriptors.

Solution:
A performance analysis does not ask, "Will the fire department extinguish the fire?" Rather, it estimates "Will the fire department extinguish the fire within 1, 2, 3, ... n rooms?"

The initial horizontal segment compares the diagnostic fire with the time to complete $M_{Part\,A}$ (MD, MN) and $M_{Part\,B}$ (MS, MR) and initial water application after first-in arrival (MA). This indicates that Building 2 has a much better detection and notification system than does Building 3. Building 1 has either a substantially delayed notification or significant difficulties in getting first water (MA) onto the fire.

Although a fire in Building 3 will fully involve Room 1 at the time of agent application (MA), the barrier is strong and will delay propagation to Room 2. This gives the fire department a good chance to stop the fire at Room 1. If the fire does extend into Room 2, there is a certainty that it will be extinguished without extension to Room 3.

The fire in Building 1 will certainly involve two rooms before initial water can be applied. This would indicate a delay in notification, particularly in comparison with Building 2. The absence of a vertical line at Room 2 would indicate that the barrier between Rooms 1 and 2 is very weak or virtual. When water is applied after two rooms are fully involved, the likelihood that the fire department can complete extinguishment before the fire grows to three rooms is very low. One would expect additional rooms to be involved with major fire damage. Fires that easily grow to multiple rooms frequently need different fire fighting tactics. Therefore, Building 1 is very weak from the perspective of manual fire extinguishment.

The fire department should be able to apply first water in Building 2 before Room 1 has reached flashover. However, the fire is nearing flashover at MA, and a slight increase in time durations would lose the room. Also, the barrier between Rooms 1 and 2 is weak. If the fire is not extinguished in Room 1, it will probably extend to Room 2. However there is a good likelihood that manual fire fighting will be able to limit the fire to two rooms. Certainly, the fire will not extend beyond three rooms.

Fires such as those described in Buildings 2 and 3 may be able to use an aggressive interior attack with rapid extinguishment. Building 1 is more likely to have major problems and require a defensive fire fighting strategy or a large first alarm force.

Although almost no information is provided about the diagnostic fire, the building, and the fire department resources, these three M Curves give a preliminary indication that Building 1 poses a very different fire suppression situation than Buildings 2 and 3.

36.7 The L Curve

The limit (L) of flame movement defines the extent to which flame-heat damage will spread. The probability for the limit (L) may be calculated using the network of Figure 36.9(a). When performance curves for I, A, and M have been constructed, the calculations can be programmed.

Often, only the active extinguishment (L_A) using A and M is evaluated, because the diagnostic fire characteristics make fire self-termination less important. From a theoretical viewpoint, self-termination is required for completeness. For practical applications, this component may be neglected because it is a part of the diagnostic fire.

During the early years of this performance analysis, the L curve was an important descriptor. Often, it was described as the "fingerprint" of the building. The analytical framework evolved by decomposing the L curve. However, as procedures to evaluate the components matured, the importance of the L curve diminished.

Conceptually, the L curve is still useful to understand the roles of fire self-termination (I), automatic suppression $\left(A|\bar{I}\right)$, and manual suppression $\left(M|\bar{I}\right)$ beyond the room of origin. Chapter 38 discusses the analytical procedures for rooms beyond the room of origin more completely. Here, we present a few observations about L curve descriptions.

Figure 36.8 shows a representative L curve for three sequential rooms. Segment AB shows the likelihood that the fire will be limited to the room of origin. The vertical line from B to C describes the barrier effectiveness between Room 1 and Room 2. A short line indicates a barrier that can provide some delay in fire propagation, but which is generally ineffective to stop the fire. A long line indicates that the barrier is effective in delaying fire propagation into the next space.

Segments CD and EF show the likelihood that, given the fire has not been limited in previous rooms, it will be limited in the room being considered. Vertical lines BC and DE indicate the probability that the barrier will stop the fire without additional propagation.

Figure 36.9(a) shows a network to describe the thought process and to calculate L curve values through successive rooms. For example:

Given EB, the probability that the fire will be limited to Room 1 is $P(L_1)$.
Given $\underline{L_1}$, the probability of barrier success is (B_{1-2}).
Given $\underline{B_{1-2}}$, the probability the fire will be limited to Room 2 is $P(L_2)$.
Given $\underline{L_2}$, the probability of barrier success is (B_{2-3})
Given $\underline{B_{2-3}}$, the probability the fire will be limited to Room 3 is $P(L_3)$.
Etc.

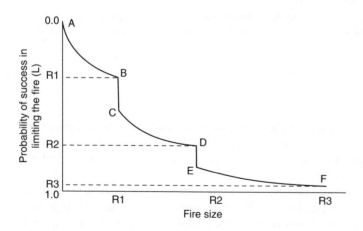

Figure 36.8 Characteristic L curve.

Figure 36.9(b) shows a SVN network to calculate the limit in a space at any selected time. Figure 36.9(c) shows the single value network that identifies the status of a barrier at any time after flashover in the room. The relationship $P(\bar{D}) + P(\bar{T}) + P(B) = 1.0$ always.

Figure 36.9(b) shows the three ways that a fire can go out within a space. The first is by self-termination. At any stage in a natural fire, the fire will continue burning or it will self-terminate. Those fires that continue burning may be put out by the sprinkler system (A) (if present) or the local fire department (M). Thus, A and M are conditional on $\bar{I}$.

$\bar{I}$ is the diagnostic fire. Because we select a room of origin and diagnostic fire scenario to represent a reasonable challenge, those room interior designs that go out easily are uncommon. Therefore, we normally neglect I (i.e. we assume I = 0). This means that the L value is based on A and M. If A is not present, L becomes the same as M.

Barriers have an important role in the L curve. The locations and types of holes to the inside of the building are a major factor in multi-room fire propagation and the diagnostic fire. They can also affect A and M beyond the room of origin.

Vertical lines that represent the strength of barriers are step functions. The time to breach the barrier becomes the time duration on each side of the vertical line. Added time becomes that of the space fire development. The IPI chart can handle this with attention to time durations. The time to breach a barrier can also be programmed. Although we normally show the abscissa of the L curve as a sequence of rooms or increasing areas of fire involvement, time is also related. The timescale is discontinuous at barriers to reflect the delay in fire spread through the barriers.

The sequence of fire defense actions is routinely handled by the calculations. Because A and M involve independent evaluations, their order is unimportant. However, A

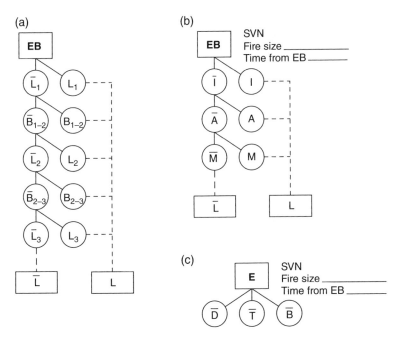

Figure 36.9 Network analysis: L curve.

normally occurs before M. Therefore, the L curve is coincident with the A curve until M occurs. The mathematics treats them as separate events to produce the L curve.

There is value in evaluating A and M as independent components. The analyses are uncomplicated by commingling actions. In practical evaluations, the fire department "mops up" the remaining fire from sprinkler operations. The L curve analysis may recognize those situations where both sprinklers and the fire department operations work together.

36.8   L Curve Illustration

The most useful application of L curves involves communication – with owners, architects, fire departments, building managers, and city officials. Although L curves may use calculated outcomes from component analysis, one may also sketch comparative L curves to describe fire paths or other features that affect fire safety.

Figure 36.10 shows three alternative design options for a group of rooms in a specific building. L curve A shows existing conditions. The fire has little chance of being extinguished within the room of origin. Barriers are weak and the fire can spread to multiple rooms very easily. L curve B shows the result when sprinklers are installed, but barriers are not changed. L curve C shows barriers that have been strengthened against fire spread. Figure 36.9(c) shows the effects of installing sprinklers and also improving the barriers.

One can relate costs, operational changes, damage expectations, and any other factors that are of interest to building management. Although the ordinate does not show a numerical scale, values could be used because the alternatives would have been the result of engineering analyses. The pictorial relationships enable decision-makers to recognize anticipated damage impacts for different alternatives.

36.9   Variability and Reliability

The window of uncertainty can be narrow or broad, depending on the knowledge base and available information. Uncertainty is expected for any individual component, and when several components must be linked, as in the M curve analysis, time durations can

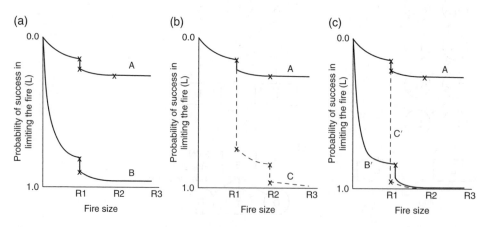

Figure 36.10 L curves.

overlap. Although a composite curve could be constructed, the product is not usually worth the effort. A better approach borrows techniques from construction project management to provide a range of answers that contribute to understanding.

Project management uses values for reasonable situations to produce analyses that are:

- Most pessimistic
- Most likely
- Most optimistic.

The bounds of uncertainty for the involved components combined with the probabilistic values for these conditions provide an insight into performance variability. The complications become tractable when these conditions give a sense of the range of performance.

36.10 Summary

The goal of a performance analysis is to understand the way a building will behave during a fire. Specifically, we want to be able to move effortlessly between details of deterministic component evaluations and holistic building behavior. Similarly, we want to transition between numerical calculations and dynamic behavior expressed by the IPI chart or performance curves. Probabilistic degrees of belief establish links among these techniques.

The framework structure, the IPI chart, deterministic evaluations, probabilistic estimates, and performance curves are tools. They help an engineer to select the most appropriate techniques for different ways to evaluate a building. These help to adapt for constraints of time, knowledge, and the technical state of the art. Each technique has its role, and each relates to the others in some way.

Graphs that depict performance can substantially enhance understanding and communication. The types of visual portrayal of component performance can vary between numerical values based on calculations and subjective descriptions based on observations and mental estimates. Visual descriptors help one to think about performance. The thinking enhances personal understanding as well as communication with others.

Sketching qualitative curves to represent component performance is fast and inexpensive. It helps an individual to understand the continuum of component performance better. Its value is increased when it becomes a focus of discussion with colleagues or when the behavior is incorporated into the effect on holistic behavior. Often, the technique is valuable for explaining performance expectations to individuals with no understanding of fire safety.

Although probability sketches are done rapidly, they require thought and understanding. Judgments are based on the best information available at the time, even though knowledge is never sufficient. Nevertheless, the exercise of identifying relevant information, envisioning expected performance, and recognizing factors that influence outcomes is the most valuable product of sketching performance curves.

37

Introduction to Risk Management

37.1   Introduction

Risk management means different things to different people. An insurance company, an investment broker, a fire officer, a building contractor, and an industrial company president all manage risk. But the ways in which they manage their own type of risk depend on the particular problem they are trying to solve. However, a common thread that is woven through each type of application is the need to make decisions under uncertainty.

One aspect of risk management involves understanding specifically what is at risk; another is to have a sense of what can happen when things go wrong; and a third is to make decisions concerning what to do about the risk before the potential hazard becomes real. Sometimes these pieces are addressed with rigor, sometimes they are addressed in a fuzzy, casual manner, and sometimes it there is a mixture of approaches.

Some view risk management exclusively as a decision about what type of insurance to purchase. Sometimes risks are ignored (e.g. a passive decision to do nothing) in the hope that misfortune will not happen. If an accident were to occur, decisions about what to do would be handled at that time. On the other hand, some risk management is formulated with a thoughtful, rational process that involves a partnership between technical experts and the team that must manage and operate the enterprise.

This introduction to risk management will suggest a way of thinking that structures a process linking technical fire performance with management of the enterprise. The goal is to make better decisions faster, easier, and more economically. Business operations must deal with a variety of day-to-day decisions to keep the company going. Fire safety often has a low priority on the list of immediate needs. Yet there are times when fire safety moves up the agenda to require decisions that can affect the physical and financial well-being of the enterprise.

Risk management is usually associated with individuals within an enterprise. Management must decide if consultants should augment corporate decision-making with specialized knowledge or whether to keep the process within the corporation. This is often described as "build it or buy it."

Although this chapter describes a way of thinking for fire risk management, many of the concepts may be adapted for natural and man-made disasters other than fire.

Fire Performance Analysis for Buildings, Second Edition. Robert W. Fitzgerald and Brian J. Meacham.
© 2017 John Wiley & Sons Ltd. Published 2017 by John Wiley & Sons Ltd.

PART ONE: THE PROCESS

37.2   Audience

The decision-maker is not the audience for this discussion. Instead, it is the individual or group that prepares recommendations, reports, and analyses to help others within a corporation or other enterprise make better decisions. Therefore, the audience for this discussion may be a private consultant that is commissioned to analyze a problem and recommend solutions. Or the audience could be a corporate risk management staff responsible for the continued safe operation of the enterprise within economic constraints. The audience might be a public service fire marshal or fire chief responsible for fire fighters or the general public. Although risk management can involve a broad range of potential applications, a general process is common to all. This chapter provides guidance for those who serve others.

37.3   Fire Safety Management

Fire safety design decisions are made with specificity. That is, the water density of a sprinkler system is x liters/min (lpm)/m^2 [gallons/min (gpm)/ft^2]; type y photoelectric smoke detectors will be installed in the following locations; fire department connections to the standpipes are located z m (ft) to the west of the main entrance; or the local fire department will respond with two pumpers, one ladder, a chief, and a staff of 12. An individual during the design and construction process or in the local community government will have made those decisions. Decisions such as these may have a significant influence on the building's fire performance or they may be relatively unimportant to the outcome.

Fire safety management decisions relate to identifying acceptable losses, selecting a deductible cost for an insurance policy, installing new fire protection equipment or upgrading existing installations. Additionally, management may decide to create a more effective fire prevention program to avoid certain expected losses. Effectively, fire safety management involves decisions that affect the complete package of loss, protection, and economics of fire.

We advocate that management decisions be based on performance knowledge having technical credibility. The primary objectives of a fire safety management program are:

- To *understand* the building fire performance and characterize associated risks.
- To *use* that understanding to make decisions easier.
- To *communicate* more effectively with others.

Decisions are made within constraints of operational functionality, financial limitations, and time.

Organized, structured procedures enable decision-makers to understand the problem, examine details, develop ways of strategic thinking, identify and evaluate alternatives, and recognize implications of decisions. The insight that is gained by a thoughtful performance analysis enables one to discuss and evaluate alternatives and recognize clearly why one course of action may be more desirable than another.

There are two major types of application that benefit from the understanding that follows a disciplined fire performance and risk characterization analysis. One addresses the management question, "How can I do better with my available resources?", and the other considers technical decisions: "Are the fire defenses appropriate for my objectives and how can I improve their performance quality?"

Some examples of resource management are:

- *Operational planning* – local fire service officers can plan appropriate and safe fire department responses to specific (target) potential incidents.
- *Performance expectations* – local fire service officers can communicate the fire suppression and life safety performance expectations more effectively to owners and occupants.
- *Fire risk management* – consultants and corporate managers can formulate risk management programs that integrate an appropriate mix of fire defense measures, insurance selection, and loss expectations into cost-effective plans that address specific needs.
- *Emergency planning* – consultants and corporate operations personnel can formulate emergency operational procedures that are tailored to the needs and activities of the specific building being studied.
- *Resource allocation* – business managers can make more informed decisions relating to allocations between accepting losses, transferring risks through the purchase of insurance, and reducing the risk by improving the building fire performance.
- *Risk discrimination* – insurance companies can evaluate the risk characterizations for different site-specific buildings more objectively to decide if they will insure and at what price.

Examples of technical decisions may include:

- *Equivalency acceptance* – local authorities having jurisdiction (AHJ) can compare building code equivalency proposals on a consistent basis.
- *Code interpretation* – local AHJ examiners can interpret the functional basis for prescriptive regulatory requirements.
- *Performance-based design approval* – local AHJ examiners can have a basis for comparing a performance-based design with the equivalent performance that would be expected from prescriptive code compliance.
- *Design decisions* – the architectural team can select features to improve performance.
- *Impairment planning* – local fire departments can document a rationale for identifying acceptable equivalency alternatives during temporary impairment of fire protection systems for servicing.
- *Cost-effectiveness comparisons* – fire safety engineers can compare cost and effectiveness among fire defense alternatives more rationally.
- *Fire reconstruction* – fire investigators can plan and conduct fire reconstructions to compare the manner in which a building actually performed with the manner in which it had been expected to perform.
- *Design alternative comparisons* – an architectural design team can compare the fire performance and associated costs that could be expected from prescribed building regulations with those expected with a desired alternative design.

37.4   Decisions and Uncertainty

A number of questions arise during fire safety deliberations. On one level they include:

- Do we have a fire problem?
- How do you know?
- If a problem exists, what kind of a problem do we have and how serious is it?
- How do you know?
- What alternatives should we consider in dealing with the problem?
- Can our staff handle the problem or should we hire a consultant?
- How do you know?
- How do we structure the request for proposal (RFP)?

Another level might consider questions of why a risk analysis may be needed? For example, initial questions about the adequacy of building fire safety may arise from occupancy changes, functional operation modifications, economic loss concerns, reports from insurance or fire department inspections, or legal suspicions.

Should the fire analysis be regulatory-based to upgrade code compliance? Should it be performance-based so we understand the real problems and can develop rational alternatives to produce the most cost-effective solutions?

Should we do the analysis in-house or hire a consultant for technical assistance? Should we prepare an RFP or should we select the consultant from our network of individual contacts? How should the RFP be worded? What do we want the consultant to do? Do we need specialists or a general practitioner? How do we identify qualifications and how do we select the consultant? How much is enough?

Often, what we don't know is more difficult to determine than what we know.

37.5   Management Applications

Both the private sector and the public sector have many applications that can be addressed by understanding the way a building will perform during a fire. The public often views fire as a regulatory issue or an insurance constraint. Awareness that a fire performance analysis can provide enough understanding to allow rational decisions among alternatives is often not recognized. In the following we present a few illustrations of decisions involving the performance analysis of buildings:

1) A private corporation has budgeted $20 000 (or $200 000) for the coming year as part of a long-term program for improving the fire risk of the company.
   a) How should they decide the initial projects on which to spend the money?
   b) What is the best long-term investment for the company? Is it better (i.e. more cost-effective) to:
      - do nothing;
      - improve the fire prevention and emergency preparedness programs;
      - purchase more insurance;
      - install fire protection equipment? (What kind? Where? How much?)
   c) How can management recognize the most effective improvement in fire safety for the money?

2) You have 10 (or 100, or 1000) buildings within your responsibility for fire safety. You do not have the resources to deal with all of these buildings.
    a) How do you identify the 10% of the buildings that have the greatest need for your attention?
    b) How do you identify the buildings that have the greatest potential for:
        – death or injury to fire fighters;
        – death or injury to occupants;
        – damage to downtime that would cause business failure?
3) How will you decide, "How much is enough?"

37.6 Comparisons

Organizational needs usually guide the best way to compare alternatives. For example, decision-making needs by an insurance company on rate-setting policy are different from those of a manufacturing company deciding what type of insurance to purchase. The former deals with a large population of insured, whereas the latter is involved with a unique, individual operation. Objective probability and statistical studies may be appropriate for an insurance company to compare alternatives. However, individual business decisions are more appropriately structured on the unique operations of the corporation. Measurements and communication should be appropriate to the needs of the application.

The risk management described here focuses on the individual, site-specific building or group of buildings. It becomes a personal analysis for the building and its operation.

37.7 Process Overview

Information has both cost and value. We advocate the principle that the value of information must be greater than its acquisition cost. Therefore, an evaluation is tailored to the needs of the problem. When a qualitative, observational evaluation is adequate to provide enough understanding of building performance and risk, no additional information is needed. When one needs a greater understanding of detailed features, we can use quantitative estimates and variation analyses to target additional information.

Figure 37.1 shows a complete risk management process. Because many potential applications can use this structure, some users will rearrange or even omit certain activities. Although activities are distinct, the information and knowledge are interactive. Our intent is to identify an organized structure. The process is holistic, and iteration is common.

Part A acquires information that becomes the basis for organizing risk management and decision-making. This group of tasks acquires knowledge that is normally provided by the fire safety engineer with some assistance from the corporate risk manager.

1) *Understand the Problem* Understanding how the building and operations work during normal usage is the most important aspect of any risk management program. This has a minor influence on risk assessment, but a major impact on risk management. In gaining an understanding of how the building works, one identifies what is at risk and

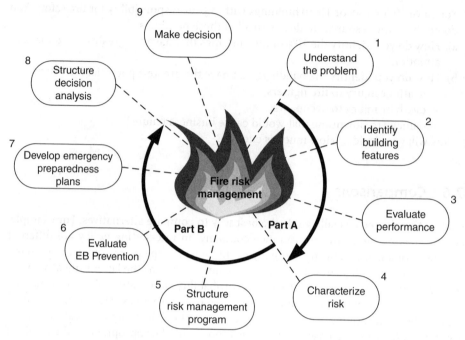

Figure 37.1 Risk management organization.

the sensitivity to fire and fire suppression products. The knowledge enables one to identify credible atypical conditions that should be examined.

2) *Identify Building Features* This activity defines the building for the performance evaluation and its associated risk characterizations. Space utilization plans can note specific features important to an evaluation.

3) *Evaluate Performance* Performance analysis is the principal content of this book. The knowledge gained from a performance evaluation provides a technical basis with which to evaluate and compare the effectiveness of proposed alternative solutions.

4) *Characterize Risk* After one understands the building's performance, threats to people, property, mission, neighbors, and the environment can be analyzed to characterize the risks.

Part B uses the knowledge from Part A to organize and evaluate ways to manage risk. Although risk managers may structure a report, fire safety engineers identify feasible alternatives, provide technical advice, evaluate performance, and characterize risk. In addition, fire safety engineers evaluate fire prevention and special hazards equipment. These activities are valuable to structure cost-effective solutions.

5) *Structure a Risk Management Program* In theory, this activity starts after characterizing the risks. However, the knowledge gained during the Part A analysis provides a clear insight into problem areas and ways to improve performance.

This activity devotes a conscious effort to structuring a comprehensive risk management program. The process usually continues in parallel with other activities, because some alternatives will depend on or be influenced by fire prevention, emergency planning, and decisions relating to alternative solutions. If a risk management program is not needed, the activity may be omitted.

6) *Evaluate Established Burning (EB) Prevention* This first step of a traditional hazard analysis is intentionally delayed until after we understand the risks, given EB. It is more useful to clearly understand the consequences of an unwanted ignition in isolation. Then, if improving or maintaining an effective fire prevention program is an option, its importance and function can be recognized more easily.

   Prevent EB has two components. One is the traditional ignition prevention involving the separation of heat energy sources and available kindling fuels. The second follows an unwanted ignition by preventing flames from growing to the size of EB. This incorporates initial fire growth and fire suppression by building occupants.

   We also examine special (spot) hazards in this location. Although spot hazards are normally protected by automatic suppression equipment, their selection, nature, and operations are not the same as an automatic sprinkler system. We find that including special hazards within fire prevention provides more systematic framework logic.

7) *Develop Emergency Preparedness Plans* Emergency preparedness plans are based on understanding the building function and operations, the fire defense system, and the associated needs of the enterprise. One may organize plans into three sets of activities that may be used before, during, and after the incident.

8) *Structure Decision Analysis* A decision analysis identifies alternative courses of action in a way that enables a manager to compare the impact of decisions. One normally packages alternatives into a very few integrated plans of action. Cost and effectiveness comparisons are included with the alternative packages.

9) *Prepare a Presentation* The risk management team normally prepares alternative solutions to enable other parts of corporate management to make a decision. A presentation should be concise and provide a clear understanding of the building's performance and associated risks. This important task is usually difficult, and we identify a separate activity to the planning.

   A clear picture of a building's present condition and its risks is challenging to develop for a short presentation. The implications of cost and value analysis of alternatives must also be succinct. Technical justification for this presentation is rarely useful and can be provided in a separate report.

10) *Make a Decision* The choice may be to do nothing, to select one of the alternatives, or to suggest additional considerations. Management decisions should be based on a clear understanding of the building's performance and associated risks. Costs and effectiveness of the alternatives are an integral part of a decision analysis structure. Although the risk manager may not make the decision, the quality of the organization and presentation often affect its outcome.

PART TWO: INFORMATION ACQUISITION

37.8 Introduction

Information acquisition includes gaining knowledge of the building and an understanding of fire performance and risk characterizations. The process is essential to gaining a clear grasp of how the building functions. Fortunately, each part of the process provides a foundation for building logical, effective ways to manage risks.

37.9 Understand the Problem

Four major areas of information collection are involved in this activity:

1) To understand how the building functions and operates.
2) To identify what is at risk.
3) To learn what managers believe is important and are willing to accept in terms of losses in a fire incident.
4) To document the problem definition. Figure 37.2 shows the major parts of this activity.

Understanding the functions, operations, and how the building works is a difficult and time-consuming, yet important, activity. This task is easier for existing buildings. Proposed new buildings may require more discussion with operations management and the architect. However, the information is essential to establish a workable, effective risk management program.

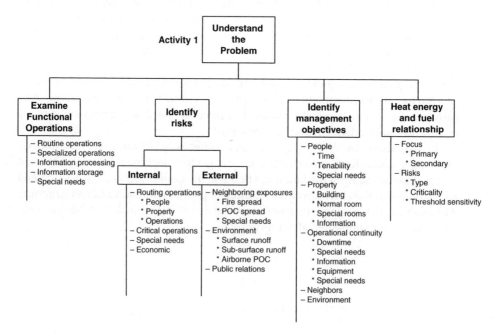

Figure 37.2 Organization chart: understand the problem.

This activity also includes identification of the sensitivity of people, contents, and operational continuity to the potential products of combustion and time frames for emergency actions. Tenable and tolerable limits of products of combustion may be identified with the building's operations management.

Often management may not know suitable values and may ask appropriate limits. One may describe tenable limits for human protection with citations and a rationale for reasonable values. Tenability limits for information technology, data storage, and equipment can be more difficult to determine, and those limits may be delayed until one gains a better understanding of the building's performance.

Neighboring properties may be sensitive to exposure fires, and the environment may be susceptible to air contamination and water runoff. A better time to incorporate this aspect of risk management may be after the performance evaluation has been completed. At that time, one has a better sense of potential problems that may exist.

Documentation for the information, results, and problems to be addressed is valuable in this activity. Space utilization plans become useful ways of documenting and retrieving the information. Broad observations may often be sufficient for the documentation. Data collection forms can be used to collect and document information for this activity.

37.10   Describe the Building

This activity identifies building features that affect performance evaluations and risk characterizations. It identifies the visible objects that link functions, risks, and performance. The activity defines the building's fire defenses and features that influence their performance and the associated risk characterizations. Figure 37.3 describes factors that are normally included.

37.11   Evaluate Performance

The fire department extinguishment (M) and automatic sprinkler suppression (A) provide an understanding of the fire size, conditions, and damage that can be expected. Although a detailed analysis may focus on a single or a few rooms of origin, extrapolation of that knowledge to envision fires in other locations enables one to describe differences in performance. Some fire locations will clearly cause less damage. Other locations may identify conditions in which the fire could cause greater loss in small, but critical pockets of operations. The single (or few) detailed analyses plus their extrapolations give an understanding of the performance variations that can occur in the building.

Figure 37.4 shows the components associated with this activity.

37.12   Characterize Risk

Performance evaluations provide knowledge of the fire, the associated structural stability, and target space smoke tenability. These evaluations provide a foundation for understanding the range of performance that can tell a credible story about the building's fire behavior. This performance understanding, combined with the

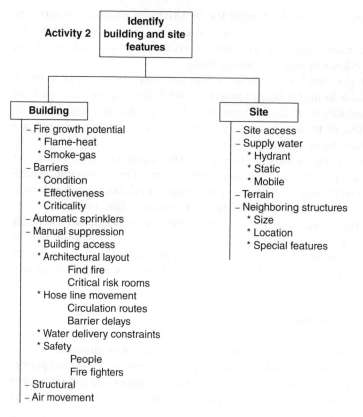

Figure 37.3 Organization chart: identify building and site features.

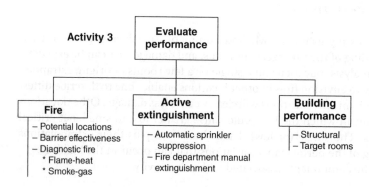

Figure 37.4 Evaluation chart: evaluate performance.

information gathered in Activity 1 ("Understand the problem"), provides the basis for characterizing risks.

The risk to people can be expressed in terms of the time the building will provide to individuals before they encounter untenable conditions. The time durations will vary with the fire location, architectural layout, and initial occupant positions.

Expected fire damage shown by a limit (L) analysis may be adapted to include the risk to important property contents, information, and functions essential to the facility's operational continuity or mission.

The impact on neighbors and the environment may also be described, if necessary. Regardless of how one measures risk, it must be described in terms that a business manager can understand quickly and accurately.

Figure 37.5 shows the ingredients of risk characterizations for a building.

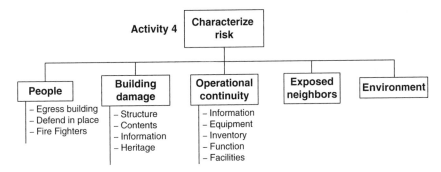

Figure 37.5 Organization chart: characterize risk.

PART THREE: DEVELOP A RISK MANAGEMENT PROGRAM

37.13   Structure a Risk Management Program

The knowledge and understanding that emerge from completing the first four parts of the risk management process in Figure 37.1 comprise a central element of almost all applications. If the purpose were to describe building performance from the perspective of code plans approval, the process would be completed because enough information is available for regulatory decisions involving acceptable performance. However, the acquired knowledge can serve a variety of other applications, such as corporate planning or in-station training for fire ground operations. A risk management program looks at the question, "How can I do better with my available resources?"

Activities 1–4 in Figure 37.1 provide a knowledge base from which to organize comprehensive services for a building or enterprise. For example, with this knowledge one may plan the general structure of a risk management program.

A risk management program is an integrated organizational strategy that establishes practices and policy with which to guide the fire risk management for the building or the enterprise. The risk management program can be developed in parallel with information from a "prevent EB" analysis and the development of emergency plans. The entire process can be incorporated within a decision analysis that packages alternative courses of action, their costs, and effectiveness.

The building's fire performance is the basis for a risk management program. This analysis can give a good perspective of the consequences of an unwanted fire anywhere in the building.

Sometimes a building's survival may rest only on the effectiveness of the "prevent EB" program. If this is the case, building management must be aware of the situation. A risk management program may give more attention to fire prevention, improving the building's fire defenses, or purchase more appropriate insurance. Regardless of the performance of the building after EB, knowledge of the quality of "prevent EB" activities becomes an important part of a comprehensive fire risk management program.

Emergency preparedness plans are another part of a fire risk management program. These plans consider actions for various fire conditions before, during, and after ignition and EB occur. Disaster plans may incorporate hazards beyond fire, such as hurricanes, earthquakes, floods, terrorism, and civil disobedience.

Figure 37.6 shows the factors of a comprehensive risk management program. The program is tailored to each specific enterprise, rather than forcing a fit into a standard format. The knowledge that is gained provides informed knowledge with which to decide an appropriate strategy.

37.14   Evaluate "Prevent EB"

The evaluation of fire prevention without spot protection involves two events. One is the traditional "prevent ignition" analysis. Here the goal is to avoid ignition by evaluating the ability of building operations to separate heat energy sources from readily ignitable fuels. Heat fluxes, piloted flames, and their time durations are a part of this

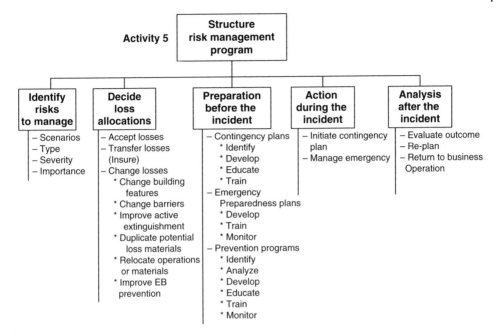

Figure 37.6 Organization chart: structure risk management program.

assessment. Qualitative analyses would use traditional hazard assessment procedures to get a sense of trouble locations for unwanted ignitions.

The second part of a "prevent EB" evaluation analyzes the fire growth from the first fragile flame to EB. This process considers the fire and early fire extinguishment. The ease and speed with which the fire grows during this period depend on the fuels and heat fluxes. Early fire extinguishment considers only occupants. Although occupant suppression may be possible for fire sizes greater than EB, from a building evaluation viewpoint, occupant extinguishment is limited to small fires. If special training or conditions warrant other definitions of occupant suppression capabilities, they can be incorporated.

Figure 37.7 shows elements of a "prevent EB" evaluation.

37.15 Evaluate Special Hazards Protection

Many industrial buildings and restaurants install special hazards equipment to control the potential for a fast fire. Most buildings that have automatic special hazards installations also have traditional sprinkler systems.

Figure 37.8 shows elements of a special hazards analysis. When these installations are present, their analysis is often an important part of understanding building performance.

37.16 Emergency Preparedness

The Latin phrase *praemontis praemuntis* "forewarned is forearmed" is appropriate for emergency planning. Although neither a specific date nor the exact scenario for a disaster may be known, contingency plans can be developed before an anticipated event.

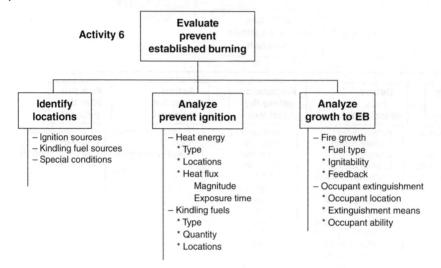

Figure 37.7 Organization chart: evaluate prevent established burning (EB).

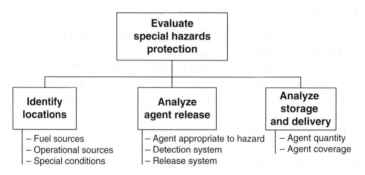

Figure 37.8 Organization chart: evaluate special hazards protection.

This provides enough time to prepare a thoughtful response strategy. This book focuses on fire emergencies. However, many contingency strategies are also appropriate for earthquakes, hurricanes, floods, terrorism, or civil disobedience.

An emergency plan is based on elements of understanding the problem and characterizing the risk, as outlined in Figures 37.2 and 37.5. An emergency plan provides a strategy to preserve the people, property, and operational continuity or mission of the building. The knowledge gained from the performance analyses provides a base to identify physical actions that can augment the building's capabilities.

The plan may be organized into actions that take place before, during, and after an event. Each action is tailored to the site-specific needs of the building. One may wish to ensure that physical and other (e.g., electronic) redundancies are separated adequately. For example, important electrical equipment often has redundancy installations. When electronically redundant systems such as those of a telephone center are separated by a

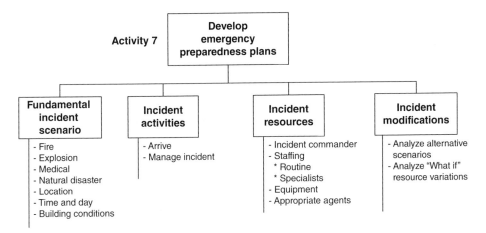

Figure 37.9 Organization chart: develop emergency preparedness plans.

few inches of air space, a fire in one will probably cause a fire in the adjacent gear. Fire is often not a consideration by other professionals in designing functional protection. Individuals sometimes tend to overlook events that are beyond the scope of their expertise. Therefore, a single risk manager must be responsible for coordinating the complete system into an integrated plan.

Figure 37.9 shows the elements of an emergency preparedness plan.

37.17 Decision Analysis

One goal of a decision analysis is to organize and compare feasible alternatives. The alternatives link advantages, disadvantages, costs, effectiveness, and potential consequences. A decision analysis helps to understand details of complex situations.

The development of feasible alternatives requires effort. It is a synergistic process that integrates performance and associated risk expectations, costs, and economic forecasting with a good understanding of the building operations and functions. Comparing a few feasible alternatives is integral to decision analysis.

One possible alternative is always the status quo. This involves no additional expenditures, and the risk characterization identifies potential losses that can occur. It is often desirable to compare alternatives with this performance. For example, the effectiveness of fire prevention procedures and emergency preparedness plans may be superimposed to provide a picture of the overall risk as well, to distinguish the role of each in the outcome. The cost and effectiveness of changes in alternative plans can be compared more easily with this base.

The mix among accepting potential losses, transferring risk by buying insurance, and changing the risk by installing fire defenses can be augmented with more effective "prevent EB" procedures and emergency preparedness plans. When one bases analysis on the understanding that evolves from performance analysis, the effect of proposed changes can be described relatively quickly and easily. Knowledge from a performance

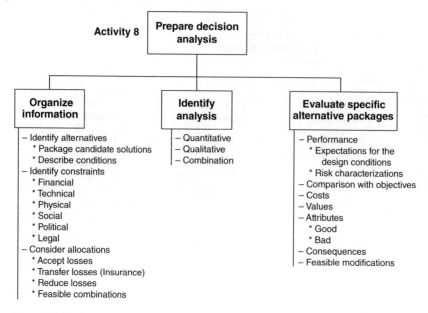

Figure 37.10 Organization chart: prepare decision analysis.

analysis enables one to recognize logical modifications for the building and their effects on performance.

Clear understanding of outcomes and costs is fundamental to a decision analysis. Implications of risk, loss aversion, cost, and affect on the operations and stability of the enterprise are a part of the process. Although engineers often think in terms of success and avoidance of failure, risk managers are more often influenced by loss considerations. While the performance framework focuses on success, business decisions are commonly expressed in terms of loss potential. The conversion is trivial, but understanding is often greatly improved.

Figure 37.10 shows the elements of a decision analysis.

37.18   Prepare the Presentation

An activity devoted to presenting a very few alternatives may be superfluous. However, its importance cannot be underestimated. The risk managers and fire safety engineers have invested much time on technical details and a thoughtful package of alternatives to decision-makers.

The decision analysis organized information and compared alternatives. The presentation must be clear enough for a decision-maker to recognize meaningful differences and implications of the alternatives. The time available to grasp the essence of the story is brief, and the story must be told succinctly and accurately. Technical justification is rarely important for a presentation to decision-makers. A clear, succinct picture of "What can we lose?," "How can we improve risks?," and "What are the costs?" enables one to make better decisions among competing demands.

37.19   Decision-Making

The final activity involves the decision to select the best solution for the corporation or building. Decision-makers are rarely the individuals who do the performance analysis or develop the risk management program. Documents that form the basis of a decision must portray the existing situation and alternatives clearly and succinctly. Management rarely has time to read reports and investigate technical reasoning for recommendations.

A decision may involve selection of a prepared alternative, or it may be a request to explore another solution not originally presented in the decision structure. Because performance is based on understanding from technical estimates, changes can often be evaluated relatively quickly. The final part of Figure 37.1 is making a decision.

38

Analytical Foundations

38.1 Historical Origins

The performance-based method described in this book evolved from a 1972 publication by Harold E. "Bud" Nelson called *Appendix D: The Interim Guide to Goals-Oriented Systems Approach to Building Firesafety*. The publication was written as a means to aid management in decision-making. Over the intervening years, the original concepts have been transformed into a logical, systematic procedure to evaluate fire performance for buildings.

In the early 1970s, Nelson was head of fire safety for the General Services Administration (GSA). In this position, he was responsible for the fire design of the Seattle Federal Building. That design was based on engineering principles rather than building code and traditional standards. The Seattle Federal Building became the "father" of performance-based design, because the thought process behind it was used to organize *Appendix D*.

The period 1960–1980 was active in the creation of new ways to examine complex systems. Fault tree analysis was originally developed in 1962 at Bell laboratories to analyze system reliability. The deductive logic to organize failure modes used Boolean mathematics. Although initial applications involved aerospace systems, its applicability became recognized for a wide variety of other complex systems. Forrester's *Systems Dynamics* was valuable for understanding time sequencing of decisions in complex systems. Computers and software were emerging to handle the mathematics.

Nelson had a remarkable ability for adapting concepts of other disciplines to fire safety applications. He recognized the strength of fault tree concepts in organizing complex systems. However, when it became evident that new designs were better understood in terms of success rather than failure, Nelson and Irwin Benjamin converted fault trees to success trees. This logic became the organization for the complex system of fire and buildings.

Initially, event logic diagrams enabled individuals to recognize relationships among major parts of the complete building–fire system. Nelson calculated fire interactions with fault tree theory. Unfortunately, with all of its attributes, the "systems approach," as it was then called, posed difficulty with practical applications. For example, Nelson's original paper used time-related descriptions whose logical transition beyond a room of origin was difficult and incorrect. Also, the logic of component evaluations was not developed. Nevertheless, even though there were many inaccuracies, this seminal paper

Fire Performance Analysis for Buildings, Second Edition. Robert W. Fitzgerald and Brian J. Meacham.
© 2017 John Wiley & Sons Ltd. Published 2017 by John Wiley & Sons Ltd.

became the foundation for modern performance-based design. "*Appendix D*," as Nelson's *Goals* paper was affectionately called, is the most significant single document in the evolution of fire safety because it was the first to organize the logic for describing the complete performance-based building–fire system.

In 1978, network diagrams were adapted to Nelson's concepts to track the analytical thought process. These networks were the first to establish a practical logic for multi-room analysis. The networks were structured to simplify calculations using probabilistic values. Event trees and fault trees were adapted into the network structure to integrate functional operations and produce the same mathematical results. Network diagram structures magnified technical capabilities that enhanced understanding and communication.

Nelson's procedures were, in reality, deterministic even though the descriptions used probabilistic concepts and terms. Today, it is easy to recognize the controversy that arose because of the probabilistic descriptors. Now, we would describe Nelson's quantification to represent a degree of belief for deterministic performance. Objections arose because frequentist or Bayesian methods were assumed, rather than degree-of-belief expressions of uncertainty for deterministic analyses.

Over the next 30 years, an analytical framework and quantification methods evolved by adapting techniques from fields such as systems analysis, risk analysis, psychology, probability and statistics, risk management, operations research, and construction project management. Care was taken to blend theoretical rigor and logic with functional behavior to evaluate performance applications. This chapter discusses selected fundamental concepts to give a sense of the techniques that are the foundation of the evolutionary development.

PART ONE: LOGIC DIAGRAMS AND NETWORKS

38.2   Event Trees

An event tree is a logic diagram that starts with a defined initial event and establishes a forward (inductive) logic that organizes sequences of future events that together describe all possible outcomes. Conditionality and sequencing are usually easy to recognize. The events show states of success or failure that can be represented by probabilistic measures.

In many ways, traditional event trees represent a scenario, analogous to a motion picture. For example, Figure 38.1(a) shows an event tree that describes the likelihood of fire extinguishment by an occupant for a building with no automatic sprinkler protection. The initiating event is ignition, and the event tree describes sequential events.

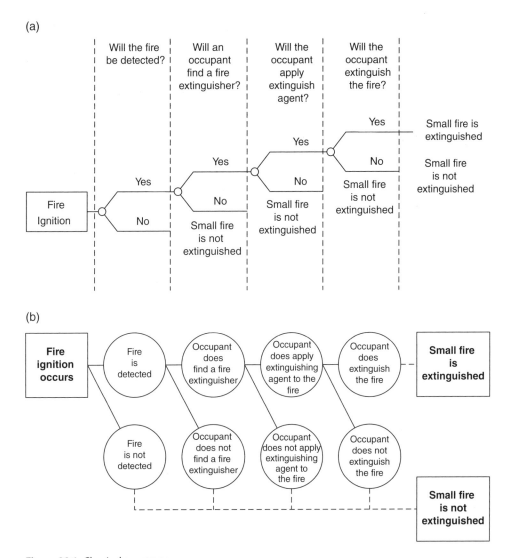

Figure 38.1 Classical event tree.

Figure 38.1(a) shows the traditional way of representing an event tree. Figure 38.1(b) illustrates the same thought process in network form. Any differences are inconsequential.

Event tree sequences give a sense of other variables, such as time, fire sizes, and intervention measures, although they are not explicitly stated. The main advantage of an event tree is its ability to describe a sequence (or thought process) of related events from the beginning to the end of a scenario. Time sequencing is usually implied, although it is rarely defined specifically. A major disadvantage of event trees is the inability to decompose events and establish a hierarchy of contributing factors. Also, it is cumbersome to incorporate concurrent paths for redundant activities.

Risk assessment event trees typically order dissimilar events to sequence a scenario, as illustrated in Figure 38.1. The scenarios consider important events that affect an outcome. The framework in this book adapts event tree concepts by structuring the forward logic of individual components explicitly with time. This enables one to examine sequential changes in the status of an event. Cells in the row of an Interactive Performance Information (IPI) chart incorporate time sequence status changes.

When probabilistic measures are used with the IPI chart, performance curves may be constructed. The IPI representation for these continuous value network curves can also structure cause–consequence analyses and failure modes and effects studies.

38.3   Fault and Success Trees

The fault tree is an important, useful tool that gives an insight into the interactions of system performance. A fault tree supplements event tree weaknesses by identifying causes that can contribute to an event failure. The tree organizes contributing events into a logical framework that enables one to deductively trace the roots of potential failures of an event.

Fault trees and their complement, success trees, show important factors and their logical relationships on a single diagram. Chapter 3 (Sections 3.4 and 3.5) discusses the construction and attributes of these event logic diagrams. Deductive logic relationships are easier to show with a fault tree. Applications for design and analysis are easier to structure with a success tree.

The success tree of Figure 38.2 shows the first several levels of events that limit a fire to a space. The events are described with the symbolic notation that has been used throughout the book. We retain the event of fire self-termination $[P(I)]$ for theoretical completeness.

Events below an OR gate must be inclusive. That is, they must be described in a way that includes all possible outcomes. Events below an AND gate need not be inclusive, although we attempt to be complete in all descriptions.

Events shown below the OR gates of Figure 38.2 describe the only (i.e. all) possible ways for a fire to go out in a room. That is, a fire can self-terminate (I), be actively suppressed by an automatic sprinkler system (A), or be extinguished by the local fire department (M).

Evaluations of A and M are dependent on fires that do not self-terminate ($\bar{I}$).

Technically, the symbols should be shown as $\left(A|\bar{I}\right)$ and $\left(M|\bar{I}\right)$, i.e. automatic sprinkler suppression, given a fire that does not self-terminate, and manual suppression, given a fire that does not self-terminate. However, a fire that does not

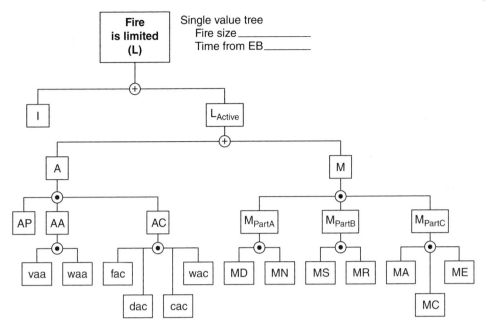

Figure 38.2 Success tree.

self-terminate ($\overline{I}$) is the diagnostic fire. Because components A and M use this condition in the evaluation, the conditionality is understood.

38.4 Fault and Success Tree Calculations

When probabilistic values are assigned to events in a fault tree or a success tree, one can calculate the outcomes. The hierarchical construction requires that a separate calculation is needed at each gate.

Logical AND and OR gates consider three conditions:

- The union of two or more mutually exclusive events, as illustrated by the Venn diagram of Figure 38.3(a).
- The union of two or more independent events that overlap, as illustrated in Figure 38.3(b).
- The intersection of two or more events, as illustrated in Figure 38.3(c).

A union is a combination of events that is represented by an OR gate. When the events are mutually exclusive, as shown by Figure 38.3(a), their union is calculated as:

$$P(E) = P(A) + P(B) \tag{38.1}$$

The intersection of two independent, mutually exclusive events is shown in Figure 38.3(b) as the shaded area of the circle overlap. This represents an AND gate whose value may be calculated as:

$$P(E) = P(A)P(B) \tag{38.2}$$

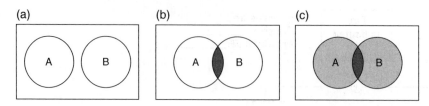

Figure 38.3 Venn diagrams.

The union of independent, mutually exclusive events in Figure 38.3(c) is the area enclosed by the circles. This area represents the probability of either OR both events. Because the shaded area has been included twice and must be deducted, the equation for this union is:

$$P(E) = P(A) + P(B) - P(A)P(B) \tag{38.3}$$

These relationships may be used with any of the event logic diagrams to calculate the probability of a higher-level event immediately above a gate. For example, assume in Figure 38.2 that $P(A)$ is the probability that the sprinkler system will extinguish a fire and $P(M)$ is the probability that the fire department will extinguish the fire. If each of these probabilities were evaluated independently, the probability of active fire extinguishment (L_{Active}) may be calculated as:

$$P(L_{Active}) = P(A) + P(M) - P(A)P(M) \tag{38.4}$$

The complexity increases with the number of events. For example, consider the three ways of limiting a fire as self-termination, $P(I)$, OR by automatic sprinklers, $P(A)$, OR by the local fire department, $P(M)$. The limit $P(L)$ may be calculated as:

$$\begin{aligned} P(L) = {} & P(I) + P(A) + P(M) - P(I)P(A) - P(I)P(M) \\ & - P(A)P(M) + P(I)P(A)P(M) \end{aligned} \tag{38.5}$$

This equation is cumbersome and may be expressed in alternate form as

$$P(L) = 1 - \left[1 - P(I)\right]\left[1 - P(A)\right]\left[1 - P(M)\right] \tag{38.6}$$

When fault (or success) trees have been constructed and probabilistic values assigned to the events, one may calculate the outcome of any event above a gate. After the probabilistic values for the lowest event levels have been estimated, the hierarchical nature of the event logic trees enables one to calculate intermediate values leading to terminal values.

38.5 Fault and Success Trees Beyond the Room of Origin

The success tree of Figure 38.2 is valid to evaluate the limit of fire in a space. Unfortunately, building fires involve a sequential propagation path of space–barrier–space–barrier… Beyond the room of origin, fire can spread in a three-dimensional array of paths that involves simultaneous involvement.

The spread is dynamic for each path and depends on the type of barrier failure, the room's interior design, automatic suppression, and local fire department operations. And some paths double back on others to complicate sequencing order. In addition, logic trees depict static relationships. The fire and sequencing of components are dynamic.

It is no wonder that Nelson's method described in *Appendix D* became unworkable beyond the room of origin. Although one could construct a fault or success tree for multiple rooms and multiple path conditions, the exercise becomes mechanical and obscures clear understanding. Fault/success trees have practical limitations beyond clearly defined conditions.

38.6   Network Organization

The complications of fault/success tree analysis for dynamic interactions involving multiple paths of fire propagation can be avoided with network analysis and modular descriptors. The networks of this book evolved from fault/success tree limitations for rooms beyond the room of origin. Evaluation networks were structured to incorporate the strengths of fault/success trees and to produce the same mathematical results. Thus, the networks in this book have a direct relationship with fault/success tree construction.

The most useful attribute of networks is the ability to integrate a thought process with event logic to produce an engineering analysis for a situation. When the network has been rigorously organized with logical relationships, the mathematics becomes compatible with performance understanding. Networks provide a good insight into analysis and performance. The IPI chart integrates the network organization of event trees and success trees.

38.7   Network Calculations

Single value networks (SVNs) have been organized around fault/success tree construction. Chapter 3 (Sections 3.6–3.8) describes the process for relating event logic diagrams to SVNs.

The network organization incorporates the same logical gates as event logic trees. Events along a continuous path represent AND gates, whereas events that terminate represent OR gates. This process enables one to recognize event dependency easier. Mutual exclusivity is naturally incorporated, although it should be recognized.

Traditional event logic trees and network diagrams use the same probabilistic measures. Although calculation procedures appear different, the correspondence may be shown to produce the same results.

Single value networks express all events in binary form, i.e. an event plus its complement must add to 1.0. Thus:

$$P(E) + P(\bar{E}) = 1.0 \tag{38.7}$$

where $P(E)$ is the probability that an event will occur and $P(\bar{E})$ is its complement, i.e. the probability the event will not occur.

Figure 38.4 shows a network for the three of events, I, A, and M that constitute the universe of ways that fire can be limited (L). When the probability is established for each event, its complement is known. Thus, when $P(\overline{I})$ is established, its complement, $P(\overline{I})=1-P(I)$. Similarly, $P(\overline{A})=1-P(A)$ and $P(\overline{M})=1-P(M)$.

The probability of Event $\overline{L}$ represents an AND gate and is:

$$P(\overline{L})=\left[(1-P(I))(1-P(A))(1-P(M))\right] \tag{38.8}$$

The probability of Event (L) represents the OR gate and is the sum of the contributory terminal events:

$$\begin{aligned}P(L)=&P(I)+P(A)+P(M)-P(I)P(A)-P(I)P(M)\\&-P(A)P(M)+P(I)P(A)P(M)\end{aligned} \tag{38.9}$$

These results become the same as the algebraic expression in equation (38.6). Combining equations (38.8) and (39.9) becomes $P(\overline{L})+P(L)=1.0$.

Thus, network calculations simplify the mathematics of traditional event logic theory. The mathematics of network calculations is the same regardless of the order in which the events is listed. When one evaluates independent events, the order in which they are shown on the network is not important. For example, the same mathematical result occurs when A and M in Figure 38.4 are reversed.

One must not confuse mathematical calculations with performance evaluations. Although networks structure calculations naturally, their most important attribute is the ability to track the thought process for functional behavior. Independent, dependent, conditional, and exclusive events are more easily recognized during an evaluation.

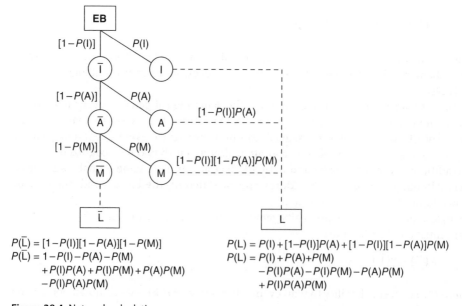

Figure 38.4 Network calculations.

38.8   Sequential Path Analysis

Fire moves through a building by propagating from a room of origin – through a barrier – into the next room, etc. Figure 38.5(a) shows a sequence of three rooms. Figure 38.5(b) shows a space–barrier network that tracks the flame movement path. This SVN is valid for event evaluations at each instant in time.

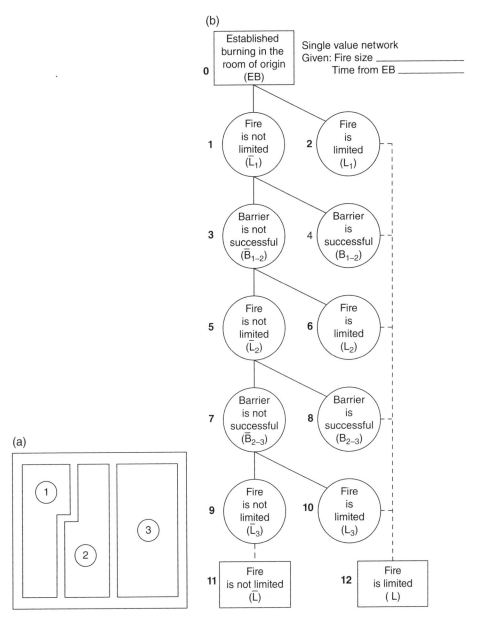

Figure 38.5 Space–barrier description.

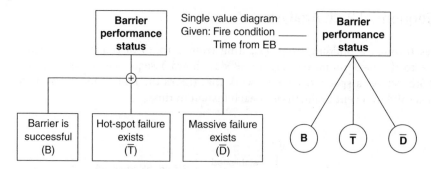

Figure 38.6 Barrier performance analysis. EB, established burning.

At any instant, a barrier can exhibit only one of three mutually exclusive conditions. Figure 38.6(a) shows the event logic diagram for the status of a barrier at any time and Figure 38.6(b) shows the equivalent network.

The size, type, and location of the barrier failure have an influence on the outcomes for I, A, and M in the adjacent room. The status of a barrier's performance changes as the barrier deteriorates due to the heat energy of the fire in the adjacent room. Thus, time is important to a barrier's performance status.

38.9 Rooms Beyond the Room of Origin

Figure 38.7 shows a decomposition of the components in the sequence of rooms of Figure 38.5 combined with barrier status conditions of Figure 38.6. Although the number of events may seem complicated, the organization actually provides a basis to understand and simplify multiple room evaluations.

38.10 Modular Analysis

The basic structure of a performance analysis is the barrier–space module. Essentially, we have a room of origin and all of the other modules that define the building's architecture. Although all parts of the complete fire safety system can be associated with barrier–space modules, the flame-heat component dominates performance.

Figure 38.8(a) shows the basic network of all flame-heat events in module. In theory, because the network is SVN, the I, A, M, and L values must be evaluated for each module and condition at each instant in time. However, practical simplifications to make analysis more tractable.

$\bar{D}$ Failure

In a module, a $\bar{D}$ failure causes a massive influx of fire gases into the adjacent room. The fire will not self-terminate [i.e. $P(I) = 0.0$]. Also, beyond the room of origin, values for $P(A)$ are normally taken to be $P(A) = 0.0$. The reasoning is based on two factors. First, if a sprinkler system in the room of origin operates, the fire is usually suppressed. A sprinkler system in a room of origin that does not suppress the fire must have significant deficiencies that would reduce performance significantly.

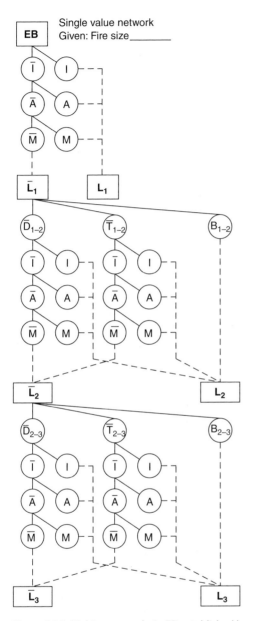

Figure 38.7 Multi-room analysis. EB, established burning.

Second, a $\bar{D}$ failure allows a fire to move very fast, and sprinkler fusing is delayed. When many sprinklers fuse, the density and water flow continuity usually become overwhelmed. Therefore, we simplify estimates beyond the room of origin by saying that sprinkler effectiveness declines significantly and may be discounted.

A $\bar{D}$ failure causes fire to spread in the adjacent room so rapidly that manual suppression is successful only when hose lines are charged and in operation at the time of the massive barrier failure. Because the likelihood of stopping the fire for this condition is very low, we assume that a $\bar{D}$ failure results in $P(M) = 0.0$ for an adjacent room.

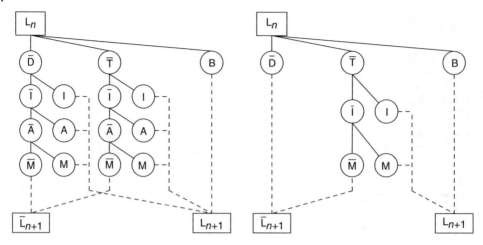

Figure 38.8 Barrier–space modular network.

$\bar{\text{T}}$ Failure

A $\bar{\text{T}}$ failure causes ignition and a "normal" fire growth, although it is faster than it would be in a room of origin. A diagnostic fire is based on $P(\text{I}) = 0.0$ and is already incorporated into the analysis. However, if room interior design conditions indicate a difficult fire growth, values of $P(\text{I})$ may be estimated. Therefore, the usual situation assumes $P(\text{I}) = 0.0$ for Figure 38.8(b). A value for $P(\text{I})$ may be estimated if conditions warrant the need.

A fire department is often successful in stopping slower-moving fires caused by a $\bar{\text{T}}$ failure. Therefore, one may examine the fire ground operations for the room to determine the likelihood of fire department extinguishment for this condition. Thus, $P(\text{M})$ may have a value.

These considerations allow one to simplify the complete array of events of Figure 38.8(a) into the less complex form of Figure 38.8(b). If special conditions are present and these assumptions are not appropriate, one can always revisit Figure 38.8(a) and estimate the appropriate events.

The "top-down" analysis described in Chapter 8 is based on the simplified module of Figure 38.8(b). Manual extinguishment evaluations are also based on this module. Thus, this network becomes the foundation for organizing a systematic analysis of performance.

These modular networks use a consistent logic. Therefore, one can use probabilistic values to calculate the limit of fire growth for any number of rooms in any configuration beyond the room of origin. The descriptions of the earlier parts of this book are based on this barrier–space modular network. Section 3A of the IPI charts can order time durations and fire propagation sequencing.

38.11 Closure

Harold E. Nelson established the foundation that created a way to understand the complex system of fire and buildings. Many individuals have contributed to its evolution in a wide variety of ways. However, this method would not exist if Bud Nelson had not had the creativity to organize his thoughts into a logical framework. Nelson said that *Appendix D* was dictated during one weekend. It took 40 years to organize Nelson's vision into a practical, rigorous procedure that enables one to integrate component behavior with holistic building performance. In many ways, Nelson's concepts organized fire safety analysis in a similar manner to that of free body analysis for structural mechanics over a century ago.

PART TWO: PROBABILITY

38.12   Meanings of Probability

Probability has had a dual meaning almost from the time it was first described. There are two extreme definitions (perhaps we can call them philosophies), as well as many positions between these extremes. One definition, called classical, objective, or frequentist, argues that probability is a measure of the frequency of occurrence of one event out of a possible set of the same events. The other approach is called personal or subjective. This procedure examines all of the evidence for a situation or problem involving uncertainty to establish a subjective judgment or belief regarding the likelihood of an occurrence.

The objective position states that probability is valid only for events that can be repeated under essentially the same conditions. This idea may be extended with limit theorems to include relative frequencies of long-run "experiments." For example, if a sprinkler manufacturer or a detector manufacturer wishes to determine the reliability of the company's product, a sample of the devices is selected from among those that were manufactured. These devices are tested and the number of failures is recorded. The probability of a bad device reaching the construction site would be very nearly the ratio of the number of bad devices out of the total sample that was selected for testing.

The objectivist likes to determine outcomes only from experiments or situations that are repeated under the same conditions. Games of chance and selecting balls out of urns are commonly used to establish concepts. To obtain the expected value of an event, statistical data are needed. The probability measure is based on the historical record from a set of situations (population) that are essentially similar. These data enable the variance, confidence limits, and other useful information to be determined.

An objective probability is viewed to be a characteristic of an identifiable physical process and is a property of the event. This definition of probability is commonly adopted in the physical and biological sciences and it has wide and successful use in a variety of industries such as medicine, pharmaceuticals, life insurance, and property insurance.

Another definition of probability is called personal or subjective. This interpretation is more commonly used for unique situations where experimental repetition under the same conditions is impossible or unfeasible. For example, the probability that the local fire department will arrive at a site-specific building before the room of origin has reached flashover, or that a detection system installation will actuate before the fire reaches 100 kW is more appropriate for personal assessments. These estimates use subjective probability to describe an expected performance.

The subjectivist can apply probability to all of the problems that an objectivist considers, and many more besides. A subjectivist views probability as a measure of the degree of belief that an individual has in a judgment or prediction. Here probability is interpreted as an intellectual process, rather than as a physical property in the objective definition. Some mathematicians view all probability as subjective, and the type and use of actuarial information distinguish the relative position of the application on the scale of interpretations.

Sometimes this definition of subjective probability is confused with Bayesian theory that updates initial subjective judgments with experimental data. Bayesian probabilities start with an initial subjective judgment of an outcome based on available information. Then those subjective values are upgraded with experimental observations to improve the quality of the subjective assessment. In theory, if enough data are available, Bayesian and classical methods will converge.

38.13   Fire Safety Applications

Which interpretation is appropriate for fire safety applications? The answer depends on the application. If one is dealing with a class of buildings (a population), the objective view is likely to be better. For example, if one wishes to compare occupant deaths, property damage, or other characteristics between sprinklered and non-sprinklered buildings, objective probability is the most appropriate. Insurance decision-making to set general policy guidelines or premiums is also more suited to objective procedures.

On the other hand, if one were interested in understanding the fire performance of a site-specific building to develop a site-specific risk management program or to develop documentation for a code equivalency proposal, subjective (degree of belief) probability is more appropriate. Similarly, a highly protected risk (HPR) insurance inspection report would be appropriate because subjective judgment is already an integral part of those reports. In cases where one needs documentation of rationale for performance selections, the subjective calibration of estimates is better.

38.14   Degree of Belief

Fire performance analyses are deterministic. That is, one views actions in terms of cause and effect or load and resistance. When fire safety engineering reaches early maturity, deterministic calculations will be available for all component evaluations. Even today the analytical framework has structured the components to a level where one can "fill the gaps" with deterministic evaluations.

Although analytical procedures are deterministic, the understanding and communication can be enriched when one incorporates probabilistic methods. The state of the art of fire science and engineering is evolving. Dynamic fire conditions and phased interaction of fire defenses make some problems difficult to describe mathematically. In addition, differences in materials and conditions can produce widely variable behavior. This creates a number of "if …, then …" situations.

Probabilistic degree-of-belief estimates expand understanding and evaluation confidence within the windows of uncertainty that exist for all fire safety components. Probabilistic estimates in the IPI chart can be converted into performance curves for better personal understanding and visual communication with others.

We characterize degree-of-belief probability estimates as encoding a state of knowledge. The assessments are based on acquired knowledge such as calculated values, computer models, physical relationships, observational facts, experimental information, failure analyses, and personal experience. Assessments use the full spectrum of information that is available and seems relevant to the problem. The existence and nature of uncertainty are clearly understood.

Degree-of-belief probability assessments describe expected performance for a site-specific condition. For example, success probabilities of 0.4 and 0.9 should not be interpreted literally as statistical outcomes of 40 times and 90 times out of 100. Rather, they should be interpreted as a calibration of performance. Although it is possible that both outcomes can occur, the event with probability 0.9 is substantially more likely to occur than the one with probability 0.4.

Probabilistic values are based on a deterministic evaluation of the situation. For example, assume that an individual evaluates a building and estimates the sprinkler system will be able to control a fire within a floor area of 2 m² (22 ft²) as $P(AC) = 0.80$. This means that, based on the information available for the evaluation, the likelihood that the sprinkler system will control the fire before it extends beyond 2 m² (22 ft²) is estimated as 80%. The numerical value expresses a degree of belief for the site-specific analysis.

38.15   Mathematics of Probability

There is no disagreement between the frequentist and the subjectivist over the mathematical foundations of probability. Differences of opinion are largely philosophical. Each definition of probability has philosophical extremists at the ends of the spectrum of opinion. Extremists often may not concede that the opposite view has validity. However, in practical evaluations for engineering disciplines, most individuals adopt the probability view that enables the best decision to be made for the specific problem under consideration.

One of the attributes of probability mathematics is that it does not consider how the numbers are determined or what they mean. Probability calculus only identifies how to use the numbers in a consistent manner. The analytical framework (i.e., network diagrams) has been carefully constructed to conform to mathematical principles. Consequently, this framework can be used regardless of the interpretation of probability.

Now, let's be careful. Just because the analytical framework enables the mathematics to be consistent, it does not relieve the individual of understanding what the numbers mean. After all, the goal of a performance evaluation is to understand the building and its behavior in a fire. Numbers are a means to an end, not the end itself. Numbers and graphs are used to assist understanding, compare alternatives, and communicate with others.

38.16   Assessment Quality

Questions may arise relating to accuracy of the numbers. In particular, when subjective estimates are used, different people may select different probabilistic values for events. One may ask, "Is this a problem?," "Who is correct?," or "Why isn't it better to have a procedure where everyone gets the same number?"

The subjective interpretation of probability accepts that different individuals may differ, even with the same evidence. This may lead to a perception that subjective probabilities are just guesses and not to be used in serious or rigorous applications. This perception is not accurate. When used appropriately, subjective judgment expressed as a probabilistic measure can be a valuable contribution to analytical rigor. Subjective probability is a valid representation of knowledge, and its strengths can be exploited to great advantage.

Degree-of-belief assessments do not view a numerical value as right or wrong. Rather, it is the quality of the assessment that becomes significant. That assessment quality may be viewed as good or bad. The quality of an estimate depends on the amount of information that is used, the skill in translating that knowledge into performance estimates, and the constraints. The time available to make an estimate is also a constraint.

The quality of an assessment requires that the event and its qualifying conditions must be identified clearly. The network structure enables conditions to be described more easily. Nevertheless, humans can have preconceived biases that affect judgment. Some of the more common influences of bias are as follows:

- Overconfidence in scientific knowledge and computer modeling is often displayed.
- Human error and anticipated human activities or capabilities are often inappropriately portrayed.
- Rare events influence judgment, particularly if they have occurred recently.
- Preconceived assumptions about the effect or value of codes and standards on performance sway opinions.
- Fear of uncertainty can produce excessive conservatism or over-adjustment of values.

Our goal is to help an individual to understand the fire performance of a site-specific building. This understanding contributes to better technical documentation of the rationale for the performance expectations.

Occasionally, differences in performance expectations for an event may arise. The framework structure enables individuals to focus on relatively narrow issues. The boundaries of event evaluations help to focus information sources and limit extraneous conditions. Discussion usually resolves the issue. Communication becomes enhanced because of the deterministic base for performance evaluations. While numerical values may differ (by relatively minor amounts in most cases), the identification of building problems and their relative importance rarely differs among individuals.

PART THREE: THE ROLE OF JUDGMENT

38.17   Introduction

Judgment is a part of all engineering evaluations. The amount of judgment needed and its role depend on the maturity of the discipline. When a discipline is in its infancy, judgment has a dominant influence in performance evaluations. As the discipline matures, judgment is gradually replaced by quantification methods.

The role of engineering is to make decisions. The decisions may relate to design, analysis, or planning. Often the decisions are incorporated into reports that allow others to understand issues and alternatives and make a final selection. The process involves understanding the problem, organizing information, and developing a logical framework for making decisions. Calculations are only tools to assist understanding. The gap between technical knowledge and performance must be filled with what has been called engineering judgment.

38.18   Building Decisions

Many practical questions in fire safety deal with the adequacy of decisions. For example, the density of the sprinkler discharge is x liters/min (lpm)/m^2 [gallons/min (gpm)/ft^2]; type y photoelectric smoke detectors will be installed in the following locations; fire department connections to the standpipes are located z m (ft) to the west of the main entrance; or the local fire department will respond with two pumpers, one ladder, a chief, and a staff of 12. Someone in the design and construction professions or in the community government makes those decisions. Those decisions may have a significant influence on the building fire performance, or they may be relatively unimportant to an outcome. A performance evaluation assesses the significance of any detail with respect to fire performance.

Sometimes decisions that influence fire performance are based on interpretations of prescriptive code or standard requirements. Sometimes they are well formulated by individuals with substantial knowledge and experience in the design, installation, and operation of the particular fire defense under consideration. Sometimes individuals with no knowledge or training make decisions in an arbitrary and uninformed manner.

38.19   Judgment in Engineering

Engineers work in a very imperfect world where they must make timely decisions for projects that must be built today. Koen defines an engineering method as "the strategy for causing the best change in a poorly understood or uncertain situation within the available resources." Available resources include areas such as knowledge, information, equipment and procedures, confidence, money, and time.

The base of understanding for any type of engineering performance uses a mix of observation, technology, and judgment to arrive at decisions. When technology is immature, judgment is a dominant influence. As technical knowledge improves, judgment is gradually supplanted by analytical capabilities that become proven with experience. However, judgment is never eliminated. Judgmental components become integrated with the evolving technological information base. The history of engineering provides many examples illustrating the transformation of judgment and technology.

38.20 Language and Culture

The architect, the different consulting engineers, contractors, code officials, and the fire service all speak different technical languages. Within each group, the technical language, the jargon, and shades of meaning are clear. Communication within any single group is easy and seldom misinterpreted. However, technical communication barriers between different professions can be substantial. Except for a few verbal and conceptual descriptors that establish an interface to conduct necessary business, a common language that ensures understanding is not present. Consequently, a structural engineer and an electrical engineer have as much difficulty in communicating the technical details of their respective disciplines as do an architect and a fire service officer.

In addition to language differences between professions, the cultures also are distinct. In many ways, the language and culture between professions involved in the design, construction, operation, and safety of a building are as distinct as the language and cultures of different nationalities. Even when technical language communication can be established, a cultural distinction and associated acceptable behavior or experiences may inhibit the easy flow of ideas and concepts between groups.

This book is about fire performance evaluations. The goal is to help fire safety professionals understand performance and communicate that understanding to others. Engineers normally think and communicate with pictures. The visual thinking of fire safety performance curves provides a universal language for different professional groups. These curves can communicate expected behavior for alternatives efficiently and clearly.

Performance curves could be constructed for the other engineering disciplines. This is unnecessary because technical experience and understanding are sufficient to make decisions and communicate behavior. Fire safety engineers do not yet have the technical base or a clear, codified description of loading conditions that are characteristic of a mature engineering discipline.

The prescriptive code is the perceived basis for fire performance. All of the building industry professions make decisions and incorporate details that affect fire performance. Some decisions or details can be insignificant to the overall fire performance. Others may have major implications. If fire performance is of interest for a building, its evaluation must be based on understanding that specific building's performance. One must be able to communicate with other disciplines regarding their decisions on fire performance.

38.21 Uncertainty and Performance

Concepts of classical safety analysis may be adapted to uncertainty information that is associated with a performance evaluation. Variability and uncertainty may be grouped into three categories:

- Stability of conditions
- Variability in physical parameters
- Professional practices.

Conditional stability is associated with the relative stability of the situation being evaluated. Stability factors reflect the influence of common arrangement changes to predicted performance. For example, how will shifting the position of the fuel items or barrier openings affect the fire growth potential of the room? How will the repositioning of movable partitions influence fire propagation or smoke movement? Conceptually, the conditional stability factor relates geometrical and dimensional variability to evaluation scenarios.

Physical parameter variability incorporates two aspects. The first involves the variability associated with the physical performance of different materials. For example, the burning characteristics of a cotton batting upholstered chair will be different from those of a functionally similar chair constructed with foam plastics. Similarly, operating and extinguishing characteristics of a common sprinkler will differ from those of an early suppression fast response sprinkler. In other words, all chairs and sprinklers are not created equal. The variability of species types introduces a form of uncertainty into performance evaluations.

Consistency of physical behavior is another type of physical variability. For example, the thermal conductivity of many materials varies with temperature and testing procedures. Similarly, the coefficient of linear expansion for materials may change with temperature. Uncertainty arises in the calculation of expected performance with many parameters.

A third category of uncertainty relates to the formulation and use of the deterministic relationships and calculation procedures used in professional practice. For example, what are the limits of validity involving flashover correlations? When does the uncertainty begin to affect the outcomes significantly? Computer programs have default values and limits of applicability incorporated into their codes. The professional practice category identifies and addresses the uncertainty and inaccuracies associated with calculation procedures.

The discussion up to this point has not included the use of statistics. Statistics and classical probabilistic analysis have a role in the complete fire safety system. For certain components, that role can be very important. However, one must be aware that a building analysis evaluates a singular, case-specific installation. Distributional information relating to a class of situations may or may not be appropriate. The goal is to understand the details that influence the performance of specific buildings so that the impact of specific alternatives can be understood and communicated clearly. Biases inherent in judgmental decisions are also present in statistical studies. Unfortunately, this is not always evident.

38.22   Summary

Professional time is a valuable commodity. The professional must conserve that commodity by reducing the time necessary to arrive at better decisions. Tailoring the analysis to needs may create or improve decision support systems to enable information to be organized, retrieved, and used more efficiently. Thus, the professional must select the type of evaluation to meet the needs of the application.

Judgment is an important part of engineering applications.

Variability evaluations use both qualitative and quantitative analyses to give a better breadth of understanding. Variability analyses may study sensitivity to expected variations in depth. Or, they may be used to give a better macro picture for "what if" alternatives to extend the knowledge base for decision-making.

The goal of an evaluation is to understand fire performance. Evaluations do not average or weight different conditions. The purpose is to tell a story. Consequently, specific scenarios are selected to represent conditions that are most appropriate to describe the risk characterizations of the building.

Judgment is an integral part of engineering practice. The role of judgment changes as technical knowledge evolves. Calculation procedures are not equally developed for all fire safety components. Even though technology may be inadequate to address the necessary performance functions, an engineering evaluation may be completed using the information that is available and reasonable engineering estimates of the loading and resistance behavior.

Judgment bridges the gap between technical capability and fire performance evaluation needs. Subjective probability is a way to introduce rigor into deterministic performance evaluations.

Evaluations and associated performance curves provide an opportunity to compare functional behavior. The greatest benefit of subjective probability evaluations is that comparisons of expected performance can be made more accurately and consistently with substantially less cost. The framework enables one to understand fire performance and portray risk characterizations.

The quality of an assessment requires that events and their qualifying conditions be identified carefully and clearly. One must understand the knowledge base and construct an interface to access it effectively.

A first-time analysis for a scenario can be time-consuming because it is an initial learning exposure. As experience is gained, comparisons are studied, methods are defined and understood, and decision-support structures are constructed, the time for any level of analysis will be reduced and results will become more consistent.

Classical safety analysis considers three types of uncertainty. Conditional stability addresses the influence of different materials or arrangements on outcomes. Changing substantive factors that influence scenario conditions can affect performance. An important part of professional practice relates to recognition of reasonable, appropriate scenarios on which to base evaluations. When it becomes necessary to recognize the "better" or "worse" conditions that may occur, variability analyses using different values can bracket performance.

Professional practice uncertainty recognizes that calculated outcomes approximate actual behavior. All fire-related equations and computer programs have limits of applicability. The professional engineer must ensure that the models are appropriate to the scenario and that results are interpreted to give reasonable performance measures. Certainly, fire models handle complex relationships that cannot be managed with other methods. However, qualitative methods are an important complement to the quantitative procedures. When deficient quantitative models must be used because other methods are not available, qualitative adjustments based on theory and technical judgment can help to reduce inaccuracies.

Physical parameter variability is often associated with mathematical models. Calculated outcomes must be adjusted when parameter variance affects performance. Professional judgment determines if variations are significant.

The way in which we deal with uncertainty has a major influence on understanding performance and the ability to communicate with others. This chapter describes techniques for managing uncertainty that are structured on information variation and its organization within the IPI chart.

Appendix A

Organizational Structure

A.1   The Organizational Framework

Building fire evaluations require dynamic analysis because the fire constantly changes and different components phase in and out of action. Concepts of systems analysis relate a building's performance for all components and events from ignition to extinguishment.

We use networks to relate interactions. The analytical framework integrates event trees to identify status changes that occur over time with success trees that decompose each component at any instant of time. The Interactive Performance Information (IPI) chart combines the continuous value event networks with single value success networks. Thus, the IPI chart provides an alternative way to depict network events and relationships.

Appendix A compiles network interactions and the associated IPI charts of the analytical framework. Levels of detail can be adjusted to meet evaluation and communication needs.

A.2   Basic Organization

Figure A.1(a) shows the major components of a performance analysis. Figure A.1(b) organizes the events in an IPI chart. The basic order and arrangement of IPI charts remain constant.

A.3   The Composite Fire

The flame-heat analysis provides core knowledge for evaluating fire performance of buildings. The composite fire (L) combines the diagnostic fire scenario with extinguishment actions to develop expectations of the building fire performance. This enables one to determine the building's response to smoke and structural behavior and also to characterize fire risks.

Figure A.2 shows network interrelations for the composite fire. The limit (L) identifies composite fire over the time until its extinguishment (i.e. limit). The subscripts indicate time durations. Subscripts can be changed to describe associated measures such as floor area, room modules, and smoke zones, or fire size measures.

Fire Performance Analysis for Buildings, Second Edition. Robert W. Fitzgerald and Brian J. Meacham.
© 2017 John Wiley & Sons Ltd. Published 2017 by John Wiley & Sons Ltd.

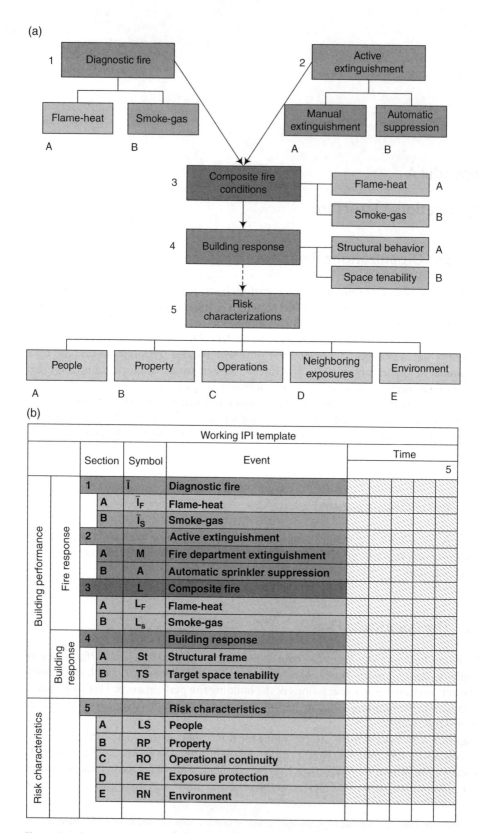

Figure A.1 System organization.

Figure A.2 Limit (L) analysis networks.

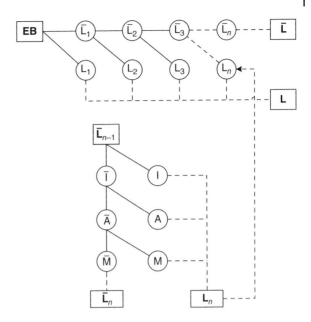

The decomposition of I (the fire goes out itself), A (the fire is controlled by the automatic sprinkler system), and M (the fire is manually extinguished by the local fire department) are shown in subsequent sections.

The term $\bar{I}$ defines the diagnostic fire. Terms A and M in Figure A.2 are conditional on the diagnostic fire not self-terminating ($\bar{I}$). Although the proper nomenclature should be (A| $\bar{I}$) and (M| $\bar{I}$) we use the symbols A and M with the conditionality being understood.

Row 3A of the IPI chart is the limit of the fire (L_F). It is the combination of I, A, and M for the fire sizes until extinguishment. For analytical purposes, we assume the fire does not self-terminate (i.e. I = 0) and use only the active extinguishment of A and M. These are shown in Sections 1 and 2 of Figure A.2. Normally, we assume I = 0 and use $\bar{I}$ as the diagnostic fire.

A.4 The Diagnostic Fire ($\bar{I}$)

Figure A.3(a) shows the network to decompose the diagnostic fire. The term ST indicates self-termination. The subscripts describe sub-areas within a building unit. Although this detail is not needed in building analyses, we show the networks for theoretical completeness for the universe of conditions.

Figure A.3(b) shows an IPI chart for the diagnostic fire. The flame-heat component examines fire propagation through a successive cluster of rooms. The designations may be called "rooms" or "modules." Here we will designate the units of fire propagation as modules to maintain the "barrier–space" modular organization. The limit (L) is associated with the flame-heat movement.

The smoke-gas component describes smoke movement through the building. We designate the building units as "zones" to distinguish them from flame-heat movement and enable one to describe contiguous spaces used for smoke analysis more conveniently.

(a)

(b)

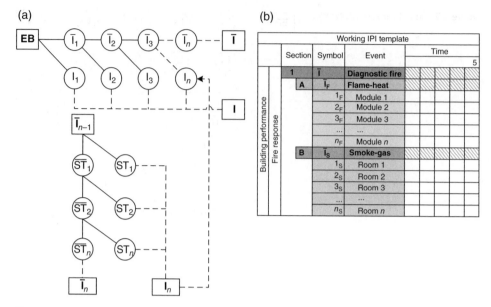

Figure A.3 Diagnostic fire.

The IPI cells can be used for many different purposes. For example, they can use shades of color to indicate fire free, pre-flashover, or post-flashover conditions. Quantitative values, qualitative descriptors, or probabilistic calibration estimates can also be noted in the cells. Additionally, digital pictures, graphs, or other information of value to understanding fire evaluations can be incorporated into IPI spreadsheet cells.

Smoke contamination $\bar{L}_s$ is dependent on the fire continuing. The cells may show contamination or visibility at each increment of time.

A.5   Fire Department Manual Extinguishment

Fire department manual extinguishment involves three major components. The first ($M_{Part\,A}$) examines the building's role in detection and fire department notification. The second ($M_{Part\,B}$) identifies the community response to the emergency. The third ($M_{Part\,C}$) examines the building's ability to work with the local fire department to extinguish the fire. Each part is discrete and conditional on completion of the preceding component. Figure A.4(a) shows the network decomposition for the major events, and Figure A.4(b) shows those events in an IPI chart.

A.6   Detection

Instruments or humans, or both, can detect a fire. Figure A.5(a) shows a network diagram that includes both forms of detection. Figure A.5(b) shows the associated IPI chart. We delete the unused method when only a single procedure is considered.

(a)

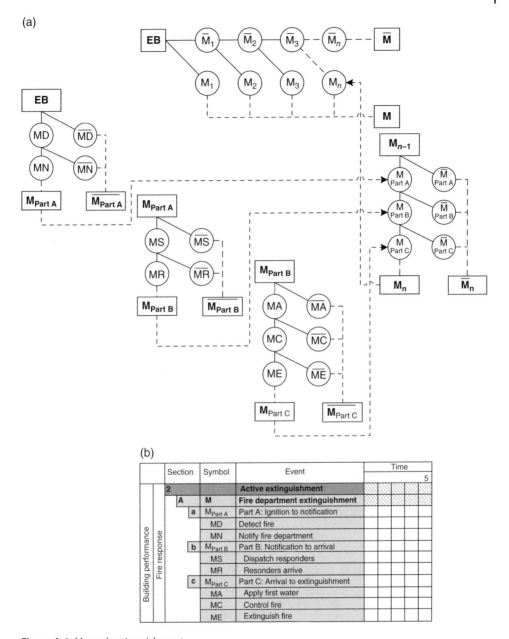

(b)

		Section	Symbol	Event	Time				
									5
		2		**Active extinguishment**					
		A	**M**	**Fire department extinguishment**					
		a	$M_{Part\,A}$	Part A: Ignition to notification					
			MD	Detect fire					
			MN	Notify fire department					
		b	$M_{Part\,B}$	Part B: Notification to arrival					
			MS	Dispatch responders					
			MR	Resonders arrive					
		c	$M_{Part\,C}$	Part C: Arrival to extinguishment					
			MA	Apply first water					
			MC	Control fire					
			ME	Extinguish fire					

Building performance — Fire response

Figure A.4 Manual extinguishment.

The outcomes can be combined if an evaluation were to use both forms of detection. The analysis can use any sequence because each evaluation is independent. Usual network calculations will give the correct mathematical outcomes when probabilistic values are used (see Chapter 35).

(a)

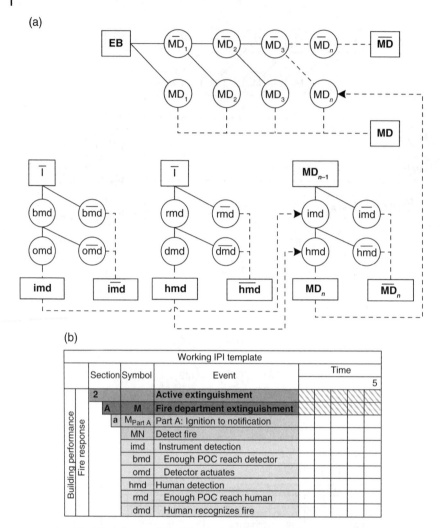

(b)

		Section	Symbol	Event	Time				5
		2		**Active extinguishment**					
		A	M	**Fire department extinguishment**					
		a	$M_{Part\ A}$	Part A: Ignition to notification					
			MN	Detect fire					
			imd	Instrument detection					
			bmd	Enough POC reach detector					
			omd	Detector actuates					
			hmd	Human detection					
			rmd	Enough POC reach human					
			dmd	Human recognizes fire					

(Left side vertical labels: Building performance / Fire response; Header: Working IPI template)

Figure A.5 Detection.

A.7 Notification

Humans or automated procedures can notify the fire department of an emergency. Although both are common, not all buildings use automated procedures. Human notification is necessary in buildings that have no automated notification systems.

Figure A.6(a) shows the analytical network. As with detection, human and automated notification procedures may be uncoupled, and then only the one appropriate for the analysis is used. When probability estimates are used, the network structures calculation procedures to combine the components and give the correct mathematical outcome.

(a)

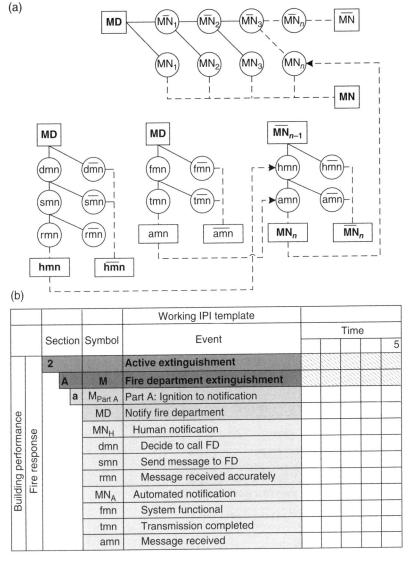

(b)

			Working IPI template					
							Time	
	Section	Symbol	Event					5
	2		**Active extinguishment**					
	A	M	**Fire department extinguishment**					
		a	$M_{Part\,A}$	Part A: Ignition to notification				
			MD	Notify fire department				
			MN_H	Human notification				
			dmn	Decide to call FD				
			smn	Send message to FD				
			rmn	Message received accurately				
			MN_A	Automated notification				
			fmn	System functional				
			tmn	Transmission completed				
			amn	Message received				

(left margin, rotated: Building performance / Fire response)

Figure A.6 Fire department notification.

A.8 Notification to Arrival

One can decompose $M_{Part\,B}$ in several ways. A network could represent the two events, MS (dispatch responding units) and MR (emergency forces respond to site). Alternatively, *fire department respond to site* could be decomposed into its components of turnout time and travel time. The important issue is the selection of appropriate time durations for these events.

The time from notification to dispatching responding units is normally a clear process that does not need additional decomposition to examine events. An analysis does not do a time and motion study, but rather makes a reasonable estimate for a defined process that has clear start and end points. Statistical analyses for dispatch center operations are useful when time durations can be identified for different staffing and operational conditions that are common for the center.

(a)

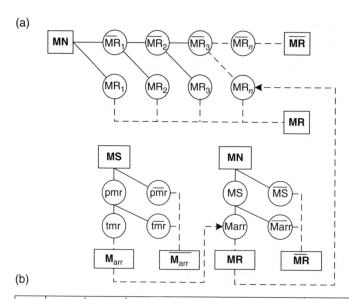

(b)

			Working IPI template						
		Section	Symbol	Event			Time		
									5
		2		**Active extinguishment**					
			A	**M**	**Fire department extinguishment**				
Building performance	Fire response			a	$M_{Part B}$	Part B: Notification to arrival			
				MR	Response arrives at site				
				MS	Dispatch response				
				M_{arr}	Company arrival				
				pmr	Turout time				
				tmr	Travel time				
				E_1	Engine E_1 arrives				
				E_2	Engine E_2 arrives				
				...	...				
				E_n	Engine E_2 arrives				
				L_1	Ladder L_1 arrives				
				L_2	Ladder L_2 arrives				
				...	...				
				L_n	Ladder L_n arrives				

Figure A.7 Fire department response.

Decomposition of the time from dispatch (MS) to arrival (MR) may be useful for special situations. Most communities with paid professional fire departments need only a single event to estimate the collective time from MS to MR. However, communities with volunteer, paid-on-call, or combination fire departments may evaluate two time durations:

1) The turnout time (mrp) from dispatch (MS) to the time when apparatus wheels cross the threshold of the fire station
2) Travel time (mrt) from the fire station to arrival at the site.

We show the more complete network in Figure A.7.

The time duration and the arrival time for each responding company are relatively easy to estimate with reasonable confidence. Consequently, the IPI chart can incorporate arrival times for each responding company. In larger communities where most companies arrive within a short period of time, this may not be necessary. In suburban communities, this information can have more importance to an evaluation. In addition, phased company arrivals are useful to establish a time of first water application and additional hose line operations.

A.9   Arrival to Extinguishment

Complicated fire ground operations can be methodically analyzed to gain a good understanding of the way a building design affects the process. The diagnostic fire size is identified at arrival of the first-in company, and the arrival times for additional companies may be calculated easily.

Phase 1 analysis determines the fire size when first water can be theoretically applied. This analysis provides much information about the general fire conditions during the times to establish supply water and advance a hose line and deliver water to the fire. Although the IPI of Figure A.8(b) shows only the first attack hose line, supplemental sheets may be used to show additional hose line evolutions. Thus, the time sequencing of all hose line operations may be calculated and described by IPI charts.

If the fire cannot be extinguished with an initial attack line or lines, one develops a good understanding of the building and fire conditions for that time frame. A Phase 1 analysis can provide information about the likely boundaries for controlling the fire and the resources needed for final extinguishment.

Figure A.8 shows the analytical networks from arrival to control and extinguishment, as well as the associated IPI chart.

Supplementary rows of the IPI chart can incorporate additional information that gives a better picture of fire suppression for the selected scenario. For example, the time and water quantity for all supply water may be calculated as described in Chapter 14. Similarly, the time to stretch and place attack lines for fire control or extinguishment can be estimated. Fire flows may be calculated. All of this information can be located with the associated event in the IPI chart. Barriers and their effectiveness as fire defenses may be located. Possible ventilation situations may be identified. A scenario analysis provides much information about the way the building and site will help or hinder fire fighting. The IPI chart serves as both a framework to understand performance and a repository for useful information and documentation.

(a)

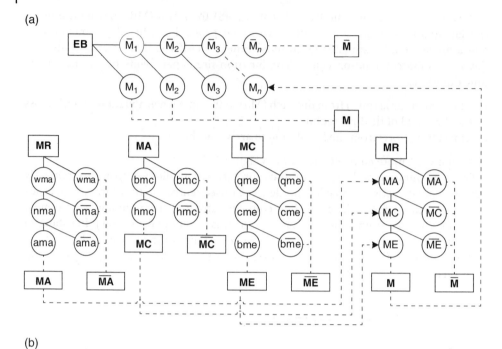

(b)

Working IPI template										
		Section	Symbol	Event	Time					
										5
Building performance	Fire response	2		**Active extinguishment**						
		A	M	**Fire department extinguishment**						
		c	M Part C	Part C: Arrival to extinguishment						
			MA	Apply first water						
			wma	Establish water supply						
			nma	Nozzle reaches fire						
			dma	Water discharges						
			MC	Control fire						
			bmc	Barrier prevent extension						
			hmc	Hose lines in position						
			ME	Extinguish fire						
			qme	Water quantity sufficient						
			cme	Water supply continuous						
			bme	Blackout occurs						

Figure A.8 Manual extinguishment analysis.

A.10   Automatic Sprinkler System

Automatic sprinkler suppression analysis compares diagnostic fire conditions with sprinkler system operations. Agent application (i.e. reliability) for the system is evaluated only once. Sprinkler effectiveness is examined for successively larger areas of fire size. Figure A.9(a) shows the network of events and Figure A.9(b) shows the associated IPI chart.

(a)

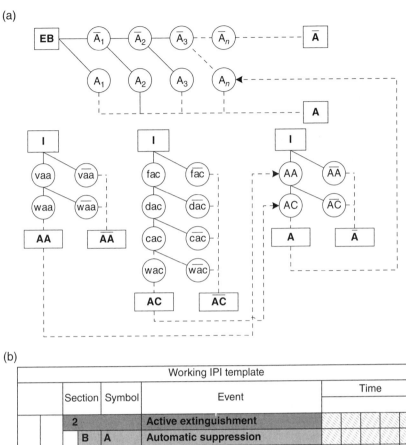

(b)

		Section	Symbol	Event	Time				5
				Working IPI template					
Building performance	Fire response	2		**Active extinguishment**					
		B	A	**Automatic suppression**					
		a	AA	Agent application					
			vaa	Valves open					
			waa	Water reaches sprinkler					
		b	AC	Sprinkler system controls fire					
			fac	Sprinklers fuse					
			dac	Dischange density sufficient					
			cac	Water quantity sufficient					
			wac	Water spray not obstructed					

Figure A.9 Automatic sprinkler suppression.

A.11   Building Response: Structural Behavior

Structural analysis for fire conditions calculates structural performance as the post-flash-over fire continues to burn. A defined structural system and protective insulation are analyzed for a specific diagnostic fire. The analysis can identify the continuum of changes in load-carrying capacity or deformations through the duration of any defined fire.

One can uncouple the dead load from the live load to consider both maximum and minimum conditions for live load. Calculations can relate to local structural members

(a)

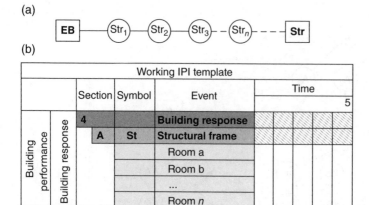

(b)

Figure A.10 Structural performance.

such as beams, slabs, and columns. Understanding local structural performance enables one to anticipate more holistic behavior.

Figure A.10(a) shows a representative linear network to describe a continuum of performance. Here, we illustrate the concept with structural load-carrying capacity. Rows indicate rooms. Structural members affected by the fire may be listed and the status may be calculated for each time duration. IPI columns relate time durations, burning conditions, and structural performance. Although one does not need a binomial description, because structural analysis has matured to the level where technology can make reasonable predictions, the "yes–no" binomial description can be used for convenience in visual perceptions.

A.12   Building Response: Space Tenability

Smoke moves through a building more rapidly and on different paths from those of flame. Because some smoke is tolerable, tenability is based on "How much is enough?" for any space. After establishing an accepted definition for tenability, one may estimate the suitability for each target space. Because forces that affect tenability can act at different times, space tenability may change from tenable to untenable and back again at different times during the fire. The IPI chart can keep track of target space changes during the fire.

A.13   Risk Characterizations

The understanding that evolves during a performance analysis enables one to recognize particular rooms of origin or fire conditions where people, property, operations, exposed facilities, or the environment may be at risk. Chapter 37 discusses ways to use performance evaluations to identify risks and manage them better. Here, we note that the IPI chart can help to identify conditions and the severity of risks.

(a)

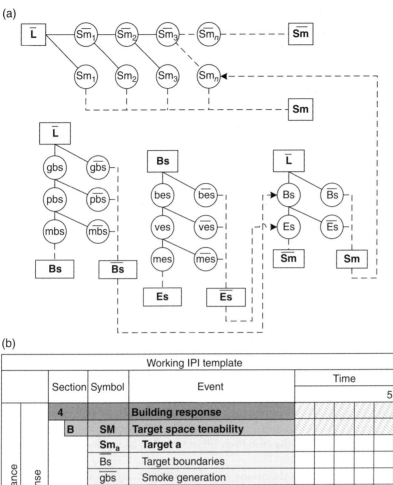

(b)

			Working IPI template		Time				
		Section	Symbol	Event					5
Building performance	Building response	4		**Building response**					
		B	SM	**Target space tenability**					
			Sm$_a$	**Target a**					
			$\overline{Bs}$	Target boundaries					
			$\overline{gbs}$	Smoke generation					
			$\overline{tbs}$	Smoke travel					
			$\overline{Es}$	Tenable environment					
			bes	Barriers successful					
			ves	Venting successful					
			mes	Pressurization successful					
			Sm$_b$	**Target b**					
			...	...					
			Sm$_n$	**Target n**					

Figure A.11 Smoke tenability analysis.

The general concept of risk characterizations is that too many of the bad products of combustion (POCs) and the exposure of interest do not occupy the same space at the same time. The locations of most exposures are fixed, and characterizations describe the ability of the POCs to affect the exposed space. However, locations for people can be variable. Rooms that house defend-in-place occupants examine the building's ability to prevent too many POCs from entering the occupied space. Movable occupants and fire fighters may occupy unsafe spaces during egress or fire fighting operations. The IPI chart helps to relate the dynamic actions of people with the changeable conditions in target spaces. This knowledge provides the basis for describing risk characterizations.

A.14   Occupant Movement

Risk characterizations for occupant egress identify conditions when occupants who attempt to leave a building encounter too many harmful POCs along the egress path. This does not indicate occupant fatality or even injury. It only identifies situations when occupants and too many harmful POCs will occupy the same space at the same time.

The IPI chart simplifies the dynamic analysis for people and POCs both moving at the same time along different paths. For a particular fire scenario, the time for POCs movement is relatively defined. However, a specific path and the time to traverse that path can vary for different occupants.

Occupant movement along an identified egress path is characterized by the time to enter and then leave target spaces along the route. An occupant movement description combines three parts:

1) Identify a specific egress path from occupant origin to egress discharge using a barrier–space modular network (Chapter 2 and 3).
2) Determine the pre-movement time for a representative occupant to cross the room of origin threshold.
3) Calculate the time the occupant enters and leaves each target space along the egress path.

Figures A.12(a) and (b) show the network and associated IPI chart to estimate the pre-movement time for an occupant to understand the existence of a fire, prepare for egress, and start moving out of the building. Figure A.12(c) identifies the egress path being evaluated.

Qualitative estimates of time durations for the different events may be selected. Detection and alert conditions can be estimated with some confidence. Time estimates between an occupant deciding to leave the building and actually leaving the room are uncertain and variable. Time estimates for people to move through egress path segments are more predictable and therefore available. Nevertheless, this analytical exercise provides much insight into the building's life safety performance.

The window of time for target spaces along the defined egress path to remain tenable is estimated in IPI Section 4B. The anticipated time for an occupant to move into each of the target spaces along the egress path is shown in the IPI chart Section 5Aa. Associating the tenability time and the occupant position for the comparable target spaces enables one to estimate the condition that people and untenable conditions can occupy the same space at the same time. Although this determination does not indicate that injury or death will occur, it does show that the building will not provide tenable conditions during the anticipated egress and fire scenario.

If the building provides an unacceptable performance for occupant egress, one may change the design to provide more tenability time in the critical target spaces, identify alternative egress paths, or provide more effective pre-fire instruction to occupants. Although a performance analysis identifies what to expect and not how to manage an outcome, the knowledge gained from the analysis often defines alternatives more clearly.

An alternative use of the IPI chart for egress identifies the maximum pre-movement time available to an individual before he or she can expect untenable conditions along a given egress path. This is relatively easy to determine. Time estimates to pass through the target spaces along the egress path (Figure A.12c) are relatively stable. One may shift these cells horizontally on the IPI chart and compare time durations (Section 4B) to recognize the maximum pre-movement time to avoid experiencing untenable conditions during egress.

A.15 Other Risks

The building response to a fire scenario (IPI chart Section 4) provides information with which to characterize other risks. Chapter 37 describes procedures to identify what is at risk in a building fire. Except for occupant egress and fire fighter safety, most of the other risks involve movement of unwanted POCs to spaces that protect defend-in-place people or important equipment. Thus, occupant defend in place, valuable property, and operational continuity are associated with the effectiveness of the building in preventing too many untenable POCs from reaching and remaining in the target space.

Fire fighter safety has become a more conscious focus of building performance. Although fire fighter risk management is associated with incident command decisions, a performance analysis can provide better knowledge with which to make informed decisions. The IPI chart Section 4 describes the building's response for structural behavior and smoke conditions that influence fire fighter safety.

A.16 Prevent Established Burning (EB): Occupant Extinguishment

Fire prevention – or, more specifically, the combination of "prevent ignition" and (given ignition) "prevent established burning" – is considered a part of risk management rather than building performance. Nevertheless, many buildings survive on the effectiveness of the prevention of ignition and EB. These components are an integral part of holistic building performance.

Given an ignition scenario location and conditions, prevent EB may be done in two ways: occupant extinguishment and automatic special (spot) hazards suppression. It is possible to blend both methods of extinguishment in a manner similar to that of human and instrument detection. However, it is more useful to portray them as separate functions.

The network and associated IPI chart of Figure A.13 guides ignition and EB analysis. Here, the ignition (IG) event may be evaluated (Chapters 34 and 35) or it may be assumed to start the remainder of the analysis. EB can be prevented by fire self-termination (IP) before it can grow to EB or by occupant extinguishment (MP) to prevent fire growth to the critical size fire (CS) that defines EB.

(a)

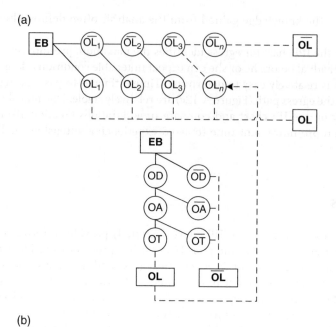

(b)

Working IPI template						Time				
		Section	Symbol	Event						5
Risk characteristics		5		**Risk characteristics**						
		A		**People**						
			a	Occupant egress						
			OL	Occupant leaves room						
			OD	Fire detection						
			OA	Occupant alerted						
			OT	Occupant crosses threshold						
				Occupant travel						
			TS_a	Target space a						
			TS_b	Target space b						
			TS_c	Target space c						
			...	...						
			TS_n	Target space n						

(c)

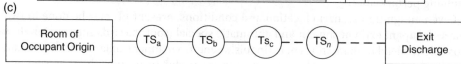

Figure A.12 Occupant egress analysis.

(a)

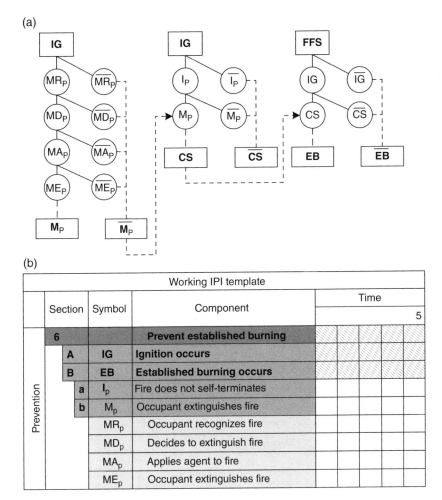

(b)

			Working IPI template						
						Time			
	Section	Symbol	Component						5
	6		**Prevent established burning**						
	A	IG	**Ignition occurs**						
	B	EB	**Established burning occurs**						
Prevention		a	I_p	Fire does not self-terminates					
		b	M_p	Occupant extinguishes fire					
			MR_p	Occupant recognizes fire					
			MD_p	Decides to extinguish fire					
			MA_p	Applies agent to fire					
			ME_p	Occupant extinguishes fire					

Figure A.13 Prevent established burning: occupant extinguishment.

A.17 Prevent EB: Special Hazards Protection

Special (spot) hazards suppression is important for many commercial and industrial occupancies. Because the role of these installations is different from that of automatic (general area) sprinkler protection, we use a separate evaluation. Figure A.14 shows the network and corresponding IPI chart for analysis.

A.18 Closure

The framework for the fire performance analysis of buildings evolved over nearly 40 years of testing and experiment. Its organization combines technical rigor with a professional engineering thought process. The structure incorporates established techniques from many associated disciplines.

(a)

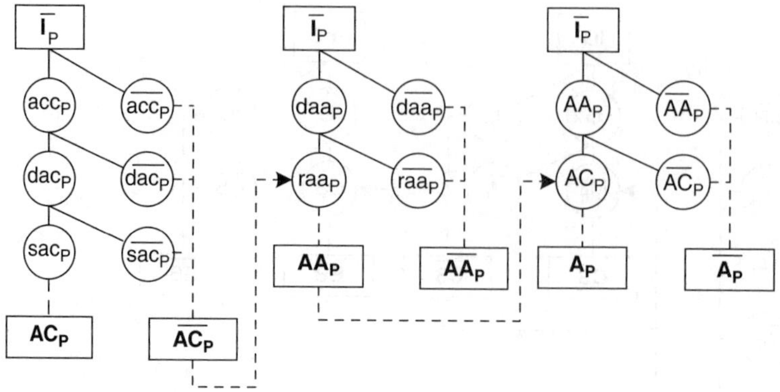

(b)

Working IPI template								
	Section	Symbol	Component	Time				
								5
Prevention	6		**Prevent established burning**					
	C	A_p	**Special hazards protection**					
		AA_p	**Agent does discharge**					
		daa_p	Detection occurs					
		raa_p	Release agent actuates					
		AC_p	**Agent controls fire**					
		acc_p	Agent is appropriate					
		dac_p	Agent discharge sufficient					
		sac_p	Storage and delivery adequate					

Figure A.14 Prevent established burning: special hazards protection.

The analytical framework guides deterministic evaluations. The IPI charts integrate the dynamic progression of fire growth with phased interactions of fire defenses and risk characterizations. The organization is analogous to that of structural engineering, because components may be examined independently and then integrated to describe holistic performance.

The networks and IPI charts can incorporate uncertainty in the form of degree-of-belief probability estimates that are based on the deterministic analyses. The network and IPI organization can structure probabilistic calculations and construct associated curves to depict building performance.

Appendix A provides a detailed compilation of the components. Although one may use all events or delete those that are irrelevant, care must be taken to maintain the basic organization. Analytical rigor for quantitative evaluations is carefully structured within the framework.

Appendix B

Model Building

Description

Plans

Building

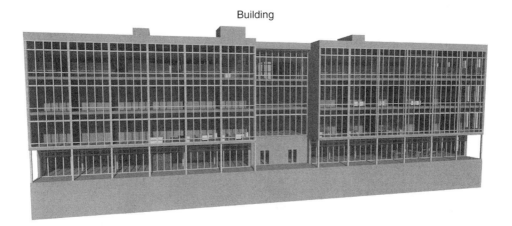

Fire Performance Analysis for Buildings, Second Edition. Robert W. Fitzgerald and Brian J. Meacham.
© 2017 John Wiley & Sons Ltd. Published 2017 by John Wiley & Sons Ltd.

Floor plan

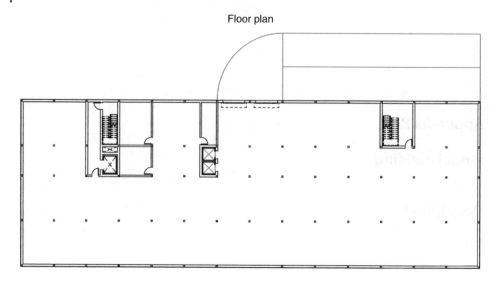

3D diagram

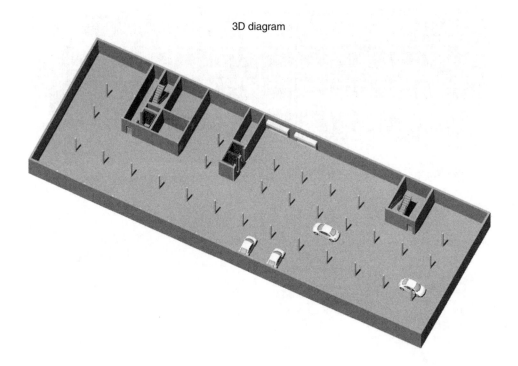

First floor

Floor plan

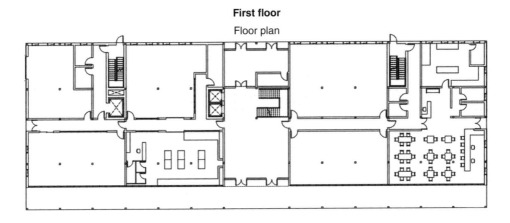

3D diagram

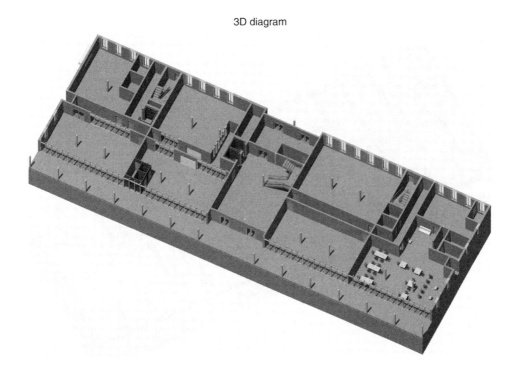

Second floor
Floor plan

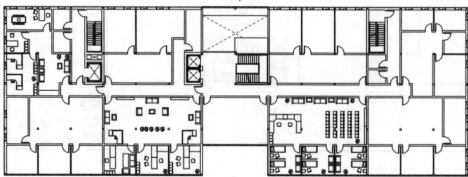

3D diagram

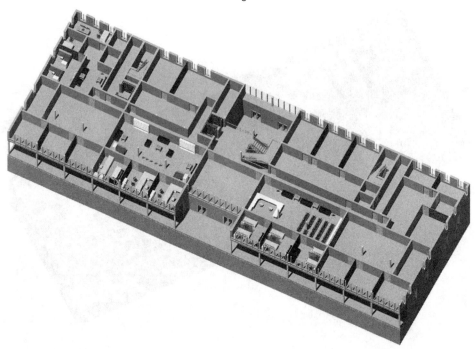

Third floor
Floor plan

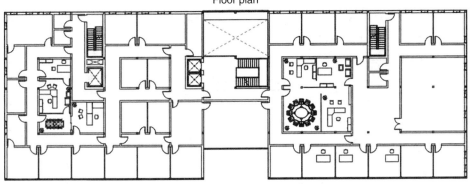

3D diagram

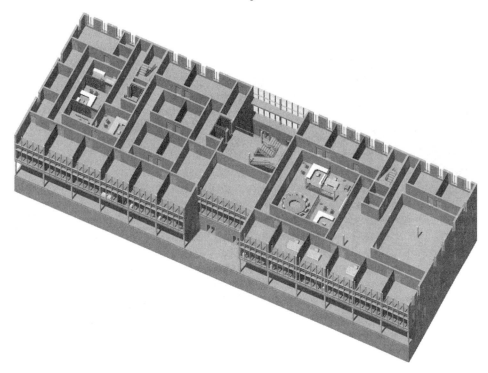

Fourth floor
Floor plan

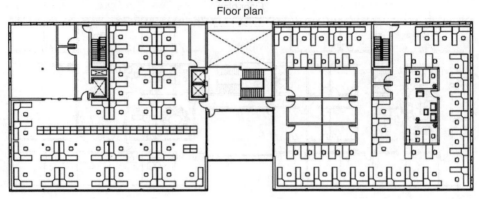

3D diagram

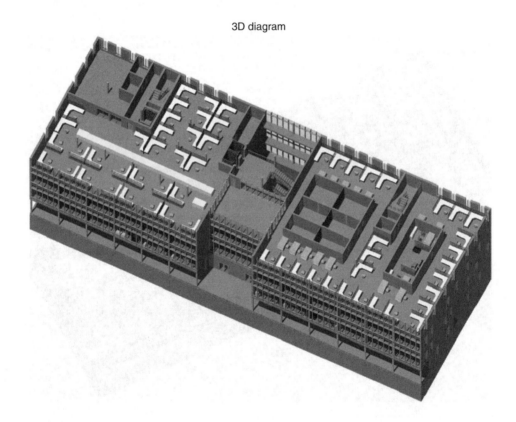

Index

Note: Figures are indicated by *italic page numbers*, Tables by **bold numbers**, and footnotes by suffix 'n'

Fire Performance Analysis for Buildings, Second Edition. Robert W. Fitzgerald and Brian J. Meacham.
© 2017 John Wiley & Sons Ltd. Published 2017 by John Wiley & Sons Ltd.